Unified Field Theory
and
Occam's Razor
Simple Solutions to Deep Questions

Unified Field Theory
and
Occam's Razor
Simple Solutions to Deep Questions

András Kovács
BroadBit Energy Technologies, Finland

Giorgio Vassallo
University of Palermo, Italy

Paul O'Hara
Sophia University Institute, Italy

Francesco Celani
INFN-LNF, Italy

Antonino Oscar Di Tommaso
University of Palermo, Italy

World Scientific

NEW JERSEY • LONDON • SINGAPORE • BEIJING • SHANGHAI • HONG KONG • TAIPEI • CHENNAI • TOKYO

Published by

World Scientific Publishing Europe Ltd.
57 Shelton Street, Covent Garden, London WC2H 9HE
Head office: 5 Toh Tuck Link, Singapore 596224
USA office: 27 Warren Street, Suite 401-402, Hackensack, NJ 07601

Library of Congress Cataloging-in-Publication Data
Names: Kovács, András (Chief technology officer), author. | Vassallo, Giorgio, author. | O'Hara, Paul (Professor of ontology), author. | Celani, Francesco, author. | Di Tommaso, Antonino Oscar, author.
Title: Unified field theory and Occam's razor : simple solutions to deep questions / András Kovács, BroadBit Energy Technologies, Finland; Giorgio Vassallo, University of Palermo, Italy; Paul O'Hara, Sophia University Institute, Italy; Francesco Celani, INFN-LNF, Italy; Antonino Oscar Di Tommaso, University of Palermo, Italy.
Description: New Jersey : World Scientific, [2022] | Includes bibliographical references.
Identifiers: LCCN 2021046878 | ISBN 9781800611290 (hardcover) | ISBN 9781800611306 (ebook for institutions) | ISBN 9781800611313 (ebook for individuals)
Subjects: LCSH: Unified field theories. | Physics--Experiments--Methodology. | Problem solving--Mathematical models.
Classification: LCC QC794.6.G7 K68 2022 | DDC 530.14/2--dc23/eng/20211130
LC record available at https://lccn.loc.gov/2021046878

British Library Cataloguing-in-Publication Data
A catalogue record for this book is available from the British Library.

For any available supplementary material, please visit
https://www.worldscientific.com/worldscibooks/10.1142/Q0333#t=suppl

Desk Editors: Soundararajan Raghuraman/Michael Beale/Shi Ying Koe

Typeset by Stallion Press
Email: enquiries@stallionpress.com

https://doi.org/10.1142/9781800611306_fmatter

Preface

This book is for all who are curious about our physical world. It builds upon the ideas of Maxwell, Dirac, and their notable contemporaries such as Einstein, Heisenberg, Pauli, and De Broglie. Indeed, we believe that our ideas formulated in Chapters 1–3, were already implicit in the theoretical foundations of physics established more than eighty years ago, but for whatever reason were never fully developed afterwards. In a plausible alternative past, our ideas might have been considered self-evident already by the mid-20th century. It is our hope that as the book becomes more widely known it will be recognized as a standard tool for understanding the concepts of modern physics.

We develop our ideas using the powerful tool of geometric algebra and the foundational concepts stemming from Maxwell and Dirac but also contribute something new in our approach. Key mathematical concepts and structures are explained in the first chapters with the aim of making the book accessible to readers who have the interest and capability for self-study. Nevertheless, it does presuppose a certain level of mathematical skill essential for developing a correct understanding of electromagnetics and its Lagrangian.

To justify our theory, we discuss a large set of validating experimental data coming from mainstream physics which we believe, in contrast to some current theories, can be explained in an insightful and simple way with our approach. We are not interested in imposing ad-hoc rules (such as imposing the Lorenz gauge rule onto electromagnetic equations, or cutting of the counting of electromagnetic field energy at a certain distance from a supposedly point-like electron) and neither in analyzing exotic fringe experiments (such as the study of yet another meson particle with 10^{-22} s lifetime), but instead, we shall refer to very fundamental data such as nuclear binding energies or the magnetic moments of elementary particles.

The history of physics may be viewed as a progression from particle-oriented to wave-oriented concepts to wave–particle duality. A thousand years ago, if one tried to talk about "fields" and "waves," they would have been met with blank stares. Only the concept of matter particles existed at the time, with direct mechanical interactions among them. Waves were only observed on the surface of water, and were neither understood mathematically nor were they thought to be related to anything else. Scientists, at the time, thought of light as a stream of small "light balls." In contrast, today, physicists describe light as waves in the electromagnetic field that also exhibits particle properties. All mechanical and chemical forces are also described as being exerted by such electromagnetic waves. Approximately, 100 years ago, the notion that elementary matter particles might also be quantum mechanical waves started to gain acceptance. In this sense, our book simply fits into the historic trend, and represents a next step in the understanding of wave-oriented concepts.

It is worth recalling that such steps in the past have usually been met by strong opposition from the scientific community. We illustrate this through the brief history of light model evolution from the original "light balls" description. Around 1687, Newton and Leibniz independently discovered differential calculus, which is needed to quantitatively describe waves. However, it took 130 more years to recognize light as a wave. In 1818, the French Academy of Sciences offered a prize for a consistent understanding of light diffraction. At that time, light diffraction experiments were considered an anomaly of the prevailing "light balls" model. One of the participants, civil engineer and optometrist Augustin-Jean Fresnel, submitted a thesis in which he explained diffraction from analysis of both the Huygens–Fresnel principle and Young's double-slit experiment. This irritated the academic "old guard," who were staunch believers in the particle theory of light and were skeptical of its alternative, the wave theory. Poisson, a member of the Académie, studied Fresnel's theory in detail and looked for a way to prove it wrong. Poisson thought that he had found a flaw when he demonstrated that Fresnel's theory predicts an on-axis bright spot in the shadow of a circular obstacle blocking a point source of light, where the particle-theory of light predicts complete darkness. Poisson argued this was absurd and Fresnel's model was wrong. The head of the committee, Dominique-François-Jean Arago, performed the experiment. To everyone's surprise, he observed the predicted bright spot, which vindicated the wave model. Fresnel won the competition, and the wave nature of light gained acceptance in mainstream scientific circles. However, the "field" part was still missing, and mainstream physicists now considered light to be a wave in the physical aether, which they believed filled the vacuum. In essence, the traveling "light balls" model was replaced by a "sea of light carrying balls" model, facilitating the waves.

Maxwell's discovery of electromagnetic field equations in 1862 spelled the beginning of the end for this "sea of light carrying balls" model. After Einstein's discovery of relativity in the early 20th century, Silberstein published a seminal book in 1914, titled *The Theory of Relativity*. In this book, he quantitatively described how light is carried by a relativistic electromagnetic field. The journal *Nature* published a vitriolic review of Silberstein's work. From the perspective of the reviewer, Silberstein's main sin was to have left the aether out of the discussion:

> There is scarcely a reference to the longings of the physicist for an objective aether ... Many will read Dr Silberstein's careful and detailed introduction to it, consider his illustrations, and follow his logic, and yet feel there is something lacking. The argument in favor of an aether is not dealt with. The reluctance that Lorentz had to abandon the aether remains. The seeker after a deeper understanding of the physical is apt to fight shy of a principle which cannot be expressed in terms of concepts to which he can give some degree of substantiality.

Eventually, the quantitative theory of electromagnetic field waves gave birth to a wide range of new radio technologies, while the theory of aether was of no use to radio engineers. The aether theory was then quietly abandoned by theoretical physicists. Starting from Fresnel's thesis, the whole process took over one hundred years.

The fundamental equations of physics appear to be remarkably straightforward, but the set of their possible solutions is anything but simple. This is not a contradiction: simple equations may allow for many classes of solutions, various symmetries, and novel interpretations. Our approach of seeing elementary particles as being akin to electromagnetic waves, although it might not initially be widely accepted, is consistent with the standard equations. We hope there will be some modern-day Aragos among readers, who will strive to resolve experimental anomalies, by carefully considering the various options, and by trying to design experiments to test the quantitative predictions of our theory.

In the last part of this introduction, we outline our scientific methodology. Scientists like to think of their respective field as an additive process, which advances through the accumulation of ever wiser ideas and ever more elaborate concepts. That is generally true, but one must be careful to retain checks and balances and correct mistakes or oversights that may slip in.

Correcting mistakes often reveals a more fundamental perspective, whereby one realizes that separate physical laws or phenomena are one and the same thing and can be unified. To give an example from a pre-Newtonian era, the celestial motion of planets was accurately and predictively described by Earth-centered epicycloid formulas. The gravitational fall of bodies was accurately and predictively described by a constant acceleration formula. These two phenomenological laws were thought to be completely unrelated, and scientists of the day had no sense of any mistaken assumption: their formulas were predictive and accurate. They were the top theoreticians of the day, drawn from the same gene pool as today's theoreticians. But their thinking was locked into a mistaken paradigm. Once Newton's efforts revealed the mistaken assumptions, the phenomena of celestial motions and falling bodies were unified, although it continued to meet opposition from some who insisted that their model of Earth-centered epicycloid motions was the correct model of reality. One must understand that phenomenological formulas may involve unrealistic models of reality, even while being experimentally correct.

Correcting mistakes is an important aspect of our methodology. We identify what we believe are some major misunderstandings in interpretation that need correcting, and also identify the time period when they were made in both actual, as well as resource-adjusted "19th century years," abbreviated as 19C-years.

(1) **Background:** In the early 20th century, mathematicians discovered that the four Maxwell equations can be written as a single differential equation of the so-called "vector potential" field. The space–time gradient of the vector potential field yields three field types: electric, magnetic, and scalar.
Mistaken assumption: The role of the scalar field was not understood at the time, and it was assumed to be always zero. This assumption shows up in current textbooks as the "Lorenz gauge" condition, and it is now referred to as a "knowledge" rather than as an assumption. There are no experimental proofs for this assumption, and it was in fact shown to lead to paradoxes [1].
Age: Over 100 years (over 1000 19C-years).
What we discover upon correcting the mistake: Electromagnetic fields and charges are not two distinct entities — they are all part of a single vector potential field. We gain a paradox-free field theory of electromagnetism, and we no longer need to ask what charges are made of.

(2) **Background:** In 1927, Dirac discovered the phenomenological equation named after him, which accurately predicts the energy levels of atomic and molecular orbitals. He noted in his initial publication that this equation is a phenomenological guess and that some of its solutions appear to yield negative energy eigenvalues.
Mistaken assumption: The negative energy eigenvalue solutions of the Dirac equation became interpreted as the physical reality of negative energy states. In the high energy limit, textbooks refer to negative energy solutions as anti-particles. But electron–positron annihilation events radiate electromagnetic energy, which would be impossible if positrons were negative energy solutions. Driven by the desire to maintain the assumption that anti-particles are negative energy solutions, some textbooks engage in wild speculation about a "Dirac-sea of electrons in the vacuum" with infinite energy density, which supposedly fills the vacuum sea with electrons. According to this hypothesis, positrons are holes in the omni-present Dirac-sea of electrons. In the low energy limit, textbooks refer to negative energy solutions as "temporary violation of energy conservation," which is supposedly allowed by the Heisenberg uncertainty, and cite quantum tunneling as an example. But the mathematics of Noether's theorem clearly states that energy conservation follows from the invariance of physical laws with respect to the flow of time. Speculation about violation of energy conservation is a radical departure from realism, as it implies that physical laws may randomly vary on microscopic time scales. Consequently, any set of quantum mechanical axioms becomes mathematically meaningless.
Age: 90 years (900 19C-years).
What we discover upon correcting the mistake: By studying the spatial geometry of Maxwell and Dirac equations' solutions, we learn that both have harmonic and exponential type solutions. Exponential solutions of the Maxwell equation correspond to the tunneling electromagnetic field associated with reflections, e.g., behind a metallic surface, while exponential solutions of the Dirac equation correspond to the quantum tunneling spinor field, e.g., behind a potential wall. No one claims that there should be negative electromagnetic energy associated with reflections, and we prove that the correct energy density calculation indeed yields a positive value. With the help of geometric algebra we offer geometric analogies of the exponential solutions of the Maxwell and Dirac equations. Once the geometry of these solutions is understood, one realizes that claims of "negative energy spinors" originate from a

mathematical misunderstanding of transient exponential solutions. Happily, there is no further need for confused speculation about temporary violation of energy conservation or about an omni-present but undetectable sea of charged electrons in the vacuum. Electrons and positrons may now be described by the same quantum mechanical wavefunction: both have positive energy, and the only remaining difference between the two particle types is the sign of their electromagnetic scalar field.

(3) **Background:** Experimental techniques for high-energy proton–electron scattering measurements became available in the 1960s. At that time, proton–electron scattering measurements performed at SLAC did not provide yet a conclusive answer about the number of subparticles comprising a proton. With the development of experimental capabilities, proton–electron scattering experiments were performed over a gradually widening energy range, which in principle, gave us experimental knowledge with ever increasing accuracy of the proton's internal structure.

Mistaken assumption: Based on theoretical reasoning, most theoretical physicists assumed by the mid-1970s that the proton comprises three sub-particles. This assumption was not quite compatible with the experimental data collected at SLAC, but theorists bridged the gap between their theory and experiment by assuming the three sub-particles are swimming in a "sea of virtual quarks" inside the proton. Such an assumption might strike the reader as being particularly extraordinary, but one must remember that its proponents considered it not to be much different from the above-discussed "Dirac-sea of electrons" model, which was already in all standard textbooks. Over the ensuing decades, textbooks started to refer to all these assumptions as "knowledge" about the three subparticle structure of the proton. Experimentally, it became possible only in the early 2000s to unambiguously count the proton's subparticles from proton–electron scattering measurements. The corresponding experimental data was collected from the SLAC, JLAB, and HERA facilities. Unexpectedly, highly trained scientists started making elementary data processing mistakes by mixing incompatible data, and published their conclusions in high-impact journals without any peer-reviewer pointing out their basic errors. Not too surprisingly, these published reports reaffirmed the already "known" three subparticle structure of the proton. As far as we know, the only mathematically correct analysis of these experimental data was published by William Stubbs. We present his results in our book, along with an

explanation of the data processing mistakes that one finds in high-impact publications.
Age: 15 years (150 19C-years).
What we discover upon correcting the mistake: Upon carefully combining the sub-particle momentum distribution function from multiple experiments, and refraining from making any unnecessary assumptions, its peak reveals the number of subparticles in a proton. For X number of subparticles, the peak is at $\frac{1}{X}$. The peak of experimental data indicates that the proton comprises 8–10 subparticles, and there is no need to hypothesize about any "sea of virtual particles." Through complementing analysis of charge radii we determine that the proton comprises nine subparticles. Investigating these protonic subparticles opens up a new frontier of particle physics.

(4) **Background:** By the 1930s, experimenters could measure the electron's magnetic moment to a rather high accuracy. These measurements revealed a slightly larger magnetic moment than what was expected from Dirac theory.
Mistaken assumption: The first phenomenological anomalous magnetic moment formula was published by Julian Schwinger, who recognized that the anomalous part is given by the $\frac{\alpha}{2\pi}$ expression, where α is the fine structure constant. However, Schwinger claimed to have a theoretical calculation, not just a phenomenological formula. But Schwinger's calculation was based on a point-particle model, which assumes that the magnetic moment of Dirac theory is an inherent property of the infinitesimal point-particle. In the book we show this assumption to be incorrect, and calculate the magnetic moment as a purely electromagnetic phenomenon. Moreover, experts who claim to understand Schwinger's calculation are debating whether his calculation implies a temporary violation of momentum conservation. Most current textbooks go with the interpretation of a temporary momentum conservation violation, which according to Noether's theorem implies that physical laws randomly fluctuate at short distances. As mentioned above, any set of quantum mechanical axioms becomes mathematically meaningless after such speculations. Nevertheless, Schwinger's formula was taken to be more than phenomenological, as physicists already became comfortable with the idea of temporary energy conservation violation, due to the assumption of negative energy Dirac spinors. Soon, Feynman suggested that a more accurate result can be obtained by

calculating higher order terms of α. A calculation based on Feynman's method fills 50 pages. One might wonder how it is possible to determine whether such calculation has any predictive power? One may check whether the calculation yields the correct result for various particle types. However, although the coefficients of Feynman's formula yield the correct electron magnetic moment, the formula fails badly for the proton. This was not considered to be a problem, because the proton and electron were thought to belong to unrelated particle families, and so it was assumed that there is a need to apply so-called "hadronic corrections" for the proton case. In other words, it was assumed to be impossible to have a universal anomalous magnetic moment formula for all particle types. By this point, hopefully, the reader sees how all the speculative assumptions are inter-related in a harmful way.
Age: 70 years (700 19C-years).
What we discover upon correcting the mistake: The solution of Maxwell's equation shows that the anomalous magnetic moments of the electron, muon, proton, and neutron can all be calculated from a universal formula. We derive this simple formula from relativistic principles, which reveals that the real control parameter is not α but the charge radius and Zitterbewegung radius parameters. There is no more need to make different calculations for each particle type, no need for 50 pages of calculations, and no need to debate whether physical laws randomly fluctuate at short distances. Based on this, we predicted in the first edition the outcome of the so-called "proton charge radius puzzle": the correct proton charge radius is in fact given by the muonic hydrogen measurements. This prediction appears to be confirmed by a recent proton charge radius measurement [2], which was not yet available at the time of publication. Our results uncover a universal internal geometry among particles which were previously considered to be unrelated, and demonstrate that particles are certainly not abstract points in a wavefunction.

(5) **Background:** Since the early days of quantum mechanics, it has been recognized that no more than two electrons may occupy any given atomic orbital. For covalent molecular orbitals, the same pairwise occupancy was observed. This pairwise rule became known as the Pauli exclusion principle and is at the heart to understanding chemistry. Without this exclusion principle, all electrons would fall into the same ground state orbital.
Mistaken assumption: Pauli was keen to derive the exclusion principle named after him, so that it would not be a new axiom. The derivation worked out by Pauli involved a new principle, which has become known

as "micro-causality." In the words of Pauli, this principle states that "all physical quantities at finite distances exterior to the light cone ... are commutable [4]. What Pauli failed to note was that this principle as formulated does not apply to entangled particle states. Entangled particles violate micro-causality by definition and for this reason, within the context of the Einstein–Podolski–Rosen (EPR) paradox, Einstein referred to spooky action at a distance. Regarding the current status of the micro-causality principle, it is still debated whether micro-causality is an axiom or a consequence of relativity, and a recent systematic review of pro and con arguments suggests that micro-causality is still in axiom status [3].
Age: 70 years (700 19C-years).
What we discover upon correcting the mistake: We first prove that *isotropically spin correlated states can only occur in pairs.* In chemistry, such states are referred to as "singlet" states, and we show that singlet states are isotropically entangled particle pair states. This essentially means that indistinguishable particle pairs are either entangled as singlets or are statistically independent of each other. We show in the book that a system of n-indistinguishable particles that contains singlet states must obey Fermi–Dirac statistics, while if there are no singlet states, then they will obey Bose–Einstein statistics, and Boltzmann statistics follows by relaxing the indistinguishability condition. In contrast to the generally assumed classification of statistics according to half-integer versus integer spin values, we find that spin value has no essential role to play. How can our theorem be experimentally distinguished from Pauli's axiom? Firstly, entangled electron states can be experimentally studied in a wide variety of physical systems, while Pauli's axiom of anti-symmetric wavefunctions is not experimentally observable, as far as we know. Secondly, when two electrons are indistinguishable in the same quantum mechanical state, a break-up of this state causes both electrons to transition pairwise out of this state. Chemists routinely observe this phenomenon for both atomic and molecular orbitals. Regarding atomic orbitals, alkali-earth metals for example are well-known to oxidize directly to +2 valence state, implying a pairwise departure of their outer electrons. Regarding molecular orbitals, the outer electrons of oxygen molecule are known to be in either triplet state or singlet state. In the triplet state, the electrons react individually, while in the singlet state they react pairwise. When chemists want to simultaneously bind two oxygens to another molecule, they start by inducing a singlet state of O_2 outer electrons. In contrast,

> there is nothing in Pauli's theory which implies that electrons in the same orbital would be pairwise transitioning. In summary, after 70 years of misunderstandings about the process behind electron statistics and chemical binding rules, we may have a chance to understand the physics of tangible matter. Our theorem is very general, and applies to all particles in an isotropically entangled state, for example also to halo nuclei which break up through a pairwise emission of the two halo nucleons.

Altogether, the above-listed realizations provide a new perspective for the advancement of physics. Since the publication of the first edition, we have received some feedback, both encouraging and discouraging. The more one reflects on the foundations, the more one understands the obstacles that stand in the way of progress. We hope readers will appreciate our quest for truth. Correcting mistakes is by no way disrespectful of the scientists who worked on the involved problems. We are grateful for the efforts of others, and recognize that without their results this book would be impossible. In particular, we would like to mention the efforts of Marcel Riesz, who pioneered the geometric algebra perspective for studying electromagnetism, and the efforts of H. E. Moses, who first proposed that Maxwell equations may be treated as proper field equations. In case some readers might be wondering why we did not cite or mention others, rest assured that we did not have the intention to ignore anyone, and will gladly add names when appropriate. We simply aim to keep the text as self-consistent and focused as possible.

Guided by the above insights, we endeavor to develop a realistic electron model. We emphasize that we seek analytic solutions of Maxwell and Dirac equations, without having to make any further assumptions or axioms about particle properties. Our book suggests that less is better and that we can have a more unified understanding of the physical world based on a few fundamental laws. As years go by, the need for a critical discussion on physicists' methods becomes ever more clear. To quote from *Lost in Math* by S. Hossenfelder [5]: "It is already clear that the old rules for theory development have run their course. Five hundred theories to explain a signal that wasn't and 193 models for the early universe are more than enough evidence that current quality standards are no longer useful to assess our theories." The principle of Occam's razor is as valid today as it was in his time, and applying it feels like a breath of fresh air.

References

[1] G. Rousseaux, The gauge non-invariance of classical electromagnetism. arXiv:physics/0506203v1.

[2] A. Grinin *et al.*, Two-photon frequency comb spectroscopy of atomic hydrogen. *Science*, **370**(6520) (2020) 1061–1066.

[3] J. Wright, Quantum field theory: Motivating the axiom of microcausality. Waterloo, Ontario, Canada, 2012.

[4] W. Pauli, The connection between spin and statistics. *Physical Review Journals A*, (1940), 716–722.

[5] S. Hossenfelder, Lost in math. *Basic Books* (2018), pp. 235–236.

https://doi.org/10.1142/9781800611306_fmatter

About the Authors

András Kovács is co-founder and CTO of Broadbit Batteries. A graduate in Physics from Columbia University, he also holds key patents in battery technology. He co-authored a book on Maxwell–Dirac theory and is coordinator of several EU co-funded research projects. He is extremely well prepared both theoretically and experimentally and it is because of his passion for Physics that he was able to assemble this group.

Giorgio Vassallo is a graduate in Physics from University of Palermo. He has a diverse background in computer science, machine learning, neural networks, electronics and physics and has published in all of these areas. He is currently professor in the Engineering Department at the University of Palermo.

Paul O'Hara has a Ph.D. in Mathematics from UCLA and has widely published in theoretical physics (about 25 papers) especially in the areas relating general relativity and quantum mechanics, as well as quantum entanglement and spin statistics. He has presented on these topics at over 30 international conferences. Currently, he is a professor of scientific methodology at the Sophia University Institute, Italy. Previously, he was full professor and chair of the Department of Mathematics at Northeast Illinois University in Chicago. He is an active member of the International Association for Relativistic Dynamics (IARD).

Francesco Celani is a graduate in Physics from the prestigious Istituto di Fisica Gugliemo Marconi (La Sapienza University, Rome). In 1976, he was promoted to Researcher at the Istituto Nazionale di Fisica Nucleare-Laboratori Nazionali di Frascati (INFN-LNF) in Italy, and in 1988, he was further promoted to First Researcher. He has over 150 published papers, holds 2 international patents and has been twice nominated for a Nobel Peace Prize (2014, 2015) as a proponent of the "Live Open Science" Project. He currently holds research grants from the EU. He was co-founder (2003) of the International Society of Condensed Matter Nuclear Science (ISCMNS) and among his many affiliations he also has an Honoris-Causa, Life Membership of the Russian Physical Society (RPS).

Antonino Oscar Di Tommaso is a professor of electrical machines and drives in the Department of Engineering at the University of Palermo from which he also holds a Ph.D. in Electrical Engineering (2004). Over the years he has been involved with 10 different national research projects in Italy related to wind generators, photovoltaic cells for integrated intelligent systems, and problems related to optimization and automation of energy systems. On four occasions (2003, 2013, 2018, 2020) he received an award for the "best presentation" at international conferences held in Italy, Germany and France. He has also published over 110 scientific papers in international journals and conference proceedings.

https://doi.org/10.1142/9781800611306_fmatter

Contents

https://doi.org/10.1142/9781800611306_fmatter01

Mathematical Preliminaries

Paul O'Hara*, Giorgio Vassallo[†,§], and András Kovács[‡]

Sophia University Institute, Loppiano, Italy

[†]*Engineering Department, University of Palermo viale delle Scienze, 90128 Palermo, Italy*

[‡]*BroadBit Energy Technologies, Metallimiehenkuja 8 C 02150 Espoo, Finland*

[§]*giorgio.vassallo@unipa.it*

We will begin by developing some of the mathematical language that will be needed for the remainder of this book. This includes the language of Clifford algebra, tensor and wedge products, groups, probability, spin measurements, and angular momentum theory. The reader is encouraged to master these mathematical aspects before reading subsequent chapters.

Mathematical Nomenclature

The following notations are used throughout the book.

- $\boldsymbol{A}_{\square}, A$**:** four-vector in Minkowski space–time;
- $\boldsymbol{A}_{\triangle}, \mathbf{A}$**:** three-vector in physical space;
- $\mathbf{e}_x \equiv \gamma_x, \mathbf{e}_{xy} \equiv \gamma_{xy}, \mathbf{e}_{xyz} \equiv \gamma_{xyz}, \mathbf{e}_{txyz} \equiv \gamma_{xyzt} \equiv I$**:** Clifford bases for vectors, bi-vectors, tri-vectors, and the pseudo-scalar;
- A^***:** a vector in the dual-space, which is defined by the exterior algebra;
- $\mathbf{v} = v^i\mathbf{e}_i$**:** representation of vector components, using Einsteinian summation convention;
- $\widetilde{A}$**:** Clifford reversion of A;
- $\bar{C}$**:** the complex conjugate of a complex number C;
- M^T**:** the transpose of a matrix M;
- $A^\dagger$**:** the adjoint of A.

0.1. Clifford Algebra Introduction

Clifford algebra is defined by the multiplication rule of its basis elements. The Clifford basis elements are defined to obey the following multiplication rules:

$$\mathbf{e}_t^2 = -1, \quad \mathbf{e}_x^2 = \mathbf{e}_y^2 = \mathbf{e}_z^2 = 1 \tag{0.1.1}$$

$$\mathbf{e}_i\mathbf{e}_j = -\mathbf{e}_j\mathbf{e}_i \; (i \neq j) \tag{0.1.2}$$

The t index represents the time coordinate, and indices x, y, z represent spatial coordinates. Thus, the Clifford algebra basis elements can be identified with the unit vectors spanning our four-dimensional space–time.

The above-defined algebra of the $3 + 1$ basis elements is referred to as $Cl_{3,1}(\mathbb{R})$ algebra type. In the context of the Dirac equation, the above base vectors are equivalent to the Dirac gamma matrices.

The multiplication of two different base vectors defines a bi-vector, which spans an oriented surface. We use the following notation for bi-vectors: $\mathbf{e}_{ij} \equiv \mathbf{e}_i\mathbf{e}_j$. Similarly, tri-vectors are volume elements, and denoted as: $\mathbf{e}_{ijk} \equiv \mathbf{e}_i\mathbf{e}_j\mathbf{e}_k$. The dimensionality of an expression goes up through multiplication of different base vector components.

Based on the above definitions, we observe the following multiplication properties:

$$\mathbf{e}_x\mathbf{e}_{xyz} = \mathbf{e}_{yz}, \quad \mathbf{e}_y\mathbf{e}_{xyz} = -\mathbf{e}_{xz}, \quad \mathbf{e}_z\mathbf{e}_{xyz} = \mathbf{e}_{xy} \tag{0.1.3}$$

$$\mathbf{e}_{yz}\mathbf{e}_{xyz} = -\mathbf{e}_x, \quad \mathbf{e}_{zx}\mathbf{e}_{xyz} = -\mathbf{e}_y, \quad \mathbf{e}_{xy}\mathbf{e}_{xyz} = -\mathbf{e}_z \tag{0.1.4}$$

$$-1 = \mathbf{e^2}_{ij} = \mathbf{e^2}_{ijk} \; (i \neq j \neq k, i \neq t, j \neq t, k \neq t) \tag{0.1.5}$$

The above multiplication examples illustrate how the dimensionality of an expression goes down through the multiplication of same base vector components.

A useful notation for a space–time vector is $Q = (Q_t\mathbf{e}_t, \mathbf{Q})$, where the bold capital notation denotes a spatial vector, i.e., $\mathbf{Q} = q_x\mathbf{e}_x + q_y\mathbf{e}_y + q_z\mathbf{e}_z$. The product of two vectors is:

$$\begin{aligned} PQ &= (P_t\mathbf{e}_t, \mathbf{P})\,(Q_t\mathbf{e}_t, \mathbf{Q}) \\ &= (-P_tQ_t + \mathbf{P}\cdot\mathbf{Q}) + (-P_t\mathbf{Q} + \mathbf{P}Q_t)\,\mathbf{e}_t + \mathbf{P}\times\mathbf{Q}\mathbf{e}_{xyz} \end{aligned} \tag{0.1.6}$$

In the above expression, the $(-P_tQ_t + \mathbf{P}\cdot\mathbf{Q})$ part is scalar, while the other terms are bi-vectors. However, we wrote the bivector terms by highlighting two spatial vectors: $(-P_t\mathbf{Q} + \mathbf{P}Q_t)$ and $\mathbf{P} \times \mathbf{Q}$. It must be kept in mind that the bold $\mathbf{P} \times \mathbf{Q}$ notation represents a spatial vector.

We define the unitary pseudoscalar as $I \equiv \mathbf{e}_{txyz}$, composed of the space–time unit vectors. Algebraically, we can identify the imaginary unit i with the unitary pseudoscalar I. We will make use of the $i = \mathbf{e}_{txyz}$ identity during the study of electromagnetism. We note that $i\mathbf{e}_t = \mathbf{e}_{txyz}\mathbf{e}_t = \mathbf{e}_{xyz}$, and that $i\mathbf{P} = -\mathbf{P}i$ because a vector contains one index same as $\mathbf{e}_{txyz}$. Because of this commutation rule, multiplying a vector by i is different from numerical multiplication. On the other hand, the multiplication of a scalar or bi-vector by i commutes the same way as numerical multiplication.

0.2. The Geometry of Clifford Algebra

Scientific knowledge is expressed mathematically, but the importance of the optimal choice of the appropriate mathematical language is often underestimated [1, 2, 4, 6]. The geometric algebra (Clifford algebra) formalism, according to Occam's razor principle, is by far the best choice for modern physics. Clifford algebra provides a simple and unifying mathematical language for coding geometric entities and operations [3, 7]. It integrates different mathematical concepts highlighting geometrical meanings that are often hidden in the ordinary algebra.

A particular real Clifford $Cl_{p,q}(\mathbb{R})$ algebra is defined in a space with $n = p + q$ dimensions with an orthonormal base of n unitary vectors. The first p vectors of this base have positive squares, whereas the remaining ones have negative squares, as shown by the following equations:

$$\gamma_i^2 = 1 \quad \text{with } 1 \le i \le p \tag{0.2.1}$$

$$\gamma_i^2 = -1 \quad \text{with } p+1 \le i \le p+q \tag{0.2.2}$$

$$\gamma_i\gamma_j = -\gamma_j\gamma_i \quad \text{with } i \ne j \tag{0.2.3}$$

where γ_i are unitary orthogonal vectors.

The geometrical product $\gamma_i\gamma_j$ represents a "segment" of an unitary "area" of undefined shape in the plane identified by unitary vectors γ_i and γ_j. The product $\gamma_1\gamma_2\ldots\gamma_n$ represents a unitary, n-dimensional volume segment identified by the unitary vectors $\gamma_1, \gamma_2, \ldots, \gamma_n$. In the n-dimensional space no more than 2^n elementary distinct "components" exist. Each of these entities corresponds to a particular subset of the orthonormal base vectors. The "grade" of these entities (called blades) is equal to the number of base vectors which are present within the subset. The blade of grade zero (empty set) is the dimensionless scalar unit. The number of blades of kth-grade in a n-dimensional space is equal to the binomial coefficient

$$N_k = \binom{n}{k} = \frac{n!}{k!\,(n-k)!} \tag{0.2.4}$$

Using Clifford algebra, there are two possible choices for the metric of space–time coordinates, namely $Cl_{1,3}(\mathbb{R})$ and $Cl_{3,1}(\mathbb{R})$. For $Cl_{1,3}(\mathbb{R})$ (signature "+ − − −"), we have

$$-\gamma_x^2 = -\gamma_y^2 = -\gamma_z^2 = \gamma_t^2 = 1$$

whereas for $Cl_{3,1}(\mathbb{R})$ (signature "+ + + −"),

$$\gamma_x^2 = \gamma_y^2 = \gamma_z^2 = -\gamma_t^2 = 1$$

In $Cl_{3,1}(\mathbb{R})$ algebra, which is used in Chapters 1–3, the spatial coordinates can be viewed as the familiar Cartesian coordinates of the Euclidean space. $Cl_{1,3}(\mathbb{R})$ signature is used however in Chapters 5 and 8. The 2^4 components of the $Cl_{3,1}(\mathbb{R})$ space–time Clifford algebra are listed in Table 0.1.

Table 0.1. Blades of space–time algebra ($Cl_{3,1}$).

Blade	Bit mask	Grade	Hex.
1	0000	0 (scalar)	0
γ_x	0001	1 (vector)	1
γ_y	0010	1 (vector)	2
$\gamma_x\gamma_y = \gamma_{xy}$	0011	2 (bivector)	3
γ_z	0100	1 (vector)	4
$\gamma_x\gamma_z = \gamma_{xz}$	0101	2 (bivector)	5
$\gamma_y\gamma_z = \gamma_{yz}$	0110	2 (bivector)	6
$\gamma_x\gamma_y\gamma_z = \gamma_{xyz} = I_\Delta$	0111	3 (pseudovector)	7
γ_t	1000	1 (vector)	8
$\gamma_x\gamma_t = \gamma_{xt}$	1001	2 (bivector)	9
$\gamma_y\gamma_t = \gamma_{yt}$	1010	2 (bivector)	A
$\gamma_x\gamma_y\gamma_t = \gamma_{xyt}$	1011	3 (pseudovector)	B
$\gamma_z\gamma_t = \gamma_{zt}$	1100	2 (bivector)	C
$\gamma_x\gamma_z\gamma_t = \gamma_{xzt}$	1101	3 (pseudovector)	D
$\gamma_y\gamma_z\gamma_t = \gamma_{yzt}$	1110	3 (pseudovector)	E
$\gamma_x\gamma_y\gamma_z\gamma_t = \gamma_{xyzt} = I$	1111	4 (pseudoscalar)	F

Each blade is associated with a real number, the blade value. The expression $a\gamma_i\gamma_j$, in which a is a real scalar, represents an area a of an undefined shape in the plane identified by vectors γ_i and γ_j.

Summing can be carried out only between coefficients of identical blades. If used for different blades, the sum must be reinterpreted as a composition, in analogy with the concept of sum between real and imaginary parts of a complex number.

A multi-vector is a generic composition of one or more blades. Within an n-dimensional space, a multi-vector is composed by no more than 2^n blades. The geometrical product between two blades gives a blade which is obtained from the application of the *bitwise exclusive OR* operation between the bit mask of blades to be multiplied. The value of the resultant blade is equal to the product of the values of the operands multiplied by a sign which is a function of the two blades. For $Cl_{3,1}(\mathbb{R})$ algebra, the sign can be easily determined by means of Table 0.2, identifying the blades to

Table 0.2. Signs of the geometrical product in the space–time algebra $Cl_{3,1}$. $I = \gamma_{xyzt} = \gamma_x\gamma_y\gamma_z\gamma_t$, $I_\Delta = \gamma_{xyz} = \gamma_x\gamma_y\gamma_z$.

Grade	0	1	1	2	1	2	2	3	1	2	2	3	2	3	3	4
Hex.	0	1	2	3	4	5	6	7	8	9	A	B	C	D	E	F
Label	1	γ_x	γ_y	γ_{xy}	γ_z	γ_{xz}	γ_{yz}	I_Δ	γ_t	γ_{xt}	γ_{yt}	γ_{xyt}	γ_{zt}	γ_{xzt}	γ_{yzt}	I
1	+	+	+	+	+	+	+	+	+	+	+	+	+	+	+	+
γ_x	+	+	+	+	+	+	+	+	+	+	+	+	+	+	+	+
γ_y	+	−	+	−	+	−	+	−	+	−	+	−	+	−	+	−
γ_{xy}	+	−	+	−	+	−	+	−	+	−	+	−	+	−	+	−
γ_z	+	−	−	+	+	−	−	+	+	−	−	+	+	−	−	+
γ_{xz}	+	−	−	+	+	−	−	+	+	−	−	+	+	−	−	+
γ_{yz}	+	+	−	−	+	+	−	−	+	+	−	−	+	+	−	−
I_Δ	+	+	−	−	+	+	−	−	+	+	−	−	+	+	−	−
γ_t	+	−	−	+	−	+	+	−	−	+	+	−	+	−	−	+
γ_{xt}	+	−	−	+	−	+	+	−	−	+	+	−	+	−	−	+
γ_{yt}	+	+	−	−	−	−	+	+	−	−	+	+	+	+	−	−
γ_{xyt}	+	+	−	−	−	−	+	+	−	−	+	+	+	+	−	−
γ_{zt}	+	+	+	+	−	−	−	−	−	−	−	−	+	+	+	+
γ_{xzt}	+	+	+	+	−	−	−	−	−	−	−	−	+	+	+	+
γ_{yzt}	+	−	+	−	−	+	−	+	−	+	−	+	+	−	+	−
I	+	−	+	−	−	+	−	+	−	+	−	+	+	−	+	−

be multiplied through their hexadecimal "label" taken from Table 0.1. In general, this product is not commutative. The sign table is obtained by the direct application of Equations (0.2.1), (0.2.2), and (0.2.3).

There are other types of products in Clifford algebra: the wedge product (symbol $\wedge$) and the scalar product (symbol $\cdot$). The result is computed following the same rules of geometric product, but is zero in some cases:

(1) the wedge product vanishes if the intersection between the set of base vectors of the first operand blade and the set of base vectors of the second operand blade is not empty, i.e., if the bitwise logical and of the operands' blades bitmasks is not zero (empty set);
(2) the scalar product is a pure number that vanishes if the blade of the first operand is different from that of the second operand.

The geometric product of two vectors in Clifford algebra can be decomposed in a scalar product and a wedge product according to the relation

$$\boldsymbol{uv} = \boldsymbol{u} \cdot \boldsymbol{v} + \boldsymbol{u} \wedge \boldsymbol{v} \tag{0.2.5}$$

It is important to note that the space–time algebra of the four γ_i vectors is isomorphic to the algebra of Majorana matrices. The Majorana matrices can be represented by the Dirac gamma matrices times the imaginary unit.

Example. Some examples of the application of products are here reported referring to $Cl_{3,1}(\mathbb{R})$:

$$\gamma_x\gamma_y\,\gamma_x\gamma_y = -\gamma_x\gamma_x\gamma_y\gamma_y = -1$$

$$\gamma_x\gamma_y \cdot \gamma_z\gamma_t = 0$$

$$a\gamma_x\gamma_y \wedge b\gamma_z = ab\gamma_x\gamma_y\gamma_z$$

$$\gamma_x\gamma_y \wedge \gamma_y = 0$$

$$\gamma_x\gamma_y \wedge \gamma_z\gamma_t = \gamma_x\gamma_y\gamma_z\gamma_t$$

$$\gamma_x \cdot \gamma_y = 0$$

$$\begin{aligned}(\gamma_x + \gamma_t)^2 &= (\gamma_x + \gamma_t)(\gamma_x + \gamma_t) = \gamma_x^2 + \gamma_x\gamma_t + \gamma_t\gamma_x + \gamma_t^2 \\ &= 1 + \gamma_x\gamma_t - \gamma_x\gamma_t - 1 = 0\,^{\text{a}}\end{aligned}$$

$$\begin{aligned}(\gamma_x + \gamma_y)^2 &= (\gamma_x + \gamma_y)(\gamma_x + \gamma_y) = \gamma_x^2 + \gamma_x\gamma_y + \gamma_y\gamma_x + \gamma_y^2 \\ &= 1 + \gamma_x\gamma_y - \gamma_x\gamma_y + 1 = 2\,^{\text{b}}\end{aligned}$$

[a]Example of a light-like vector, its square is 0.
[b]Example of a space-like vector, its square is >0.

$$
\begin{aligned}
&(a\gamma_t)^2 = -a^2 \text{ [c]}\\
&(a\gamma_x + b\gamma_y)^2 = (a\gamma_x + b\gamma_y)(a\gamma_x + b\gamma_y) = a^2\gamma_x^2 + ab\gamma_x\gamma_y + ba\gamma_y\gamma_x + b^2\gamma_y^2\\
&\qquad = a^2 + ab\gamma_x\gamma_y - ab\gamma_x\gamma_y + b^2 = a^2 + b^2 \text{ [d]}\\
&(a\gamma_x + b\gamma_t)^2 = (a\gamma_x + b\gamma_t)(a\gamma_x + b\gamma_t) = a^2\gamma_x^2 + ab\gamma_x\gamma_t + ba\gamma_t\gamma_x + b^2\gamma_t^2\\
&\qquad = a^2 + ab\gamma_x\gamma_t - ab\gamma_x\gamma_t - b^2 = a^2 - b^2
\end{aligned}
$$

In these examples, a and b are generic real scalars.

0.2.1. *Reflection and rotation of vectors*

In order to perform a mirror reflection of a vector with respect to a plane, the following formula holds in Clifford algebra:

$$\boldsymbol{a}' = -\boldsymbol{mam} \tag{0.2.6}$$

where $\boldsymbol{m}$ is the unitary vector orthogonal to surface α, as shown in Figure 0.1. If $\boldsymbol{a}_\perp$ and $\boldsymbol{a}_\parallel$ are the components of vector $\boldsymbol{a}$ orthogonal and parallel to m, then

$$
\begin{aligned}
\boldsymbol{a} &= \boldsymbol{a}_\perp + \boldsymbol{a}_\parallel\\
\boldsymbol{a}' &= -\boldsymbol{m}\left(\boldsymbol{a}_\perp + \boldsymbol{a}_\parallel\right)\boldsymbol{m} = -\boldsymbol{m}\boldsymbol{a}_\perp\boldsymbol{m} - \boldsymbol{m}\boldsymbol{a}_\parallel\boldsymbol{m}\\
&= \boldsymbol{a}_\perp\boldsymbol{m}^2 - \boldsymbol{a}_\parallel\boldsymbol{m}^2 = \boldsymbol{a}_\perp - \boldsymbol{a}_\parallel
\end{aligned}
$$

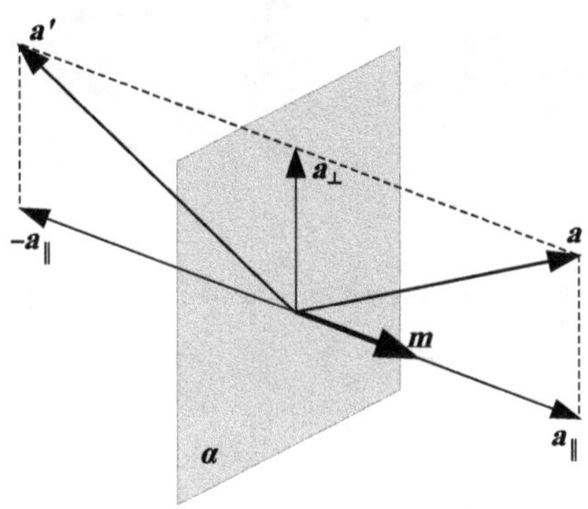

Figure 0.1. Reflection of a vector with respect to plane α.

[c] Example of a time-like vector, its square is <0.

[d] Always a space-like vector.

This operation is justified by the fact that the product between parallel vectors commutes, i.e.,

$$\boldsymbol{m}\boldsymbol{a}_{\parallel} = \boldsymbol{a}_{\parallel}\boldsymbol{m}$$

whereas the product between orthogonal vectors anti-commutes, i.e.,

$$\boldsymbol{m}\boldsymbol{a}_{\perp} = -\boldsymbol{a}_{\perp}\boldsymbol{m}$$

Now we take a second unitary vector $\boldsymbol{n}$, which is located in the plane spanned by $\boldsymbol{a}$ and $\boldsymbol{m}$, and is rotated with respect to $\boldsymbol{m}$ by an angle θ. The placement of $\boldsymbol{a}$, $\boldsymbol{m}$, and $\boldsymbol{n}$ is shown in Figure 0.2. As before, we reflect $\boldsymbol{a}$ with respect to the unitary vector $\boldsymbol{m}$: $\boldsymbol{a}' = -\boldsymbol{m}\boldsymbol{a}\boldsymbol{m}$. If vector $\boldsymbol{a}'$ in now reflected again with respect to the unitary vector $\boldsymbol{n}$, we obtain:

$$\boldsymbol{a}'' = -\boldsymbol{n}\boldsymbol{a}'\boldsymbol{n} = \boldsymbol{n}\boldsymbol{m}\boldsymbol{a}\boldsymbol{m}\boldsymbol{n} \tag{0.2.7}$$

Interestingly, vector $\boldsymbol{a}''$ becomes rotated, with respect to vector $\boldsymbol{a}$, by an angle 2θ on the common plane of the two-vector $\boldsymbol{m}$ and $\boldsymbol{n}$ as shown in Figure 0.2. This result can be seen from the angle decomposition shown in Figure 0.3. The angle between $\boldsymbol{m}$ and $\boldsymbol{n}$ is $\theta = \alpha + \beta$, and the obtained angle between $\boldsymbol{a}$ and $\boldsymbol{a}''$ is $2\theta = 2\alpha + 2\beta$.

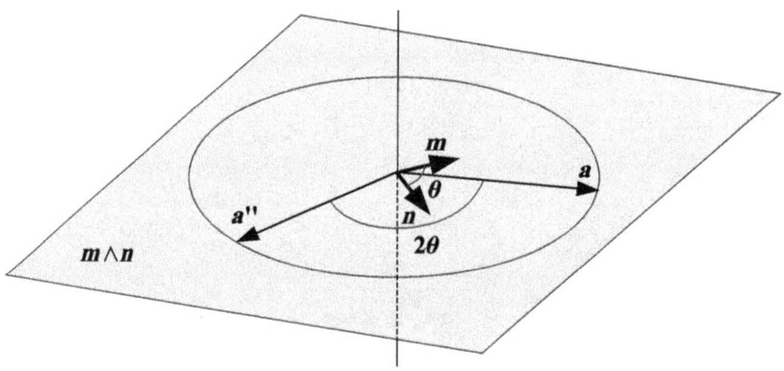

Figure 0.2. Rotation of vector $\boldsymbol{a}$ due to two subsequent reflections.

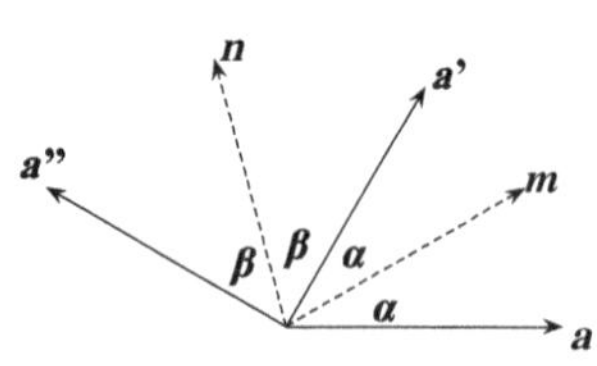

Figure 0.3. Decomposition of the rotation angle 2θ, due to two subsequent reflections.

Consider the $\gamma_x\gamma_y$ plane spanned by the orthogonal unit vectors γ_x and γ_y. A rotation in this plane by an angle 2θ can of course be understood as a linear map r determined by

$$\begin{aligned}\gamma_x \mapsto r(\gamma_x) &= \cos(2\theta)\,\gamma_x + \sin(2\theta)\,\gamma_y \\ \gamma_y \mapsto r(\gamma_y) &= -\sin(2\theta)\,\gamma_x + \cos(2\theta)\,\gamma_y\end{aligned} \tag{0.2.8}$$

We saw in the multiplication examples that $(\gamma_x\gamma_y)^2 = \gamma_x\gamma_y\,\gamma_x\gamma_y = -\gamma_x\gamma_x\gamma_y\gamma_y = -1$, which means that the unit blade $\gamma_x\gamma_y$ behaves like the complex imaginary unit i in this regard, and we can rewrite Equation (0.2.8) as

$$\begin{aligned}r(\gamma_x) &= \cos(2\theta)\,\gamma_x + \sin(2\theta)\,\gamma_x\gamma_x\gamma_y \\ &= \gamma_x\left(\cos(2\theta) + \sin(2\theta)\,\gamma_x\gamma_y\right) = \gamma_x e^{2\theta\gamma_x\gamma_y} \\ r(\gamma_y) &= -\sin(2\theta)\,\gamma_y\gamma_y\gamma_x + \cos(2\theta)\,\gamma_y \\ &= \gamma_y\left(\sin(2\theta)\,\gamma_x\gamma_y + \cos(2\theta)\right) = \gamma_y e^{2\theta\gamma_x\gamma_y}\end{aligned} \tag{0.2.9}$$

We further arrange the above equation as follows:

$$\begin{aligned}r(\gamma_x) &= \gamma_x e^{\theta\gamma_x\gamma_y} e^{\theta\gamma_x\gamma_y} = \gamma_x\left(\cos(\theta) + \sin(\theta)\,\gamma_x\gamma_y\right) e^{\theta\gamma_x\gamma_y} \\ &= \left(\cos(\theta) - \sin(\theta)\,\gamma_x\gamma_y\right)\gamma_x e^{\theta\gamma_x\gamma_y} = e^{-\theta\gamma_x\gamma_y}\gamma_x e^{\theta\gamma_x\gamma_y}\end{aligned} \tag{0.2.10}$$

$$r(\gamma_y) = e^{-\theta\gamma_x\gamma_y}\gamma_y e^{\theta\gamma_x\gamma_y} \tag{0.2.11}$$

The point with the above rewriting is that the resulting expression also applies to the basis vector γ_x, which is orthogonal to the rotation plane and thus unaffected by the rotation:

$$r(\gamma_z) = \gamma_z e^{-\theta\gamma_x\gamma_y} e^{\theta\gamma_x\gamma_y} = e^{-\theta\gamma_x\gamma_y}\gamma_z e^{\theta\gamma_x\gamma_y} \tag{0.2.12}$$

where we used that $\gamma_z\gamma_x\gamma_y = \gamma_x\gamma_y\gamma_z$. By linearity, the rotation r acting on any vector $\boldsymbol{a}$ can be written as

$$\boldsymbol{a}" = r(\boldsymbol{a}) = e^{-\theta\gamma_x\gamma_y}\boldsymbol{a}e^{\theta\gamma_x\gamma_y} = \boldsymbol{R}\boldsymbol{a}\tilde{\boldsymbol{R}} \tag{0.2.13}$$

where the object $\boldsymbol{R} = e^{-\theta\gamma_x\gamma_y}$ is called a rotor, and it encodes in a compact way both the plane of rotation and the angle. Also note that by the generality of the construction, and by choosing a basis accordingly, any rotation can be expressed in this form by some rotor $\boldsymbol{R} = e^{-\theta\boldsymbol{m}\wedge\boldsymbol{n}}$, where the blade $\boldsymbol{m}\wedge\boldsymbol{n}$ represents the plane in which the rotation is performed, and the angle of rotation is 2θ.

At the same time, we remember that the rotor can also be expressed as a product of unitary vectors:

$$\boldsymbol{R} = \boldsymbol{nm} = \boldsymbol{m} \cdot \boldsymbol{n} - \boldsymbol{m} \wedge \boldsymbol{n}$$

and

$$\tilde{\boldsymbol{R}} = \boldsymbol{mn} = \boldsymbol{m} \cdot \boldsymbol{n} + \boldsymbol{m} \wedge \boldsymbol{n}$$

It is important to note that these rules are independent from the signature, and for this reason they are also valid for four-vectors of the space–time algebra. In particular, rotors with pure spatial bivector parts (such as $\gamma_x\gamma_y$) generate ordinary rotations, whereas rotors containing bivectors with the term γ_t (such as $\gamma_z\gamma_t$) generate hyperbolic rotations which are the Lorentz boosts. Rotor operations are a very powerful geometric tool for numerous applications [5].

In the above analysis, we identified the exponential form of a rotor with a unitary complex number. In a fixed coordinate system, any spatial rotation can be decomposed into two consecutive rotations along two blades, e.g., along $\gamma_x\gamma_y$ and $\gamma_y\gamma_z$. The group of three-dimensional rotations is the $SO\,(3)$ group, which may thus be represented by two unitary complex numbers. The group of two unitary complex numbers is the $SU\,(2)$ group. The important point here is that a rotor with angle θ generates physical rotation with angle 2θ. Thus, one full circle in the rotor space corresponds to two full circles of physical rotation. This is what mathematicians mean by saying that the $SU\,(2)$ group is a double cover of the $SO\,(3)$ group.

0.2.2. *Clifford reversion*

We define Clifford reversion as the operator which reverses the order of base vectors. Scalars and vectors thus remain unchanged upon reversion: $\widetilde{S} = S$ and $\widetilde{\gamma_i} = \gamma_i$. Reversion acts as follows on bi-vectors and tri-vectors: $\widetilde{(\gamma_i\gamma_j)} = \gamma_j\gamma_i = -\gamma_i\gamma_j$ and $\widetilde{(\gamma_i\gamma_j\gamma_k)} = \gamma_k\gamma_j\gamma_i = -\gamma_i\gamma_j\gamma_k$.

0.3. Exterior Algebra and Wedge Products

The meaning of exterior algebra can be intuitively grasped with a Clifford algebra approach. Given two vectors $\mathbf{m}$ and $\mathbf{n}$, their wedge product $\mathbf{A} = \mathbf{m} \wedge \mathbf{n}$ can be thought of as an oriented area element. In three-dimensional space, the oriented area element can be associated with a normal vector to this area, with the length of the vector representing the area size. As illustrated in Figure 0.4, this results in a one-to-one mapping between vectors

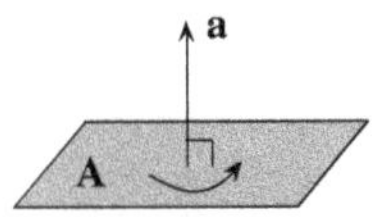

Figure 0.4. A vector $\mathbf{a}$ and its dual bivector $\mathbf{A} = \mathbf{a}\mathbf{e}_{xyz}$.

and bivectors. This mapping is called the Hodge dual, and is denoted by the $*$ symbol:

$$\mathbf{a} \mapsto \mathbf{A} = \mathbf{a}^* = a_x\mathbf{e}_y \wedge \mathbf{e}_z + a_y\mathbf{e}_z \wedge \mathbf{e}_x + a_z\mathbf{e}_x \wedge \mathbf{e}_y \tag{0.3.1}$$

In three dimensions, this dual mapping is essentially an association between a wedge product and a cross product:

$$\begin{aligned} \mathbf{m} \wedge \mathbf{n} &= (\mathbf{m} \times \mathbf{n})^* = (\mathbf{m} \times \mathbf{n})\,\mathbf{e}_{xyz} \\ \mathbf{m} \times \mathbf{n} &= (\mathbf{m} \wedge \mathbf{n})^* = (\mathbf{m} \wedge \mathbf{n})\,\mathbf{e}_{xyz} \end{aligned} \tag{0.3.2}$$

Such dual mapping works on the basis of same dimensionality between the mapped spaces. In the four-dimensional Minkowski space, the analogous mapping is between vectors and oriented volume elements: $\mathbf{a} \mapsto \mathbf{a}^* = \mathbf{a}\mathbf{e}_{txyz}$, and it frequently shows up in electromagnetism and quantum mechanics.

The basis elements of the dual space are denoted as $\mathbf{e}_1 \wedge \cdots \wedge \mathbf{e}_r$. The wedge product multiplication rules define the exterior algebra:

$$\mathbf{e}_i \wedge \mathbf{e}_i = 0 \tag{0.3.3}$$

$$\mathbf{e}_i \wedge \mathbf{e}_j = -\mathbf{e}_j \wedge \mathbf{e}_i \; (i \neq j) \tag{0.3.4}$$

Compare the above exterior algebra rules with the multiplication rules of Clifford algebra bases given by Equations (0.1.1) and (0.1.2). Using the above rules, and recalling the Clifford product rule given in Equation (0.2.5), we can express a wedge product in terms of antisymmetric Clifford products:

$$\mathbf{m} \wedge \mathbf{n} = \frac{1}{2}(\mathbf{mn} - \mathbf{nm}) \tag{0.3.5}$$

We now make ordinary algebraic definitions of wedge product spaces. We assume the reader is familiar with the language of vector spaces and linear transformations. Einsteinian notation will be used throughout unless ambiguity occurs, in which case the summation symbol will be included.[a]

[a]If $\mathbf{v} \in V$, where V is a vector space, such that $\mathbf{v} = \sum_{i=1}^{n} v^i e_i$, then in Einsteinian notation we will write this as $\mathbf{v} = v^i e_i$. In general, when there is no ambiguity, we will always take the presence of a double index to represent summation.

Let V be a vector space of dimension n. An exterior r-form on V is an r-linear map ω such that

$$\omega(X_1, \ldots X_i, \ldots X_j, \ldots X_r) = -\omega(X_1, \ldots X_j, \ldots X_i, \ldots X_r) \tag{0.3.6}$$

for all $(X_1, \ldots X_r) \in V$ and for all $i,\ j = 1, \ldots, r$ and $r \leq n$.

In terms of a local dual basis $\{e^i\}$ for V^*, we can write

$$\omega = \omega_{[i_1 \ldots i_r]} e^{i_1} \otimes \cdots \otimes e^{i_r}$$

with the understanding that the $\omega_{[i_1 \ldots i_r]}$ is skew symmetric. In other words,

$$\omega_{[i_1 \ldots i_r]} = \frac{1}{r!} \delta^{j_1 \ldots j_r}_{i_1 \ldots i_r} \omega_{j_1 \ldots j_r} \tag{0.3.7}$$

where

$$\begin{aligned} \delta^{j_1 \ldots j_r}_{i_1 \ldots i_r} &= +1, \text{ if } j_1 \ldots j_r \text{ is an even permutation of } i_1 \ldots i_r \\ &= -1, \text{ if } j_1 \ldots j_r \text{ is an odd permutation of } i_1 \ldots i_r \\ &= 0 \text{ otherwise} \end{aligned}$$

For example, in the case of $r = 2$, we write

$$\omega = \omega_{[i_1 i_2]} e^{i_1} \otimes e^{i_2} = \omega_{12} e^1 \otimes e^2 - \omega_{21} e^1 \otimes e^2$$

Next we define an exterior product of r 1-forms:

Definition. Let $\omega^1, \ldots, \omega^r \in V^*$. Then their exterior product is defined by

$$\omega^1 \wedge \cdots \wedge \omega^r \equiv \frac{1}{r!} \delta^{1 \ldots r}_{i_1 \ldots i_r} \omega^{i_1} \otimes \cdots \otimes \omega^{i_r}$$

For example, in the case of $r = n = 3$ it is an easy exercise to show that for $\mathbf{a} = a_i e^i, \mathbf{b} = b_i e^i, \mathbf{c} = c_i e^i$,

$$\begin{aligned} \mathbf{a} \wedge \mathbf{b} \wedge \mathbf{c} = & \frac{1}{3!}(\mathbf{a} \otimes \mathbf{b} \otimes \mathbf{c} + \mathbf{b} \otimes \mathbf{c} \otimes \mathbf{a} + \mathbf{c} \otimes \mathbf{a} \otimes \mathbf{b} \\ & - \mathbf{a} \otimes \mathbf{c} \otimes \mathbf{b} - \mathbf{b} \otimes \mathbf{a} \otimes \mathbf{c} - \mathbf{c} \otimes \mathbf{b} \otimes \mathbf{a}) \\ = & \ det[\mathbf{a}, \mathbf{b}, \mathbf{c}] e^1 \wedge e^2 \wedge e^3 \end{aligned}$$

and in general it follows from the definition that

$$\omega^1 \wedge \cdots \wedge \omega^r = \det[\omega^1, \ldots, \omega^r] e^1 \wedge \cdots \wedge e^r$$

It is also important to note that from the defining Equation (0.3.6), if $X_i = X_j$ for some $i \neq j$, then $\omega_{[i_1 \ldots i_r]} = 0$ by skew symmetry. Indeed, it is easy to check in the example above that if $\mathbf{a} = \mathbf{b}$, then $\mathbf{a} \wedge \mathbf{b} \wedge \mathbf{c} = 0$.

0.4. Tensors

The language of tensor and wedge products will permit us to work with invariants of the system. Let $\mathbf{v} \in V$, where V is a vector space, such that

$$\mathbf{v} = v^i e_i \tag{0.4.1}$$

where $\{e_i\}$ is a basis for the system. Under a linear map A, the vector will transform into

$$A\mathbf{v} = a^i_j v^j e_i = (v')^i e_i \tag{0.4.2}$$

in the same basis. From now on, we abuse this notation slightly and write $(v')^i = v^{i'}$. For example, if A represents a rotation, then $A\mathbf{v}$ is the new vector that comes from rotating $\mathbf{v}$. However, it is also possible to choose a basis $\{e_{i'}\}$ such that

$$\mathbf{v} = v^i e_i = v^{i'} e_{i'}, \quad \text{where } v^{i'} = a^i_j v^j \tag{0.4.3}$$

Viewed from this perspective, $\mathbf{v}$ is independent of the choice of basis and can be considered an invariant of the system. Moreover, a^i_j is independent of the choice of $\mathbf{v}$ and depends only on the change of basis. We call the set $\{\mathbf{v}\}$ a (contravariant) tensor space V of order (1,0).

Similarly, we can define a (covariant) tensor by considering the vector space V^*, which is the dual space of V. In other words, let $\phi \in V^*$ be a function on V such that

$$V \to \mathbb{R} : \mathbf{v} \mapsto \phi(\mathbf{v}) \tag{0.4.4}$$

then the set $\{\phi\}$ defines a vector space. In particular, if $\{e_i\}$ define a basis for V, then there exists a set of maps $\{\phi^i\}$ such that

$$\phi^i(e_j) = \delta^i_j \tag{0.4.5}$$

As shown in Section 0.3, the set of maps with this characteristic defines a basis for V^* which from here on we will denote by $\{e^i\}$. Consequently, for an arbitrary $\phi \in V^*$, we can write

$$\phi = \phi_i e^i \tag{0.4.6}$$

The set $\{e^i\}$ is called the dual basis of $\{e_i\}$. Also, under a change of basis from $\{e^i\}$ to $\{e^{i'}\}$ we can write

$$\phi = \phi_i e^i = \phi_{i'} e^{i'} \tag{0.4.7}$$

We call the set of $\{\phi\}$ a (covariant) tensor space of order (0,1).

Equivalently, we can define a tensor of type (1,0) as a set of n quantities (v^i) associated with a basis $\{e_i\}$ of a vector space, which under a transformation

$$e_{i'} = a^j_{i'} e_j \tag{0.4.8}$$

transforms as

$$v^{i'} = a^{i'}_j v^j \tag{0.4.9}$$

In the particular case of coordinate transformations on a differential manifold, this can be expressed in the form

$$v^{i'} = v^j \frac{\partial x^{i'}}{\partial x^j} \tag{0.4.10}$$

or in differential notation as

$$dx^{i'} = dx^j \frac{\partial x^{i'}}{\partial x^j} \tag{0.4.11}$$

An analogous definition can also be used to define a tensor of type (0,1). It is important to note that although dx^j are variables in Equation (0.4.11), for an affine transformation each coefficient $\frac{\partial x^{i'}}{\partial x^j}$ is a constant for all i and j. Geometrically, this can be interpreted to mean that under an affine change of basis the two frames remain fixed relative to each other.

Now consider a vector space which we symbolically represent by $V_1 \otimes V_2$, constructed from the vector spaces V_1 and V_2, respectively. If $T \in V_1 \otimes V_2$, then we can write T in basis form as

$$T = T^{ij} e_i \otimes e_j \tag{0.4.12}$$

with the transformation properties

$$T = T^{ij} e_i \otimes e_j = T^{i'j'} e_{i'} \otimes e_{j'} \tag{0.4.13}$$

where

$$T^{i'j'} = a^{i'}_i a^{j'}_j T^{ij} \tag{0.4.14}$$

In this case, we can say that T^{ij} is the component of a $(2,0)$ tensor. Similarly, in the case of a dual map we say that T_{ij} is the component of a $(0,2)$ tensor, and that T^i_j is the component of a mixed tensor. In both of these cases, we can write

$$T = T_{ij}e^i \otimes e^j \tag{0.4.15}$$

with the transformation properties

$$T = T_{ij}e^i \otimes e^j = T_{i'j'}e^{i'} \otimes e^{j'} \tag{0.4.16}$$

where

$$T_{i'j'} = a^i_{i'}a^j_{j'}T_{ij} \tag{0.4.17}$$

and in the case of the mixed tensor

$$T = T^i_j e_i \otimes e^j \tag{0.4.18}$$

with the transformation properties

$$T = T^i_j e_i \otimes e^j = T^{i'}_{j'} e_{i'} \otimes e^{j'} \tag{0.4.19}$$

where

$$T^{i'}_{j'} = a^{i'}_i a^j_{j'} T^i_j \tag{0.4.20}$$

In general, we define a tensor product as follows: Let $V_1^* \otimes V_2^* \otimes \cdots \otimes V_r^* \otimes V_{r+1} \otimes \cdots \otimes V_n$ be a vector space whose components $\mathbf{v_1^*} \otimes \mathbf{v_2^*} \otimes \cdots \otimes \mathbf{v_n}$ transform as a tensor under coordinate transformations, then $V_1^* \otimes V_2^* \otimes \cdots \otimes V_n$ is called a tensor product of the n vector spaces.

Specifically, if $e^1 \otimes e^2 \otimes \cdots \otimes e^r \otimes e_{r+1} \dots e_n$ defines a basis for $V_1^* \otimes V_2^* \otimes \cdots \otimes V_n$, then for any $T \in V_1^* \otimes V_2^* \otimes \cdots \otimes V_n$ we can write

$$T = T^{i_{r+1}\dots i_n}_{i_1\dots i_r} e^1 \otimes e^2 \otimes \cdots \otimes e^r \otimes e_{r+1} \dots e_n \tag{0.4.21}$$

such that

$$T^{i_{r+1}\dots i'_n}_{i'_1\dots i'_r} = a^1_{i'_1} \dots a^r_{i'_r} a^{i'_{r+1}}_{r+1} \dots a^{i'_n}_n T^{i_{r+1}\dots i_n}_{i_1\dots i_r} \tag{0.4.22}$$

for the transformation of the various components of the tensor under a change of basis.

For example, in the case of a differential manifold this change of basis permits us to write for a contravariant tensor

$$T^{i'_1 i'_2 \dots i'_n} = \frac{\partial x^{i'_1}}{\partial x^1}\frac{\partial x^{i'_2}}{\partial x_2} \cdots \frac{\partial x^{i'_n}}{\partial x_n} T^{i_1 i_2 \dots i_n} \tag{0.4.23}$$

Another useful example is the antisymmetric tensor of type (0,2) given by

$$alt(\mathbf{T}) = T_{[i_1,i_2]}e^{i_1} \otimes e^{i_2} \quad \text{where } T_{[i_1,i_2]} \equiv \frac{1}{2!}(T_{i_1 i_2} - T_{i_2 i_1})$$

The reader may notice that $T_{[i_1,i_2]}e^{i_1} \otimes e^{i_2}$ is an exterior 2-form, as defined in Section 0.3.

0.5. Singlet State

One of the purposes of this book is to relate spin statistics and entanglement, or more specifically to relate Fermi–Dirac statistics to singlet states. Physicists refer to a singlet state as a system in which particles are paired and which has net angular momentum zero. This definition, while useful for our intuition, needs to be refined if we are to do more analytical work. With this in mind, we present a more formal definition in terms of wedge products.

Definition. Let **a** and **b** represent two quantum states of a vector space V. Then the state $\mathbf{a} \wedge \mathbf{b}$ is called a singlet state.

Singlet states will be at the center of discussion in Chapter 9: from one perspective, that entire chapter pertains to the properties of singlet states as they relate to the Pauli exclusion principle and Fermi–Dirac statistics.

In the case that V is a two-dimensional space, the singlet state takes on the explicit form

$$\mathbf{a} \wedge \mathbf{b} \equiv \frac{1}{\sqrt{2}}\left[\begin{pmatrix} a_1 \\ a_2 \end{pmatrix}\begin{pmatrix} b_1 \\ b_2 \end{pmatrix} - \begin{pmatrix} b_1 \\ b_2 \end{pmatrix}\begin{pmatrix} a_1 \\ a_2 \end{pmatrix}\right] \tag{0.5.1}$$

$$= \frac{a_1 b_2 - a_2 b_1}{\sqrt{2}}\left[\begin{pmatrix} 1 \\ 0 \end{pmatrix}\begin{pmatrix} 0 \\ 1 \end{pmatrix} - \begin{pmatrix} 0 \\ 1 \end{pmatrix}\begin{pmatrix} 1 \\ 0 \end{pmatrix}\right] \tag{0.5.2}$$

where we applied a tensor product between the **a** and **b** vectors. To see the above result, one may expand $\begin{pmatrix} a_1 \\ a_2 \end{pmatrix}\begin{pmatrix} b_1 \\ b_2 \end{pmatrix}$ as follows:

$$a_1 b_2 \begin{pmatrix} 1 \\ 0 \end{pmatrix}\begin{pmatrix} 0 \\ 1 \end{pmatrix} - a_2 b_1 \left[-\begin{pmatrix} 0 \\ 1 \end{pmatrix}\begin{pmatrix} 1 \\ 0 \end{pmatrix}\right] + a_1 b_1 \begin{pmatrix} 1 \\ 0 \end{pmatrix}\begin{pmatrix} 1 \\ 0 \end{pmatrix} + a_2 b_2 \begin{pmatrix} 0 \\ 1 \end{pmatrix}\begin{pmatrix} 0 \\ 1 \end{pmatrix}$$

Similarly, one may expand $-\binom{b_1}{b_2}\binom{a_1}{a_2}$ as follows:

$$-b_1a_2\begin{pmatrix}1\\0\end{pmatrix}\begin{pmatrix}0\\1\end{pmatrix}+b_2a_1\left[-\begin{pmatrix}0\\1\end{pmatrix}\begin{pmatrix}1\\0\end{pmatrix}\right]-b_1a_1\begin{pmatrix}1\\0\end{pmatrix}\begin{pmatrix}1\\0\end{pmatrix}$$
$$-b_2a_2\begin{pmatrix}0\\1\end{pmatrix}\begin{pmatrix}0\\1\end{pmatrix}$$

The result of Equation (0.5.2) follows by adding the above expressions. This result holds for both real-valued and complex-valued spaces.

The advantage of this definition is that it not only captures the explicit pairing process as can be seen from Equation (0.5.2) but also implicitly captures the isotropy associated with the pairing. Singlet states do not simply refer to a polarization in a plane but to a polarization in every plane at once. In others words, if two particles are in a spin-singlet state, it means that the total spin angular momentum is isotropically spin correlated and the resulting state is rotationally invariant. We will prove in Section 0.8.5 that $\mathbf{a} \wedge \mathbf{b}$ is invariant under any possible spatial orientation of the $\mathbf{a}$ and $\mathbf{b}$ vectors, which is encoded by their complex-valued coordinates.

0.6. Group Actions

For a (finite or infinite) set G, with a binary operation $*$ such that

$$G \times G \to G : (g_1, g_2) \mapsto g_1 * g_2$$

$(G, *)$ is a group if and only if

(1) For all $g_1, g_2 \in G$, $g_i * g_2 \in G$;
(2) For all $g_1, g_2, g_3 \in G$ then $g_1 * (g_2 * g_3) = (g_1 * g_2) * g_3$;
(3) There exists $e \in G$ such that for all $g \in G$, $e * g = g * e = g$;
(4) For all $g \in G$, there exists a $h \in G$ such that $g * h = h * g = e$. h is usually denoted by g^{-1}.

Later in the book, our focus will be on the action of linear groups of $n \times n$ matrices acting on the tensor product of vector spaces. A group representation is defined as follows [11]:

Definition. A representation of G assigns to each element $g \in G$ a unique linear transformation L:

$$D(g) \to L(V) : D(g_1)D(g_2) = D(g_1 g_2), \quad \forall g_1, g_2 \in G$$

For example, if $G = \{e^{i\theta} | \theta \in [0, 2\pi)\}$, then a map of the form

$$\exp(i\theta) \to D(\theta) = \begin{pmatrix} \cos\theta & \sin\theta \\ -\sin\theta & \cos\theta \end{pmatrix}$$

defines a group representation of G. Of particular interest will be the continuous linear group of $n \times n$ matrices of determinant 1, defined over the complex numbers. It is denoted by $SL(n, \mathbb{C})$. The most important of these continuous linear groups and their properties are listed in the following table:

Symbol	Name of group	Properties of matrix		No. of free parameters
$GL(n,\mathbb{C})$	General linear	complex,	$det(A) \neq 0$	$2n^2$
$GL(n,\mathbb{R})$	General linear	real,	$detA \neq 0$	n^2
$SL(n,\mathbb{C})$	Special linear	complex,	$detA = 1$	$2n^2 - 2$
$SL(n,\mathbb{R})$	Special linear	real,	$detA = 1$	$n^2 - 1$
$U(n)$	unitary	complex,	$A^T\bar{A} = E$	n^2
$SU(n)$	Special unitary	complex,	$A^T\bar{A} = E$	
			$detA = 1$	$n^2 - 1$
$O(n)$	Orthogonal	real,	$A^TA = E$	$\frac{1}{2}n(n-1)$
$SO(n)$	Special orthogonal	real,	$A^TA = E$	
	(proper orthogonal)		$detA = 1$	$\frac{1}{2}n(n-1)$

The action of a group G on a set X is defined to be a map

$$\phi : G \times X \to X : (g, x) \mapsto g(x)$$

In the case of group representations acting on a vector space, this means

$$\phi : D(G) \times V \to V : (D(g), x) \mapsto D(g)(x)$$

In particular, the general linear group $GL(n, \mathbb{C})$ acts on the vector space $\mathbb{C}^n$. Of interest to us will be the invariant action of the elements of $SL(n, \mathbb{C})$ on subsets of $\mathbb{C}^n \otimes \cdots \otimes \mathbb{C}^n$. For example, in the case of $\mathbf{a} \wedge \mathbf{b}$ as given in Equation (0.5.1), it is easy to check that $S(\mathbf{a} \wedge \mathbf{b}) = \mathbf{a} \wedge \mathbf{b}$ for all $S \in SL(2, \mathbb{C})$. In fact, it is the only invariant element of $SL(2, \mathbb{C})$.

0.7. Harmonic Functions

In this book, harmonic functions are defined as functions that satisfy Laplace's equation over Minkowski space. That is, if f is a harmonic function, it satisfies the following equation:

$$\frac{\partial^2 f}{\partial x^2} + \frac{\partial^2 f}{\partial y^2} + \frac{\partial^2 f}{\partial z^2} - \frac{\partial^2 f}{\partial t^2} = 0$$

We are interested in harmonic functions because fundamental physical entities, for example, the electromagnetic vector potential, are described by such functions. In Section 0.9, we will show how spinors are involved in solving a quadratic polynomial equation, with analogous structure to Laplace's equation.

0.8. Notation for Quantum Mechanics

0.8.1. *Definition of quantum states*

In this section, we introduce some mathematical definitions which are used in quantum mechanics. Quantum mechanics is essentially a probability theory of states, where a state is defined as an element of a so-called Hilbert space. A Hilbert space $\mathbb{S}$ is a vector space with an inner product defined on it. When we turn to the language of quantum mechanics, we usually denote contravariant vector of the form $\mathbf{v} \in \mathbb{S}$ by $|v\rangle$ and its corresponding dual (covariant) vector $\phi \in \mathbb{S}^*$ by $\langle u|$. We also note that $\langle u|v\rangle$ defines an inner product and is invariant under coordinate transformations. Let $|v\rangle = |\psi(\lambda)\rangle \in \mathbb{S}$ represent a quantum mechanical state associated with a particle, where λ are some coordinates of interest. For example, λ may stand for position, momentum, or spin coordinates, and $\psi(\lambda)$, maps these coordinates of interest to the Hilbert space coordinates. A main difficulty with quantum mechanics is the abstractness of its formalism, which may lead to a confusion about its physical interpretation. Avoiding such confusion is a main theme of this book.

For a system of n independent particles whose state space is defined on the Hilbert Space $\mathbb{S}_1 \otimes \mathbb{S}_2 \otimes \cdots \otimes \mathbb{S}_n$, we denote the n-particle state by $|\psi_1(\lambda_1)\rangle \ldots |\psi_n(\lambda_n)\rangle$. This definition expresses that the states of independent particles evolve independently of each other.

As an example of this notation being used in conjunction with tensors, consider a two-particle system with wave function $|\psi(\lambda_1, \lambda_2)\rangle = |\psi(\lambda_1)\rangle\, |\psi(\lambda_2)\rangle \in \mathbb{S}_1 \otimes \mathbb{S}_2$. In general, the particle state may be factorized into spin-independent and spin-representing components: $\mathbb{S}_i = |\psi(x)\rangle\, s_i$, where $s_i = \{|+\rangle, |-\rangle\}$ represents the set of two possible spin-states of a particle (for example, an electron). The binary outcome of the spin-state measurement was historically first noticed in the Stern–Gerlach experiment, which was a surprising result at the time. The physical background of $|\psi(x)\rangle\, s_i$ factorization will be discussed in Chapter 7.

We can write the two-particle state as

$$|\psi(\lambda_1, \lambda_2)\rangle = |\psi(\lambda_1)\rangle \otimes |\psi(\lambda_2)\rangle \tag{0.8.1}$$

$$= |\psi(x)\rangle\, s_1 \otimes |\psi(y))\rangle\, s_2 \tag{0.8.2}$$

$$= |\psi(x_1)\rangle\, |\psi(x_2)\rangle\, s_1 \otimes s_2 \tag{0.8.3}$$

0.8.2. *Probability and quantum states*

If in classical mechanics we conceive of the dynamics of a particle as being determined by Newton's laws of motion or Maxwell's laws of electrodynamics, in quantum theory we prescind from asking about the particle's motion and instead try to classify the possible states permitted to the particle in accordance with the laws of probability. The dynamics of the particle per se are replaced by the evolutionary dynamics of the possible states associated with a linear operator defined on the space. Such an operator is usually referred to as an "observable." In general, for the Hilbert space $\mathbb{S}$, an observable O defined on $\mathbb{S}$ is a linear map

$$O : \mathbb{S} \to \mathbb{S}$$

In principle, the theory applies to all linear operators. In practice, we focus on specific operators of interest to us. Operator Q usually represents position, P momentum, $L = Q \wedge P$ angular momentum, S spin, and $H(Q, P)$ a Hamiltonian operator. Also, if an operator O has a complete set of eigenvectors $|e_i\rangle \in \mathbb{S}$ with corresponding eigenvalues λ_i then every $|\psi\rangle \in \mathbb{S}$ can be expressed as

$$|\psi\rangle = c_i\, |e_i\rangle, \quad \text{where } O\, |e_i\rangle = \lambda_i\, |e_i\rangle$$

Again, in practice $|\psi\rangle$ usually determines the state of a particle for a given experiment, where the c_i^2 are the probabilities that a given eigenstate will be observed when the experiment is carried out. For example, the eigenstates of the non-relativistic Hamiltonian operator $H = H(Q, P)$ can be determined from solving the Schrödinger equation

$$(-\hbar^2 \nabla^2 + V)\, |\psi_i\rangle = \mathcal{E}_i\, |\psi_i\rangle \tag{0.8.4}$$

In the language of probability, we can say that the set of eigenstates for the Schrödinger operator defines the sample space for the normalized probability distribution associated with energy level $\mathcal{E}_i$ and probability $\mathcal{E}_i^2$. Note that normalized means that $\sum_{i=1}^{\infty} \mathcal{E}_i^2 = 1$.

In a steady state, an elementary particle state is described via the above-defined state vector assignment. How to describe the particle during $|e_1\rangle \to |e_2\rangle$ state transitions? During the transition process, we must take into account the simultaneous presence of $|e_1\rangle$ and $|e_2\rangle$, which are treated under the same mathematical formalism as a two-particle system: $|e_1\rangle|e_2\rangle \equiv \left(\begin{smallmatrix} |e_1\rangle \\ |e_2\rangle \end{smallmatrix}\right) \in \mathbb{S}_1 \otimes \mathbb{S}_2$. In this case, the Hamiltonian operator acting on $\left(\begin{smallmatrix} |e_1\rangle \\ |e_2\rangle \end{smallmatrix}\right)$ becomes a matrix operator: $H = \left(\begin{smallmatrix} H_{11} & H_{12} \\ H_{21} & H_{22} \end{smallmatrix}\right)$. Working with such a matrix operator is essential for a mathematically consistent description of a state transition process. In Chapter 2, we will use this formalism to describe the essential dynamics of quantum mechanical state transitions. Our insistence on a mathematically consistent description of state transition processes deviates from quantum mechanics textbooks, which treat state transitions as an instantaneous jump process.

0.8.3. *Spin*

Explaining the electron spin in physical terms is one goal of this book. Mathematically, the electron spin is simply a two-dimensional rotation state. The $s_i = \{|+\rangle, |-\rangle\}$ set distinguishes left-handed versus right-handed rotation chirality.

As explained in Section 0.2.1, a rotation angle θ is generated in two dimensions by a unitary rotor $\boldsymbol{R} = e^{-\frac{\theta}{2}\boldsymbol{m}\wedge\boldsymbol{n}}$, where $\boldsymbol{m}$ and $\boldsymbol{n}$ are two nonparallel vectors and $|\boldsymbol{m} \wedge \boldsymbol{n}| = 1$ is the unit area element. The unitary rotor remains invariant under the action of the $SL(2,\mathbb{R})$ group, which generates various $\boldsymbol{m}$ and $\boldsymbol{n}$ representations of the base elements while keeping $|\boldsymbol{m} \wedge \boldsymbol{n}| = 1$. The chirality of rotation is defined by the $\mathbf{e}_i \wedge \mathbf{e}_j$ versus $\mathbf{e}_j \wedge \mathbf{e}_i$ ordering. Switching the ordering of these vectors switches the orientation of the rotation plane.

Since s_i takes on binary values, what we really want to encode is the direction of rotation. Quantum theoreticians make the following choice: let $S_z = \left(\begin{smallmatrix} \frac{1}{2} & 0 \\ 0 & -\frac{1}{2} \end{smallmatrix}\right)$ be a matrix representation of $\mathbf{e}_x \wedge \mathbf{e}_y$, and let the $\mathbf{e}_x \wedge \mathbf{e}_y$ versus $\mathbf{e}_y \wedge \mathbf{e}_x$ ordering be encoded by the $|+\rangle \equiv \left(\begin{smallmatrix} 1 \\ 0 \end{smallmatrix}\right)$ versus $|-\rangle \equiv \left(\begin{smallmatrix} 0 \\ 1 \end{smallmatrix}\right)$ eigenvector choice. The corresponding $\frac{1}{2}$ and $-\frac{1}{2}$ eigenvalues are meant to encode the $\frac{1}{2}$ factor in the exponent of the rotor which corresponds to angle θ rotation. S_z is referred to as the spin operator in the z direction. This way of encoding

Table 0.3. Spin operators and eigenvectors.

Rotor exponents ($\theta = \pm 1$)	Spin operator	Eigenvectors	Eigenvalues
$\frac{1}{2}\mathbf{e}_y \wedge \mathbf{e}_z, \frac{1}{2}\mathbf{e}_z \wedge \mathbf{e}_y$	$S_x = \begin{pmatrix} 0 & \frac{1}{2} \\ \frac{1}{2} & 0 \end{pmatrix}$	$\begin{pmatrix} \frac{1}{\sqrt{2}} \\ \frac{1}{\sqrt{2}} \end{pmatrix}, \begin{pmatrix} \frac{-1}{\sqrt{2}} \\ \frac{1}{\sqrt{2}} \end{pmatrix}$	$\frac{1}{2}, -\frac{1}{2}$
$\frac{1}{2}\mathbf{e}_z \wedge \mathbf{e}_x, \frac{1}{2}\mathbf{e}_x \wedge \mathbf{e}_z$	$S_y = \begin{pmatrix} 0 & \frac{-i}{2} \\ \frac{i}{2} & 0 \end{pmatrix}$	$\begin{pmatrix} \frac{1}{\sqrt{2}} \\ \frac{i}{\sqrt{2}} \end{pmatrix}, \begin{pmatrix} \frac{i}{\sqrt{2}} \\ \frac{1}{\sqrt{2}} \end{pmatrix}$	$\frac{1}{2}, -\frac{1}{2}$
$\frac{1}{2}\mathbf{e}_x \wedge \mathbf{e}_y, \frac{1}{2}\mathbf{e}_y \wedge \mathbf{e}_x$	$S_z = \begin{pmatrix} \frac{1}{2} & 0 \\ 0 & -\frac{1}{2} \end{pmatrix}$	$\begin{pmatrix} 1 \\ 0 \end{pmatrix}, \begin{pmatrix} 0 \\ 1 \end{pmatrix}$	$\frac{1}{2}, -\frac{1}{2}$

the $\mathbf{e}_x \wedge \mathbf{e}_y$ versus $\mathbf{e}_y \wedge \mathbf{e}_x$ ordering is due to the above-discussed method of using operators, eigenvectors, and eigenvalues in quantum mechanics.

The quantum mechanical definitions for the three spin operators and corresponding eigenvalues are shown in Table 0.3, where the eigenvectors are normalized to unit length. These spin operators correspond to the $\mathbf{e}_x \wedge \mathbf{e}_y \to S_z$, $\mathbf{e}_y \wedge \mathbf{e}_z \to S_x$ and $\mathbf{e}_z \wedge \mathbf{e}_x \to S_y$ assignments: the spin operators are a matrix representation of the exterior algebra discussed in Section 0.3, obeying the same commutation rules. Note that there is no inherent need to use complex valued matrices in quantum mechanics: it is simply a choice to use a complex valued matrix representation of Clifford basis vectors.

The usefulness of this quantum mechanical representation is seen when we try to calculate measurements and probabilities of spin states. The electron spin can be measured along any direction, and orthogonal measurements give independent results. Firstly, we measure the electron spin state along the z direction. Before the measurement, a general spin state is written as

$$|s\rangle = a\,|+\rangle_z + b\,|-\rangle_z = a\begin{pmatrix} 1 \\ 0 \end{pmatrix} + b\begin{pmatrix} 0 \\ 1 \end{pmatrix}$$

Without any measurement and filtering, we have $a = b = \frac{1}{\sqrt{2}}$, which means $\frac{1}{2}$ probability for each spin eigenstate. Acting with the operator S_z on s means observing the average spin state along the z axis, and results in:

$$S_z\,|s\rangle = \frac{a}{2}\begin{pmatrix} 1 \\ 0 \end{pmatrix} - \frac{b}{2}\begin{pmatrix} 0 \\ 1 \end{pmatrix}$$

Suppose that we want to know just the coefficient b of the spin state. This is obtained by the following expression:

$$_z\langle -|\, s\rangle = a\,(0\;1)^\dagger \begin{pmatrix} 1 \\ 0 \end{pmatrix} + b\,(0\;1)^\dagger \begin{pmatrix} 0 \\ 1 \end{pmatrix} = b$$

The above method can be checked for all other directions, and shows that our eigenvectors are orthogonal for any measurement direction.

Suppose that we also want to know the spin along the x direction. We write the above eigenvectors as follows:

$$|+\rangle_z = \begin{pmatrix} 1 \\ 0 \end{pmatrix} = \frac{1}{\sqrt{2}}\left[\begin{pmatrix} \frac{1}{\sqrt{2}} \\ \frac{1}{\sqrt{2}} \end{pmatrix} - \begin{pmatrix} \frac{-1}{\sqrt{2}} \\ \frac{1}{\sqrt{2}} \end{pmatrix}\right] = \frac{1}{\sqrt{2}}\left[|+\rangle_x - |-\rangle_x\right]$$

$$|-\rangle_z = \begin{pmatrix} 0 \\ 1 \end{pmatrix} = \frac{1}{\sqrt{2}}\left[\begin{pmatrix} \frac{1}{\sqrt{2}} \\ \frac{1}{\sqrt{2}} \end{pmatrix} + \begin{pmatrix} \frac{-1}{\sqrt{2}} \\ \frac{1}{\sqrt{2}} \end{pmatrix}\right] = \frac{1}{\sqrt{2}}\left[|+\rangle_x + |-\rangle_x\right]$$

The above result means that if we filter the electron to be in the $|+\rangle_z$ state, a subsequent spin measurement along the x axis yields equal probability of finding the electron in any of the two spin states. Similarly, upon filtering the electron to be in the $|-\rangle_z$ state, a subsequent spin measurement along the x axis yields again equal probability of finding the electron in any of the two spin states. Therefore, our mathematical notation captures the independence of spin measurements in orthogonal directions.

0.8.4. *Indistinguishability*

Indistinguishability is a very important concept in modern physics. However, at times it is not always clear which states are indistinguishable and why. Moreover, between indistinguishable states and fully distinguishable states there is a variety of intermediary or mixed states.

We begin our inquiry by asking what is the correct way to write down the spin of an electron. When we measure spin with a Stern–Gerlach device, we are free to choose our direction of measurement arbitrarily. As discussed in the previous section, the measured spin is denoted by $|+\rangle$ (spin-up) and $|-\rangle$ (spin-down) or equivalently as $\binom{1}{0}$ and $\binom{0}{1}$, respectively. We will use both of these notations interchangeably.

Until the spin is first measured in a given direction or some attempt is made to detect it, we can assume both that the spin is unknown and that the state is symmetrical. Consequently, we write $\frac{1}{\sqrt{2}}(|+\rangle + |-\rangle)$ for the spin state. Technically, we could have assumed the state to be of the form

$$\left|\tilde{\psi}\right\rangle = \frac{1}{\sqrt{2}}(|+\rangle - |-\rangle) \tag{0.8.5}$$

However, this state is not permutable in that if we switch the + and −, then $|\tilde{\psi}\rangle$ becomes $-|\tilde{\psi}\rangle$, in contrast to $|\psi\rangle$ where

$$|\psi\rangle = \frac{1}{\sqrt{2}}(|+\rangle + |-\rangle) = \frac{1}{\sqrt{2}}(|-\rangle + |+\rangle) \tag{0.8.6}$$

In this case, we say that the state $|\psi\rangle$ is indistinguishable under a permutation S_1, and we write $S_1|\psi\rangle = |\psi\rangle$.

Now consider the spin state of two arbitrary electrons. If nothing is known, then the state function should be written as

$$|\psi\rangle = \frac{1}{2}(|+\rangle|+\rangle + |+\rangle|-\rangle + |-\rangle|+\rangle + |-\rangle|-\rangle) \tag{0.8.7}$$

which is completely invariant under permutations. Symbolically, this can be written as $S_2|\psi\rangle = |\psi\rangle$, where S_2 corresponds to a permutation on the product space associated with the pair of particles. Also note that implicit in this equation is the assumption that each component occurs with a probability of $\frac{1}{4}$. This leads us to the following methodological question: is it reasonable to assume that each state has the same probability of occurring? One thing that can be stated is that regardless of our (reasonable) assumptions, the actual outcome needs to be ultimately confirmed by experiment.

This brings us to the next question. Do there exist other permutable states besides those of Equation (0.8.7)? For example, one might magnetically distinguish the following three states (triplet states, with combined spins 1, 0, −1, respectively):

$$|1\rangle = |+\rangle|+\rangle \tag{0.8.8}$$

$$|0\rangle = \frac{1}{\sqrt{2}}(|+\rangle|-\rangle + |-\rangle|+\rangle) \tag{0.8.9}$$

$$|-1\rangle = |-\rangle|-\rangle \tag{0.8.10}$$

In this case, we have partially separated the four states into the two distinguishable states $|1\rangle$ and $|-1\rangle$ and an **indistinguishable** state $|0\rangle = \frac{1}{\sqrt{2}}(|+\rangle|-\rangle + |-\rangle|+\rangle)$. What is the nature of this last indistinguishable state? Is it something intrinsic to the two-particle system or does it stem from our inability to perform an adequate experiment which could enable us to further distinguish the terms?

To answer these questions, one might consider each electron to be in a $1s$ state of two different hydrogen atoms, not forming a molecule. In this latter state, we again obtain a $|1\rangle, |0\rangle, |-1\rangle$, with probability distribution $\frac{1}{4}, \frac{1}{2}, \frac{1}{4}$.

It is useful to distinguish between extrinsically indistinguishable states and intrinsically indistinguishable states. An extrinsically indistinguishable spin state means that the spin of each particle is statistically independent of the spin of the other, as in the above example.

An intrinsically indistinguishable spin state means that the particle spins are correlated in each measurement. Two particles in an intrinsically indistinguishable spin state are said to be "entangled into opposite chirality" if the two spin rotors are given by $\boldsymbol{R}_1 = e^{-\frac{\theta}{2}\boldsymbol{e}_i \wedge \boldsymbol{e}_j}$ and $\boldsymbol{R}_2 = e^{-\frac{\theta}{2}\boldsymbol{e}_j \wedge \boldsymbol{e}_i}$, respectively. We look more closely into entangled states in Section 0.8.5.

0.8.5. *Spin entanglement into opposite chirality*

Let's consider two electrons being entangled, where the individual electron spins always have opposite chirality. This means that each and every measurement of their combined spin yields zero value. We don't know the individual spin values before measuring them, but we know that they will be opposite.

We can write the two-particle state as

$$|\psi(\lambda_1, \lambda_2)\rangle = |\psi(x_1, x_2)\rangle\, s_1 \otimes s_2 \tag{0.8.11}$$

Because the two electrons have opposite chirality, and recalling Equation (0.3.4), we can write the following identity for their rotor exponent:

$$(\mathbf{e}_i \wedge \mathbf{e}_j) \wedge (\mathbf{e}_j \wedge \mathbf{e}_i) = 0 \tag{0.8.12}$$

The above identity remains true for any choice of i, j and therefore it expresses isotropic spin correlation. Using the singlet state definitions made in Section 0.5, we can write the left-hand side of Equation (0.8.12) with quantum mechanical notation:

$$(s_1 \otimes s_2)_{\text{singlet}} = s_1 \wedge s_2 = \frac{1}{\sqrt{2}}\left(|+\rangle|-\rangle - |-\rangle|+\rangle\right) \tag{0.8.13}$$

The above equation means that the spin-singlet state of two electrons is antisymmetric, and the two electrons are entangled into opposite chirality.

Let's see what this spin-singlet state means for various measurement directions. For a spin measurement in the z direction, Equation (0.8.13) is written as follows:

$$s_1 \wedge s_2 = \frac{1}{\sqrt{2}}\left(\begin{pmatrix}1\\0\end{pmatrix}\begin{pmatrix}0\\1\end{pmatrix} - \begin{pmatrix}0\\1\end{pmatrix}\begin{pmatrix}1\\0\end{pmatrix}\right)$$

where we have tensor multiplication between the spin state vectors. For a spin measurement in the y direction, Equation (0.8.13) is written as follows:

$$\begin{aligned} s_1 \wedge s_2 &= \frac{1}{\sqrt{2}}\left(\begin{pmatrix}\frac{1}{\sqrt{2}}\\\frac{i}{\sqrt{2}}\end{pmatrix}\begin{pmatrix}\frac{i}{\sqrt{2}}\\\frac{1}{\sqrt{2}}\end{pmatrix} - \begin{pmatrix}\frac{i}{\sqrt{2}}\\\frac{1}{\sqrt{2}}\end{pmatrix}\begin{pmatrix}\frac{1}{\sqrt{2}}\\\frac{i}{\sqrt{2}}\end{pmatrix}\right) \\ &= \frac{1}{\sqrt{2}}\left(\frac{1}{2} - \frac{i^2}{2}\right)\left(\begin{pmatrix}1\\0\end{pmatrix}\begin{pmatrix}0\\1\end{pmatrix} - \begin{pmatrix}0\\1\end{pmatrix}\begin{pmatrix}1\\0\end{pmatrix}\right) \end{aligned}$$

where we used the results of Equation (0.5.1). We get exactly the same result for the spin measurement in the z and y directions. The same result can be worked out also for the x direction. Therefore, Equation (0.8.13) indeed expresses isotropic spin entanglement in any direction.

If the particles form a spin-singlet state but are otherwise independent, then we can write the joint $|\psi\rangle = |\psi(\lambda_1, \lambda_2)\rangle$ wave function in different ways according to where we want to place the emphasis

$$|\psi\rangle = \frac{1}{\sqrt{2}}|\psi(\lambda_1)\rangle\,|\psi(\lambda_2)\rangle - \frac{1}{\sqrt{2}}|\psi(\lambda_2)\rangle\,|\psi(\lambda_1)\rangle \tag{0.8.14}$$

$$= \frac{1}{\sqrt{2}}|\psi(x_1)\rangle\, s_1 \otimes |\psi(x_2)\rangle\, s_2 - \frac{1}{\sqrt{2}}|\psi(x_2)\rangle\, s_2 \otimes |\psi(x_1)\rangle\, s_1 \tag{0.8.15}$$

$$= det(|\psi(x_1)\rangle\,, |\psi(x_2)\rangle) s_1 \wedge s_2 \tag{0.8.16}$$

$$= \delta^{12}_{i_1 i_2}\,|\psi\,(\lambda_{i_1})\rangle\,|\psi\,(\lambda_{i_2})\rangle \tag{0.8.17}$$

Apart from the notation, all of these expressions are equivalent to the singlet state (0.5.1) defined previously.

The wedge product of n 1-forms is given by: $a_1 \wedge \cdots \wedge a_n = \frac{1}{n!}\delta^{1\ldots n}_{i_1\ldots i_r} a^{i_1} \otimes \cdots \otimes a^{i_n}$

Specifically,

$$\begin{aligned} a^1 \wedge a^2 \wedge a^3 &= \frac{1}{3!}(a^1 \otimes a^2 \otimes a^3 + a^2 \otimes a^3 \otimes a^1 + \\ &\quad + a^3 \otimes a^1 \otimes a^2 - a^2 \otimes a^1 \otimes a^3 - a^1 \otimes a^3 \otimes a^2 - a^3 \otimes a^2 \otimes a^1) \\ &= det(a_1, a_2, a_3) e^1 \otimes e^2 \otimes e^3 \end{aligned}$$

Thus, for a system of n indistinguishable particles forming singlet states, the wavefunction generalizes to:

$$|\psi(\lambda_1, \lambda_2, \dots \lambda_n)\rangle = det(|\psi(x_1)\rangle, \dots |\psi(x_n)\rangle)s_1 \wedge \cdots \wedge s_n \tag{0.8.18}$$

$$= \frac{1}{n!}\delta^{1\dots n}_{i_1\dots i_n} |\psi(\lambda_{i_1}\rangle \dots |\psi(\lambda_{i_n}\rangle \tag{0.8.19}$$

Equation (0.8.19) defines a Fermi–Dirac state and obeys Fermi–Dirac statistics. If the antisymmetry condition is dropped and

$$|\psi(\lambda_1, \lambda_2), \dots \lambda_n\rangle = \frac{1}{n!}\sum_{i_1 i_2 \dots i_n} |\psi(\lambda_{i_1}\rangle \dots |\psi(\lambda_{i_n}\rangle \tag{0.8.20}$$

then it is called a Bose–Einstein state and obeys Bose–Einstein statistics. We will discuss the origin of Fermi–Dirac statistics in depth in Chapter 9.

Note that in all of this discussion we did not rely on any particular spin rotation angle θ value, nor any specific frequency value. Therefore, these results hold true for all particle-specific spin values, i.e., for spin 1/2, spin 1, as well as for spin 2 particle types.

0.9. Spinors

0.9.1. *Introduction to spinors*

To understand spinors, we should first ask what is the difference, if any, between the complex number $z = x_1 + ix_2$ and the vector $v = x_1\mathbf{e}_1 + x_2\mathbf{e}_2$, especially since both are usually represented by the coordinates (x_1, x_2). Note that $|z|^2 = \bar{z}z = x_1^2 + x_2^2$ and $|v|^2 = v \cdot v = x_1^2 + x_2^2$ and therefore $|z| = |v|$. In other words, both $|z|$ and $|v|$ can be considered to be the square root of $x_1^2 + x_2^2$. However, there is a fundamental difference in that $|z|^2$ is the product of the two different (although conjugate) complex numbers $\bar{z}$ and z, and indeed $|z| = \sqrt{\bar{z}z}$, while $|v|^2$ is obtained as the dot product of v with itself. It is this two-to-one relationship that is at the heart of understanding spinors, and it also gives us an alternative perspective on Clifford algebra. Given the special role of $\sqrt{z}$ and $\sqrt{\bar{z}}$ in defining $|z|$, we let $\xi = \sqrt{z}$ and refer to ξ and $\bar{\xi}$ as spinors associated with the vector v in that $|v| = \bar{\xi}\xi$.

More specifically, given $\bar{z} = x_1 - ix_2$ (equivalently $v = (x_1, x_2)$) and $w = y_1 + iy_2$ (equivalently $\omega = (y_1, y_2)$), then their product

$$\bar{z}w = (x_1y_1 + x_2y_2) + i(x_1y_2 - y_1x_2)$$

has the form of a Clifford algebra, while in contrast $v{\cdot}\omega = (x_1y_1{+}x_2y_2) \neq \bar{z}w$. We can overcome this difficulty by defining a Clifford product

$$v\omega = v \cdot \omega + v \wedge \omega$$

such that

$$v\omega = (x_1y_1 + x_2y_2) + (x_1y_2 - x_2y_1)\mathbf{e}_1\mathbf{e}_2$$

The above definition is essentially the same as defining the following multiplication rules for $\mathbf{e}_1$ and $\mathbf{e}_2$, analogously to the Equations (0.1.1) and (0.1.2):

$$\mathbf{e}_1^2 = 1, \mathbf{e}_2^2 = 1 \tag{0.9.1}$$

$$\mathbf{e}_1\mathbf{e}_2 = -\mathbf{e}_2\mathbf{e}_1 \tag{0.9.2}$$

Note that when $v = \omega$, then $v\omega = v \cdot \omega = |v|^2$, which again allows us to interpret v as a proper square root with respect to the Clifford product. Moreover, we may denote the unit vector $\mathbf{e}_i$ by the following matrix representation:

$$\sigma_1 = \begin{pmatrix} 0 & 1 \\ 1 & 0 \end{pmatrix}, \quad \sigma_2 = \begin{pmatrix} 0 & -i \\ i & 0 \end{pmatrix} \tag{0.9.3}$$

The reader may verify that the multiplication of σ_1 and σ_2 obeys analogous rules to Equations (0.9.1) and (0.9.2).

We can write a matrix representation of v in two equivalent ways:

(1) $v = x_1\sigma_1 + x_2\sigma_2 = \sigma_1(x_1 + x_2\sigma_1\sigma_2)$,
(2) $v = x_1\sigma_1 + x_2\sigma_2 = (x_1 - x_2\sigma_1\sigma_2)\sigma_1$.

We can identify $x_1 + x_2\sigma_1\sigma_2$ with the complex number $x_1 + ix_2$, and $x_1 - x_2\sigma_1\sigma_2$ with $x_1 - ix_2$. We may write the norm of v in the following way:

$$|v| = \sqrt{x_1^2 + x_2^2} = \sqrt{(x_1 + ix_2)(x_1 - ix_2)} \equiv \xi\bar{\xi}$$

where $\xi = \sqrt{x_1 + ix_2}$ and its conjugate are called spinors associated with the vector v.

Given a spinor we can construct v, and given v we can derive the spinors. For example, if $v = x_1\sigma_1 + x_2\sigma_2$, then $z = |v|\exp(i\theta)$ and therefore $\xi = \sqrt{|v|}\exp(i\theta/2)$ and conversely given $\xi = \sqrt{|v|}\exp(i\theta/2)$, we can define $v = (|v|\cos(\theta), |v|\sin(\theta))$.

Unfortunately, the exact definition of spinors varies in the literature. Elie Cartan's work essentially defines a spinor as a solution over a ($2n+1$ dimensional) vector space of the equation $|v| = \bar{\xi}\xi$, where ξ is an n-dimensional complex vector. If this is our starting point, then both v and ξ are not unique in that $\xi = \sqrt{|v|}\exp(i\theta/2)$ satisfies $|v| = \bar{\xi}\xi$ for all θ. Consequently, v is any vector defined over the $2n$-dimensional sphere such that $|v|^2$ is a quadratic polynomial. For the above-discussed two-dimensional case, a given $|v|$ corresponds to a circle of complex-valued ξ.

Given this arbitrariness, Cartan refers to such vectors v as isotropic vectors. These isotropic vectors can be written as quadratic polynomials. To see this, let $|v| = ix_0$; meaning that x_0 is complex valued, and points along the imaginary axis. Then $|v|^2 = -x_0^2 = x_1^2 + x_2^2$ and therefore

$$x_0^2 + x_1^2 + x_2^2 = 0 \tag{0.9.4}$$

In fact, Cartan refers to $\mathbf{x} = (x_0, x_1, x_2)$ as the isotropic vector which can be expressed in terms of the σ matrix basis as $\boldsymbol{X} = x_i\sigma_i$. In order for $\mathbf{x} = (x_0, x_1, x_2)$ to be an isotropic vector, we need to extend the two-dimensional space spanned by $x_1\sigma_1$ and $x_2\sigma_2$ to a three-dimensional space having the same structure. We extend the multiplication rules (0.9.1) and (0.9.2) to three dimensions:

$$\mathbf{e}_0^2 = 1, \quad \mathbf{e}_1^2 = 1, \quad \mathbf{e}_2^2 = 1 \tag{0.9.5}$$

$$\mathbf{e}_0\mathbf{e}_1 = -\mathbf{e}_1\mathbf{e}_0, \mathbf{e}_0\mathbf{e}_2 = -\mathbf{e}_2\mathbf{e}_0, \mathbf{e}_1\mathbf{e}_2 = -\mathbf{e}_2\mathbf{e}_1 \tag{0.9.6}$$

The matrix representation of the above-defined $\mathbf{e}_i$ bases is known in the literature as the Pauli spin matrices:

$$\sigma_0 = \begin{pmatrix} 1 & 0 \\ 0 & -1 \end{pmatrix}, \quad \sigma_1 = \begin{pmatrix} 0 & 1 \\ 1 & 0 \end{pmatrix}, \quad \sigma_2 = \begin{pmatrix} 0 & -i \\ i & 0 \end{pmatrix} \tag{0.9.7}$$

Comparing the above definition against the spin operators listed in Table 0.3, one recognizes why these matrices are called "spin matrices."

Let's consider the following matrix equation, where ϕ is defined to be non-zero:

$$\boldsymbol{X}\phi = 0 \tag{0.9.8}$$

The above equation requires that $\det(\boldsymbol{X}) = 0$. This can be seen from the following argument. Suppose $\det(\boldsymbol{X}) \neq 0$, then $\boldsymbol{X}^{-1}$ exists. Upon left-multiplying the above equation by $\boldsymbol{X}^{-1}$ we get $\phi = 0$, but that contradicts our definition of ϕ. Therefore, $\det(\boldsymbol{X}) = 0$, and we explicitly calculate this determinant:

$$0 = \left(-x_0^2 - (x_1 - ix_2)(x_1 + ix_2)\right) = -\left(x_0^2 + x_1^2 + x_2^2\right)$$

On the one hand, Equation (0.9.8) leads to Cartan's $x_0^2 + x_1^2 + x_2^2 = 0$ polynomial. On the other hand, Equation (0.9.8) means that ϕ is an eigenvector of $\boldsymbol{X}$ with eigenvalue 0.

The last step is to solve for ϕ. Using the above matrix basis, we can easily express $\boldsymbol{X}$ in terms of the spinor ξ:

$$\boldsymbol{X} = \begin{pmatrix} -i\xi\bar{\xi} & \bar{\xi}^2 \\ \xi^2 & i\xi\bar{\xi} \end{pmatrix}$$

We find that Equation (0.9.8) is solved by $\phi = \left(\bar{\xi}, i\xi\right)$:

$$\boldsymbol{X}\phi = \begin{pmatrix} -i\xi\bar{\xi}\bar{\xi} + i\bar{\xi}^2\xi \\ \xi^2\bar{\xi} - \xi\bar{\xi}\xi \end{pmatrix} = 0$$

This result means that a spinor, which is a given vector's "square root," can be used to construct the eigenvector of the linear Equation (0.9.8).

Cartan had a preference to write his quadratic polynomials as elements of a Euclidean space. Instead, we could have chosen to define the quadratic polynomial in Minkowski space given by $x_0^2 - x_1^2 - x_2^2 = 0$. In that case, the Minkowski null metric also emerges naturally in Hermann Weyl's approach (see in what follows).

0.9.2. *An alternative construction of the eigenvector equation*

In the above paragraphs, we established the equivalence between the zero-valued polynomial Equation (0.9.4) and the eigenvector Equation (0.9.8), which has a zero eigenvalue. However, we could have chosen to write Equation (0.9.4) as

$$x_1^2 + x_2^2 = -x_0^2 \tag{0.9.9}$$

and to then construct our eigenvector equation as

$$\boldsymbol{X}\phi = s\phi \tag{0.9.10}$$

where $\boldsymbol{X} = x_1\sigma_1 + x_2\sigma_2$ and $s = ix_0$ is a non-zero eigenvalue. The eigenvector equation has a solution only if $\det(\boldsymbol{X} - s\mathbf{I}) = 0$, where $\mathbf{I}$ is the identity matrix.[b] We calculate this determinant:

$$0 = \left(-x_0^2 - (x_1 - ix_2)(x_1 + ix_2)\right) = -\left(x_0^2 + x_1^2 + x_2^2\right)$$

Like before, we find an equivalence between Equation (0.9.4) and our eigenvector equation.

Our eigenvector equation is now solved by $\phi = \left(\bar{\xi}, \xi\right)$:

$$\boldsymbol{X}\phi = \begin{pmatrix} 0 & \bar{\xi}^2 \\ \xi^2 & 0 \end{pmatrix} \begin{pmatrix} \bar{\xi} \\ \xi \end{pmatrix} = \begin{pmatrix} \bar{\xi}^2\xi \\ \xi^2\bar{\xi} \end{pmatrix}$$

$$s\phi = \xi\bar{\xi} \begin{pmatrix} \bar{\xi} \\ \xi \end{pmatrix} = \begin{pmatrix} \xi\bar{\xi}\bar{\xi} \\ \xi\bar{\xi}\xi \end{pmatrix}$$

This method of eigenvector construction is preferable in some physics applications. In particular, we will use the $\boldsymbol{X}\phi = s\phi$ eigenvector construction for the analysis of the Dirac equation.

0.9.3. *Spacetime vectors under Cartan's approach*

Historically, the notion of spinors was introduced into mathematics by Elie Cartan in 1913 and then further developed in his book titled *The Theory of Spinors* [1]. Cartan begins with a three-dimensional Euclidean space defined over the complex numbers.

We can extend the results of Section 0.9.1 to Minkowski space by defining a four-vector $X = x_\mu \sigma^\mu$, where $\sigma^0 = \mathbf{I}$ and $\{\sigma^1, \sigma^2, \sigma^3\}$ are the Pauli spin matrices. The σ^3 matrix of this section has the same value as the zero-indexed Pauli spin matrix of Equation (0.9.7). We then define the corresponding set of spinors ϕ such that

$$X\phi = 0 \iff |X| = x_0^2 - x_1^2 - x_2^2 - x_3^2 = 0 \tag{0.9.11}$$

where $|X|$ is the determinant of X. One may check that a solution for spinor $\phi = (\xi_1, \xi_2)$ is

$$\xi_1 = \sqrt{x_1 - ix_2}\sqrt{x_3 - x_0}, \quad \xi_2 = i\sqrt{x_1 + ix_2}\sqrt{x_3 + x_0}$$

Note that the spinors define vectors on the null cone. Also when $x_0 = 0$, then ϕ is the same as in the previously discussed three-dimensional case.

---

[b]We use the bold **I** symbol for the identity matrix in order to distinguish it from the Clifford pseudoscalar notation.

Moreover, the Lorentz transformation of Equation (0.9.11) can be easily calculated. Suppose X is Lorentz transformed to $X' = SXS^\dagger$, where $S \in SL(2,\mathbb{C})$. Our Lorentz transformed spinor equation is now:

$$X'\phi' = 0 \iff |X'| = x_0^2 - x_1^2 - x_2^2 - x_3^2 = 0$$

Since $X\phi = 0$, the solution to the above equation is $\phi' = (S^\dagger)^{-1}\phi$.

The above results work for null-vectors, such as light propagation or neutrino propagation, where $x_0^2 - x_1^2 - x_2^2 - x_3^2 = 0$.

For massive particles, we have non-zero spacetime vectors, i.e.,

$$|X| = x_0^2 - x_1^2 - x_2^2 - x_3^2 \neq 0$$

Therefore, we write our eigenvalue equation as

$$X\phi = s\phi \tag{0.9.12}$$

where $X = x_\mu\sigma^\mu$ is our four-vector and $s = im$ is our eigenvalue. Analogously to the case discussed in Section 0.9.2, the equivalent polynomial equation[c] is

$$x_0^2 - x_1^2 - x_2^2 - x_3^2 = m^2 \tag{0.9.13}$$

When x_0 measures time and $\mathbf{x} = (x_1, x_2, x_3)$ measures spatial coordinates, the above polynomial equation resembles the Klein–Gordon equation.

Our eigenvector equation has a solution only if $\det(X - s\mathbf{I}) = 0$. Analogously to the example in Section 0.9.2, we cannot use the identity matrix $\mathbf{I}$ as a basis element of X because $s\mathbf{I}$ plays the role of an independent coordinate dimension. It turns out that it is not possible to find a 2×2 matrix representation with four non-commutative basis elements, and therefore we use a 4×4 matrix representation. A popular matrix representation choice is to use the Dirac γ_μ matrices, defined as

$$\gamma_0 = \begin{pmatrix} \mathbf{I} & 0 \\ 0 & -\mathbf{I} \end{pmatrix}, \quad \gamma_1 = \begin{pmatrix} 0 & -\sigma_1 \\ \sigma_1 & 0 \end{pmatrix}, \quad \gamma_2 = \begin{pmatrix} 0 & -\sigma_2 \\ \sigma_2 & 0 \end{pmatrix}, \quad \gamma_3 = \begin{pmatrix} 0 & -\sigma_3 \\ \sigma_3 & 0 \end{pmatrix} \tag{0.9.14}$$

where $\mathbf{I}$ is the 2×2 identity matrix and σ_i are the Pauli spin-matrices. Our four-vector is now $X = \sum x_\mu\gamma_\mu$. The chosen γ_μ notation is analogous to the notation which we used in Section 0.2 to represent the basis elements of $Cl_{1,3}(\mathbb{R})$-type Clifford algebra. That is not a coincidence, as the Dirac

[c]The sign of m depends on whether $Cl_{1,3}(\mathbb{R})$ or $Cl_{3,1}(\mathbb{R})$ is used.

γ_μ matrices have the same commutation relations as the basis elements of $Cl_{1,3}(\mathbb{R})$ algebra, and are therefore a representation of that algebra.

The solution of the eigenvector Equation (0.9.12) becomes more complex than the null-vector case, and we will not dwell on its further details here. The main goal of this section was to show that the quadratic Equation (0.9.13), which defines the length of a four-vector in Minkowski space, can be always linearized into the spinor eigenvalue Equation (0.9.12). This result will be used for understanding the Dirac equation in chapter 5.

0.9.4. *Weyl's approach to spinors*

Hermann Weyl took a different approach to spinor theory. He begins by constructing a vector with respect to the Pauli spin matrices. More explicitly, let $\xi = (\xi_1, \xi_2)$ and define

$$y_i = \bar{\xi}\sigma_i\xi \tag{0.9.15}$$

where $\{\sigma_1, \sigma_2, \sigma_3\}$ are the Pauli spin matrices.

We find that $y_1 = \bar{\xi}_1\xi_2 + \bar{\xi}_2\xi_1$, $y_2 = i(-\bar{\xi}_1\xi_2 + \bar{\xi}_2\xi_1)$, $y_3 = \bar{\xi}_1\xi_1 - \bar{\xi}_2\xi_2$ and their squares sum up to

$$y_1^2 + y_2^2 + y_3^2 = (\bar{\xi}_1\xi_1 + \bar{\xi}_2\xi_2)^2 \tag{0.9.16}$$

Although this approach is different from Cartan's, the vector (ξ_1, ξ_2) — which comprises two spinors — can be again interpreted as a square root of a quadratic polynomial. Also, as before, we can extend this approach to Minkowski space by defining a four-vector $Y = y_\mu\sigma^\mu$, where $\sigma^0 = \mathbf{I}$ and $\{\sigma_1, \sigma_2, \sigma_3\}$ are the Pauli spin matrices. Recalling that σ_i is the matrix representation of a unit vector $\mathbf{e}_i$, and recalling the definition of rotors from Section (0.2.1), it becomes clear from Equation (0.9.15) that a Weyl spinor ξ is in fact analogous to a rotor. Since any Lorentz transformation can be encoded by a rotor, a Weyl spinor may be used to represent a Lorentz transformation.

The reader might be wondering what is the relationship between a null-vector X in the Cartan approach and the four-vector Y in the above construction. As it turns out, the vector X defined by Cartan can be generated by the Weyl spinor ξ operating on $C\sigma_i$, where

$$C = \begin{pmatrix} 0 & 1 \\ -1 & 0 \end{pmatrix}$$

Specifically, the x_i components of X are generated from the Weyl spinor $\xi = (\xi_1, \xi_2)$ as follows:

$$x_i = \xi^T C \sigma_i \xi$$

Although, Cartan's and Weyl's original definitions are different, in reality both definitions of spinors are related to the square roots of quadratic polynomials. This is the unifying factor in both approaches.

0.10. Angular Momentum Theory

In classical mechanics, the angular momentum of a particle is defined to be a vector $\mathbf{L} = \mathbf{r} \times \mathbf{p}$ in $\mathbb{R}^3$, where $\mathbf{r}$ and $\mathbf{p}$ represent the particle's position and momentum respectively.

For the relativistic angular momentum calculation one must consider the space–time position four-vector $\mathbf{r}_\square = (t\mathbf{e}_t + \mathbf{r})$ and energy–momentum four-vector $\mathbf{P}_\square = (E\mathbf{e}_t + \mathbf{p})$. We used natural units to express these four-vectors. Their Clifford product becomes

$$\mathbf{r}_\square \mathbf{P}_\square = ([\mathbf{r} \cdot \mathbf{p} - tE] + (\mathbf{r}E - t\mathbf{p})\,\mathbf{e}_t + \mathbf{r} \times \mathbf{p}\mathbf{e}_{xyz}) \tag{0.10.1}$$

The $\mathbf{r} \times \mathbf{p}\mathbf{e}_{xyz} = \mathbf{r} \wedge \mathbf{p}$ term gives the orbital angular momentum, and it is part of a larger object. The scalar $\mathbf{r} \cdot \mathbf{p} - tE$ term gives the center-of-mass motion.

Generally, quantum mechanical calculations take the non-relativistic approximation and work with the $\mathbf{L} = \mathbf{r} \wedge \mathbf{p}$ term only, where $\mathbf{r}$ is the position operator and $\mathbf{p}$ is the momentum operator. In this non-relativistic case, for particles with angular momenta $\mathbf{L_1} = \mathbf{r} \wedge \mathbf{p_1}$ and $\mathbf{L_2} = \mathbf{r} \wedge \mathbf{p_2}$, the joint angular momentum is given by $\mathbf{L_1} + \mathbf{L_2}$.

In the context of an electron orbital, $\mathbf{L} = \mathbf{r} \wedge \mathbf{p}$ represents the orbital angular momentum, which is generated by a unitary rotor. Thus, we may directly use the mathematical formalism of Section 0.8.3 also for the angular momentum calculation: we do not need to adjust anything in Table 0.3 values because it contains nothing specific about the rotation frequency or angular momentum value.

As mentioned in Section 0.8.2, atomic electron orbitals are eigenstates of the Hamiltonian operator. The more exact relativistic solution is given by the Dirac equation, which we will discuss in detail in the following chapters. It is a well-studied problem to find the solutions of Equation (0.8.4), where the potential field V is the electrostatic potential field of the nucleus. Its solutions are of the following form: $|\psi(x)\rangle\,\mathrm{s} = Y_l^m\,|\psi(r)\rangle\,\mathrm{s}$, where Y_l^m are the

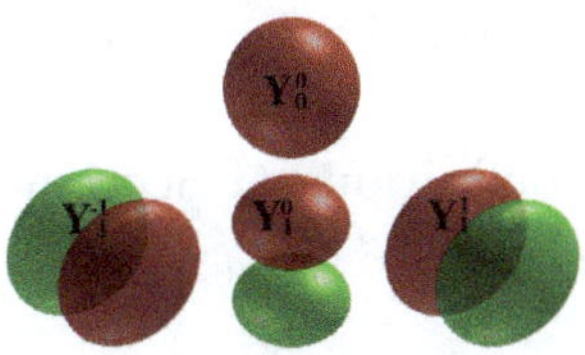

Figure 0.5. An illustration of spherical harmonics for s-orbitals (Y_0^0) and p-orbitals ($Y_1^{-1} \cdots Y_1^1$).

spherical harmonic functions, $|\psi(r)\rangle$ is the radial wavefunction, and s is the spin state.

The spherical harmonic functions are illustrated in Figure 0.5. In terms of (θ, ϕ) spherical coordinates, Y_l^m take the following values in various electron orbitals:

- s-orbital: $Y_0^0 = \frac{1}{\sqrt{4\pi}}$,
- p-orbitals: $Y_1^1 = \sqrt{\frac{3}{8\pi}} e^{-i\phi} \cos\theta$, $Y_1^0 = \sqrt{\frac{3}{4\pi}} \cos\theta$, $Y_1^1 = -\sqrt{\frac{3}{8\pi}} e^{i\phi} \cos\theta$.

Note that the s-orbital wavefunction is rotationally invariant, i.e., isotropic. The question arises whether the electron has any angular momentum in the s-orbital state. One mathematical possibility is that the constant Y_0^0 factor implies that the electron has exactly zero angular momentum in the s-orbital state, and thereby its kinetic energy and momentum are derived from a purely radial oscillation. Mainstream quantum theorists insist on this possibility, and consider the electron to have exactly zero angular momentum in the s-orbital state: they claim it's a pure coincidence that highly excited s-orbitals converge to the classical Kepler-orbit solution around the nucleus. By convergence at high quantum numbered s-orbitals, we mean the following: (i) at the same mean radius values of $\psi(r)$ and the Kepler-orbit, the binding energies are equal in both cases, and (ii) the rate of quantum mechanical radius change via state transitions is the same as the rate at which the Kepler-orbit spirals into the nucleus. One should be suspicious of coincidence theories.

Another mathematical possibility is that the constant Y_0^0 factor means that the s-state electron's orbital angular momentum L_o is isotropically entangled with some other angular momentum L_c, and measurements detect only the combined angular momentum $L_o \otimes L_c$. Regarding L_o, the "o" index stands for "orbital," and regarding L_c the "c" index stands for a yet undefined "companion" angular momentum. This mathematical model is exactly analogous to the isotropic spin entanglement between two particles, which was discussed in Section 0.8.5. We discuss this possibility in the following paragraphs.

Using the singlet state definitions of Section 0.5, we write the isotropically entangled orbital-singlet state of $L_e \otimes L_c$:

$$(L_o \otimes L_c)_{\text{singlet}} = L_o \wedge L_c = \frac{1}{\sqrt{2}}(|+\rangle|-\rangle - |-\rangle|+\rangle) \tag{0.10.2}$$

As discussed in Section 0.8.5, the above expression yields isotropically zero net sum. Thus we identify Equation (0.10.2) with the measured angular momentum of s-orbitals: $\mathbf{L}_{s\text{-orbital}} = L_o \wedge L_c$.

The other possibility for $L_o \otimes L_c$ states is an extrinsically indistinguishable coupling, which we denote by $(L_o \otimes L_c)_{\text{triplet}}$ and we write the resulting orbital-triplet states as

$$|1\rangle = |+\rangle|+\rangle \tag{0.10.3}$$

$$|0\rangle = \frac{1}{\sqrt{2}}(|+\rangle|-\rangle + |-\rangle|+\rangle) \tag{0.10.4}$$

$$|-1\rangle = |-\rangle|-\rangle \tag{0.10.5}$$

Equations (0.10.3)–(0.10.5) give the measured angular momentum of p-orbitals: $\mathbf{L}_{p\text{-orbital}} = (L_o \otimes L_c)_{\text{triplet}}$, with probability distribution $\frac{1}{4}, \frac{1}{2}, \frac{1}{4}$. This also matches the experimentally measured values.

So far, we found that one may use the exact same mathematical model for describing electron spins and for describing the angular momentum electron orbitals. As noted in Section 0.8, this math is independent of the rotation frequency, i.e., does not tell the amount of measured angular momentum which the $|\pm\rangle$ symbol stands for. However, spin measurements reveal that $|\pm\rangle$ represents $\hbar/2$ angular momentum, while orbital angular momentum measurements reveal that $|1\rangle$ represents $\hbar$ angular momentum for the lowest quantum numbered state. Therefore, the $|\pm\rangle$ symbol has exactly the same meaning and same magnitude in Section 0.8 as in this section!

Thus, we have two possible mathematical models of the s-orbital: a purely radial oscillation versus an orbital-singlet type circulation. Both models give the same angular momentum, but which one leads to a more correct understanding of the wavefunction? This issue is mathematically investigated in Ref. [10]: its author finds that the radially oscillating s-orbital is the solution in flat space–time, while the non-zero L_o value arises in curved spacetime. While quantum theorists neglect any space–time curvature effect, in Chapter 8 we carefully calculate the s-orbital solution by not neglecting the space–time curvature effect of electromagnetic energy density.

We find that the electron's orbital angular momentum is non-zero even in the s-orbital state. Thus, the $L_o \wedge L_c$ solution is the correct one.

Under the mainstream interpretation of quantum mechanics, an s-orbital has zero angular momentum while a p-orbital has non-zero angular momentum. In that interpretation, the measurable $|-1\rangle, |0\rangle, |1\rangle$ angular momentum values along some measurement axis are the quantum mechanically allowed measurement values of the non-zero angular momentum. In contrast, our mathematical model implies that at a given main quantum number, the electron has exactly the same angular momentum in either s-type or p-type orbitals. The measured value of the electron's angular momentum is represented by the $|\pm\rangle$ symbol for either orbital type. In Chapter 6, we show how it follows from angular momentum conservation that the electron must indeed have the same angular momentum in either s-type or p-type orbitals. Our result proves from yet another perspective that the $L_o \wedge L_c$ orbital-singlet expression is the correct mathematical model of s-orbitals.

So far, mathematics only tells us that orbital eigenstates must have a companion angular momentum L_c, but we do not know yet the physical meaning of L_c. The interpretation of L_c will be discussed in Section 6.9.

In practice, quantum mechanical angular momentum is measured by placing the investigated atom under a magnetic field, which creates an angular momentum dependent Zeeman-splitting of various electron states' energy levels. We can classify the Zeeman-splitting interaction into three classes, as shown in Table 0.4. These classes were named by experimentalists who pioneered these measurements. In reality, there is nothing "anomalous" about the anomalous Zeeman effect. Quantum theorists also refer to the columns of Table 0.4 as doublet, singlet, and triplet states.

The angular momentum of a charged particle defines its magnetic moment. In order to understand what is really being measured under an applied magnetic field, one must remember the magnetic moment vector is in Larmor precession around the applied magnetic field lines. Since the particle

Table 0.4. Classification of Zeeman effect types.

Anomalous Zeeman effect	No Zeeman effect	Normal Zeeman effect
$\|+\rangle$		$\|+\rangle\|+\rangle$
	$\frac{1}{\sqrt{2}}(\|+\rangle\|-\rangle - \|-\rangle\|+\rangle)$	$\frac{1}{\sqrt{2}}(\|+\rangle\|-\rangle + \|-\rangle\|+\rangle)$
$\|-\rangle$		$\|-\rangle\|-\rangle$

Note: The table values show the structure of measured states.

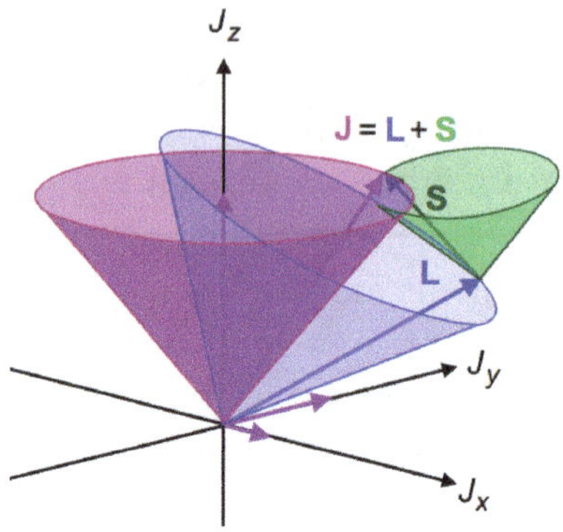

Figure 0.6. The Larmor precession of $\mathbf{J} = \mathbf{L} + \mathbf{S}$ in a magnetic field, with magnetic field lines along the z-direction.

charge is constant, we can equivalently talk about the Larmor precession of the angular momentum vector. Figure 0.6 illustrates the Larmor precession of the total angular momentum $\mathbf{J}$. The measured J_z value is the projection of $\mathbf{J}$ vector onto the z axis.

Considering the electron's $|\psi(x)\rangle$ s wavefunction, let the $\mathbf{L}$ vector represent angular momentum of $|\psi(x)\rangle$ and let the $\mathbf{S}$ vector represent angular momentum of the spin s. Both rotations interact with magnetic field: Figure 0.6 illustrates that each vector has its own Larmor precession. For any individual electron, the Zeeman split is generated by total angular momentum $\mathbf{J} = \mathbf{L} + \mathbf{S}$, and the measured value reveals only the projection of this angular momentum vector onto some measurement axis. We emphasize that the mathematics is exactly the same for both $\mathbf{L}$ and $\mathbf{S}$, and will revisit this Larmor precession effect in Chapters 3 and 4.

It is not surprising that Zeeman split-related calculations can get quite complicated, especially when considering the presence of multiple electrons. Furthermore, spectroscopy measurements reveal only the energy difference during state transitions, i.e., do not directly measure the energy of any electron state. But we do not need to get lost in the details of such calculations: our focus here is on establishing a correct understanding of angular momentum related concepts.

In summary, we adhere to a contradiction-free and mathematically consistent angular momentum theory in this book. We find the mathematics of electron spin and angular momentum to be completely analogous. Our approach somewhat deviates from the angular momentum theory of quantum mechanics textbooks, which assign zero angular momentum to s-type orbitals, claim that the mathematics of $\mathbf{L}$ and $\mathbf{S}$ would be fundamentally different due to a supposedly "different topology" of **S**, and deny applying the Larmor precession effect to **S**.

References

[1] G. Bettini, Clifford algebra, 3 and 4-dimensional analytic functions with applications. Manuscripts of the last century. *Quantum Physics*, (2011), 1–63.

[2] J.M. Chappell, S.P. Drake, C.L. Seidel, L.J. Gunn, A. Iqbal, A. Allison, and D. Abbott. Geometric algebra for electrical and electronic engineers, *Proceedings of the IEEE*, **102**(9) (2014), 1340–1363.

[3] D. Hestenes, Spacetime physics with geometric algebra, *American Journal of Physics*, **71**(3) (2003), 691–714.

[4] J. Dressel, K.Y. Bliokh, and F. Nori, Spacetime algebra as a powerful tool for electromagnetism, *Physics Reports*, **589** (2015), 1–71.

[5] S. Franchini, A. Gentile, F. Sorbello, G. Vassallo, and S. Vitabile, Conformal ALU: A conformal geometric algebra coprocessor for medical image processing, *IEEE Transactions on Computers*, **64**(4) (2015), 955–970.

[6] D. Hestenes, *Reforming the Mathematical Language of Physics, Oersted Medal Lecture*, 2002.

[7] A.M. Shaarawi, Clifford algebra formulation of an electromagnetic charge-current wave theory, *Foundations of Physics*, **30**(11) (2000) 1911–1941.

[8] A. Aspect, J. Dalibard, and G. Roger, Experimental test of Bell's inequalities using time-varying analyzers. *Physics Review Letters*, **49**(25) (1982) 1804–1807.

[9] J.S. Bell, On the Einstein Podolsky Rosen paradox. *Physics*, **1** (1964), 195–200.

[10] A. Lopez-Castillo, *Exact intrinsic half angular momentum from the Schrödinger equation, Theoretical Chemistry Accounts*, **139** (2020), 51.

[11] A. Fassler and E. Stiefel *Group Theoretical Methods and Their Applications.* Birkhauser, Boston, Basel, Berlin, 1991.

[12] G. Horne, Shimony and Zeilinger, *American Journal of Physics*, **58**(12) (1990), 1139.

[13] H.E. Hall, *Solid State Physics.* John Wiley, (1974), 142–146.

[14] H.F. Hameka, *Quantum Theory of the Chemical Bond*, Vol. 93. Hafner Press, New York and London, 1975.

[15] P. Offenhartz, *Atomic and Molecular Orbital Theory.* McGraw-Hill, 1970, pp. 118–119.

[16] P. O'Hara, *Hadronic Journal*, **25**(4) (2002), 407–429.

[17] P. O'Hara, *Hadronic Journal*, **25**(5) (2002), 529–542.

[18] P. O'Hara, *Spin 96 Proceedings.* World Scientific, Singapore, 1997.

[19] W. Pauli, The connection between spin and statistics. *Physics Review Journals A*, **58** (1940), 716–722.

[20] C. Quigg, *Gauge Theories of the Strong, Weak, and Electromagnetic Interactions*, Vol. 12. Benjamin Cummins, 1983.

[21] E.P. Wigner, On hidden variables and quantum mechanical probabilities. *American Journal of Physics*, **38**(8) (1970) 1005.

[22] E. Cartan, *The Theory of Spinors.* Dover, Mineola, 1981, p. 49.

Part 1

Foundations: Electromagnetism, General Relativity, and Quantum Mechanics

https://doi.org/10.1142/9781800611306_0001

Chapter 1

Maxwell's Equations and Occam's Razor

Francesco Celani[*,†,§], Antonino Oscar Di Tommaso[‡,¶], and Giorgio Vassallo[‡,||]

[*] *National Institute of Nuclear Physics (INFN-LNF) Via E. Fermi 56, 00044 Frascati, Roma, Italy*

[†] *International Society for Condensed Matter Nuclear Science (ISCMNS), UK*

[‡] *Engineering Department, University of Palermo viale delle Scienze, 90128 Palermo, Italy*

[§] *francesco.celani@lnf.infn.it*

[¶] *antoninooscar.ditommaso@unipa.it*

[||] *giorgio.vassallo@unipa.it*

Nomenclature

Symbol, name, SI units, natural units (NU).

$\boldsymbol{A}_{\square}$: electromagnetic four-potential $(\mathrm{V\cdot s\cdot m^{-1}})$, (eV);
$\boldsymbol{r}_{\square}$: four-position vector (m), $(\mathrm{eV^{-1}})$;
$\boldsymbol{G}$: electromagnetic field $(\mathrm{V\cdot s\cdot m^{-2}})$, $(\mathrm{eV^2})$;
$\boldsymbol{F}$: electromagnetic field bivector $(\mathrm{V\cdot s\cdot m^{-2}})$, $(\mathrm{eV^2})$;
$\boldsymbol{B}$: flux density field $(\mathrm{V\cdot s\cdot m^{-2}}) = (\mathrm{T})$, $(\mathrm{eV^2})$;
$\boldsymbol{E}$: electric field $(\mathrm{V\cdot m^{-1}})$, $(\mathrm{eV^2})$;
S: scalar field $(\mathrm{V\cdot s\cdot m^{-2}})$, $(\mathrm{eV^2})$;
$\boldsymbol{J}_{\square e}$: four-current density field $(\mathrm{A\cdot m^{-2}})$, $(\mathrm{eV^3})$;
$\boldsymbol{v}_{\square}$: four-velocity vector $(\mathrm{m\cdot s^{-1}})$, (1);
$\boldsymbol{A}'$: electromagnetic eight-potential $(\mathrm{V\cdot s\cdot m^{-1}})$, (eV);
P: pseudoscalar field $(\mathrm{V\cdot s\cdot m^{-2}})$, $(\mathrm{eV^2})$;
$\boldsymbol{J}_{\square m}$: magnetic four-current density field $(\mathrm{A\cdot s\cdot m^{-3}})$, $(\mathrm{eV^3})$;
ρ: charge density $(\mathrm{A\cdot s\cdot m^{-3}} = \mathrm{C\cdot m^{-3}})$, $(\mathrm{eV^3})$;
ρ_m: magnetic charge density $(\mathrm{A\cdot m^{-2}})$, $(\mathrm{eV^3})$;

x, y, z: space coordinates (m), (eV^{-1}), ($1.973\,270\,5 \cdot 10^{-7}$ m $= 1$ eV^{-1});
t: time variable (s), (eV^{-1}), ($6.582\,122 \cdot 10^{-16}$ s $= 1$ eV^{-1});
c: light speed in vacuum ($2.997\,924\,58 \cdot 10^{8}$ $\mathrm{m \cdot s^{-1}}$), (1);
μ_0: permeability of vacuum ($4\pi \cdot 10^{-7}$ $\mathrm{V \cdot s \cdot A^{-1} \cdot m^{-1}}$), ($4\pi$);
ϵ_0: dielectric constant of vacuum ($8.854\,187\,817 \cdot 10^{-12}$ $\mathrm{A \cdot s \cdot V^{-1} \cdot m^{-1}}$), ($\frac{1}{4\pi}$);
$\boldsymbol{P}_\square$: electromagnetic four-momentum ($\mathrm{kg \cdot m \cdot s^{-1}}$), (eV);
$\mathfrak{S}$: generalized Poynting vector ($\mathrm{W \cdot m^{-2}}$), (eV^4).
w: specific energy ($\mathrm{J \cdot m^{-3}}$), (eV^4).

1.1. Introduction

Science is based on the creation and validation of models of abstract concepts and experimental data. For this reason, it is important to examine the rules used to evaluate the quality of a model. Occam's razor principle emphasizes the simplicity and conciseness of the model: among different models that fit experimental data, the simplest one must be preferred, i.e., the model that does not introduce concepts or entities that are not strictly necessary. The following sentences in Latin [2] briefly illustrate this principle:

Pluralitas non est ponenda sine necessitate
Frustra fit per plura quod potest fieri per pauciora
Entia non sunt multiplicanda praeter necessitatem

which can be translated respectively as "plurality should not be posited without necessity," "it is futile to do with more things that which can be done with fewer," and "entities must not be multiplied beyond necessity."

According to this principle, the model quality can be measured by means of two fundamental parameters:

(1) Good agreement of model's predictions with experimental data and/or with other expected results.
(2) The model's simplicity: a value that is inversely related to the amount of information, concepts, entities, exceptions, postulates, parameters, and variables used by the model itself.

These rules are universal, and can be applied in many contexts [16]. In this chapter, we apply Occam's razor principle to Maxwell's equations.

We introduce and use the space–time Clifford algebra, showing that only one fundamental physical entity is sufficient to describe the origin of electromagnetic fields, charges, and currents: the electromagnetic four-potential. The vector potential should not be viewed only as a mathematical tool but as a real physical entity, as suggested by the Aharonov–Bohm effect, a quantum mechanical phenomenon in which a charged particle is affected by the electromagnetic potentials in regions in which the electromagnetic fields are null [1]. Actually, many papers deal with the application of geometric algebra to Maxwell's equation (see [3–5, 7, 19] and many others), but few of them deal with the concept of scalar field. Among the most interesting works we can find a paper by Giuliano Bettini [3], two papers written by van Vlaenderen [21, 22], and two papers of Lee M. Hively [14, 15].

In this work, we propose a reinterpretation of Maxwell's equations which does not use any gauge: the unique constraint is that the electromagnetic four-potential must be represented by a *harmonic* function, as proposed by Bettini [3]. A mistaken application of the so-called "Lorenz gauge" in Maxwell's equations *denies the status of real physical entity* to a scalar field that, although not directly observable, has a gradient in space-time with clear physical meaning: the microscopic four-current density field. As a consequence of our approach, a spinor field that encodes the electromagnetic fields and the derivative of a scalar field emerges from Maxwell's equations. The scalar field will be investigated here and it will be shown that its existence has many implications and consequences on the microscopic structure of electrical charges and currents. It points towards a particular Zitterbewegung model for charged elementary particles.

This chapter is composed of the following parts: Section 1.2 illustrates how Maxwell's equations can be derived only from the electromagnetic four potential; Section 1.3 deals with the main properties of the electromagnetic field, the derivation of Maxwell's equations from the Lagrangian density, the generalized Poynting vector, the symmetrical Maxwell's equations and, finally, in Section 1.4 some essential points are summarized.

1.2. The Electromagnetic Field and the Wave Function

The behavior of electromagnetic waves was described in 1865 by James Clerk Maxwell in his work *Dynamical Theory of the Electromagnetic Field.* Maxwell's equations are a system of partial differential equations, where different concepts are employed: electric field, flux density (or magnetic) field, charge density, and current density [3, 4, 7].

In order to study the undulatory behavior of particles, the concept of wave function was introduced. Following the interpretation of Born, the "square" of this function represents the probability density to find a particle in a point of space, just like the undulatory theory of light, whose intensity is given by the square of the electromagnetic wave amplitude. Now, following the principle of Occam's razor, which suggests carefulness in the introduction of new concepts, we consider two interesting possibilities:

(1) find a common origin of the conceptual entities used in Maxwell's equations;
(2) consider the wave function as a particular reformulation of concepts/entities already present in Maxwell's equations.

1.2.1. *The electromagnetic four potential*

Maxwell's equations can be reinterpreted by means of a unique entity, namely, the electromagnetic four-potential, as defined by the following equation:

$$\boldsymbol{A}_{\square}\left(\boldsymbol{r}_{\square}\right)=\gamma_x A_x\left(\boldsymbol{r}_{\square}\right)+\gamma_y A_y\left(\boldsymbol{r}_{\square}\right)+\gamma_z A_z\left(\boldsymbol{r}_{\square}\right)+\gamma_t A_t\left(\boldsymbol{r}_{\square}\right) \tag{1.2.1}$$

where each of the vector potential components A_x, A_y, A_z, and A_t are functions of the space–time coordinates and $\boldsymbol{r}_{\square}(x,y,z,t)=\gamma_x x+\gamma_y y+\gamma_z z-\gamma_t ct=\boldsymbol{r}_{\triangle}-\gamma_t ct$ is the position vector in space-time. The γ_i unit vectors are the basis elements of $Cl_{3,1}$ Clifford algebra. From now on, in the four-potential and in other field quantities the variable $\boldsymbol{r}_{\square}$ will be omitted for simplicity. The four-potential has dimensions in SI units equal to $[V\cdot s\cdot m^{-1}]$. Two basic assumptions are made:

(1) the vector potential field $\boldsymbol{A}_{\square}$ is represented by a *harmonic* function;
(2) the space is homogeneous, linear, and isotropic.

Therefore, we assume a function that links a vector of four components to each point of the space–time as the unique source of Maxwell's equations entities.

We use the following definition of the operator $\boldsymbol{\partial}$ in space–time algebra

$$\boldsymbol{\partial}=\gamma_x\frac{\partial}{\partial x}+\gamma_y\frac{\partial}{\partial y}+\gamma_z\frac{\partial}{\partial z}+\gamma_t\frac{1}{c}\frac{\partial}{\partial t}=\nabla+\gamma_t\frac{1}{c}\frac{\partial}{\partial t} \tag{1.2.2}$$

where $\nabla=\gamma_x\frac{\partial}{\partial x}+\gamma_y\frac{\partial}{\partial y}+\gamma_z\frac{\partial}{\partial z}$ and $c=1/\sqrt{\epsilon_0\mu_0})$.

If $\boldsymbol{A}_{\square}$ is the vector potential defined by (1.2.1), the following expression can be written:

$$\boldsymbol{\partial}\boldsymbol{A}_{\square}=\boldsymbol{\partial}\cdot\boldsymbol{A}_{\square}+\boldsymbol{\partial}\wedge\boldsymbol{A}_{\square}=S+\boldsymbol{F}=\boldsymbol{G} \tag{1.2.3}$$

Table 1.1. Products $\boldsymbol{\partial A}_{\square}$.

$\boldsymbol{\partial}\mathbf{A}_{\square}$	$\gamma_x A_x$	$\gamma_y A_y$	$\gamma_z A_z$	$\gamma_t A_t$
$\gamma_x \frac{\partial}{\partial x}$	$\frac{\partial A_x}{\partial x}$	$\gamma_x\gamma_y \frac{\partial A_y}{\partial x}$	$\gamma_x\gamma_z \frac{\partial A_z}{\partial x}$	$\gamma_x\gamma_t \frac{\partial A_t}{\partial x}$
$\gamma_y \frac{\partial}{\partial y}$	$-\gamma_x\gamma_y \frac{\partial A_x}{\partial y}$	$\frac{\partial A_y}{\partial y}$	$\gamma_y\gamma_z \frac{\partial A_z}{\partial y}$	$\gamma_y\gamma_t \frac{\partial A_t}{\partial y}$
$\gamma_z \frac{\partial}{\partial z}$	$-\gamma_x\gamma_z \frac{\partial A_x}{\partial z}$	$-\gamma_y\gamma_z \frac{\partial A_y}{\partial z}$	$\frac{\partial A_z}{\partial z}$	$\gamma_z\gamma_t \frac{\partial A_t}{\partial z}$
$\gamma_t \frac{1}{c}\frac{\partial}{\partial t}$	$-\gamma_x\gamma_t \frac{1}{c}\frac{\partial A_x}{\partial t}$	$-\gamma_y\gamma_t \frac{1}{c}\frac{\partial A_y}{\partial t}$	$-\gamma_z\gamma_t \frac{1}{c}\frac{\partial A_z}{\partial t}$	$-\frac{1}{c}\frac{\partial A_t}{\partial t}$

where

$$\boldsymbol{G}(x,y,z,t) = S + \gamma_x\gamma_t F_{xt} + \gamma_y\gamma_t F_{yt} + \gamma_z\gamma_t F_{zt} + \gamma_y\gamma_z F_{yz} + \gamma_x\gamma_z F_{xz} + \gamma_x\gamma_y F_{xy} \tag{1.2.4}$$

Expanding (1.2.3), by considering the products as shown in Table 1.1 and by collecting all terms with the same blade, the following set of equations is found:

$$\boldsymbol{\partial}\cdot\boldsymbol{A}_{\square} = S = \frac{\partial A_x}{\partial x} + \frac{\partial A_y}{\partial y} + \frac{\partial A_z}{\partial z} - \frac{1}{c}\frac{\partial A_t}{\partial t} \tag{1.2.5}$$

$$\gamma_x\gamma_t F_{xt} = \gamma_x\gamma_t \frac{1}{c} E_x = \gamma_x\gamma_t\left(\frac{\partial A_t}{\partial x} - \frac{1}{c}\frac{\partial A_x}{\partial t}\right) \tag{1.2.6}$$

$$\gamma_y\gamma_t F_{yt} = \gamma_y\gamma_t \frac{1}{c} E_y = \gamma_y\gamma_t\left(\frac{\partial A_t}{\partial y} - \frac{1}{c}\frac{\partial A_y}{\partial t}\right) \tag{1.2.7}$$

$$\gamma_z\gamma_t F_{zt} = \gamma_z\gamma_t \frac{1}{c} E_z = \gamma_z\gamma_t\left(\frac{\partial A_t}{\partial z} - \frac{1}{c}\frac{\partial A_z}{\partial t}\right) \tag{1.2.8}$$

$$\gamma_y\gamma_z F_{yz} = \gamma_y\gamma_z B_x = \gamma_y\gamma_z\left(\frac{\partial A_z}{\partial y} - \frac{\partial A_y}{\partial z}\right) \tag{1.2.9}$$

$$\gamma_x\gamma_z F_{xz} = -\gamma_x\gamma_z B_y = \gamma_x\gamma_z\left(\frac{\partial A_z}{\partial x} - \frac{\partial A_x}{\partial z}\right) \tag{1.2.10}$$

$$\gamma_x\gamma_y F_{xy} = \gamma_x\gamma_y B_z = \gamma_x\gamma_y\left(\frac{\partial A_y}{\partial x} - \frac{\partial A_x}{\partial y}\right) \tag{1.2.11}$$

where $S = S_1 + S_2 + S_3 + S_4$ is a scalar field, whose meaning will be clarified later. It is to be noted that equating (1.2.5) to zero, i.e., $S = 0$, gives an expression that takes the form of the "Lorenz gauge" condition if $A_t = -\frac{\varphi}{c}$, where φ is the scalar potential of the electric field [4, 20, 22].

Equation (1.2.5) can be rewritten as

$$S = \nabla \cdot \boldsymbol{A}_{\triangle} - \frac{1}{c}\frac{\partial A_t}{\partial t} \tag{1.2.12}$$

where $\boldsymbol{A}_{\triangle} = \gamma_x A_x + \gamma_y A_y + \gamma_z A_z$ is the usual three components vector potential.

Using the so-called "Lorenz gauge" the scalar field S is considered zero everywhere, *denying its status of a real physical entity* [3]. Similar considerations can be done for the "Coulomb gauge" that assigns zero value to each addendum S_i. We simply do not apply any "gauge," apart from defining $\boldsymbol{A}_{\square}$ as a harmonic function. According to our point of view, both Lorenz and Coulomb "gauges" should be considered just as boundary conditions and the scalar field S, although not directly observable, has a gradient in space–time with a clear physical meaning. Similar considerations are normally presented in electromagnetism to introduce the concept of vector potential, which is not a directly measurable field. The components of the geometric product $\boldsymbol{\partial A}_{\square}$ are shown in Table 1.1. An electromagnetic field $\boldsymbol{G}$ with seven components emerges, composed by one scalar and six bivectors.

The set of equations from (1.2.6) to (1.2.11) can be rewritten also in the following way:

$$E_x = c\frac{\partial A_t}{\partial x} - \frac{\partial A_x}{\partial t} \tag{1.2.13}$$

$$E_y = c\frac{\partial A_t}{\partial y} - \frac{\partial A_y}{\partial t} \tag{1.2.14}$$

$$E_z = c\frac{\partial A_t}{\partial z} - \frac{\partial A_z}{\partial t} \tag{1.2.15}$$

$$B_x = \frac{\partial A_z}{\partial y} - \frac{\partial A_y}{\partial z} \tag{1.2.16}$$

$$B_y = -\frac{\partial A_z}{\partial x} + \frac{\partial A_x}{\partial z} \tag{1.2.17}$$

$$B_z = \frac{\partial A_y}{\partial x} - \frac{\partial A_x}{\partial y} \tag{1.2.18}$$

where

$$\boldsymbol{E} = \gamma_x E_x + \gamma_y E_y + \gamma_z E_z = c\nabla A_t - \frac{\partial \boldsymbol{A}_{\triangle}}{\partial t} \tag{1.2.19}$$

$$\boldsymbol{B} = \gamma_x B_x + \gamma_y B_y + \gamma_z B_z = \nabla \times \boldsymbol{A}_{\triangle} \tag{1.2.20}$$

The sum of all diagonal elements in Table 1.1 represents the scalar product

$$S = \boldsymbol{\partial} \cdot \boldsymbol{A}_{\square} \tag{1.2.21}$$

whereas the sum of all extra-diagonal elements gives the six components of electromagnetic bivector F

$$\boldsymbol{F} = \boldsymbol{\partial} \wedge \boldsymbol{A}_{\square} \tag{1.2.22}$$

Referring to the function $\boldsymbol{G}$, it is possible to note that the "electromagnetic field" is characterized by seven values: three for the electric field, three for the flux density field, and one for the scalar field S.

Table 1.2 represents the relation between the fundamental electromagnetic entities and the space–time components of the vector potential $\boldsymbol{A}_{\square}$.

With reference to Table 1.2, the electromagnetic field $\boldsymbol{G}$ can also be expressed as

$$\begin{aligned}\boldsymbol{G}(x,y,z,t) = S + \boldsymbol{F} = S + \gamma_x\gamma_t\frac{E_x}{c} + \gamma_y\gamma_t\frac{E_y}{c} + \gamma_z\gamma_t\frac{E_z}{c} + \gamma_y\gamma_z B_x \\ - \gamma_x\gamma_z B_y + \gamma_x\gamma_y B_z = S + \frac{1}{c}\boldsymbol{E}\gamma_t + I\boldsymbol{B}\gamma_t \\ = S + \frac{1}{c}(\boldsymbol{E} + Ic\boldsymbol{B})\gamma_t\end{aligned} \tag{1.2.23}$$

where

$$I = \gamma_x\gamma_y\gamma_z\gamma_t \tag{1.2.24}$$

is the unitary pseudoscalar and

$$\boldsymbol{F} = \frac{1}{c}\boldsymbol{E}\gamma_t + I\boldsymbol{B}\gamma_t = \frac{1}{c}(\boldsymbol{E} + Ic\boldsymbol{B})\gamma_t \tag{1.2.25}$$

Table 1.2. Relation between electromagnetic entities and the vector potential $\boldsymbol{A}_{\square}$.

$\boldsymbol{\partial A}_{\square}$	$\gamma_x A_x$	$\gamma_y A_y$	$\gamma_z A_z$	$\gamma_t A_t$
$\gamma_x \frac{\partial}{\partial x}$	S_1	B_{z1}	$-B_{y1}$	$\frac{1}{c} \cdot E_{x1}$
$\gamma_y \frac{\partial}{\partial y}$	B_{z2}	S_2	B_{x1}	$\frac{1}{c} \cdot E_{y1}$
$\gamma_z \frac{\partial}{\partial z}$	$-B_{y2}$	B_{x2}	S_3	$\frac{1}{c} \cdot E_{z1}$
$\gamma_t \frac{1}{c}\frac{\partial}{\partial t}$	$\frac{1}{c}E_{x2}$	$\frac{1}{c}E_{y2}$	$\frac{1}{c}E_{z2}$	S_4

On the other hand, with reference to Table 1.1, the electromagnetic field $\boldsymbol{G}$ can be expressed in the following compact form:

$$\boldsymbol{G}(x,y,z,t) = \nabla \cdot \boldsymbol{A}_{\triangle} - \frac{1}{c}\frac{\partial A_t}{\partial t} + \nabla A_t \gamma_t - \frac{1}{c}\frac{\partial \boldsymbol{A}_{\triangle}}{\partial t}\gamma_t + I\nabla \times \boldsymbol{A}_{\triangle}\gamma_t \tag{1.2.26}$$

which again results in Equations (1.2.12), (1.2.19), and (1.2.20) by taking (1.2.23) into account.

1.2.2. *Maxwell's equations*

Now, by applying the operator $\boldsymbol{\partial}$ to the multivector $\boldsymbol{G}$ (1.2.3) and equating it to zero, a new expression is found,

$$\boldsymbol{\partial G} = \boldsymbol{\partial}^2 \boldsymbol{A}_{\square} = 0 \tag{1.2.27}$$

whose components are shown in Table 1.3. The equation $\boldsymbol{\partial G} = 0$ *can be seen as an extension in four dimensions of the Cauchy–Riemann conditions for analytic functions of a complex (two-dimensional) variable* [3, 11]. In [11], Hestenes writes: "Members of this audience will recognize $\square\psi_0 = 0$ as a generalization of the Cauchy-Riemann equations to spacetime, so we can expect it to have a rich variety of solutions. The problem is to pick out those

Table 1.3. Products $\boldsymbol{\partial G} = \boldsymbol{\partial}(\boldsymbol{\partial A}_{\square})$. $\gamma_{ij} = \gamma_i\gamma_j$, $\gamma_{ijk} = \gamma_i\gamma_j\gamma_k$.

$\boldsymbol{\partial}^2\boldsymbol{A}_{\square}$	S	$\gamma_{xt}\frac{1}{c}E_x$	$\gamma_{yt}\frac{1}{c}E_y$	$\gamma_{zt}\frac{1}{c}E_z$
$\gamma_x\frac{\partial}{\partial x}$	$\gamma_x\frac{\partial S}{\partial x}$	$\gamma_t\frac{1}{c}\frac{\partial E_x}{\partial x}$	$\gamma_{xyt}\frac{1}{c}\frac{\partial E_y}{\partial x}$	$\gamma_{xzt}\frac{1}{c}\frac{\partial E_z}{\partial x}$
$\gamma_y\frac{\partial}{\partial y}$	$\gamma_y\frac{\partial S}{\partial y}$	$-\gamma_{xyt}\frac{1}{c}\frac{\partial E_x}{\partial y}$	$\gamma_t\frac{1}{c}\frac{\partial E_y}{\partial y}$	$\gamma_{yzt}\frac{1}{c}\frac{\partial E_z}{\partial y}$
$\gamma_z\frac{\partial}{\partial z}$	$\gamma_z\frac{\partial S}{\partial z}$	$-\gamma_{xzt}\frac{1}{c}\frac{\partial E_x}{\partial z}$	$-\gamma_{yzt}\frac{1}{c}\frac{\partial E_y}{\partial z}$	$\gamma_t\frac{1}{c}\frac{\partial E_z}{\partial z}$
$\gamma_t\frac{1}{c}\frac{\partial}{\partial t}$	$\gamma_t\frac{1}{c}\frac{\partial S}{\partial t}$	$\gamma_x\frac{1}{c^2}\frac{\partial E_x}{\partial t}$	$\gamma_y\frac{1}{c^2}\frac{\partial E_y}{\partial t}$	$\gamma_z\frac{1}{c^2}\frac{\partial E_z}{\partial t}$
$\boldsymbol{\partial}^2\boldsymbol{A}_{\square}$	S	$\gamma_{yz}B_x$	$-\gamma_{xz}B_y$	$\gamma_{xy}B_z$
$\gamma_x\frac{\partial}{\partial x}$	$\gamma_x\frac{\partial S}{\partial x}$	$\gamma_{xyz}\frac{\partial B_x}{\partial x}$	$-\gamma_z\frac{\partial B_y}{\partial x}$	$\gamma_y\frac{\partial B_z}{\partial x}$
$\gamma_y\frac{\partial}{\partial y}$	$\gamma_y\frac{\partial S}{\partial y}$	$\gamma_z\frac{\partial B_x}{\partial y}$	$\gamma_{xyz}\frac{\partial B_y}{\partial x}$	$-\gamma_x\frac{\partial B_z}{\partial y}$
$\gamma_z\frac{\partial}{\partial z}$	$\gamma_z\frac{\partial S}{\partial z}$	$-\gamma_y\frac{\partial B_x}{\partial z}$	$\gamma_x\frac{\partial B_y}{\partial z}$	$\gamma_{xyz}\frac{\partial B_z}{\partial z}$
$\gamma_t\frac{1}{c}\frac{\partial}{\partial t}$	$\gamma_t\frac{1}{c}\frac{\partial S}{\partial t}$	$\gamma_{yzt}\frac{1}{c}\frac{\partial B_x}{\partial t}$	$-\gamma_{xzt}\frac{1}{c}\frac{\partial B_y}{\partial t}$	$\gamma_{xyt}\frac{1}{c}\frac{\partial B_z}{\partial t}$

solutions with physical significance." In fact, if $\boldsymbol{A}_{\Box}$ is harmonic, then

$$\boldsymbol{\partial}^2 \boldsymbol{A}_{\Box} = \nabla^2 \boldsymbol{A}_{\Box} - \frac{1}{c^2}\frac{\partial^2 \boldsymbol{A}_{\Box}}{\partial t^2} = 0 \tag{1.2.28}$$

which represents the wave equation of the four-potential and where

$$\boldsymbol{\partial}^2 = \frac{\partial^2}{\partial x^2} + \frac{\partial^2}{\partial y^2} + \frac{\partial^2}{\partial z^2} - \frac{1}{c^2}\frac{\partial^2}{\partial t^2} = \nabla^2 - \frac{1}{c^2}\frac{\partial^2}{\partial t^2}$$

It should be noted that in our case, considering the scalar field $S \neq 0$ and $\boldsymbol{A}_{\Box}$ harmonic, (1.2.28) is always homogeneous.

By collecting all common factors contained in Table 1.3, the following equations are derived:

$$\gamma_x \left(\frac{\partial S}{\partial x} - \frac{\partial B_z}{\partial y} + \frac{\partial B_y}{\partial z} + \frac{1}{c^2}\frac{\partial E_x}{\partial t} \right) = 0 \tag{1.2.29}$$

$$\gamma_y \left(\frac{\partial B_z}{\partial x} + \frac{\partial S}{\partial y} - \frac{\partial B_x}{\partial z} + \frac{1}{c^2}\frac{\partial E_y}{\partial t} \right) = 0 \tag{1.2.30}$$

$$\gamma_z \left(-\frac{\partial B_y}{\partial x} + \frac{\partial B_x}{\partial y} + \frac{\partial S}{\partial z} + \frac{1}{c^2}\frac{\partial E_z}{\partial t} \right) = 0 \tag{1.2.31}$$

$$\gamma_t \frac{1}{c} \left(\frac{\partial E_x}{\partial x} + \frac{\partial E_y}{\partial y} + \frac{\partial E_z}{\partial z} + \frac{\partial S}{\partial t} \right) = 0 \tag{1.2.32}$$

$$\gamma_y \gamma_z \gamma_t \frac{1}{c} \left(\frac{\partial E_z}{\partial y} - \frac{\partial E_y}{\partial z} + \frac{\partial B_x}{\partial t} \right) = 0 \tag{1.2.33}$$

$$\gamma_x \gamma_z \gamma_t \frac{1}{c} \left(\frac{\partial E_z}{\partial x} - \frac{\partial E_x}{\partial z} - \frac{\partial B_y}{\partial t} \right) = 0 \tag{1.2.34}$$

$$\gamma_x \gamma_y \gamma_t \frac{1}{c} \left(\frac{\partial E_y}{\partial x} - \frac{\partial E_x}{\partial y} + \frac{\partial B_z}{\partial t} \right) = 0 \tag{1.2.35}$$

$$\gamma_x \gamma_y \gamma_z \left(\frac{\partial B_x}{\partial x} + \frac{\partial B_y}{\partial y} + \frac{\partial B_z}{\partial z} \right) = 0 \tag{1.2.36}$$

Rearranging all equations from (1.2.29) to (1.2.36), the following are derived:

$$\frac{\partial B_z}{\partial y} - \frac{\partial B_y}{\partial z} = \frac{\partial S}{\partial x} + \frac{1}{c^2}\frac{\partial E_x}{\partial t} \tag{1.2.37}$$

$$\frac{\partial B_x}{\partial z} - \frac{\partial B_z}{\partial x} = \frac{\partial S}{\partial y} + \frac{1}{c^2}\frac{\partial E_y}{\partial t} \tag{1.2.38}$$

$$\frac{\partial B_y}{\partial x} - \frac{\partial B_x}{\partial y} = \frac{\partial S}{\partial z} + \frac{1}{c^2}\frac{\partial E_z}{\partial t} \tag{1.2.39}$$

$$\frac{\partial E_x}{\partial x} + \frac{\partial E_y}{\partial y} + \frac{\partial E_z}{\partial z} = -\frac{\partial S}{\partial t} \tag{1.2.40}$$

$$\frac{\partial E_z}{\partial y} - \frac{\partial E_y}{\partial z} = -\frac{\partial B_x}{\partial t} \tag{1.2.41}$$

$$\frac{\partial E_x}{\partial z} - \frac{\partial E_z}{\partial x} = -\frac{\partial B_y}{\partial t} \tag{1.2.42}$$

$$\frac{\partial E_y}{\partial x} - \frac{\partial E_x}{\partial y} = -\frac{\partial B_z}{\partial t} \tag{1.2.43}$$

$$\frac{\partial B_x}{\partial x} + \frac{\partial B_y}{\partial y} + \frac{\partial B_z}{\partial z} = 0 \tag{1.2.44}$$

which are coincident with Maxwell's equations if

$$\frac{\partial S}{\partial x} = \mu_0 J_{ex} = \mu_0 \frac{\partial q}{\partial y \partial z \partial t} = \mu_0 \frac{\partial q \partial x}{\partial x \partial y \partial z \partial t} = \mu_0 \rho v_x \tag{1.2.45}$$

$$\frac{\partial S}{\partial y} = \mu_0 J_{ey} = \mu_0 \frac{\partial q}{\partial x \partial z \partial t} = \mu_0 \frac{\partial q \partial y}{\partial x \partial y \partial z \partial t} = \mu_0 \rho v_y \tag{1.2.46}$$

$$\frac{\partial S}{\partial z} = \mu_0 J_{ez} = \mu_0 \frac{\partial q}{\partial x \partial y \partial t} = \mu_0 \frac{\partial q \partial z}{\partial x \partial y \partial z \partial t} = \mu_0 \rho v_z \tag{1.2.47}$$

$$\frac{1}{c}\frac{\partial S}{\partial t} = \mu_0 J_{et} = -\mu_0 c \frac{\partial q}{\partial x \partial y \partial z} = -\mu_0 c \rho \tag{1.2.48}$$

where ∂q is the differential of a generic charge [4, 20]. Equation (1.2.48) can also be written as

$$\frac{\partial S}{\partial t} = c\mu_0 J_{et} = -\mu_0 c \frac{\partial q}{\partial x \partial y \partial z} = -\mu_0 c^2 \rho = -\frac{\rho}{\epsilon_0} \tag{1.2.49}$$

By taking into account (1.2.45), (1.2.46), (1.2.47), and (1.2.48), the following relation holds for the current density field,

$$\frac{1}{\mu_0}\boldsymbol{\partial} S = \frac{1}{\mu_0}\left(\gamma_x \frac{\partial S}{\partial x} + \gamma_y \frac{\partial S}{\partial y} + \gamma_z \frac{\partial S}{\partial z} + \gamma_t \frac{1}{c}\frac{\partial S}{\partial t}\right) = \boldsymbol{J}_{\Box e} \tag{1.2.50}$$

where

$$\begin{aligned}\boldsymbol{J}_{\Box e} &= \gamma_x J_{ex} + \gamma_y J_{ey} + \gamma_z J_{ez} + \gamma_t J_{et} = \gamma_x J_{ex} + \gamma_y J_{ey} + \gamma_z J_{ez} - \gamma_t c\rho \\ &= \boldsymbol{J}_{\triangle} - \gamma_t c\rho = \rho\left(\boldsymbol{v}_{\triangle} - \gamma_t c\right)\end{aligned} \tag{1.2.51}$$

is the four-current vector,

$$\boldsymbol{v}_{\square} = \gamma_x v_x + \gamma_y v_y + \gamma_z v_z - \gamma_t c = \boldsymbol{v}_{\Delta} - \gamma_t c \tag{1.2.52}$$

is a four-velocity vector and $\boldsymbol{v}_{\Delta}$ is the speed in the ordinary space.

In this formulation, the partial derivatives of the scalar field S with respect to time and space coordinates can be interpreted as charge density and current density, respectively. As a matter of fact, (1.2.37), (1.2.38), and (1.2.39) represent the spatial components of Ampere's law, i.e.,

$$\nabla \times \boldsymbol{B} = \mu_0 \boldsymbol{J}_{\Delta} + \frac{1}{c^2}\frac{\partial \boldsymbol{E}}{\partial t} \tag{1.2.53}$$

where $\boldsymbol{J}_{\Delta} = \gamma_x J_{ex} + \gamma_y J_{ey} + \gamma_z J_{ez}$ is the three component vector of current density, (1.2.40) is Gauss's law for the electric field

$$\nabla \cdot \boldsymbol{E} = \frac{\rho}{\epsilon_0} \tag{1.2.54}$$

(1.2.41), (1.2.42) and (1.2.43) represent the spatial components of the Faraday–Neumann–Maxwell–Lenz law

$$\nabla \times \boldsymbol{E} = -\frac{\partial \boldsymbol{B}}{\partial t} \tag{1.2.55}$$

and (1.2.44) Gauss's law for the flux density field

$$\nabla \cdot \boldsymbol{B} = 0 \tag{1.2.56}$$

Finally, by applying the $\boldsymbol{\partial}\cdot$ operator to (1.2.50) and setting the result to zero, the equation representing the law of electric charge conservation is obtained

$$\frac{1}{\mu_0}\boldsymbol{\partial} \cdot (\boldsymbol{\partial} S) = \boldsymbol{\partial} \cdot \boldsymbol{J}_{\square e} = \frac{\partial J_{ex}}{\partial x} + \frac{\partial J_{ey}}{\partial y} + \frac{\partial J_{ez}}{\partial z} + \frac{\partial \rho}{\partial t} = 0 \tag{1.2.57}$$

It is important to note that the wave equation of the scalar field S can be deduced from the charge-current conservation law:

$$\boldsymbol{\partial} \cdot (\boldsymbol{\partial} S) = \boldsymbol{\partial}^2 S = \frac{\partial^2 S}{\partial x^2} + \frac{\partial^2 S}{\partial y^2} + \frac{\partial^2 S}{\partial z^2} - \frac{1}{c^2}\frac{\partial^2 S}{\partial t^2} = \nabla^2 S - \frac{1}{c^2}\frac{\partial^2 S}{\partial t^2} = 0 \tag{1.2.58}$$

Now, by applying the time derivative to (1.2.58) and remembering (1.2.49), the wave equation of a massless charge field $\rho(\boldsymbol{r}_{\square})$ can also be deduced, i.e.,

$$\frac{\partial}{\partial t}\left(\boldsymbol{\partial}^2 S\right) = \boldsymbol{\partial}^2 \left(\frac{\partial S}{\partial t}\right) = \boldsymbol{\partial}^2\left(-\mu_0 c^2 \rho\right) = -\mu_0 c^2 \boldsymbol{\partial}^2 \rho = 0 \tag{1.2.59}$$

which gives

$$\partial^2 \rho = \nabla^2 \rho - \frac{1}{c^2}\frac{\partial^2}{\partial t^2}\rho = 0 \tag{1.2.60}$$

Clearly, both (1.2.58) and (1.2.60) represent, respectively, fields (S and ρ) that must necessarily propagate at the speed of light [17, 20]. Equation (1.2.50) means also that the four-vector current density field can be derived directly from the scalar field S. The hypothesis of existence of scalar waves has been recently explored at the Oak Ridge laboratories: "The new theory predicts a new charge-fluctuation-driven scalar wave, having energy but not momentum for zero magnetic and electric fields. The scalar wave can co-exist with a longitudinal-electric wave, having energy and momentum. The new theory in four-vector form is relativistically covariant. New experimental tests are needed to confirm this theory" [15].

The proposed reinterpretation of Maxwell's equations in this work is in agreement with the principle of Occam's razor: the concepts of charge and current density are not inserted "*ad hoc*" but are deduced from a single more fundamental entity, the four-dimensional vector potential field $\boldsymbol{A}_\square(\boldsymbol{r}_\square) = \boldsymbol{A}_\square(x, y, z, t)$.

In fact, a big advantage of our perspective on Maxwell's equations is the ability to simply specify both current density and charge density distributions and then see what fields result. Nevertheless, the added constraint on the charge and current density seems to imply that one is no longer free to specify charge and current density distributions at will, because this information is indeed included within the electromagnetic four potential $\boldsymbol{A}_\square$.

Equation (1.2.60) imposes a precise condition on charge dynamics, describing only distributions of charge density moving in vacuum at the speed of light c. At first glance, this result seems to be incompatible with experimental observations, with the usual concepts of charge and current, and with the traditional way of working with Maxwell's equations. A skeptic might ask: "What does a light-speed moving and massless charge field have to do with the electron?" In the following chapters, we precisely derive how this light-speed moving $\rho(\boldsymbol{r}_\square)$ field is perceived as a massive electron particle and as a quantum mechanical wavefunction. To reach this discovery, we use nothing else than Maxwell's equations and General Relativity.

As will be shown later, we can interpret Equation (1.2.60) as a constraint for the definition of models of elementary charged particles. This constraint, however, can be removed when considering macroscopic electromagnetic systems or even the dynamics of a single elementary charge at a spatial scale greater than the particle Compton wavelength λ_c and at a time scale greater than the Compton period $\frac{\lambda_c}{c}$. In this case, "static" elementary charges can be seen as charge density distributions moving at the speed of light on a closed trajectory but with a zero average speed (this generalization would be consistent with static charge densities, electrets, dielectrics), whereas currents can be considered as ordered motions of charge density distributions moving with an absolute velocity equal to the speed of light but with an arbitrary average speed lower than c. As an example, referring to Figure 1.1, the electromagnetic effects generated by an elementary charge q, moving at instantaneous speed c in a helical motion of radius $\leq \frac{\lambda_c}{2\pi}$ with average velocity v_z along the helix axis z and tangential velocity $v_\perp$, can be approximated, on a spatial scale $\gg \lambda_c$ and a temporal scale $\gg \frac{\lambda_c}{c}$, to those produced by the same elementary charge q moving at uniform velocity v_z, creating the current density

$$\boldsymbol{J}_z = J_z\gamma_z \approx \frac{q}{\delta x \delta y \delta z}\frac{dz}{dt}\gamma_z = \frac{q}{\delta V}\frac{dz}{dt}\gamma_z = \rho v_z \gamma_z = \rho \boldsymbol{v}_z \tag{1.2.61}$$

where $\delta V = \delta x \delta y \delta z \approx \lambda_c^3$. In this view and at a macroscopic level, the new interpretation of Maxwell's equations proposed here remains compatible with the traditional way of working with them, i.e., by assigning the sources and determining, as a consequence, both the electric and the flux density (magnetic) field.

The new formulation of Maxwell's equations expressed by (1.2.27) is quite similar to the Dirac–Hestenes equation for $m = 0$ (Weyl equation). In all cases, the solution is a *spinor* field. A spinor is a mathematical object

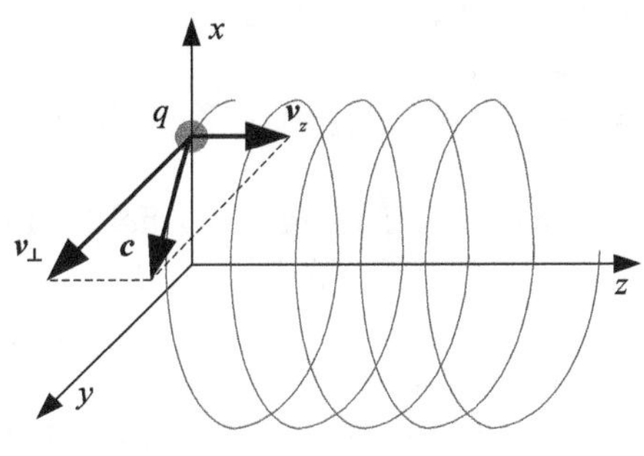

Figure 1.1. Helical motion of an elementary charge q moving at the speed of light, with $v_z^2 + v_\perp^2 = c^2$.

that in space–time algebra is simply a multivector of even grade components. The motion of a massless charge that moves at speed of light can be described using a composition of a rotation in the $\gamma_x\gamma_y$ plane followed by a scaled hyperbolic rotation in the $\gamma_z\gamma_t$ plane and can be encoded in real $Cl_{3,1}$ algebra with a single spinor.

At this point, we encourage the reader by an interesting sentence of P. A. M. Dirac. In fact, in his Nobel lecture [6], held in 1933, Dirac proposed an electron model in which a charge moves at the speed of light: "It is found that an electron which seems to us to be moving slowly, must actually have a very high frequency oscillatory motion of small amplitude superposed on the regular motion which appears to us. As a result of this oscillatory motion, the velocity of the electron at any time equals the velocity of light."

1.3. Properties of the Electromagnetic Field

In this section, the main properties of the electromagnetic field will be presented and discussed by means of $Cl_{3,1}$ Clifford algebra.

1.3.1. *Derivation of Maxwell's equations from Lagrangian density*

Maxwell's equations can be derived considering the following Lagrangian density, in form of a composition of a scalar and a pseudoscalar part:

$$\begin{aligned} \boldsymbol{L} &= \frac{1}{2\mu_0}\boldsymbol{\partial A}_{\square}\widetilde{\boldsymbol{\partial A}_{\square}} = \frac{1}{2\mu_0}\boldsymbol{G}\widetilde{\boldsymbol{G}} = \frac{1}{2\mu_0}\|\boldsymbol{G}\|^2 = \frac{1}{2\mu_0}(S+\boldsymbol{F})(S-\boldsymbol{F}) \\ &= \frac{1}{2\mu_0}\left(S^2-\boldsymbol{F}^2\right) = \frac{1}{2\mu_0}\left(-\frac{E^2}{c^2}+B^2+S^2-\frac{2}{c}\boldsymbol{I}\boldsymbol{E}\cdot\boldsymbol{B}\right) \end{aligned} \tag{1.3.1}$$

where, bearing (1.2.25) in mind,

$$\boldsymbol{F} = \frac{1}{c}\boldsymbol{E}\gamma_t + \boldsymbol{I}\boldsymbol{B}\gamma_t = \frac{1}{c}(\boldsymbol{E}+\boldsymbol{I}c\boldsymbol{B})\gamma_t \tag{1.3.2}$$

is the bivector part of the electromagnetic field and $\widetilde{\ }$ represents the Clifford reversion operator, as defined in the Mathematical Preliminaries chapter. Essentially, Equation (1.3.1) equates Lagrangian density with the electromagnetic field's norm-square.

Expanding (1.3.1), and taking equations from (1.2.12) to (1.2.18) into account, we obtain the Lagrangian density as a function of the derivatives of the electromagnetic four-potential components, i.e.,

$$\begin{aligned}
\boldsymbol{L} = \frac{1}{2\mu_0}\Bigg\{ &-\left(\frac{\partial A_t}{\partial x} - \frac{1}{c}\frac{\partial A_x}{\partial t}\right)^2 - \left(\frac{\partial A_t}{\partial y} - \frac{1}{c}\frac{\partial A_y}{\partial t}\right)^2 - \left(\frac{\partial A_t}{\partial z} - \frac{1}{c}\frac{\partial A_z}{\partial t}\right)^2 \\
&+ \left(\frac{\partial A_z}{\partial y} - \frac{\partial A_y}{\partial z}\right)^2 + \left(\frac{\partial A_x}{\partial z} - \frac{\partial A_z}{\partial x}\right)^2 + \left(\frac{\partial A_y}{\partial x} - \frac{\partial A_x}{\partial y}\right)^2 \\
&+ \left(\frac{\partial A_x}{\partial x} + \frac{\partial A_y}{\partial y} + \frac{\partial A_z}{\partial z} - \frac{1}{c}\frac{\partial A_t}{\partial t}\right)^2 \\
&- 2I\left[\left(\frac{\partial A_t}{\partial x} - \frac{1}{c}\frac{\partial A_x}{\partial t}\right)\left(\frac{\partial A_z}{\partial y} - \frac{\partial A_y}{\partial z}\right) + \left(\frac{\partial A_t}{\partial y} - \frac{1}{c}\frac{\partial A_y}{\partial t}\right)\right. \\
&\times \left.\left(\frac{\partial A_x}{\partial z} - \frac{\partial A_z}{\partial x}\right) + \left(\frac{\partial A_t}{\partial z} - \frac{1}{c}\frac{\partial A_z}{\partial t}\right)\left(\frac{\partial A_y}{\partial x} - \frac{\partial A_x}{\partial y}\right)\right]\Bigg\}
\end{aligned} \tag{1.3.3}$$

In $Cl_{3,1}$ algebra, the Euler–Lagrange equations can be expressed, considering as variables the electromagnetic four-potential components $A_x(x, y, z, t)$, $A_y(x, y, z, t)$, $A_z(x, y, z, t)$, and $A_t(x, y, z, t)$, in the following way:

$$\sum_{j=x,y,z,t}\left(\sum_{i=x,y,z,t}\gamma_i\frac{\partial}{\partial i}\left(\frac{\partial \boldsymbol{L}}{\gamma_i\gamma_j\partial\left(\frac{\partial A_j}{\partial i}\right)}\right) - \frac{\partial \boldsymbol{L}}{\gamma_j\partial A_j}\right) = 0 \tag{1.3.4}$$

which reduces itself to

$$\sum_{j=x,y,z,t}\left(\sum_{i=x,y,z,t}\gamma_i\frac{\partial}{\partial i}\left(\frac{\partial \boldsymbol{L}}{\gamma_i\gamma_j\partial\left(\frac{\partial A_j}{\partial i}\right)}\right)\right) = 0 \tag{1.3.5}$$

considering that in this case

$$\sum_{j=x,y,z,t}\left(\frac{\partial \boldsymbol{L}}{\gamma_j\partial A_j}\right) = 0 \tag{1.3.6}$$

because in (1.3.3) only the derivative terms of the four-potential $(\frac{\partial A_j}{\partial i})$ appear. By expanding (1.3.5), for example, with $j = t$, we achieve, after some trivial calculation steps,

$$
\begin{aligned}
-\gamma_t \frac{\partial \boldsymbol{L}}{\partial A_t} &= -\gamma_t \frac{\partial}{\partial x} \frac{\partial \boldsymbol{L}}{\partial \left(\frac{\partial A_t}{\partial x}\right)} - \gamma_t \frac{\partial}{\partial y} \frac{\partial \boldsymbol{L}}{\partial \left(\frac{\partial A_t}{\partial y}\right)} - \gamma_t \frac{\partial}{\partial z} \frac{\partial \boldsymbol{L}}{\partial \left(\frac{\partial A_t}{\partial z}\right)} - \gamma_t \frac{\partial}{\partial t} \frac{\partial \boldsymbol{L}}{\partial \left(\frac{\partial A_t}{\partial t}\right)} \\
&= \gamma_t \frac{1}{\mu_0} \left(\frac{1}{c} \frac{\partial E_x}{\partial x} + I \frac{\partial B_x}{\partial x} + \frac{1}{c} \frac{\partial E_y}{\partial y} + I \frac{\partial B_y}{\partial y} \right. \\
&\quad \left. + \frac{1}{c} \frac{\partial E_z}{\partial z} + I \frac{\partial B_z}{\partial z} + \frac{1}{c} \frac{\partial S}{\partial t} \right) = 0
\end{aligned}
\tag{1.3.7}
$$

and this equation returns Gauss's laws for the electric field (see Equation (1.2.32)) and for the flux density field (see Equation (1.2.36)), respectively:

$$
\gamma_t \frac{1}{c} \left(\frac{\partial E_x}{\partial x} + \frac{\partial E_y}{\partial y} + \frac{\partial E_z}{\partial z} + \frac{\partial S}{\partial t} \right) = 0
$$

$$
\gamma_x \gamma_y \gamma_z \left(\frac{\partial B_x}{\partial x} + \frac{\partial B_y}{\partial y} + \frac{\partial B_z}{\partial z} \right) = 0
$$

Now, if we expand (1.3.5) with $j = x$, we obtain

$$
\begin{aligned}
\gamma_x \frac{\partial \boldsymbol{L}}{\partial A_x} &= \gamma_x \frac{\partial}{\partial x} \frac{\partial \boldsymbol{L}}{\partial \left(\frac{\partial A_x}{\partial x}\right)} + \gamma_x \frac{\partial}{\partial y} \frac{\partial \boldsymbol{L}}{\partial \left(\frac{\partial A_x}{\partial y}\right)} + \gamma_x \frac{\partial}{\partial z} \frac{\partial \boldsymbol{L}}{\partial \left(\frac{\partial A_x}{\partial z}\right)} + \gamma_x \frac{\partial}{\partial t} \frac{\partial \boldsymbol{L}}{\partial \left(\frac{\partial A_x}{\partial t}\right)} \\
&= \gamma_x \frac{1}{\mu_0} \left(\frac{\partial S}{\partial x} - \frac{\partial B_z}{\partial y} + \frac{I}{c} \frac{\partial E_z}{\partial y} + \frac{\partial B_y}{\partial z} - \frac{I}{c} \frac{\partial E_y}{\partial z} + \frac{1}{c^2} \frac{\partial E_x}{\partial t} + \frac{I}{c} \frac{\partial B_x}{\partial t} \right) \\
&= 0
\end{aligned}
\tag{1.3.8}
$$

Equation (1.3.8) gives (1.2.29) and (1.2.33):

$$
\gamma_x \left(\frac{\partial B_y}{\partial z} - \frac{\partial B_z}{\partial y} + \frac{\partial S}{\partial x} + \frac{1}{c^2} \frac{\partial E_x}{\partial t} \right) = 0
$$

$$
\gamma_y \gamma_z \gamma_t \frac{1}{c} \left(\frac{\partial E_z}{\partial y} - \frac{\partial E_y}{\partial z} + \frac{\partial B_x}{\partial t} \right) = 0
$$

If we carry on the above procedures with $j = y$ and $j = z$, the other remaining components of Maxwell's equation can be determined, i.e.,

(1.2.30), (1.2.34), (1.2.31), and (1.2.35):

$$\gamma_y \frac{\partial \boldsymbol{L}}{\partial A_y} = \gamma_y \frac{\partial}{\partial x} \frac{\partial \boldsymbol{L}}{\partial \left(\frac{\partial A_y}{\partial x}\right)} + \gamma_y \frac{\partial}{\partial y} \frac{\partial \boldsymbol{L}}{\partial \left(\frac{\partial A_y}{\partial y}\right)} + \gamma_y \frac{\partial}{\partial z} \frac{\partial \boldsymbol{L}}{\partial \left(\frac{\partial A_y}{\partial z}\right)} + \gamma_y \frac{\partial}{\partial t} \frac{\partial \boldsymbol{L}}{\partial \left(\frac{\partial A_y}{\partial t}\right)}$$

$$= \gamma_y \frac{1}{\mu_0} \left(\frac{\partial B_z}{\partial x} - \frac{I}{c} \frac{\partial E_z}{\partial x} + \frac{\partial S}{\partial y} - \frac{\partial B_x}{\partial z} \right.$$

$$\left. + \frac{I}{c} \frac{\partial E_x}{\partial z} + \frac{1}{c^2} \frac{\partial E_y}{\partial t} + \frac{I}{c} \frac{\partial B_y}{\partial t} \right) = 0 \tag{1.3.9}$$

$$\gamma_z \frac{\partial \boldsymbol{L}}{\partial A_z} = \gamma_z \frac{\partial}{\partial x} \frac{\partial \boldsymbol{L}}{\partial \left(\frac{\partial A_z}{\partial x}\right)} + \gamma_z \frac{\partial}{\partial y} \frac{\partial \boldsymbol{L}}{\partial \left(\frac{\partial A_z}{\partial y}\right)} + \gamma_z \frac{\partial}{\partial z} \frac{\partial \boldsymbol{L}}{\partial \left(\frac{\partial A_z}{\partial z}\right)} + \gamma_z \frac{\partial}{\partial t} \frac{\partial \boldsymbol{L}}{\partial \left(\frac{\partial A_z}{\partial t}\right)}$$

$$= \gamma_z \frac{1}{\mu_0} \left(- \frac{\partial B_y}{\partial x} + \frac{I}{c} \frac{\partial E_y}{\partial x} + \frac{\partial B_x}{\partial y} - \frac{I}{c} \frac{\partial E_x}{\partial y} \right.$$

$$\left. + \frac{\partial S}{\partial z} + \frac{1}{c^2} \frac{\partial E_z}{\partial t} + \frac{I}{c} \frac{\partial B_z}{\partial t} \right) = 0 \tag{1.3.10}$$

that give, as expected, respectively,

$$\gamma_y \left(\frac{\partial B_z}{\partial x} + \frac{\partial S}{\partial y} - \frac{\partial B_x}{\partial z} + \frac{1}{c^2} \frac{\partial E_y}{\partial t} \right) = 0$$

$$\gamma_x \gamma_z \gamma_t \frac{1}{c} \left(\frac{\partial E_z}{\partial x} - \frac{\partial E_x}{\partial z} - \frac{\partial B_y}{\partial t} \right) = 0$$

$$\gamma_z \left(\frac{\partial B_x}{\partial y} + \frac{\partial S}{\partial z} - \frac{\partial B_y}{\partial x} + \frac{1}{c^2} \frac{\partial E_z}{\partial t} \right) = 0$$

$$\gamma_x \gamma_y \gamma_t \frac{1}{c} \left(\frac{\partial E_y}{\partial x} - \frac{\partial E_x}{\partial y} + \frac{\partial B_z}{\partial t} \right) = 0$$

By analyzing the above-reported equations it is possible to reach some conclusions. First of all, the Lagrangian density, as defined in (1.3.1), can be divided in the sum of two parts

$$\boldsymbol{L} = L_{\text{field}} + L_{\text{int}} \tag{1.3.11}$$

The first part

$$L_{\text{field}} = \frac{1}{2\mu_0} \left(- \frac{E^2}{c^2} + B^2 \right) = - \frac{1}{2\mu_0} (\boldsymbol{F} \cdot \boldsymbol{F}) \tag{1.3.12}$$

represents the "field part" of the Lagrangian density, as known in the literature, and the second

$$L_{\text{int}} = \frac{1}{2\mu_0}\left(S^2 - \frac{2}{c}I\boldsymbol{E}\cdot\boldsymbol{B}\right) \tag{1.3.13}$$

represents the "interaction term" of the Lagrangian density, that takes the interaction of the electromagnetic field with the sources into account, remembering, in addition, that the derivatives of the scalar field S, with respect to the four-dimensional space coordinates x, y, z, and t, are bounded, respectively, to the sources J_{ex}, J_{ey}, J_{ez}, and $J_{et} = -c\rho$ (see Equation (1.2.50)). Indeed, by deriving only the interaction terms of the Lagrangian density with respect to the four-potential, i.e., by performing the operation $\frac{\partial L_{\text{int}}}{\partial A_j}$, it is possible to derive the term $\boldsymbol{J}_{\Box e}\cdot\boldsymbol{A}_{\Box}$. In fact, for the component along γ_t we find

$$\begin{aligned}
-\gamma_t\frac{\partial L_{\text{int}}}{\partial A_t} &= -\gamma_t\frac{\partial}{\partial x}\frac{\partial L_{\text{int}}}{\partial\left(\frac{\partial A_t}{\partial x}\right)} - \gamma_t\frac{\partial}{\partial y}\frac{\partial L_{\text{int}}}{\partial\left(\frac{\partial A_t}{\partial y}\right)} - \gamma_t\frac{\partial}{\partial z}\frac{\partial L_{\text{int}}}{\partial\left(\frac{\partial A_t}{\partial z}\right)} - \gamma_t\frac{\partial}{\partial t}\frac{\partial L_{\text{int}}}{\partial\left(\frac{\partial A_t}{\partial t}\right)}\\
&= \frac{\gamma_t}{\mu_0}\left(I\frac{\partial B_x}{\partial x} + I\frac{\partial B_y}{\partial y} + I\frac{\partial B_z}{\partial z} + \frac{1}{c}\frac{\partial S}{\partial t}\right) = \frac{\gamma_t}{\mu_0}\left(I\nabla\cdot\boldsymbol{B} + \frac{1}{c}\frac{\partial S}{\partial t}\right)\\
&= \frac{\gamma_t}{\mu_0 c}\frac{\partial S}{\partial t} = \gamma_t J_{et} = -\gamma_t c\rho
\end{aligned} \tag{1.3.14}$$

Integration of (1.3.14) yields

$$\begin{aligned}
L_{\text{int}}|_t &= \int\frac{\partial L_{\text{int}}}{\partial A_t}dA_t = \int\frac{1}{\mu_0 c}\frac{\partial S}{\partial t}dA_t = \frac{1}{\mu_0 c}\frac{\partial S}{\partial t}A_t = -\frac{1}{\mu_0}\mu_0 c\rho A_t\\
&= -c\rho A_t = J_{et}A_t
\end{aligned} \tag{1.3.15}$$

For the component along γ_x, we find

$$\begin{aligned}
\gamma_x\frac{\partial L_{\text{int}}}{\partial A_x} &= \gamma_x\frac{\partial}{\partial x}\frac{\partial L_{\text{int}}}{\partial\left(\frac{\partial A_x}{\partial x}\right)} + \gamma_x\frac{\partial}{\partial y}\frac{\partial L_{\text{int}}}{\partial\left(\frac{\partial A_x}{\partial y}\right)} + \gamma_x\frac{\partial}{\partial z}\frac{\partial L_{\text{int}}}{\partial\left(\frac{\partial A_x}{\partial z}\right)} + \gamma_x\frac{\partial}{\partial t}\frac{\partial L_{\text{int}}}{\partial\left(\frac{\partial A_x}{\partial t}\right)}\\
&= \frac{\gamma_x}{\mu_0}\left(\frac{\partial S}{\partial x} + \frac{I}{c}\frac{\partial E_z}{\partial y} - \frac{I}{c}\frac{\partial E_y}{\partial z} + \frac{I}{c}\frac{\partial B_x}{\partial t}\right) = \frac{\gamma_x}{\mu_0}\frac{\partial S}{\partial x} = \gamma_x J_{ex}
\end{aligned} \tag{1.3.16}$$

Integration of (1.3.16) yields

$$L_{\text{int}}|_x = \int\left(\frac{\partial L_{\text{int}}}{\partial A_x}\right)dA_x = \int\frac{1}{\mu_0}\frac{\partial S}{\partial x}dA_x = \frac{1}{\mu_0}\frac{\partial S}{\partial x}A_x = J_{ex}A_x \tag{1.3.17}$$

The same procedure is clearly valid also for the components in γ_y and γ_z. Finally, by integration of (1.3.6), we get the Lagrangian density interaction term as

$$L_{\text{int}} = \sum_{j=x,y,z,t} \int \left(\frac{\partial L_{\text{int}}}{\partial A_j} \right) dA_j = J_{ex}A_x + J_{ey}A_y + J_{ez}A_z - c\rho A_t = \boldsymbol{J}_{\square e} \cdot \boldsymbol{A}_{\square} \tag{1.3.18}$$

which is the usual "source" term that is added in traditional Lagrangian theory for classical electricity and magnetism in order to obtain the complete set of Maxwell equations [5, 7, 20]. The scalar product $\boldsymbol{J}_{\square e} \cdot \boldsymbol{A}_{\square}$ has a dimension of energy per volume ($J \cdot m^{-3}$); in particular, the contribution of the spatial components of vectors $\boldsymbol{J}_{\square e}$ and $\boldsymbol{A}_{\square}$ (the scalar product $\boldsymbol{J}_{\triangle} \cdot \boldsymbol{A}_{\triangle}$) can be considered as the specific "kinetic" energy of the electromagnetic field, whereas the term $J_{et}A_t = -c\rho A_t$, the "potential" energy. By virtue of (1.3.13), (1.3.18) becomes

$$L_{\text{int}} = \frac{1}{2\mu_0} \left(S^2 - \frac{2}{c} I \boldsymbol{E} \cdot \boldsymbol{B} \right) = \boldsymbol{J}_{\square e} \cdot \boldsymbol{A}_{\square} \tag{1.3.19}$$

The pseudoscalar term $\frac{2}{c} I \boldsymbol{E} \cdot \boldsymbol{B}$ is clearly null as the electric and the magnetic flux density fields are always orthogonal with respect to each other: indeed, this term contains information about (1.2.55). A direct consequence of (1.3.19) is, therefore, the following relation between the scalar field, the electromagnetic four-potential, and the four-current density:

$$S^2 = 2\mu_0 \boldsymbol{J}_{\square e} \cdot \boldsymbol{A}_{\square} \tag{1.3.20}$$

By inspection of (1.3.14) and (1.3.16), and generalizing, it is possible to define the four-current vector $\boldsymbol{J}_{\square e}$ from the interaction Lagrangian term:

$$\begin{aligned} \sum_{j=x,y,z,t} \left(\frac{\partial L_{\text{int}}}{\gamma_j \partial A_j} \right) &= \gamma_x \frac{\partial L_{\text{int}}}{\partial A_x} + \gamma_y \frac{\partial L_{\text{int}}}{\partial A_y} + \gamma_z \frac{\partial L_{\text{int}}}{\partial A_z} - \gamma_t \frac{\partial L_{\text{int}}}{\partial A_t} \\ &= \gamma_x J_{ex} + \gamma_y J_{ey} + \gamma_z J_{ez} + \gamma_t J_{et} = \boldsymbol{J}_{\square e} \end{aligned} \tag{1.3.21}$$

and, again, by virtue of Noether's theorem, the law of current and charge conservation

$$\boldsymbol{\partial} \cdot \left[\sum_{j=x,y,z,t} \left(\frac{\partial L_{\text{int}}}{\gamma_j \partial A_j} \right) \right] = \boldsymbol{\partial} \cdot \boldsymbol{J}_{\square e} = 0 \tag{1.3.22}$$

which returns, consequently, the wave equations (1.2.58) and (1.2.60), respectively.

As can be seen, the definition of the electromagnetic field $\boldsymbol{G}$ is complete and it includes in itself the information of both action and interaction, without the need of any additional term: this is in full accordance with the principle of Occam's razor.

Thanks to the $Cl_{3,1}$ Clifford algebra, the Euler–Lagrange equations can be conveniently defined in a very compact form:

$$\boldsymbol{\partial}\left(\frac{\partial L}{\partial\left(\boldsymbol{\partial}\wedge\boldsymbol{A}_{\square}\right)}\right)-\frac{\partial L}{\partial \boldsymbol{A}_{\square}}=\boldsymbol{\partial}\left(\frac{\partial L}{\partial \boldsymbol{F}}\right)-\frac{\partial L}{\partial \boldsymbol{A}_{\square}}=0 \tag{1.3.23}$$

where, now, the scalar Lagrangian density is

$$L=L_{\text{field}}+L_{\text{int}}=-\frac{1}{2\mu_0}\boldsymbol{F}\cdot\boldsymbol{F}+\boldsymbol{J}_{\square e}\cdot\boldsymbol{A}_{\square} \tag{1.3.24}$$

Substituting (1.3.24) in (1.3.23), one can achieve directly Maxwell's equations in $Cl_{3,1}$ in the form shown in the previous sections (see Equation (1.2.27)), i.e.,

$$\begin{aligned}
&\boldsymbol{\partial}\left(\frac{\partial\left(-\frac{1}{2\mu_0}\boldsymbol{F}\cdot\boldsymbol{F}+\boldsymbol{J}_{\square e}\cdot\boldsymbol{A}_{\square}\right)}{\partial\boldsymbol{F}}\right)-\frac{\partial\left(-\frac{1}{2\mu_0}\boldsymbol{F}\cdot\boldsymbol{F}+\boldsymbol{J}_{\square e}\cdot\boldsymbol{A}_{\square}\right)}{\partial\boldsymbol{A}_{\square}}\\
&=-\frac{1}{\mu_0}\boldsymbol{\partial}\boldsymbol{F}-\boldsymbol{J}_{\square e}=0
\end{aligned} \tag{1.3.25}$$

which yields

$$\boldsymbol{\partial}\boldsymbol{F}+\mu_0\boldsymbol{J}_{\square e}=\boldsymbol{\partial}\boldsymbol{F}+\boldsymbol{\partial}S=\boldsymbol{\partial}\left(\boldsymbol{F}+S\right)=\boldsymbol{\partial}\boldsymbol{G}=0 \tag{1.3.26}$$

1.3.2. *Energy of the electromagnetic field*

The light-like energy-momentum density vector of the field $\boldsymbol{G}$ can be represented by a rotation (see Section 0.2.1) of the base vector γ_t:

$$\begin{aligned}
\boldsymbol{w}&=\frac{1}{2\mu_0}\left(\boldsymbol{G}\gamma_t\widetilde{\boldsymbol{G}}\right)\\
\boldsymbol{G}&=S+\boldsymbol{F}=S+\frac{1}{c}\left(\boldsymbol{E}+Ic\boldsymbol{B}\right)\gamma_t\\
\boldsymbol{G}\gamma_t&=S\gamma_t+\boldsymbol{F}\gamma_t=S\gamma_t-\frac{1}{c}\left(\boldsymbol{E}+Ic\boldsymbol{B}\right)\\
\boldsymbol{w}&=\frac{1}{2\mu_0}\left[S\gamma_t-\frac{1}{c}\left(\boldsymbol{E}+Ic\boldsymbol{B}\right)\right]\left[S-\frac{1}{c}\left(\boldsymbol{E}+Ic\boldsymbol{B}\right)\gamma_t\right]
\end{aligned}$$

$$\boldsymbol{w} = \frac{S^2}{2\mu_0}\gamma_t + \frac{\epsilon_0\boldsymbol{E}^2}{2}\gamma_t + \frac{\boldsymbol{B}^2}{2\mu_0}\gamma_t - \frac{1}{c\mu_0}I\boldsymbol{E}\wedge\boldsymbol{B}\gamma_t + \frac{S\boldsymbol{E}}{c\mu_0}$$

$$\boldsymbol{w} = \left(\frac{S^2}{2\mu_0} + \frac{\epsilon_0\boldsymbol{E}^2}{2} + \frac{\boldsymbol{B}^2}{2\mu_0}\right)\gamma_t - \frac{1}{c\mu_0}(\boldsymbol{E}\times\boldsymbol{B} - S\boldsymbol{E}) \tag{1.3.27}$$

or using natural units:

$$\left[\boldsymbol{w} = \frac{1}{4\pi}\left(\frac{1}{2}(S^2 + E^2 + B^2)\gamma_t - (\boldsymbol{E}\times\boldsymbol{B} - S\boldsymbol{E})\right)\right]_{NU}$$

$$\boldsymbol{w} = (w_s + w_e + w_m)\gamma_t - \frac{1}{c}\cdot\mathfrak{S} \tag{1.3.28}$$

where $w_s = \frac{S^2}{2\mu_0} = \boldsymbol{J}_{\square e}\cdot\boldsymbol{A}_{\square}$, $w_e = \epsilon_0\frac{\boldsymbol{E}^2}{2}$, and $w_m = \frac{\boldsymbol{B}^2}{2\mu_0}$ are the specific energies of the scalar, the electric, and the magnetic flux density fields, and where

$$\mathfrak{S} = \frac{1}{\mu_0}(\boldsymbol{E}\times\boldsymbol{B} - S\boldsymbol{E}) \tag{1.3.29}$$

is the generalized Poynting vector. It contains the usual $\boldsymbol{E}\times\boldsymbol{B}$ term, regardless of the scalar field presence. The poynting vector's sign depends on the field polarization type. Another geometric derivation of energy-momentum vector will be addressed in Chapter 2.

1.3.3. *The scalar field and the Feynman concept of unworldliness*

Richard Feynman writes in Chapter 25 of the second volume of his lectures:

> *Let us show you something interesting that we have recently discovered. All of the laws of physics can be contained in one equation. That equation is $U = 0$. This "law" means, of course, that the sum of the squares of all the individual mismatches is zero, and the only way the sum of a lot of squares can be zero is for each one of the terms to be zero. So the "beautifully" simple law is equivalent to the whole series of equations that you originally wrote down. It is therefore absolutely obvious that a simple notation that just hides the complexity in the definitions of symbols is not real simplicity. It is just a trick.*

Introducing the concept of "error function" or "unworldliness function" in physical equations can be a useful and illuminating "trick." Let's start with a very simple example: a massive body in a parabolic potential well.

E_p is the potential energy, k and C are generic constants, m is the mass, and v the speed:

$$E_p(x(t)) = \frac{1}{2}kx^2(t) - C$$

with the following boundary conditions the total energy (kinetic + potential) is always zero:

$$x(0) = \sqrt{\frac{2C}{k}}$$
$$v(0) = 0$$
$$E_p(x(0)) = E = 0$$

E_k is the kinetic energy:

$$E_k\left(x\left(t\right)\right) = \frac{1}{2}mv^2\left(x\left(t\right)\right)$$
$$E_p\left(x\left(t\right)\right) + E_k\left(x\left(t\right)\right) = E_p\left(x\left(0\right)\right) = E = 0$$
$$\left[\frac{1}{2}kx^2\left(t\right) - C\right] + \left[\frac{1}{2}mv^2\left(x\left(t\right)\right)\right] = E = 0$$

Now we introduce the "trick," an unworldliness function $U\left(x\left(t\right)\right)$ that "measures" the local "violation of physical laws": only when $U\left(x\left(t\right)\right) = 0$ the energy conservation law is respected:

$$E_p\left(x\left(t\right)\right) + E_k\left(x\left(t\right)\right) = U\left(x\left(t\right)\right)$$

It's important to note that a change in $U\left(x\left(t\right)\right)$ implies a local "violation" of a physical law and consequently the time derivative of U must be equal to zero along the physical trajectory of the body:

$$\frac{dU}{dt} = \frac{dU}{dx}\frac{dx}{dt} = \frac{dU}{dx}v = 0$$

or

$$\frac{dU}{dx} = 0$$

Setting to zero the derivative gives Newton's Second Law of Motion (f=ma)

$$\frac{dU}{dx} = kx + mv\frac{dv}{dx} = kx + mv\frac{dv}{vdt} = kx + m\frac{dv}{dt} = kx + ma = 0$$

Another example deals with the Lorenz gauge:

$$\boldsymbol{\partial} \cdot \boldsymbol{A}_{\square} = S = \frac{\partial A_x}{\partial x} + \frac{\partial A_y}{\partial y} + \frac{\partial A_z}{\partial z} - \frac{1}{c}\frac{\partial A_t}{\partial t} = 0$$

The trick here consists *in using the scalar field S as the unworldliness function and setting its total time derivative to zero*:

$$S(x, y, z, t) = U(x, y, z, t)\,.$$

Using natural units, we can write:

$$\frac{dS}{dt} = \frac{\partial S}{\partial x}\frac{dx}{dt} + \frac{\partial S}{\partial y}\frac{dy}{dt} + \frac{\partial S}{\partial z}\frac{dz}{dt} + \frac{\partial S}{\partial t} = 0$$

$$\frac{dS}{dt} = \frac{\partial S}{\partial x}v_x + \frac{\partial S}{\partial y}v_y + \frac{\partial S}{\partial z}v_z + \frac{\partial S}{\partial t} = 0$$

$$\frac{dS}{dt} = \mu_0 J_x v_x + \mu_0 J_y v_y + \mu_0 J_z v_z - \mu_0 \rho = 0$$

$$\mu_0 \rho v_x^2 + \mu_0 \rho v_y^2 + \mu_0 \rho v_z^2 - \mu_0 \rho = 0$$

This equation is satisfied only if $v^2 = v_x^2 + v_y^2 + v_z^2 = c^2 = 1$, i.e., if the charges move at speed of light, enforcing the plausibility of the Zitterbewegung model. Consequently, in this approach the scalar field can be interpreted as a useful function, with a zero average value, that "measures" a local "violation" of the Lorenz gauge. The derivatives of the scalar field *have a well-defined physical meaning*, and therefore is important to avoid the error of always setting everywhere S to zero and ignoring it.

1.3.4. *Electrostatic field and vector potential*

In the case of non-time-varying potential $\frac{\partial A_t}{\partial t} = 0$, the scalar field S becomes the divergence of the classical 3D vector potential $\boldsymbol{A}_{\triangle}$:

$$S = \nabla \cdot \boldsymbol{A}_{\triangle}$$

The time derivative of both sides gives

$$\frac{\partial S}{\partial t} = \frac{\partial(\nabla \cdot \boldsymbol{A}_{\triangle})}{\partial t} = \nabla \cdot \frac{\partial \boldsymbol{A}_{\triangle}}{\partial t}$$

Considering that $\frac{\partial \boldsymbol{A}_{\triangle}}{\partial t} = -\boldsymbol{E}$ (see Equations (1.2.13), (1.2.14), and (1.2.15)) and that $\frac{\partial S}{\partial t} = \frac{-\rho}{\epsilon_0}$ we rediscover Gauss's law,

$$\nabla \cdot \boldsymbol{E} = \frac{\rho}{\epsilon_0}$$

showing that even the electrostatic field may be seen as generated from time derivatives of a vector potential field!

1.3.5. *Electric charge, antimatter, and time direction*

Richard Feynman proposed to interpret the positron (a particle which is identical to the electron but with a positive charge) as an electron traveling back in time. Such an interpretation seems to be perfectly compatible with the definition of electric charge density given in (1.2.48):

$$\frac{\partial S}{\partial t} = \frac{-\rho}{\epsilon_0} \tag{1.3.30}$$

By multiplying both sides of (1.3.30) by −1, we obtain

$$-\frac{\partial S}{\partial t} = \frac{\rho}{\epsilon_0}$$

or, equivalently

$$\frac{\partial S}{\partial(-t)} = \frac{\rho}{\epsilon_0} \tag{1.3.31}$$

The positron traveling back in time is represented in the annihilation reaction diagram proposed by Feynman and shown in Figure 1.2 [9].

"I did not take the idea that all the electrons were the same one from him as seriously as I took the observation that positrons could simply be represented as electrons going from the future to the past in a back section of their world lines" [8].

While the obtained equations certainly facilitate such a "traveling back in time" interpretation, we shall look further into this question in the next chapter, to determine whether it is the correct interpretation.

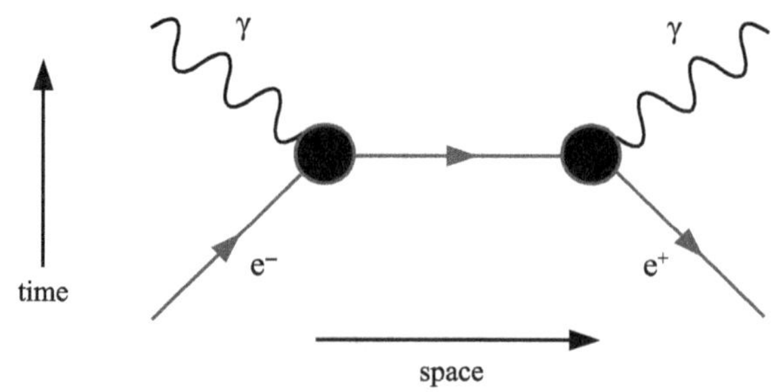

Figure 1.2. Feynman's diagram of proton–electron annihilation.

1.3.6. *Magnetic charges and currents*

Starting from a hypothetical eight component "vector potential" that includes the four pseudovectors (trivectors) $\boldsymbol{T}$ of space–time algebra, symmetrical Maxwell's equations emerge. This new set of equations now include the magnetic charge and magnetic current densities that are the time and spatial derivatives of a pseudoscalar field P. By considering (1.2.1) and the four pseudovectors defined as

$$\boldsymbol{T} = \gamma_y\gamma_z\gamma_t T_x + \gamma_x\gamma_z\gamma_t T_y + \gamma_x\gamma_y\gamma_t T_z + \gamma_x\gamma_y\gamma_z T_t \tag{1.3.32}$$

a new vector potential can be defined as

$$\begin{aligned}\boldsymbol{A}' = {} & \gamma_x A_x + \gamma_y A_y + \gamma_z A_z + \gamma_t A_t + \gamma_y\gamma_z\gamma_t T_x + \gamma_x\gamma_z\gamma_t T_y \\ & + \gamma_x\gamma_y\gamma_t T_z + \gamma_x\gamma_y\gamma_z T_t\end{aligned} \tag{1.3.33}$$

from which we obtain

$$\boldsymbol{\partial}(\boldsymbol{A}') = \boldsymbol{\partial}\left(\boldsymbol{A}_{\square} + \boldsymbol{T}\right) = S + \boldsymbol{F} + P \tag{1.3.34}$$

Using SI units and following the same procedure as shown in Section 1.2, we can write:

$$\begin{aligned}
S &= \frac{\partial A_x}{\partial x} + \frac{\partial A_y}{\partial y} + \frac{\partial A_z}{\partial z} - \frac{1}{c}\frac{\partial A_t}{\partial t} \\
\gamma_x\gamma_t\frac{1}{c}E_x &= \gamma_x\gamma_t\left(\frac{\partial A_t}{\partial x} - \frac{\partial T_z}{\partial y} - \frac{\partial T_y}{\partial z} - \frac{1}{c}\frac{\partial A_x}{\partial t}\right) \\
\gamma_y\gamma_t\frac{1}{c}E_y &= \gamma_y\gamma_t\left(\frac{\partial T_z}{\partial x} + \frac{\partial A_t}{\partial y} - \frac{\partial T_x}{\partial z} - \frac{1}{c}\frac{\partial A_y}{\partial t}\right) \\
\gamma_z\gamma_t\frac{1}{c}E_z &= \gamma_z\gamma_t\left(\frac{\partial T_y}{\partial x} + \frac{\partial T_x}{\partial y} + \frac{\partial A_t}{\partial z} - \frac{1}{c}\frac{\partial A_z}{\partial t}\right) \\
\gamma_x\gamma_y\gamma_z\gamma_t P &= \gamma_x\gamma_y\gamma_z\gamma_t\left(\frac{\partial T_x}{\partial x} - \frac{\partial T_y}{\partial y} + \frac{\partial T_z}{\partial z} - \frac{1}{c}\frac{\partial T_t}{\partial t}\right) \\
\gamma_y\gamma_z B_x &= \gamma_y\gamma_z\left(\frac{\partial T_t}{\partial x} + \frac{\partial A_z}{\partial y} - \frac{\partial A_y}{\partial z} - \frac{1}{c}\frac{\partial T_x}{\partial t}\right) \\
\gamma_x\gamma_z B_y &= \gamma_x\gamma_z\left(-\frac{\partial A_z}{\partial x} + \frac{\partial T_t}{\partial y} + \frac{\partial A_x}{\partial z} + \frac{1}{c}\frac{\partial T_y}{\partial t}\right)
\end{aligned}$$

$$\gamma_x\gamma_y B_z = \gamma_x\gamma_y\left(\frac{\partial A_y}{\partial x} - \frac{\partial A_x}{\partial y} + \frac{\partial T_t}{\partial z} - \frac{1}{c}\frac{\partial T_z}{\partial t}\right)$$

By applying again the $\boldsymbol{\partial}$ operator to (1.3.34) and equating to zero:

$$\boldsymbol{\partial}^2 \boldsymbol{A}' = \boldsymbol{\partial}\left(S + \boldsymbol{F} + P\right) = \boldsymbol{0} \tag{1.3.35}$$

Here

$$\boldsymbol{\partial F} = -\boldsymbol{\partial} S - \boldsymbol{\partial} P = \mu_0 \boldsymbol{J}_{\Box e} + \frac{1}{\epsilon_0}\boldsymbol{J}_{\Box m} \tag{1.3.36}$$

where $\boldsymbol{J}_{\Box e}$ is the four-current as defined in (1.2.50), $\boldsymbol{J}_{\Box m} = \gamma_x J_{mx} + \gamma_y J_{my} + \gamma_z J_{mz} + \gamma_t J_{mt} = \gamma_x J_{mx} + \gamma_y J_{my} + \gamma_z J_{mz} - \gamma_t \frac{1}{c}\rho_m$ is the magnetic four-current vector and ρ_m the magnetic charge. By carrying out all calculation in (1.3.35), the following set of equations is obtained:

$$\gamma_x\left(\frac{\partial S}{\partial x} - \frac{\partial B_z}{\partial y} + \frac{\partial B_y}{\partial z} + \frac{1}{c^2}\frac{\partial E_x}{\partial t}\right) = 0$$

$$\gamma_y\left(\frac{\partial B_z}{\partial x} + \frac{\partial S}{\partial y} - \frac{\partial B_x}{\partial z} + \frac{1}{c^2}\frac{\partial E_y}{\partial t}\right) = 0$$

$$\gamma_z\left(-\frac{\partial B_y}{\partial x} + \frac{\partial B_x}{\partial y} + \frac{\partial S}{\partial z} + \frac{1}{c^2}\frac{\partial E_z}{\partial t}\right) = 0$$

$$\gamma_t\frac{1}{c}\left(\frac{\partial E_x}{\partial x} + \frac{\partial E_y}{\partial y} + \frac{\partial E_z}{\partial z} + \frac{\partial S}{\partial t}\right) = 0$$

$$\gamma_y\gamma_z\gamma_t\frac{1}{c}\left(\frac{\partial P}{\partial x} + \frac{\partial E_z}{\partial y} - \frac{\partial E_y}{\partial z} + \frac{\partial B_x}{\partial t}\right) = 0$$

$$\gamma_x\gamma_z\gamma_t\frac{1}{c}\left(\frac{\partial E_z}{\partial x} - \frac{\partial P}{\partial y} - \frac{\partial E_x}{\partial z} - \frac{\partial B_y}{\partial t}\right) = 0$$

$$\gamma_x\gamma_y\gamma_t\frac{1}{c}\left(\frac{\partial E_y}{\partial x} - \frac{\partial E_x}{\partial y} + \frac{\partial P}{\partial z} + \frac{\partial B_z}{\partial t}\right) = 0$$

$$\gamma_x\gamma_y\gamma_z\left(\frac{\partial B_x}{\partial x} + \frac{\partial B_y}{\partial y} + \frac{\partial B_z}{\partial z} + \frac{\partial P}{\partial t}\right) = 0$$

This set of equations represents the symmetrical Maxwell equations considering the hypothesis of existing magnetic currents and charges. This hypothesis was experimentally proved by several independent experiments, which we discuss in a later chapter on magnetic monopoles.

1.4. Conclusions

Simplicity is an important and concrete value in scientific research. Connections between very different concepts in physics can be evidenced if we use the language of geometric algebra. The application of Occam's Razor principle to Maxwell's equations highlights some essential concepts:

(1) Clifford algebra is by far the most appropriate, simple, and intuitive mathematical language for encoding in general the laws of physics and here, particularly, for the laws of electromagnetism;
(2) a scalar field derives from the definition of "harmonic" electromagnetic four-potential and is at the origin of charges and currents;
(3) the charge density derived from the scalar field follows the wave equation with a propagation speed equal to the speed of light;
(4) the Feynman model of the positron, seen as an electron traveling back in time, seems to be compatible with the definition of electric charge density as the time derivative of a scalar field.

In particular, the important element emerging from the present work is that (1.2.60) imposes a precise condition on charge dynamics, describing distributions of charge density moving in vacuum at the speed of light.

In the model proposed here, the added constraint on the charge and current density seems to imply that one is no longer free to specify charge and current density distributions at will, because this information is indeed included within the definition of the four-potential $\boldsymbol{A}_{\square}$. However, this constraint can be removed when considering macroscopic electromagnetic systems or even the dynamics of a single elementary charge at a spatial scale greater than the particle Compton wavelength λ_c and at a time scale greater than the Compton period $\frac{\lambda_c}{c}$. In this case, static elementary charges can be visualized as charge density distributions moving at the speed of light on a closed trajectory but with a zero average speed (this generalization would be consistent with static charge densities, electrets, dielectrics), whereas currents can be considered as ordered motions of charge density distributions moving with an absolute velocity equal to the speed of light but with an arbitrary absolute average speed lower than c. *This observation favors a pure electromagnetic model of elementary particles based on a particular Zitterbewegung interpretation of quantum mechanics* [10, 12]. Therefore, the free electron, and perhaps all other elementary charged particles, can be viewed as a charge distribution that rotates at the speed of light along a circumference whose length is equal to its Compton wavelength [18].

It is our opinion that the Zitterbewegung interpretation of quantum mechanics may be an important contribution for understanding the structure of ultra-dense hydrogen and the origin of anomalous heat in some metal-hydrogen systems. A Zitterbewegung electron model and a preliminary hypothesis for the structure of ultradense deuterium will be treated more deeply in later chapters.

Finally, a Lagrangian density equal to the square module of the seven component electromagnetic field reveals an energy density formula for both fields and currents. Moreover, it has been demonstrated that Maxwell's equations can be explicitly derived in a simple way directly from a Lagrangian density of the electromagnetic bivector and the scalar field. An interesting consequence is also that the specific energy of the scalar field is deeply connected to the interaction term of the Lagrangian density and, therefore, both to the electromagnetic four-potential and the four-current density.

References

[1] Y. Aharonov and D. Bohm, Significance of electromagnetic potentials in the quantum theory. *Physical Review*, **115** (1959) 485–491.

[2] L. Bauer, *The Linguistics Student's Handbook*. Edinburgh University Press, 2007.

[3] G. Bettini, Clifford algebra, 3 and 4-dimensional analytic functions with applications. Manuscripts of the Last Century. viXra.org, *Quantum Physics*, (2011) 1–63. http://vixra.org/abs/1107.0060.

[4] J.M. Chappell, S.P. Drake, C.L. Seidel, L.J. Gunn, A. Iqbal, A. Allison, and D. Abbott, Geometric algebra for electrical and electronic engineers. *Proceedings of the IEEE*, **102**(9) (2014) 1340–1363.

[5] J.M. Chappell, A. Iqbal, and D. Abbott, A simplified approach to electromagnetism using geometric algebra, (2010), arxiv e-prints, https://arxiv.org/pdf/1010.4947.pdf.

[6] P.A.M. Dirac, Nobel lecture: Theory of Electrons and positrons. Nobel lectures, *Physics*, (1965) 1922–1941.

[7] J. Dressel, K.Y. Bliokh, and F. Nori, Spacetime algebra as a powerful tool for electromagnetism. *Physics Reports*, **589** (2015) 1–71.

[8] R.P. Feynman, Nobel lecture: The development of the space-time view of quantum electrodynamics. Nobel lectures, *Physics*, (1972) 1963–1970.

[9] R.P. Feynman, *QED: The Strange Theory of Light and Matter*. Penguin Books. Penguin, 1990.

[10] D. Hestenes, *Quantum Mechanics from Self-interaction. Foundations of Physics*, **15**(1) (1985) 63–87.

[11] D. Hestenes, *Clifford Algebra and the Interpretation of Quantum Mechanics*, Springer Netherlands, Dordrecht, 1986, pp. 321–346.

[12] D. Hestenes, Zitterbewegung Modeling. Foundations of Physics, **23**(3) (1993) 365–387.

[13] D. Hestenes, *Reforming the Mathematical Language of Physics*, Oersted Medal Lecture 2002.

[14] L.M. Hively, Implications of a new electrodynamic theory, (2015), https://www.researchgate.net.

[15] L.M. Hively and G.C. Giakos, Toward a more complete electrodynamic theory. *International Journal of Signals and Imaging Systems Engineering*, **5**(1) (2012).

[16] G. Pilato and G. Vassallo, TSVD as a statistical estimator in the latent semantic analysis paradigm. *IEEE Transactions on Emerging Topics in Computing*, **3**(2) (2015) 185–192.

[17] A. Rathke, A critical analysis of the hydrino model. *New Journal of Physics*, **7**(127) (2005).

[18] F. Santandrea and P. Cirilli, Unificazione elettromagnetica, concezione elettronica dello spazio, dell'energia e della materia. *Atlante di numeri e lettere, Laboratorio di ricerca industriale*, (2006) 1–8, http://www.atlantedinumerielettere.it/energia2006/pdf/labor.pdf.

[19] A.M. Shaarawi, Clifford algebra formulation of an electromagnetic charge-current wave theory. *Foundations of Physics*, **30**(11) (2000) 1911–1941.

[20] K. Simonyi, *Theoretische Elektrotechnik. Hochschulbücher für Physik*, 9th edition. VEB Deutscher Verlag der Wissenschaften, Berlin, Germany, 1989.

[21] K.J. van Vlaenderen, A generalisation of classical electrodynamics for the prediction of scalar field effects. (2003), arxiv Physics e-prints, https://arxiv.org/pdf/physics/0305098.pdf.

[22] K.J. van Vlaenderen and A. Waser, Generalization of classical electrodynamics to admit a scalar field and longitudinal waves. *Hadronic Journal*, **24** (2001), 609–628.

https://doi.org/10.1142/9781800611306_0002

Chapter 2

Electromagnetic and Quantum Mechanical Waves

András Kovács*,‡ and Paul O'Hara†
*BroadBit Energy Technologies, Metallimiehenkuja 8 C
02150 Espoo, Finland
†Sophia University Institute, Loppiano, Italy
‡andras.kovacs@broadbit.com

2.1. Introduction

This chapter will serve as a bridge between the deeper understanding of Maxwell's equation, which was developed in Chapter 1, and the quantum mechanics related discussions in subsequent chapters. The following study of electromagnetic wave modes lays the groundwork for the exploration of quantum mechanical waves, including the relationship between mass and electric charge.

Quantum mechanics was initiated over 100 years ago, upon the discovery that electromagnetic energy is emitted and absorbed in quantized energy units, colloquially called photons. The photon wave is presented in introductory textbooks as a trivial solution of the electromagnetic wave equation. One may presume that 100 years ago physicists firstly investigated everything that there is to know about the quantum mechanics of photons before moving onto more complex matters, but that is not the case. The historical overview of what has been written about the photon structure is well described in Refs. [1, 2], which we briefly summarize.

The first serious analysis of the photon structure was undertaken by De Broglie, who proposed in 1934 that the photon is a composite particle constituted by two waves, each having an angular momentum value of $\frac{\hbar}{2}$

and satisfying the Dirac equation; together these waves yield an angular momentum of $\pm\hbar$. De Broglie denoted the wave structure by $\psi \otimes \varphi$, where ψ and φ are spinors. Despite Heisenberg's support for this idea, there was no consensus for de Broglie's solution and it was mostly forgotten. Subsequently, quantum mechanics textbooks referred to the issue of writing down a quantum mechanical photon wavefunction as either an impossible task, or else a trivial one. Those who claimed the task to be trivial, generally wrote down Maxwell's free-space equations for left- and right-circularly polarized radiation, without any further connection to quantum mechanics. Clearly, this is not yet a well-understood topic.

By the end of the chapter, we hope that the reader will gain a precise understanding of the principal electromagnetic wave modes. Our second goal is to lay the foundations for quantum mechanics in the light of Chapter 1. Such an approach means that ultimately one only needs to know how to calculate the evolution of the electromagnetic vector potential field in spacetime. The key is to clearly understand two factors: the presence of an electromagnetic vacuum fluctuation and the curvature of the spacetime metric, which is created by the electromagnetic energy present in it. These two factors will enable us to solve Maxwell's equation in the microscopic limit and introduce the concept of mass. This introductory chapter on quantum mechanics contains several pointers to subsequent chapters, where specifics are discussed.

2.2. Maxwell's Equation Revisited

Using the basis elements of Clifford algebra together with natural units, we define the differentiation operator as follows:

$$D \equiv -i\frac{\partial}{\partial t}\mathbf{e}_t + \frac{\partial}{\partial x}\mathbf{e}_x + \frac{\partial}{\partial y}\mathbf{e}_y + \frac{\partial}{\partial z}\mathbf{e}_z \equiv (-i\partial_t \mathbf{e}_t, \nabla) \tag{2.2.1}$$

where $\nabla \equiv \frac{\partial}{\partial x}\mathbf{e}_x + \frac{\partial}{\partial y}\mathbf{e}_y + \frac{\partial}{\partial z}\mathbf{e}_z$.

We define the following representation of the electromagnetic vector potential: $A \equiv (iA_t\mathbf{e}_t, \mathbf{A})$. The reader may note that these definitions slightly differ from the one used in Chapter 1; the way we measure time is different. This difference amounts to a change of variable along the time axis, i.e., it is just a different representation of the same physics. This representation is selected because it is helpful for making connections with quantum mechanics.

We evaluate the spacetime differential of A:

$$\begin{aligned} DA &= (-\partial_t A_t + \nabla \cdot \mathbf{A}) + (i\partial_t \mathbf{A} + i\nabla A_t)\, \mathbf{e}_t + \nabla \times \mathbf{A}\mathbf{e}_{xyz} \\ &= (-\partial_t A_t + \nabla \cdot \mathbf{A}) + \mathbf{B}\mathbf{e}_{xyz} - i\mathbf{E}\mathbf{e}_t \end{aligned} \tag{2.2.2}$$

In Chapter 1, we defined the scalar field $S \equiv -\partial_t A_t + \nabla \cdot \mathbf{A}$ which was found to be describing the flow of charges. In a charge-free region of space, we therefore require $-\partial_t A_t + \nabla \cdot \mathbf{A} = 0$. Then in the free-space region, the electromagnetic vector potential derivative takes the following form:

$$DA = F \tag{2.2.3}$$

where $F = (0 + \mathbf{B}\mathbf{e}_{xyz} - i\mathbf{E}\mathbf{e}_t)$ by definition. Note, the zero term emphasizes the absence of a scalar field. The above equation conveys the geometric meaning of Maxwell's equations. Reapplying the differential operator D to the above equation results in the following magnetic and electric fields:

$$D\mathbf{B}\mathbf{e}_{xyz} = -i\partial_t \mathbf{B}\mathbf{e}_{txyz} + (\nabla \cdot \mathbf{B})\mathbf{e}_{xyz} + \nabla \times \mathbf{B}\mathbf{e}_{xyz}^2 \tag{2.2.4}$$

$$D\mathbf{E}\mathbf{e}_t = i\partial_t \mathbf{E} + (\nabla \cdot \mathbf{E})\mathbf{e}_t + \nabla \times \mathbf{E}\mathbf{e}_{txyz} \tag{2.2.5}$$

Combining terms in the above two equations, we get

$$\begin{aligned} DF &= D\,(0 + \mathbf{B}\mathbf{e}_{xyz} - i\mathbf{E}\mathbf{e}_t) \\ &= -i\nabla \cdot \mathbf{E}\mathbf{e}_t + \nabla \cdot \mathbf{B}\mathbf{e}_{xyz} - i\,(\nabla \times \mathbf{E} + \partial_t \mathbf{B})\, \mathbf{e}_{txyz} - (\nabla \times \mathbf{B} - \partial_t \mathbf{E}) \end{aligned} \tag{2.2.6}$$

Note that in a source-free region $\nabla \cdot \mathbf{E} = 0$ and $\nabla \cdot \mathbf{B} = 0$, while $(\nabla \times \mathbf{E} + \partial_t \mathbf{B}) = 0$ by Faraday's law and $(\nabla \times \mathbf{B} - \partial_t \mathbf{E})$ yields zero according to Ampere's law of induction. Therefore, in a source-free region Maxwell's equations take the following form:

$$DF = 0 \tag{2.2.7}$$

This result means that the electromagnetic wave equation can be written in the following form:

$$D^2 A = 0 \tag{2.2.8}$$

In the above definition, the $A = (iA_t, \mathbf{A})$ choice defines a particular chirality of the electromagnetic wave, which means a particular handedness of its circular polarization. The $A_r = (-iA_t, \mathbf{A})$ choice together with $D_r = (i\partial_t \mathbf{e}_t, \nabla)$ reverses the direction of the electric field, while leaving the magnetic field unchanged. This allows us to define $D_r A_r = F_r =$

$(0, \mathbf{B}\mathbf{e}_{xyz} + i\mathbf{E}\mathbf{e}_t)$, which is the opposite chirality. The Maxwell equations for the two circular polarizations can be written as follows:

$$D_l A_{l+} = F_{l+} = (0 + \mathbf{B}\mathbf{e}_{xyz} - i\mathbf{E}\mathbf{e}_t) \tag{2.2.9}$$

$$D_r A_{r+} = F_{r+} = (0 + \mathbf{B}\mathbf{e}_{xyz} + i\mathbf{E}\mathbf{e}_t) \tag{2.2.10}$$

where we used the following definitions:

- $A_{l+} = (iA_t\mathbf{e}_t, \mathbf{A})$, $D_l = (-i\partial_t\mathbf{e}_t, \nabla)$
- $A_{r+} = (-iA_t\mathbf{e}_t, \mathbf{A})$, $D_r = (i\partial_t\mathbf{e}_t, \nabla)$

It is critical to use the correct physical interpretation of mathematical equations. As the above defined D_l and D_r operators have opposite signs for time-wise differentiation, one might naively think that left-handed and right-handed polarization travel in opposite time-wise directions. However, this should only be taken in a metaphorical way. There is no physical reason for such an interpretation, as the arrow of time is defined by the direction of entropy increase, which acts the same way for both left-handed and right-handed polarization. The more appropriate interpretation is that the basis vector along which the passage of time is measured points into $-\mathbf{e}_t$ and $\mathbf{e}_t$ directions for the two polarizations, respectively. A physical quantity is therefore measured along different scales for different types of electromagnetic fields; this idea is an important and recurring theme in this chapter.

2.3. Two Different Time Representations

The meaning of the "+" index, which was introduced above, will now be explored in depth. The definitions of the electromagnetic vector potential imply a choice of A_0 as imaginary and $\mathbf{A}$ as real valued. However, if Equations (2.2.9) and (2.2.10) are multiplied by $i = \mathbf{e}_{txyz}$, an alternative interpretation is suggested from a physics point of view. Mathematically, multiplication by the number i essentially means to rotate a vector in the plane by $\pi/2$ radians. From a physics perspective, in the context of the above equations we can take it as describing an electromagnetic vector orthogonal to the initial one, which in the notation of Clifford algebra means that we are multiplying the base units by $\mathbf{e}_{txyz}$. The resulting equation obtained by multiplying by i will be denoted by the "-" index as follows:

$$D_l A_{l-} = F_{l-} = (0 + i\mathbf{B}\mathbf{e}_{xyz} + \mathbf{E}\mathbf{e}_t) \tag{2.3.1}$$

$$D_r A_{r-} = F_{r-} = (0 + i\mathbf{B}\mathbf{e}_{xyz} - \mathbf{E}\mathbf{e}_t) \tag{2.3.2}$$

where we used the following definitions:

- $A_{l-} = (-A_t\mathbf{e}_t, i\mathbf{A})$, $D_l = (-i\partial_t\mathbf{e}_t, \nabla)$
- $A_{r-} = (A_t\mathbf{e}_t, i\mathbf{A})$, $D_r = (i\partial_t\mathbf{e}_t, \nabla)$

Playing with the $i = \mathbf{e}_{txyz}$ equivalence, we can write the electromagnetic field expressions in the following equivalent ways:

- $F_{l+} = (0 + \mathbf{B}\mathbf{e}_{xyz} - i\mathbf{E}\mathbf{e}_t) = (0 - i\mathbf{B}\mathbf{e}_t + \mathbf{E}\mathbf{e}_{xyz})$
- $F_{l-} = (0 + i\mathbf{B}\mathbf{e}_{xyz} + \mathbf{E}\mathbf{e}_t) = (0 + \mathbf{B}\mathbf{e}_t + i\mathbf{E}\mathbf{e}_{xyz})$
- $F_{r+} = (0 + \mathbf{B}\mathbf{e}_{xyz} + i\mathbf{E}\mathbf{e}_t) = (0 - i\mathbf{B}\mathbf{e}_t - \mathbf{E}\mathbf{e}_{xyz})$
- $F_{r-} = (0 + i\mathbf{B}\mathbf{e}_{xyz} - \mathbf{E}\mathbf{e}_t) = (0 + \mathbf{B}\mathbf{e}_t - i\mathbf{E}\mathbf{e}_{xyz})$

We have four distinguishable electromagnetic field types, each of which obeys the Maxwell equation. The following section describes how to evaluate some important physical properties of these various field types.

It should be noted that scientists, when using a complex number rather than Clifford algebra notation, often write the electromagnetic field as $F = B + iE$, which also means that complex conjugation becomes a well-defined operation in that case. However, complex conjugation is not well defined under the Clifford algebra framework, as the $i\mathbf{E}\mathbf{e}_t$ and $-\mathbf{E}\mathbf{e}_{xyz}$ expressions are equivalent. With this in mind, we introduce the "electromagnetic conjugate" expression, which reverses the electric field but leaves the magnetic field unchanged. Physically, this conjugation corresponds to space inversions because $\mathbf{E}$ transforms as a polar vector and $\mathbf{B}$ transforms as an axial vector under space inversions. The conjugated electromagnetic fields are written as follows:

- $\bar{F}_{l+} = (0 + \mathbf{B}\mathbf{e}_{xyz} + i\mathbf{E}\mathbf{e}_t) = (0 - i\mathbf{B}\mathbf{e}_t - \mathbf{E}\mathbf{e}_{xyz})$
- $\bar{F}_{l-} = (0 + i\mathbf{B}\mathbf{e}_{xyz} - \mathbf{E}\mathbf{e}_t) = (0 + \mathbf{B}\mathbf{e}_t - i\mathbf{E}\mathbf{e}_{xyz})$
- $\bar{F}_{r+} = (0 + \mathbf{B}\mathbf{e}_{xyz} - i\mathbf{E}\mathbf{e}_t) = (0 - i\mathbf{B}\mathbf{e}_t + \mathbf{E}\mathbf{e}_{xyz})$
- $\bar{F}_{r-} = (0 + i\mathbf{B}\mathbf{e}_{xyz} + \mathbf{E}\mathbf{e}_t) = (0 + \mathbf{B}\mathbf{e}_t + i\mathbf{E}\mathbf{e}_{xyz})$

The $D = (-i\partial_t\mathbf{e}_t, \nabla)$ and $A = (iA_t\mathbf{e}_t, \mathbf{A})$ definitions, which we use in this chapter, yield the same Maxwell equations as the vector potential definition of Chapter 1. These two ways of defining the electromagnetic vector potential can be regarded as a change of variable along the time axis. In other words, the vector potential definitions of Chapter 1 and the present chapter involve two different ways to measure time, both representing the same physics.

2.4. The Energy and Lagrangian of the F_+ and F_- Fields

In the previous section, we defined $A_{l+} = iA_t\mathbf{e}_t + A_x\mathbf{e}_x + A_y\mathbf{e}_y + A_z\mathbf{e}_z$ and $A_{l-} = -A_t\mathbf{e}_t + iA_x\mathbf{e}_x + iA_y\mathbf{e}_y + iA_z\mathbf{e}_z$ for one polarization, while for the other polarization the vector potential is $A_{r+} = -iA_t\mathbf{e}_t + A_x\mathbf{e}_x + A_y\mathbf{e}_y + A_z\mathbf{e}_z$ and $A_{r-} = A_t\mathbf{e}_t + iA_x\mathbf{e}_x + iA_y\mathbf{e}_y + iA_z\mathbf{e}_z$.

Throughout this section, we use the left-handed notation, and leave it to the reader to work out the right-handed analogues in the same way. Squaring A_+ give

$$A_{l+}A_{l+} = (iA_t)^2\,\mathbf{e}_t^2 + A_x^2\mathbf{e}_x^2 + A_y^2\mathbf{e}_y^2 + A_z^2\mathbf{e}_z^2 = A_t^2 + A_x^2 + A_y^2 + A_z^2 \tag{2.4.1}$$

Similarly, the square of A_- becomes

$$\begin{aligned} A_{l-}A_{l-} &= (iA_t)^2\,(i\mathbf{e}_t)^2 + A_x^2\,(i\mathbf{e}_x)^2 + A_y^2\,(i\mathbf{e}_y)^2 + A_z^2\,(i\mathbf{e}_z)^2 \\ &= -\,(iA_t)^2\,\mathbf{e}_t^2 - A_x^2\mathbf{e}_x^2 - A_y^2\mathbf{e}_y^2 - A_z^2\mathbf{e}_z^2 = -A_t^2 - A_x^2 - A_y^2 - A_z^2 \end{aligned} \tag{2.4.2}$$

Comparing the squares A_+A_+ and A_-A_-, we see that both expressions have opposite signs. The act of multiplying Maxwell's equation by i thus yields a vector potential field with a negative norm square.

We now calculate the energy and momentum of the electromagnetic field. Recall from Section 1.3.2 that the electromagnetic energy–momentum density is $\frac{1}{2}G\mathbf{e}_t\widetilde{G}$. In the scalar-free case, this formula reduces to $\frac{1}{2}F\mathbf{e}_t\widetilde{F} = -\frac{1}{2}F\mathbf{e}_tF$. The same formula is also derived in [5]. Evaluating this expression, we obtain

$$N_{l+} = -\frac{1}{2}F_{l+}\mathbf{e}_tF_{l+} = \frac{1}{2}\left(E^2 + B^2\right)\mathbf{e}_t + i\mathbf{B}\times\mathbf{E} \tag{2.4.3}$$

The equivalent calculation for the right-handed polarization yields

$$N_{r+} = -\frac{1}{2}F_{r+}\mathbf{e}_tF_{r+} = \frac{1}{2}\left(E^2 + B^2\right)\mathbf{e}_t - i\mathbf{B}\times\mathbf{E} \tag{2.4.4}$$

Exercise 1. Prove Equations (2.4.3) and (2.4.4) by writing them out explicitly, in terms of $F_{l+} = (0+\mathbf{B}\mathbf{e}_{xyz} - i\mathbf{E}\mathbf{e}_t)$ and $F_{r+} = (0+\mathbf{B}\mathbf{e}_{xyz} + i\mathbf{E}\mathbf{e}_t)$.

Comparing the $F_+\mathbf{e}_tF_+$ and $F_-\mathbf{e}_tF_-$ expressions in terms of energy, an exactly analogous i^2 multiplication and sign reversal situation arises as in the above-described A_+A_+ versus A_-A_- comparison. It therefore appears that the F_- field has negative energy. Does such an object have a physical meaning, or is it just a mathematical construction? This question will be investigated in Section 2.6.

We now consider multiplying the electromagnetic field by its electromagnetic conjugate, as follows:

$$\begin{aligned}\frac{1}{2}F_{l+}\bar{F}_{l+} &= \frac{1}{2}(0 + \mathbf{B}\mathbf{e}_{xyz} - i\mathbf{E}\mathbf{e}_t)(0 + \mathbf{B}\mathbf{e}_{xyz} + i\mathbf{E}\mathbf{e}_t)\\ &= \frac{1}{2}(0 + \mathbf{B}\mathbf{e}_{xyz} + \mathbf{E}\mathbf{e}_{xyz})(0 + \mathbf{B}\mathbf{e}_{xyz} - \mathbf{E}\mathbf{e}_{xyz}) \qquad (2.4.5)\end{aligned}$$

Upon evaluating the $F\mathbf{e}_tF$ expression, the cross-product terms sum up, while the dot-product terms cancel out. In contrast, upon evaluating the above expression, the cross-product terms cancel out, while the dot-product terms sum up:

$$\frac{1}{2}F_{l+}\bar{F}_{l+} = \frac{1}{2}\left(B^2 - E^2\right) - \mathbf{B}\cdot\mathbf{E} \qquad (2.4.6)$$

$$\frac{1}{2}F_{r+}\bar{F}_{r+} = \frac{1}{2}\left(B^2 - E^2\right) + \mathbf{B}\cdot\mathbf{E} \qquad (2.4.7)$$

From Chapter 1, we recognize Equations (2.4.6) and (2.4.7) as the electromagnetic Lagrangian. In the absence of a scalar field, the above-introduced electromagnetic conjugation is equivalent to the reversion operation of Clifford algebra. Moreover, the $\mathbf{B}^2 - \mathbf{E}^2$ and $\mathbf{B}\cdot\mathbf{E}$ terms are Lorentz-invariant and therefore the electromagnetic Lagrangian is Lorentz invariant.

Exercise 2. Use the well-known electromagnetic Lorentz transformation formulas to prove the Lorentz invariance of B^2-E^2 and $\mathbf{B}\cdot\mathbf{E}$. These formulas are the following:

$$\mathbf{E}_{\perp}^{'} = \frac{1}{\sqrt{1-\frac{v^2}{c^2}}}\left(\mathbf{E}_{\perp} + \mathbf{v}\times\mathbf{B}\right), \mathbf{E}_{\|}^{'} = \mathbf{E}_{\|}$$

$$\mathbf{B}_{\perp}^{'} = \frac{1}{\sqrt{1-\frac{v^2}{c^2}}}\left(\mathbf{B}_{\perp} - \frac{\mathbf{v}\times\mathbf{E}}{c^2}\right), \mathbf{B}_{\|}^{'} = \mathbf{B}_{\|}$$

Using the same logic as in the case of evaluating the total electromagnetic energy, the $F_+\bar{F}_+$ and $F_-\bar{F}_-$ Lagrangians also have positive versus negative values.

The above equation conveys an important lesson. Naively, it may appear that one could just represent electromagnetic fields with the complex-valued $F = B + iE$ expression, without considering the interpretation from the perspective of geometric algebra. The complex analysis approach works either for F_+ or F_- type functions. But this complex number-based method fails when both the F_+ and F_- type fields are present. The correct signs of the above equations are obtained only when geometric algebra is correctly used to represent these electromagnetic fields.

Exercise 3. Prove that $F_+\bar{F}_+$ and $F_-\bar{F}_-$ have opposite signs.

2.5. What is the Quantum Mechanical Wavefunction?

We mathematically defined the wavefunction concept in the preliminaries chapter. One physical interpretation of a quantum-mechanical wavefunction Ψ is that its complex norm squared $\Psi\bar{\Psi}$ yields the particle's probability density. Given that $\Psi\bar{\Psi}$ specifies the particle location, it must be related to the particle's Lagrangian density since the Lagrangian density determines a particle's path.

In the preceding section, we noted that the $F\bar{F}$ expression defines the Lagrangian of a scalar-free electromagnetic field. At this point, the reader may start to suspect that F_+ or F_- has analogous geometry to the quantum mechanical wavefunction, especially if we were to write it as $\Psi = (e + ib)$. We note that there are no source terms generating the quantum mechanical e and b fields, otherwise Ψ would have an r^{-2} type dependence on the radial distance from its source. In other words, both Ψ and F are scalar-free fields.

The analogy between the F and Ψ fields is more direct in the so-called "Proca model" of quantum mechanics, where the quantum mechanical wavefunction is considered to be a mass-carrying electromagnetic wave. As we will show in Chapter 4, the Proca model indeed applies to all quantum mechanical wavefunctions, including the electron.

In the light of the preceding sections, it would seem to be better to treat $\Psi = (e + ib)$ not as a complex-valued field but rather to treat i as a Clifford pseudo-scalar. This, in turn, allows us to derive the correct Lagrangian and energy–momentum values of the quantum-mechanical wavefunction. This important point will be further explained in the next section.

After stating similarities, we now consider their differences. While F propagates at the speed of light, the $\Psi = (e + ib)$ field's propagation speed is always lower than the speed of light. What is the mechanism which slows down a quantum mechanical wave? As we shall see in Chapter 5, this reduction in velocity is related to mass. It also appears strange that the Ψ field is, on the one hand, a source-free wave, but, on the other hand, it is somehow generated by an indivisible elementary particle. What is the mechanism behind this wave–particle duality? Both of these questions will be answered in Section 5.2 of Chapter 5, where the propagation speed of the quantum-mechanical wavefunction will also be derived within the context of spinor fields and general relativity.

2.6. Spacetime Solutions of Wave Equations

We noted that there is an analogy between F and ψ fields. In this section, this analogy is further investigated to help us better understand the meaning of negative energy levels, the positron wave, and quantum tunneling. Hopefully, it will help the reader develop a more intuitive understanding of quantum mechanics. So far, we were developing Maxwell's equations in the local field equation form. The next step is to more fully understand the meaning of their spacetime solutions and from there to better understand the quantum wavefunction.

The solutions corresponding to the F_+ field are sinusoidal, and are described extensively in electromagnetism textbooks. For a spatially uniform wave propagating along the z axis, circularly polarized solutions are given by $E_x = E\sin(\omega t - kz)$, $E_y = E\cos(\omega t - kz)$, $B_x = B\cos(\omega t - kz)$, $B_y = -B\sin(\omega t - kz)$. Besides its amplitude and direction of propagation, the wave is characterized by its frequency ω or wave number k.

The solutions corresponding to the F_- field are given by exponentially decaying functions. In the simplest case, such exponential decay of the electric field arises during the discharge of capacitor plates; the electric field between these plates goes exponentially to zero. The F_- field appears to be a transient solution during an exponential decay. Is the energy density of the F_- field negative during the discharge? With respect to the electric energy density of a charged capacitor, the F_- field indeed embodies a negative energy density. The key point is that our starting condition was a static condition of positive energy density between the supercapacitor plates, locally given by $\frac{1}{2}E^2$. **The F_- wave represents a negative energy only with respect to a positive-valued static reference.** With respect to the $A = 0$ reference of an empty space, an F_- field would be of course unphysical.

In the above example, the F_- field is comprised only of an electric field, and thus differs from the F_+ field of a transversal light wave which comprises an equal amount of electric and magnetic field energy. Consider a more sophisticated example of a cylindrical capacitor, which is placed into a magnetic field. The magnetic field lines point along the cylindrical axis, and can be generated by solenoid coils placed along the axis of the capacitor. This yields a static configuration which has both electric and magnetic field energy between the capacitor plates. Moreover, the electric field points radially while the magnetic field points axially, and thus $\mathbf{B} \times \mathbf{E} \neq 0$, implying a static circulation of positive electromagnetic momentum. At a given moment in time, we short-circuit both the cylindrical capacitor plates and the solenoid terminals. This causes an exponential decay of both electric and magnetic fields. The F_- field of this scenario is now more analogous to the F_+ field of a light wave.

In Section 2.5, we suggested that the quantum-mechanical wavefunction Ψ may be viewed as a source-free $\Psi = (e + ib)$ type field. Based on this analogy, we distinguish Ψ_+ and Ψ_- solution types. In the Ψ_+ case, the generic solution is a $\Psi_+ = e^{-i(\Omega t - \mathbf{kr})}$ wave, and applying the $i\hbar\partial_t$ energy operator yields the $\hbar\Omega$ energy eigenvalue. In the Ψ_- case, the generic solution has $\Psi_- = \Psi e^{-\Omega t}$ time dependence, and applying the same $i\hbar\partial_t$ energy operator would yield an imaginary $-i\hbar\Omega$ energy eigenvalue. The correct energy operator to apply onto Ψ_- is therefore $\hbar\partial_t = -\hbar\partial_t$, and it gives the $-\hbar\Omega$ energy eigenvalue. The first point here is that the operator-based formalism of quantum mechanics must apply the appropriate operator to the appropriate field type. The second point is that in so far as the F_- field is analogous to the Ψ_- solution, the Ψ_- field arises during a "discharge" of the quantum mechanical wavefunction. When would such a discharge take place? It is well known that the quantum mechanical wavefunction disappears during an electron–positron interaction. This suggests that the Ψ_- field arises during electron–positron annihilation. However, it is clear that the Ψ_- field is then a transient wavefunction of a disappearing electron. It would be a mistake to claim that the Ψ_- field "describes the positron." This important point will be further examined in Section 7.2.

In the above examples, we saw that the timewise termination of a static field configuration is always accompanied by an exponential transient tail. In the same way, the spatial reflection of a uniform wave is always accompanied by a spatially exponential tail. Such an exponential tail is known to arise at the vacuum–metal interface, where the vacuum side contains an F_+ field with spatially uniform energy density while the electromagnetic field in the metal has a spatially exponential amplitude. Mathematically, the exponential tail can be described by a complex wavenumber $k = k_r + ik_i$, where k_r and k_i are the real and imaginary components. In most metals $k_i \gg k_r$, and therefore the electric field of the exponential tail may be written as $E_x = E\sin(\omega t)\, e^{-kz}$, $E_y = E\cos(\omega t)\, e^{-kz}$, $B_x = B\cos(\omega t)\, e^{-kz}$, $B_y = -B\sin(\omega t)\, e^{-kz}$. The need for such an exponential tail can be easily seen from Maxwell's equation: since $|A|$ is non-zero and spatially uniform in the incoming electromagnetic wave, its reflection without an exponential tail would imply $\frac{\partial A}{\partial z} \to \infty$ at the metal interface, which requires an infinitely large electromagnetic field. In practice, k_i is determined by the electromagnetic permittivity and permeability parameters of the reflecting

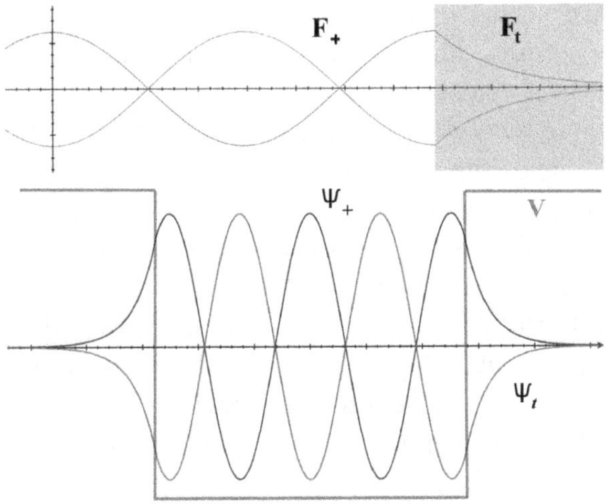

Figure 2.1. A coupling of the F_+/F_t type electromagnetic fields and of the Ψ_+/Ψ_t type quantum-mechanical fields. Top: light reflection from a metal. Bottom: quantum-mechanical wavefunction of a charged particle in a potential well. The green and blue colors represent snapshots at different times.

material. Figure 2.1 illustrates this scenario of light reflection from a metallic surface.

We denote the exponential electromagnetic tail by the F_t symbol. In the above example, the F_t field behind a metal surface is generated by a coupling between electric charges and the incoming F_+ type field. However, it is also possible to find an example where the F_t field arises on the vacuum side. Light waves traveling in a glass medium undergo total internal reflection at its boundary when they strike it at an angle greater than the so-called critical angle. The exponentially decaying F_t field arises on the vacuum side for the above stated reason of electromagnetic continuity, i.e., $\frac{\partial A}{\partial z}$ must remain a finite value.

For elementary particles in a potential well, an electromagnetic potential creates similar coupling between a sinusoidally oscillating Ψ_+ type quantum mechanical wave and its exponential and Ψ_t tail. This phenomenon is well known in quantum mechanics, and is referred to as quantum tunneling. The bottom part of Figure 2.1 illustrates quantum tunneling for an elementary particle wavefunction, placed into a potential well. The analogy with the macroscopic scenario can be seen from the figure. Mainstream theorists consider the quantum mechanical wavefunction to be in a "negative energy state" outside the potential wall region, and claim that such "negative energy" quantum tunneling is caused by Heisenberg uncertainty. The first problem with such as interpretation is the manifest unphysicality of "negative energy" states. The second problem is that a quantum mechanical state is defined as an eigenstate of the whole electron wavefunction, which corresponds to one energy eigenvalue. In other words, the electron is not supposed to have different energy values at different locations of its wavefunction. To resolve this confusion, we can use the analogy from electromagnetism, and consider the Ψ_t tail to have an imaginary k value of its wavefunction, in contrast to the real-valued wavenumber of its Ψ_+ part. The Ψ_+ and Ψ_t parts both have $\Psi e^{-i\Omega t}$ time dependence, and applying the $i\hbar\partial_t$ energy operator yields $\hbar\Omega$ energy eigenvalue for both parts. The sinusoidal Ψ_+ wavefunction is in essence reflecting back and forth between the two potential walls. This quantum tunneling phenomenon will be revisited in Sections 7.6 and 8.7 of Chapters 7 and 8, respectively.

With the above background, we progress to explore the origin of elementary particles' internal oscillation.

2.7. From Vacuum Fluctuations to Heisenberg Uncertainty

The magnitude of quantum mechanical uncertainty is characterized by the Heisenberg uncertainty relation: $\sigma_u \sigma_v \geq \frac{\hbar}{2}$, where u and v are two conjugate variables and σ represents their standard deviation. Such uncertainty in our ability to measure physical observables suggests the presence of an electromagnetic noise background, which is present everywhere in the vacuum.

Taking energy and periodicity time as conjugate variables, Heisenberg uncertainty means that the electromagnetic vacuum fluctuation energy at a given angular frequency is $\frac{\hbar\omega}{2}$. This fluctuation energy is colloquially referred to as "zero-point vacuum energy," and the $\frac{\hbar\omega}{2}$ distribution is Lorentz invariant because the vacuum has the same electromagnetic properties in any reference frame. It is important to note that this vacuum fluctuation has been directly measured in multiple experiments: it manifests as the so-called Casimir force, which acts on close-by metallic surfaces. By identifying this vacuum fluctuation with the Heisenberg uncertainty relation, the following results emerge:

(1) An explanation of the Lamb shift effect. Indeed, one of the achievements of this book is the derivation of the experimentally measured Lamb shift values as will be shown in Chapter 6, where we will give a more detailed description of this background noise.

(2) The bound electron state is quantum mechanically modeled as a harmonic oscillator, which contains spectral information in the radial direction. The expectation value of this bound wavefunction's ground state Hamiltonian is determined by the noise floor. Consequently, assuming the laws of thermodynamics, "the Heisenberg uncertainty postulate" allows us to state in very concrete physical terms that the energy of a ground state oscillation cannot be lower than the vacuum fluctuation energy which is the thermodynamic ground state. This suggests that the quantum-mechanical wavefunction is only that part of physical reality which is measurable relative to the background noise, and that this background could serve as the physical substratum for defining a standard or universal time parameter, as suggested by Prof. Larry Horwitz and members of the International Association of Relativistic Dynamics (IARD). When one considers a highly accelerated elementary particle, it is experimentally perceived as a tiny spherical charge moving along a straight trajectory. The experimental aspect in this case means scattering experiments. The amount of spectral information describing this straight-moving spherical charge approaches infinity. Obviously, this new information could not have been produced merely by the accelerating forces; instead of accelerating the particle, one could have accelerated the scattering target. This newly perceived spectral information about the elementary particle must have been present all along, but hidden by the noise floor. The act of accelerating an elementary particle allows one to measure more information about its already existing frequency spectrum.

(3) A quantum-mechanical unit of electromagnetic energy exchange is known as a photon. This unit of energy exchange is related to the photon frequency by $\epsilon = \hbar\omega$. Recognizing that an electromagnetic frequency ω corresponds to an energy exchange unit $\hbar\omega$, it is logical to guess that the internal electromagnetic oscillation frequency of an elementary particle is related to its mass according to $\hbar\omega = mc^2$. De Broglie made this guess already 100 years ago. To avoid confusion, we emphasize that a bound elementary particle is characterized by two frequencies: its electromagnetic oscillation frequency ω and the kinetic oscillation frequency of its quantum-mechanical wavefunction Ψ. In the ground state, the energy of this quantum-mechanical wavefunction Ψ coincides with the electromagnetic noise background and in accordance with the Heisenberg uncertainty relation, its energy is $\frac{\hbar\Omega}{2}$. We discuss this in more detail in Section 8.9.

(4) An oscillating particle cannot be point-like. Point-like particles should only be seen as the classical limit of the quantum-mechanical reality. From the perspective of electromagnetism, the electric field energy of a "point particle" is infinite. In contrast, by applying the logical and coherent principles of electromagnetism developed so far, we prove in Chapter 3 that the electron contains a finite amount of electromagnetic field energy, and that this electromagnetic field energy is exactly the mc^2 energy associated with its mass. This might be seen as the electromagnetic equivalent of Mach's principle that was so much liked by Einstein. Also, the question of the electron's indivisibility leads to the question of its charge quantization. Since the electromagnetic fine structure constant is related to the elementary charge by the formula $\alpha = \frac{\mu_0}{4\pi}\frac{ce^2}{\hbar}$, asking "why is the elementary charge indivisible?" is equivalent to asking "why is α constant?" We revisit this question too in Chapter 4.

2.8. The Electromagnetic Frequency of a Massive Particle

When an electron and a positron annihilate, the produced electromagnetic radiation is perceived as two photons, each carrying $\hbar\omega = mc^2$ energy and $\hbar$ angular momentum. By conservation of energy and angular momentum, an electron must comprise then an electromagnetic field which contains the same amount of energy and angular momentum, $\hbar\omega$ and $\hbar$, respectively. What is the electromagnetic topology difference between a transversal electromagnetic wave and an elementary particle?

In the above discussed transversal electromagnetic wave, the electric and magnetic fields have locally the same magnitude, induce each other via rotating field orientation, and the wave has zero rest mass. The electron however carries an elementary charge, therefore its scalar field is non-zero. The circularly polarized electromagnetic field is now described by $G = S + F = (S + \mathbf{0}) + (0 + \mathbf{B}\mathbf{e}_{xyz} - i\mathbf{E}\mathbf{e}_t)$. As derived in Chapter 1, both the scalar and transversal field components move at the speed of light. In an analogous way to the existence of F_+ and F_- electromagnetic fields, there must be two types of scalar-comprising electromagnetic fields. The corresponding G_+ and G_- electromagnetic fields take the following form for fields with electric charges:

- $F_{l+} = (0 + \mathbf{B}\mathbf{e}_{xyz} - i\mathbf{E}\mathbf{e}_t) \rightarrow G_{l+} = (S + \mathbf{B}\mathbf{e}_{xyz} - i\mathbf{E}\mathbf{e}_t)$
- $F_{l-} = (0 + i\mathbf{B}\mathbf{e}_{xyz} + \mathbf{E}\mathbf{e}_t) \rightarrow G_{l-} = (iS + i\mathbf{B}\mathbf{e}_{xyz} + \mathbf{E}\mathbf{e}_t)$

- $F_{r+} = (0 + \mathbf{B}\mathbf{e}_{xyz} + i\mathbf{E}\mathbf{e}_t) \rightarrow G_{r+} = (-S + \mathbf{B}\mathbf{e}_{xyz} + i\mathbf{E}\mathbf{e}_t)$
- $F_{r-} = (0 + i\mathbf{B}\mathbf{e}_{xyz} - \mathbf{E}\mathbf{e}_t) \rightarrow G_{r-} = (-iS + i\mathbf{B}\mathbf{e}_{xyz} + \mathbf{E}\mathbf{e}_t)$

In the above expressions, the sign of the field S determines whether the charge is positive or negative; i.e., it has opposite sign for a particle and its anti-particle.

Using the results of Chapter 1, Maxwell's equations retain their usual structure:

$$D_l G_{l+} = 0,\ D_r G_{r+} = 0 \tag{2.8.1}$$

$$D_l G_{l-} = 0,\ D_r G_{r-} = 0 \tag{2.8.2}$$

The energy and momentum densities of a massive particle are now given by the $N_+ = \frac{1}{2} G_+ \mathbf{e}_t \widetilde{G_+}$ expression, which replaces the $\frac{1}{2} F_+ \mathbf{e}_t \widetilde{F_+}$ expression of the scalar-free field. We evaluate the electromagnetic energy–momentum:

$$\begin{aligned}
N_{l+} &= \frac{1}{2}\left(G_{l+}\mathbf{e}_t\right)\widetilde{G_{l+}} = -\frac{1}{2}(S\mathbf{e}_t + i\mathbf{B} + i\mathbf{E})(-S + \mathbf{B}\mathbf{e}_{xyz} + \mathbf{E}\mathbf{e}_{xyz}) \\
&= \frac{1}{2}S^2\mathbf{e}_t - \frac{1}{2}S\left(\mathbf{e}_t\left(\mathbf{B}+\mathbf{E}\right)\mathbf{e}_{xyz} - i\left(\mathbf{B}+\mathbf{E}\right)\right) - \frac{1}{2}i\left(\mathbf{B}+\mathbf{E}\right)\left(\mathbf{B}+\mathbf{E}\right)\mathbf{e}_{xyz} \\
&= \frac{1}{2}\left(S^2 + E^2 + B^2\right)\mathbf{e}_t - \frac{1}{2}S\left(-\left(\mathbf{B}+\mathbf{E}\right)i - i\left(\mathbf{B}+\mathbf{E}\right)\right) + \frac{1}{2}i2\mathbf{B}\times\mathbf{E} \\
&= \frac{1}{2}\left(S^2 + E^2 + B^2\right)\mathbf{e}_t + i\mathbf{B}\times\mathbf{E}
\end{aligned} \tag{2.8.3}$$

$$\begin{aligned}
N_{r+} &= \frac{1}{2}\left(G_{r+}\mathbf{e}_t\right)\widetilde{G_{r+}} = -\frac{1}{2}(S\mathbf{e}_t + i\mathbf{B} - i\mathbf{E})(-S + \mathbf{B}\mathbf{e}_{xyz} - \mathbf{E}\mathbf{e}_{xyz}) \\
&= \frac{1}{2}\left(S^2 + E^2 + B^2\right)\mathbf{e}_t - i\mathbf{B}\times\mathbf{E}
\end{aligned} \tag{2.8.4}$$

The above equations correspond to the energy and momentum densities of the G_+ electromagnetic field. Analogously, the energy and momentum of the G_- electromagnetic field is given by the $N_- = -\frac{1}{2} G_- \mathbf{e}_t \widetilde{G_-}$ expression.

Equations (2.8.3) and (2.8.4) imply that the Poynting-vector part of the electromagnetic field is the well-known $\mathbf{B}\times\mathbf{E}$ formula, regardless of the scalar field's presence or absence. This result deviates from the $\mathbf{B}\times\mathbf{E} + S\mathbf{E}$ formula obtained in Chapter 1. This difference stems from the different time variable choice in the $A = (iA_t\mathbf{e}_t, \mathbf{A})$ definition, which we use in this chapter. The $A = (iA_t\mathbf{e}_t, \mathbf{A})$ choice is remarkable because (i) the Poynting-vector formula remains the same as in the scalar-free case, and (ii) the energy and momentum density formulas are symmetric in terms of E and B fields.

According to the Heisenberg uncertainty principle, when the G_+ or G_- field oscillates at a certain frequency ω, it has at least $\frac{\hbar\omega}{2}$ energy and $\frac{\hbar}{2}$ angular momentum. However, the electron–positron annihilation process demonstrates that the electron carries $\hbar\omega$ energy and $\hbar$ angular momentum. In this way, the electron–positron pair satisfies energy and angular momentum conservation with respect to the transversal wave from which it is created or into which it is annihilated. These values are a starting point in the search for an analytic solution of a particle's internal fields.

Since S, E, and B are all derived from the same electromagnetic vector potential, they all oscillate at the same frequency. This internal frequency is related to the particle mass via the $\hbar\omega = mc^2$ formula. In Chapter 3, we shall explicitly derive the relativistic increase of particle mass under Lorentz boost, i.e., we will prove the equivalence between the relativistic increase of particle mass and the relativistic shift in the particle's electromagnetic oscillation frequency. Here, we just note a pertinent point about mass conservation. In a thought experiment, we place an electron and a positron into an empty box, and initially both particles are at rest. As the two particles speed toward each other due to electric attraction, they gain kinetic energy. Where is this energy coming from? There is no other source for it than the electric field of the particles. As the two particles get closer, there is more and more cancellation between the electric fields of the two particles, and the overall electric field energy is reduced. In other words, the energy of the charges' extended electric field is gradually converted into the kinetic energy of the particles. As the particles gain kinetic energy, their relativistic mass increases. From the perspective of the whole box, there was no outside influence, so its total mass must remain invariant. Therefore the mass gained by the kinetic boost must be exactly counterbalanced by the mass lost by electric field cancellations. This thought experiment is a direct illustration of the electromagnetic field origin of particle mass. One must refrain from thinking about a particle being at a certain point in space, since the particle is represented by the whole electromagnetic wave.

The main result from this section is the $\hbar\omega = mc^2$ relationship between the massive particle and its internal electromagnetic frequency. If the particle's electromagnetic field solution involves a circular rotation of its charge, the angular frequency of this rotation must be $\omega = \frac{mc^2}{\hbar}$. This angular frequency is the so-called "de Broglie frequency." Although the $\omega = \frac{mc^2}{\hbar}$ angular frequency is very high, the authors of [7] succeeded in experimentally measuring it a few years ago.

2.9. A New "Rotation" Axis

In this section, we look into the meaning of using real versus i-multiplied time coordinate. In Chapter 1, we defined the electromagnetic field and the spacetime differential using real coordinates:

- $A_{l+} = (A_t\mathbf{e}_t, \mathbf{A})$, $\partial_l = (-\partial_t\mathbf{e}_t, \nabla)$
- $A_{r+} = (-A_t\mathbf{e}_t, \mathbf{A})$, $\partial_r = (\partial_t\mathbf{e}_t, \nabla)$
- $A_{l-} = T_l = (iA_t\mathbf{e}_t, i\mathbf{A})$, $\partial_l = (-\partial_t\mathbf{e}_t, \nabla)$
- $A_{r-} = T_r = (-iA_t\mathbf{e}_t, i\mathbf{A})$, $\partial_r = (\partial_t\mathbf{e}_t, \nabla)$

In the above expression, the T symbol refers to the vector potential which generates magnetic charges and currents, as introduced in Section 1.3.6 of Chapter 1. In this chapter, we introduced the $\mathbf{e}_t \to i\mathbf{e}_t$ transformation of the time-wise coordinate:

- $A'_{l+} = (iA_t\mathbf{e}_t, \mathbf{A})$, $\partial'_l = D_l = (-i\partial_t\mathbf{e}_t, \nabla)$
- $A'_{r+} = (-iA_t\mathbf{e}_t, \mathbf{A})$, $\partial'_r = D_r = (i\partial_t\mathbf{e}_t, \nabla)$
- $A'_{l-} = (-A_t\mathbf{e}_t, i\mathbf{A})$, $\partial'_l = D_l = (-i\partial_t\mathbf{e}_t, \nabla)$
- $A'_{r-} = (A_t\mathbf{e}_t, i\mathbf{A})$, $\partial'_r = D_r = (i\partial_t\mathbf{e}_t, \nabla)$

We verified explicitly that Maxwell's equations remain the same regardless of how we choose the time-wise basis. Electromagnetism has a symmetry with respect to this coordinate transformation. In fact, it is not a discrete, but a continuous symmetry because we can generalize the choice of time coordinate transformation to any angle according to the $\mathbf{e}_t \to e^{i\theta}\mathbf{e}_t$ choice. Even though i is the Clifford pseudo-scalar, this generalization is still valid because the exponentiation of the Clifford pseudo-scalar is a well-defined operation in Clifford algebra.

By comparing Equation (1.3.27) with Equations (2.8.3) and (2.8.4), we observe the same expression for the electromagnetic energy density: $\frac{1}{2}\left(S^2 + E^2 + B^2\right)\mathbf{e}_t$. Although the $\mathbf{e}_t \to e^{i\theta}\mathbf{e}_t$ transformation changes the momentum density part, it preserves the electromagnetic energy density. Considering θ as yet another rotation angle, electromagnetic energy density has a $U(1)$ type rotation-like symmetry with respect to this angle.

Physical rotations of space also preserve electromagnetic energy density. Therefore, the above-defined $\mathbf{e}_t \to e^{i\theta}\mathbf{e}_t$ rotation plus spatial rotations form the $U(1) \times SU(2)$ group of transformations, which preserves the electromagnetic energy density.

The study of mathematical symmetries is a useful tool for identifying conserved quantities. Rotational symmetries of space correspond to angular momentum conservation. The $\mathbf{e}_t \to e^{i\theta}\mathbf{e}_t$ rotational symmetry also corresponds to some conserved current. Such current may or may not be present in a given elementary particle state. Since the nuclear capture and emission of electrons is always accompanied by the capture or emission of neutrinos, it is possible that neutrinos carry a conserved current, corresponding to the $\mathbf{e}_t \to e^{i\theta}\mathbf{e}_t$ symmetry. This hypothesis will be investigated step-by-step throughout the book.

2.10. The Longitudinal Electromagnetic Wave

In Section 2.6, we described the simplest spatial solutions of Maxwell's equations, which correspond to scalar-free transversal waves. We wrote the equations for circularly polarized modes, which carry the conserved angular momentum of space rotation symmetry. Considering the $\mathbf{e}_t \to e^{i\theta}\mathbf{e}_t$ symmetry, can we find a wave solution which is the corresponding electromagnetic oscillation mode? A continuous $\mathbf{e}_t \to e^{i\theta}\mathbf{e}_t$ type "rotation" means an oscillation between the electric and magnetic type scalar field. The corresponding spatial solution of Maxwell's equations is a longitudinal wave:

$$\nu_+ \begin{cases} E_z = E_0 \sin(\omega t - kz), \; B_z = \pm B_0 \cos(\omega t - kz) \\ S = S_0 \sin(\omega t - kz) \pm i S_0 \cos(\omega t - kz) \end{cases} \tag{2.10.1}$$

where $S_0 = E_0 = B_0$ in natural units, and the wave is propagating into the z direction. In this simple plane wave case, the electric and magnetic fields have no $x - y$ components. The chosen ν_+ symbol indicates that this is the electromagnetic field equation of neutrino waves. This basic longitudinal wave solution is illustrated in Figure 2.2.

Equation (2.10.1) essentially describes a circular polarization of the propagating scalar field. The choice of $\pm$ sign in Equation (2.10.1) indicates

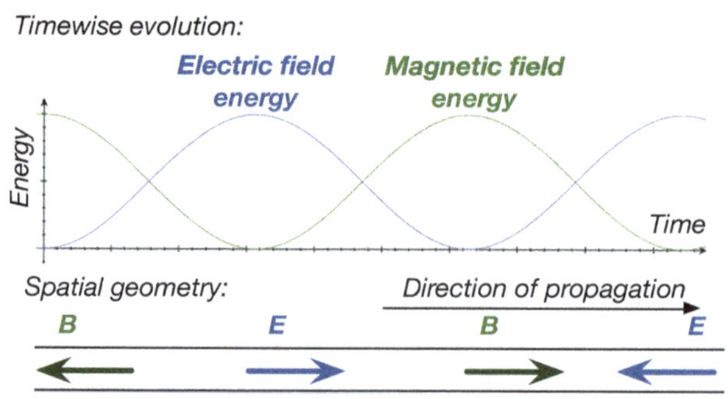

Figure 2.2. An illustration of the longitudinal electromagnetic wave. The blue color corresponds to the real-valued scalar field, while the green color corresponds to the i-multiplied scalar field.

a chirality choice. Depending on the $\pm$ sign choice, we may thus talk about "left-handed" ν_{l+} or "right-handed" ν_{r+}, even though this wave is longitudinal.

Considering the $S_0 = E_0 = B_0$ condition of the longitudinal wave solution, the $\frac{1}{2}\left(S^2 + E^2 + B^2\right)\mathbf{e}_t$ electromagnetic energy density formula implies that $\frac{1}{2}$ of the electromagnetic field energy is carried by the scalar field, while $\frac{1}{4}$ of the field energy is carried by the electric field and $\frac{1}{4}$ of the field energy is carried by the magnetic field. This is in contrast to the transversal wave solution, which is scalar-free.

Upon the reflection of a longitudinal wave, an evanescent tail appears on the other side of the reflecting interface: this phenomenon is completely analogous to the other wave reflections illustrated in Figure 2.1. However, it is difficult to identify the corresponding equation of an exponentially decaying ν_t tail. This problem can be solved after the $\mathbf{e}_{t'} = i\mathbf{e}_t$ transformation of the time coordinate. In the transformed system, the exponential tail formula can be easily found:

$$\nu_t' \begin{cases} E_z' = E_0 e^{\omega t' - kz}, \; B_z' = \pm B_0 e^{\omega t' - kz} \\ S' = S_0 e^{\omega t' - kz} \pm i S_0 e^{\omega t' - kz} \end{cases} \tag{2.10.2}$$

Exercise 4. Find the expression for ν_t upon transforming the time coordinate back to $\mathbf{e}_t$.

Analogous to the linear polarization of a transversal wave, a longitudinal electromagnetic wave may also be linearly polarized. In that case, the scalar field is given by $S = S_0 \sin(\omega t - kz)$, and the corresponding electric field is given by $E_z = E_0 \sin(\omega t - kz)$. Such polarization does not have a magnetic component. It is interesting to note that such a linearly polarized longitudinal wave appears to be realized by the acoustic phonon oscillations of condensed matter. Although microscopically an acoustic phonon wave comprises many individual positively and negatively charged particles, from the macroscopic continuum perspective this wave solution is given by the above-stated simple formulas for S and E–while keeping in mind that the wave propagation speed is very different than in vacuum. From the microscopic perspective, the individually moving particles of a phonon wave generate magnetic fields as well, but these magnetic fields cancel each other out from the macroscopic perspective. We will revisit this type of electromagnetic wave in Chapter 15.

We conclude this section by a summary of electromagnetic wave polarization modes. We identified several principal electromagnetic wave modes in this chapter, as listed in Table 2.1. Each polarization mode has a trivial

Table 2.1. A classification of electromagnetic wave polarization modes.

	A_+ type transversal wave	A_t type transversal wave
Trivial spatial solution	Sinusoidal wave (see Section 2.6)	Exponential function (see Section 2.6)
Scalar field in the trivial solution	$S = 0$	$S = 0$
Where is it observed?	Lamps, radar, radio, QM state transitions	Evanescent fields e.g., at vacuum—metal interfaces
	ν_+ type longitudinal wave	ν_t type longitudinal wave
Trivial spatial solution	Sinusoidal wave see Eq. (2.10.1)	Exponential function see Eq. (2.10.2)
Scalar field in the trivial solution	$S \neq 0$ everywhere, $S = S_0 e^{it}$ rotation	$S \neq 0$ everywhere
Where is it observed?	Neutrino radiation	See Chapter 15

spatial solution, which may be further classified, e.g., according to left/right-handed chirality. Historically, the trivial solution of the A_+ principal mode was the sole focus of electromagnetism studies.

Despite its apparent simplicity, Maxwell's equation admits rather complex spatial solutions even under the $S = 0$ restriction, such as Laguerre–Gaussian beams or cylindrical Bessel beams. The mathematical form of spatial solutions also depends on the boundary conditions: electromagnetic waves in spherical waveguides, for example, are described by spherical harmonic functions. By removing the $S = 0$ restriction, we may begin to study the full range of Maxwell equation solutions. The complete set of Maxwell equation solutions might still be very under-explored.

Our main interest in this book is to find those solutions which describe the electromagnetic field of massive particles and to explore neutrino-type longitudinal wave solutions.

Acknowledgments

The authors thank Jan von Pfaler for clarifications about Clifford algebra.

References

[1] G. Lochak, A new electromagnetism based on 4 photons: electric, magnetic, with spin 1 and spin 0. *Annales de la Fondation Louis de Broglie*, 35 (2010).
[2] M. K.-H. Kiessling *et al.*, On the quantum-mechanics of a single photon. *Journal of Mathematical Physics*, 59(11) (2018).
[3] A.K. Jha *et al.*, Fourier relationship between the angle and angular momentum of entangled photons. *Physical Review A*, 78 (2008).
[4] I. Bialynicki-Birula, Photon wave function. *Progress in Optics*, 36 (1996), 245–294.
[5] D. Hestenes, *Space-Time Algebra*, 2nd edition, Gordon and Breach Science Publishers, 1966.
[6] R. Feynman, *The Feynman Lectures on Physics*, Volume II, Chapter 28.
[7] S.Y. Lan *et al.*, A clock directly linking time to a particle's mass. *Science*, 339 (2013), 6119.

https://doi.org/10.1142/9781800611306_0003

Chapter 3

The Electron and Occam's Razor

Francesco Celani*,†,§, Antonino Oscar Di Tommaso‡,¶, and Giorgio Vassallo‡,‖

**National Institute of Nuclear Physics (INFN-LNF) Via E. Fermi 56, 00044 Frascati, Roma, Italy*

†*International Society for Condensed Matter Nuclear Science (ISCMNS), UK*

‡*Engineering Department, University of Palermo viale delle Scienze, 90128 Palermo, Italy*

§*francesco.celani@lnf.infn.it*

¶*antoninooscar.ditommaso@unipa.it*

‖*giorgio.vassallo@unipa.it*

Nomenclature

Symbol, name, SI units, natural units (NU).

$\boldsymbol{A}_\square$: electromagnetic four-potential $(\mathrm{V \cdot s \cdot m^{-1}})$, (eV);
$\boldsymbol{A}_\triangle$: electromagnetic vector potential $(\mathrm{V \cdot s \cdot m^{-1}})$, (eV);
$\boldsymbol{G}$: electromagnetic field $(\mathrm{V \cdot s \cdot m^{-2}})$, $(\mathrm{eV^2})$;
$\boldsymbol{F}$: electromagnetic field bivector $(\mathrm{V \cdot s \cdot m^{-2}})$, $(\mathrm{eV^2})$;
$\boldsymbol{B}$: flux density field $(\mathrm{V \cdot s \cdot m^{-2}}) = (\mathrm{T})$, $(\mathrm{eV^2})$;
$\boldsymbol{E}$: electric field $(\mathrm{V \cdot m^{-1}})$, $(\mathrm{eV^2})$;
S: scalar field $(\mathrm{V \cdot s \cdot m^{-2}})$, $(\mathrm{eV^2})$;
$\boldsymbol{J}_{\square e}$: four-current density field $(\mathrm{A \cdot m^{-2}})$, $(\mathrm{eV^3})$;
$\boldsymbol{J}_\triangle$: current density field $(\mathrm{A \cdot m^{-2}})$, $(\mathrm{eV^3})$;
$\boldsymbol{v}_\square$: four-velocity vector $(\mathrm{m \cdot s^{-1}})$, (1);
$\boldsymbol{v}_\triangle$: velocity vector $(\mathrm{m \cdot s^{-1}})$, (1);
ρ: charge density $(\mathrm{A \cdot s \cdot m^{-3} = C \cdot m^{-3}})$, $(\mathrm{eV^3})$;
x, y, z: space coordinates (m), $(\mathrm{eV^{-1}})$, $(1.9732705 \cdot 10^{-7}\ \mathrm{m} \approx 1\ \mathrm{eV^{-1}})$;
t: time variable [s], $(\mathrm{eV^{-1}})$, $(6.5821220 \cdot 10^{-16}\ \mathrm{s} \approx 1\ \mathrm{eV^{-1}})$;

c: light speed in vacuum ($2.997\,924\,58 \cdot 10^8$ m $\cdot$ s^{-1}), (1);
$\hbar$: reduced Planck constant ($\hbar = \frac{h}{2\pi}$) ($1.054\,571\,726 \cdot 10^{-34}$ J $\cdot$ s), (1);
μ_0: permeability of vacuum ($4\pi \cdot 10^{-7}$ V $\cdot$ s $\cdot$ A^{-1} $\cdot$ m^{-1}), (4π);
ϵ_0: dielectric constant of vacuum ($8.854\,187\,817 \cdot 10^{-12}$ A $\cdot$ s $\cdot$ V^{-1} $\cdot$ m^{-1}), ($\frac{1}{4\pi}$);
e: electron charge ($1.602\,176\,565 \cdot 10^{-19}$ A $\cdot$ s), (0.085 424 546);
m_e: mass of the electron at rest ($9.109384 \cdot 10^{-31}$ kg), ($0.510\,998\,946 \cdot 10^6$ eV);
λ_c: electron Compton wavelength ($2.426\,310\,2389 \cdot 10^{-12}$ m), ($1.229\,588\,259 \cdot 10^{-5}$ eV^{-1});
r_e: reduced Compton wavelength of electron (Compton radius) $r_e = \frac{\lambda_c}{2\pi}$;
$\omega_e = \frac{c}{r_e}$: *Zitterbewegung* angular frequency;
T: Zitterbewegung period $T = \frac{2\pi}{\omega_e}$;
$\boldsymbol{P}_\square$: energy–momentum four-vector (kg $\cdot$ m $\cdot$ s^{-1}), (eV);
$\boldsymbol{P}_\triangle$: momentum vector (kg $\cdot$ m $\cdot$ s^{-1}), (eV);
U, W: energy ($J = $ kg $\cdot$ m^2 $\cdot$ s^{-2}), (eV).

3.1. Introduction

The application of Occam's razor principle to Maxwell's equations suggests a simple electromagnetic model for charge, mass, and inertia. Building onto the results of Chapters 1 and 2, a new and particularly simple model of the electron is proposed in this chapter.

One of the most detailed and interesting Zitterbewegung electron models to date has been proposed by David Hestenes [1]. He rewrote the Dirac equation for the electron using the four-dimensional real Clifford algebra $Cl_{1,3}(R)$ of space–time with *Minkowski signature* "$+---$," eliminating unnecessary complexities and redundancies arising from the traditional use of matrices. The Dirac gamma matrices γ_μ and the associated algebra can be seen as an isomorphism of the four-basis vector of space–time geometric algebra. This simple isomorphism allows a full encoding of the geometric properties of the Dirac algebra, and a rewriting of Dirac equation that does not require complex numbers or matrix algebra. In this context, the wave function ψ is characterized by the eight real values of the even grade multivectors of space–time algebra $Cl_{1,3}$ (STA). Even grade multivectors of STA can encode ordinary rotations as well as Lorentz transformations in the six planes of space–time. Hestenes associates the rotations encoded by the wave function with an intrinsic very rapid rotation of the electron, the "Zitterbewegung," that is considered to generate the electron spin and magnetic moment. The word Zitterbewegung was originally used by Schrödinger to indicate a fast movement attributed to an hypothetical interference between "positive" and "negative" energy states. Kerson Huang later, more realistically, interpreted the Zitterbewegung as a circular motion [4].

In particular, B. Sidharth states that "The well-known Zitterbewegung may be looked upon as a circular motion about the direction of the electron spin with radius equal to the Compton wavelength (divided by 2π) of the electron. The intrinsic spin of the electron may be looked upon as the orbital angular momentum of this motion. The current produced by the Zitterbewegung is seen to give rise to the intrinsic magnetic moment of the electron" [5].

Hestenes considers the complex phase of the wave function solution of the traditional Dirac equation as the phase of the Zitterbewegung rotation, showing "the inseparable connection between quantum mechanical phase and spin" consequently rejecting the "conventional wisdom that phase is an essential feature of quantum mechanics, while spin is a mere detail that can often be ignored" [6]. Using the space–time algebra in Ref. [7], Hestenes defines the *"canonical form"* of *the real wave function* ψ:

$$\psi(x) = \left(\varrho e^{i\beta}\right)^{\frac{1}{2}} R$$

In the above equation, x is a generic space–time point, $\varrho = \varrho(x)$ is a scalar function interpreted as a probability density proportional to charge density, i is the spatial bivector $i = \gamma_2\gamma_1$, $\beta = \beta(x)$ is a function representing the value of a rotation phase in the plane $\gamma_2\gamma_1$ and R is a rotor valued function that encodes a Lorentz transformation. In the STA canonical form for Dirac's wave function, the imaginary unit i is replaced by a bivector that generates rotation in a well-defined space-like plane and not in a generic undefined "complex plane." In Chapter 2, we demonstrated that quantum mechanics' imaginary unit i is in fact the Clifford pseudo-scalar. The canonical form introduced by Hestenes is another way of expressing the same geometry, and we will work with this spinor form in Chapter 7.

Based on the results of Chapters 1 and 2, the electron characteristics may be explained by a massless charge distributed on the surface of a sphere that rotates at the speed of light along a circumference with a radius equal to the reduced electron Compton wavelength (≈0.386159 pm). This radius corresponds to speed of light rotation at the $\omega = \frac{mc^2}{\hbar}$ de Broglie frequency. We pointed out in the previous chapter that the physical reality of this frequency is experimentally validated. The electron mass-energy, expressed in natural units, is then equal to the angular speed of the Zitterbewegung rotation and to the inverse of the orbit radius (i.e., ≈511 keV), whereas the angular momentum is equal to the reduced

Planck constant $\hbar$. Consequently, unlike the Hestenes prediction, our model proposes a relativistic contraction of the Zitterbewegung radius and a corresponding instantaneous Zitterbewegung angular speed that increases as the electron speed increases.

The inter-nuclear distance in UDD of ≈2.3 pm, found by Holmlid [2], seems to be compatible with proton–electron structures at the Compton scale [8] where the Zitterbewegung phases of neighbor electrons are correlated. These structures may generate unusual nuclear reactions and transmutations, considering the different sizes, time-scale, and energies of these composites with respect to the dimension of the particles (such as neutrons) normally used in nuclear experiments.

By using the electromagnetic four-potential as a *Materia Prima*, a natural connection between electromagnetic concepts and Newtonian and relativistic mechanics seems to be possible. The vector potential should not be viewed only as a pure mathematical tool to evaluate spatial electromagnetic field distributions but as a real physical entity, as suggested by the Aharonov–Bohm effect, a quantum mechanical phenomenon in which a charged particle is affected by the vector potential in regions in which the electromagnetic fields are null [9].

The present chapter is structured in the following manner: Section 3.2 deals with a brief presentation of Maxwell's equations that does not use Lorenz gauge; Section 3.3 presents a new simple Zitterbewegung model of the electron with a list of the main parameters that can be deduced by applying this model; Section 3.4 describes an original method to easily derive the Lorentz force law from the electromagnetic field; Section 3.5 consists of a short introduction to the concept of "quanta current"; and Section 3.6 presents some hypotheses on Compton scale aggregates.

In this chapter, all equations enclosed in square brackets with subscript "NU" have dimensions expressed in natural units.

3.2. Maxwell's Equations in $Cl_{3,1}$

The space–time algebra is a four dimensions Clifford algebra with *Minkowski signature* $Cl_{1,3}$ ("west coast metric") or $Cl_{3,1}$ ("east coast metric") [10, 11].

In $Cl_{3,1}$ algebra, used in this work, calling $\{\gamma_x, \gamma_y, \gamma_z, \gamma_t\}$ the four unitary vectors of an orthonormal base, the following rules apply:

$$\gamma_i\gamma_j = -\gamma_j\gamma_i \text{ with } i \neq j \text{ and } i, j \in \{x, y, z, t\} \tag{3.2.1}$$

$$\gamma_x^2 = \gamma_y^2 = \gamma_z^2 = -\gamma_t^2 = 1 \tag{3.2.2}$$

Maxwell's equations can be rewritten considering all the derivatives of the electromagnetic four-potential $\boldsymbol{A}_{\square}$:

$$\boldsymbol{A}_{\square}(x, y, z, t) = \gamma_x A_x + \gamma_y A_y + \gamma_z A_z + \gamma_t A_t \tag{3.2.3}$$

Each of the vector potential components A_x, A_y, A_z, and A_t is a function of space and time coordinates and has dimension in SI units equal to $[\mathrm{V} \cdot \mathrm{s} \cdot \mathrm{m}^{-1}]$. $\boldsymbol{A}_{\square}$ is a *harmonic* function that can be seen as the unique source of all concepts-entities in Maxwell's equations. We recall from Chapter 1 that using the following definition of the operator $\boldsymbol{\partial}$ in space–time algebra (where $\nabla = \gamma_x \frac{\partial}{\partial x} + \gamma_y \frac{\partial}{\partial y} + \gamma_z \frac{\partial}{\partial z}$ and $c = 1/\sqrt{\epsilon_0 \mu_0}$)

$$\boldsymbol{\partial} = \gamma_x \frac{\partial}{\partial x} + \gamma_y \frac{\partial}{\partial y} + \gamma_z \frac{\partial}{\partial z} + \gamma_t \frac{1}{c} \frac{\partial}{\partial t} = \nabla + \gamma_t \frac{1}{c} \frac{\partial}{\partial t} \tag{3.2.4}$$

the following expression can be written (see Table 3.1):

$$\boldsymbol{\partial} \boldsymbol{A}_{\square} = \boldsymbol{\partial} \cdot \boldsymbol{A}_{\square} + \boldsymbol{\partial} \wedge \boldsymbol{A}_{\square} = S + \boldsymbol{F} = \boldsymbol{G} \tag{3.2.5}$$

where

$$S = \nabla \cdot \boldsymbol{A}_{\triangle} - \frac{1}{c} \frac{\partial A_t}{\partial t} \tag{3.2.6}$$

is the scalar field,

$$\boldsymbol{F} = \frac{1}{c} \boldsymbol{E} \gamma_t + I \boldsymbol{B} \gamma_t = \frac{1}{c} \left(\boldsymbol{E} + I c \boldsymbol{B} \right) \gamma_t \tag{3.2.7}$$

the electromagnetic field, and

$$I = \gamma_x \gamma_y \gamma_z \gamma_t \tag{3.2.8}$$

is the pseudoscalar unit.

Table 3.1. Relation between electromagnetic entities and the vector potential.

$\boldsymbol{\partial} \boldsymbol{A}_{\square}$	$\gamma_x A_x$	$\gamma_y A_y$	$\gamma_z A_z$	$\gamma_t A_t$
$\gamma_x \frac{\partial}{\partial x}$	S_1	B_{z1}	$-B_{y1}$	$\frac{1}{c} E_{x1}$
$\gamma_y \frac{\partial}{\partial y}$	B_{z2}	S_2	B_{x1}	$\frac{1}{c} E_{y1}$
$\gamma_z \frac{\partial}{\partial z}$	$-B_{y2}$	B_{x2}	S_3	$\frac{1}{c} E_{z1}$
$\gamma_t \frac{1}{c} \frac{\partial}{\partial t}$	$\frac{1}{c} E_{x2}$	$\frac{1}{c} E_{y2}$	$\frac{1}{c} E_{z2}$	S_4

The electromagnetic field $\boldsymbol{G}$ can be expressed in the following compact form:

$$\boldsymbol{G}(x,y,z,t) = \nabla \cdot \boldsymbol{A}_{\Delta} - \frac{1}{c}\frac{\partial A_t}{\partial t} + \nabla A_t \gamma_t - \frac{1}{c}\frac{\partial \boldsymbol{A}_{\Delta}}{\partial t}\gamma_t + I\nabla \times \boldsymbol{A}_{\Delta}\gamma_t \quad (3.2.9)$$

and the expression

$$\boldsymbol{\partial G} = \boldsymbol{\partial}^2 \boldsymbol{A}_{\square} = 0 \quad (3.2.10)$$

represents the four Maxwell equations.

By applying, now, the $\boldsymbol{\partial}$ operator to the scalar field S, we obtain the expression of the four-current as

$$\frac{1}{\mu_0}\boldsymbol{\partial} S = \frac{1}{\mu_0}\left(\gamma_x \frac{\partial S}{\partial x} + \gamma_y \frac{\partial S}{\partial y} + \gamma_z \frac{\partial S}{\partial z} + \gamma_t \frac{1}{c}\frac{\partial S}{\partial t}\right) = \boldsymbol{J}_{\square e} \quad (3.2.11)$$

where $\boldsymbol{J}_{\square e} = \gamma_x J_{ex} + \gamma_y J_{ey} + \gamma_z J_{ez} - \gamma_t c\rho = \boldsymbol{J}_{\Delta} - \gamma_t c\rho = \rho(\boldsymbol{v} - \gamma_t c)$ is the four-current vector and $\boldsymbol{v}_{\square} = \gamma_x v_x + \gamma_y v_y + \gamma_z v_z - \gamma_t c = \boldsymbol{v} - \gamma_t c$ is a four-velocity vector. The $\boldsymbol{\partial}$ operator applied to the four-current gives the charge-current conservation law

$$\frac{1}{\mu_0}\boldsymbol{\partial} \cdot (\boldsymbol{\partial} S) = \boldsymbol{\partial} \cdot \boldsymbol{J}_{\square e} = \frac{\partial J_{ex}}{\partial x} + \frac{\partial J_{ey}}{\partial y} + \frac{\partial J_{ez}}{\partial z} + \frac{\partial \rho}{\partial t} = 0 \quad (3.2.12)$$

which can be written alternatively as

$$\boldsymbol{\partial} \cdot (\boldsymbol{\partial} S) = \boldsymbol{\partial}^2 S = \frac{\partial^2 S}{\partial x^2} + \frac{\partial^2 S}{\partial y^2} + \frac{\partial^2 S}{\partial z^2} - \frac{1}{c^2}\frac{\partial^2 S}{\partial t^2} = \nabla^2 S - \frac{1}{c^2}\frac{\partial^2 S}{\partial t^2} = 0 \quad (3.2.13)$$

The charge is related to the scalar field according to

$$\frac{1}{c}\frac{\partial S}{\partial t} = \mu_0 J_{et} = -\mu_0 c \frac{\partial q}{\partial x \partial y \partial z} = -\mu_0 c\rho \quad (3.2.14)$$

so that, by applying the time derivative to (3.2.13) and remembering (3.2.14), the wave equation of the charge density field $\rho(x,y,z,t)$ can be deduced:

$$\frac{\partial}{\partial t}\left(\boldsymbol{\partial}^2 S\right) = \boldsymbol{\partial}^2 \left(\frac{\partial S}{\partial t}\right) = \boldsymbol{\partial}^2 \left(-\mu_0 c^2 \rho\right) = -\mu_0 c^2 \boldsymbol{\partial}^2 \rho = 0 \quad (3.2.15)$$

whose last equality gives

$$\boldsymbol{\partial}^2 \rho = \nabla^2 \rho - \frac{1}{c^2}\frac{\partial^2}{\partial t^2}\rho = 0 \quad (3.2.16)$$

A more detailed development of the above equations was discussed in Chapter 1.

3.3. Electron Zitterbewegung Model

The concept of charge that emerges from this rewriting of Maxwell's equations, has a non-trivial implication: the analysis of (3.2.13) and (3.2.16) shows that the time derivative of a field S which propagates at the speed of light must necessarily represent charges that are also moving at the speed of light.

This observation advises a pure electromagnetic model of elementary particles based on the Zitterbewegung interpretation of quantum mechanics [1, 12]. According to this interpretation, the electron structure consists of a massless charge that rotates at the speed of light along a circumference equal to electron Compton wavelength λ_c [13, 14]. Calling r_e the Zitterbewegung radius, ω_e the angular speed, and T its period, we have:

$$r_e = \frac{\lambda_c}{2\pi} \approx 3.861\,593 \cdot 10^{-13}\ \mathrm{m} \tag{3.3.1}$$

$$\omega_e = \frac{c}{r_e} = 2\pi \cdot \frac{c}{\lambda_c} \approx 7.763\,440 \cdot 10^{20}\ \mathrm{rad \cdot s^{-1}} \tag{3.3.2}$$

$$T = \frac{2\pi}{\omega_e} = \frac{2\pi r_e}{c} \approx 8.093\,300 \cdot 10^{-21}\ \mathrm{s} \tag{3.3.3}$$

The value of the electron mass, expressed in SI units, can be derived from the following energy equations [1]:

$$W_{\mathrm{tot}} = m_e c^2 = \hbar\omega_e = \frac{\hbar c}{r_e} \tag{3.3.4}$$

from which

$$m_e = \frac{\hbar\omega_e}{c^2} = \frac{\hbar}{c r_e} = \frac{h}{c\lambda_c} \approx 9.109\,383 \cdot 10^{-31}\ \mathrm{kg} \tag{3.3.5}$$

is obtained. Using natural units with $\hbar = c = 1$, the electron mass (in eV) is equal to the angular speed ω_e and to the inverse of r_e:

$$\left[m_e = \omega_e = \frac{1}{r_e} \approx 0.511 \cdot 10^{6}\ \mathrm{eV} \right]_{NU}$$

3.3.1. *Simple electron model*

A charge rotating at the speed of light generates a current I_e that is equal to the ratio of the elementary charge e and its rotation period T [15]:

$$I_e = \frac{e}{T} = \frac{ec}{2\pi r_e} = \frac{e\omega_e}{2\pi} \approx 19.796\,331 \text{ A} \tag{3.3.6}$$

Neglecting the electron's charge radius, the electron magnetic moment μ_B (Bohr magneton) is equal to the product between the current I_e and the enclosed area $\mathcal{A}_e$

$$\mu_B = I_e\mathcal{A}_e = \frac{e\omega_e}{2\pi}\pi r_e^2 = \frac{ec}{2}r_e = \frac{ec^2}{2\omega_e} = \frac{e\hbar}{2m_e} \approx 9.274\,010 \cdot 10^{-24} \text{ A} \cdot \text{m}^2 \tag{3.3.7}$$

The angular momentum effects of the non-zero electron charge radius will be investigated in Chapter 6.

Occam's razor is an effective epistemological instrument that imposes to avoid as much as possible the introduction of exceptions. Following this rule, a pure electromagnetic origin of the electron's "intrinsic" angular momentum should be found. Consequently, the canonical momentum P_t of the rotating massless charge may be seen as the cause of the intrinsic angular momentum:

$$\Omega = P_t r_e$$

where the canonical momentum P_t of e, in presence of a vector potential A, generated by the current I_e, is

$$P_t = eA$$

Imposing the constraint that $\Omega = \hbar$, we can compute A as function of I_e

$$\Omega = eAr_e = \frac{eAc}{\omega_e} = \frac{e^2cA}{2\pi I_e} = \hbar \tag{3.3.8}$$

from which it is possible to calculate the vector potential A seen by the spinning charge

$$A = \| \boldsymbol{A}_\square \| = \frac{2\pi\hbar}{e^2c}I_e = \frac{\hbar}{er_e} = \frac{\hbar\omega_e}{ec} = \frac{m_ec}{e} \approx 1.704\,509 \cdot 10^{-3} \text{ V} \cdot \text{s} \cdot \text{m}^{-1} \tag{3.3.9}$$

From (3.3.9) it is possible to derive the fine structure constant (FSC)

$$\alpha = \frac{\mu_0}{4\pi} \cdot \frac{ce^2}{\hbar} = \frac{\mu_0}{4\pi} \cdot \frac{e\omega_e}{A} \approx 7.297\,352 \cdot 10^{-3} \tag{3.3.10}$$

Using natural units we get these simple relations:

$$\left[A = 2\pi\alpha^{-1}I_e\right]_{NU}$$

$$\left[eA = \omega_e = r_e^{-1} = m_e = P_t\right]_{NU}$$

where α^{-1}, the inverse of the FSC, is a pure number and e is the elementary charge expressed in natural units

$$\left[\alpha^{-1} = e^{-2} \approx 137.035989\right]_{NU}$$

3.3.2. *Spin and intrinsic angular momentum*

The intrinsic angular momentum $\hbar$ of the electron model (see (3.3.8)) is compatible with the spin value $\frac{\hbar}{2}$ if we consider the electron interaction with the external flux density field $\boldsymbol{B}_E$, as in the Stern–Gerlach experiment. We can interpret the spin value $\pm\frac{\hbar}{2}$ as the component of the intrinsic angular momentum $\boldsymbol{\Omega} = \hbar$ aligned with the external flux density field $\boldsymbol{B}_E$. In this case, the angle between the $\boldsymbol{B}_E$ vector and the angular momentum have only two possible values, namely $\frac{\pi}{3}$ and $\frac{2\pi}{3}$, while the electron is subjected to a Larmor precession with angular frequency $\omega_p = \frac{d\vartheta_p}{dt}$. The Larmor precession is generated by the mechanical momentum

$$\tau = |\boldsymbol{\mu}_B \times \boldsymbol{B}_E| = B_E\mu_B \sin\left(\frac{\pi}{3}\right) \tag{3.3.11}$$

But

$$d\Omega = \Omega_\perp d\vartheta_p = \Omega \sin\left(\frac{\pi}{3}\right) d\vartheta_p$$

where $\Omega_\perp$ is the component of the intrinsic angular momentum orthogonal to $\boldsymbol{B}_E$ and, therefore, it is possible to write

$$\tau = \frac{d\Omega}{dt} = \Omega \sin\left(\frac{\pi}{3}\right) \frac{d\vartheta_p}{dt} \tag{3.3.12}$$

By equating (3.3.11) and (3.3.12), we get

$$B_E\mu_B = \Omega\omega_p$$

from which it is possible to determine the precession angular frequency

$$\omega_p = \frac{B_E\mu_B}{\Omega} = \frac{B_E\mu_B}{\hbar} \tag{3.3.13}$$

3.3.3. Value of the vector potential, cyclotron resonance, and flux density field

The pure electromagnetic momentum eA of the spinning charge of an electron at rest can be seen as if it were the momentum of a particle of mass m_e and speed c in classical Newtonian mechanics. Considering ω_e as the cyclotron angular frequency (which is coincident with the Zitterbewegung angular speed) given by the flux density field B generated by the current I_e

$$\omega_e = \frac{eB}{m_e} = \frac{eBc^2}{\hbar\omega_e}$$

it is possible to deduce the magnetic flux density produced by the electron

$$B_e = \frac{\hbar\omega_e^2}{ec^2} \approx 4.414\,004 \cdot 10^9 \text{ V} \cdot \text{s} \cdot \text{m}^{-2} \tag{3.3.14}$$

This very high flux density value (also known as *Landau critical value*) seems to be related to the physics of neutron stars and pulsars [16–18] or to that of superconductivity [19, 20].

It is also possible to calculate the flux density at the center of the electron orbit by the following expression derived from the Biot–Savart law

$$B_o = \frac{\mu_0}{2} \cdot \frac{I_e}{r_e} \approx 32.210\,548 \cdot 10^6 \text{ V} \cdot \text{s} \cdot \text{m}^{-2} \tag{3.3.15}$$

Considering that

$$dA = Ad\vartheta \Rightarrow \frac{dA}{dt} = A\omega_e \tag{3.3.16}$$

where $d\vartheta = \omega_e dt$ is the differential of the Zitterbewegung phase, and considering that the magnetic force F_B must be equal to the time derivative of the canonical momentum, it is possible to write

$$F_B = B_e ec = e\frac{dA}{dt} = eA\frac{d\vartheta}{dt} = eA\omega_e \approx 0.212\,014 \text{ N} \tag{3.3.17}$$

Finally, by manipulating the previous equation, it is possible to recompute by another method the norm of the vector potential

$$A = \frac{B_e c}{\omega_e} = \frac{\hbar\omega_e}{ec} = \frac{\hbar}{er_e}$$

3.3.4. *Value of magnetic and electrostatic energy, magnetic flux quantization, and radius of the elementary charge*

Once we obtain the expression for the vector potential, it is possible to determine the magnetic flux produced by the rotating elementary charge by applying the circulation of the vector potential A:

$$\phi_e = \oint_{\lambda_c} A d\lambda = \int_0^{2\pi} \frac{\hbar}{er_e} r_e d\vartheta = 2\pi\frac{\hbar}{e} = \frac{h}{e} \approx 4.135\,667 \cdot 10^{-15}\ \text{V} \cdot \text{s} \tag{3.3.18}$$

i.e., the magnetic flux crossing the surface described by the charge trajectory is quantized (*flux quantum*). This flux quantum, though different from the value given in CODATA 2014, is compatible with $\frac{h}{2e}$ for the same reasons explained in Section 3.3.2, with reference to $\hbar$. Now it is possible to calculate the magnetic energy stored in the field produced by the spinning charge

$$W_m = \frac{1}{2}\phi_e I_e = \frac{1}{2} \cdot 2\pi\frac{\hbar}{e} \cdot \frac{ec}{2\pi r_e} = \frac{\hbar c}{2r_e} \approx 4.093\,553 \cdot 10^{-14}\ \text{J} \tag{3.3.19}$$

which is equal to half the electron rest energy W_{tot} as can be seen from (3.3.4). The other half part can be attributed to electrostatic energy, i.e.,

$$W_{\text{tot}} - W_m = W_e = \iiint_V w_e \, dV \tag{3.3.20}$$

where w_e is the electrostatic energy per unit of volume and V is the volume in which the whole energy W_e is stored, and whose expression is given by

$$w_e = \frac{1}{2}\epsilon_0 E^2 = \frac{\epsilon_0}{2}\left(\frac{1}{4\pi\epsilon_0} \cdot \frac{e}{r^2}\right)^2 = \frac{1}{32\pi^2\epsilon_0} \cdot \frac{e^2}{r^4} \tag{3.3.21}$$

By expanding and integrating (3.3.20), with $dV = 4\pi r^2 dr$ (the generic elementary volume of a spherical thin shell centered in the middle of the electron trajectory) we obtain

$$W_e = \frac{e^2}{32\pi^2\epsilon_0}\int_{r_0}^{\infty} \frac{1}{r^4} \cdot 4\pi r^2 dr = \frac{e^2}{8\pi\epsilon_0}\int_{r_0}^{\infty} \frac{1}{r^2} dr = -\frac{e^2}{8\pi\epsilon_0} \frac{1}{r}\bigg|_{r_0}^{\infty} = \frac{e^2}{8\pi\epsilon_0 r_0} \tag{3.3.22}$$

Now, by taking into account that $W_e = W_m$, we get the radius

$$r_0 = \frac{e^2}{8\pi\epsilon_0 W_e} = \frac{e^2}{8\pi\epsilon_0 W_m} = \frac{e^2}{8\pi\epsilon_0} \cdot \frac{2r_e}{\hbar c} = \frac{e^2 r_e}{4\pi\epsilon_0 \hbar c} \approx 2.817\,940 \cdot 10^{-15}\ \text{m} \tag{3.3.23}$$

whose value is coincident with the classical electron radius [21]. The above equation states that the rotating charge must have a finite dimension, in particular it may be visualized as a sphere with charge equal to e uniformly distributed over its surface. The charge cannot be concentrated in a point in order to exhibit a finite electrostatic energy. It is interesting and at the same time very important to note that the ratio $\frac{r_e}{r_0}$ is exactly equal to the inverse of the FSC, i.e.,

$$\frac{r_e}{r_0} = \frac{4\pi\epsilon_0\hbar c}{e^2 r_e} r_e = \frac{4\pi\epsilon_0\hbar c}{e^2} = \alpha^{-1} \approx 137.035\,999 \tag{3.3.24}$$

Equations (3.3.19) and (3.3.22) describe the magnetic and electric energy calculation of a static scenario, considering a continuous current loop and a stationary charge surface. At first glance, neither of these conditions holds: the electron charge is not stationary and the current loop is not continuous. One may ask then: why do these static calculations work, i.e., why do they yield the total electron energy? The answer to this question cannot be found from Maxwell's equation alone. At the Compton-scale, one must consider that the electric charge and magnetic flux are quantized. The physical effects which cause this quantization are effectively creating a static scenario.

A well-known analogy is the electron orbital around a nucleus, which is given by an eigenstate solution of the Dirac equation. Without the physical effects which cause orbital quantization, the electron would just keep spiraling into the nucleus. However, quantization effects drive the electron into a static scenario, which is manifested as an eigenstate solution. Similarly, the effects which cause electric charge quantization allow us to calculate the electric energy on the basis of a stationary charge surface, and the effects which cause magnetic flux quantization allow us to calculate the magnetic energy on the basis of a static current loop. The appropriateness of this electron orbital analogy is indicated by the $\frac{r_{Bohr}}{r_e}$ ratio, which is α^{-1}; exactly the same value as the $\frac{r_e}{r_0}$ ratio. A detailed analysis of the elementary charge and flux quantization process will be given in Chapters 4 and 6.

The expression of the ratio $\frac{\phi_e}{T}$ has in SI the dimension of a voltage:

$$V_e = \frac{\phi_e}{T} = \frac{h}{e} \cdot \frac{c}{2\pi r_e} = \frac{\hbar c}{e r_e} \approx 5.109\,989 \cdot 10^5 \text{ V} \tag{3.3.25}$$

where T is defined by (3.3.3). Now, dividing the above voltage by the current generated by the rotating charge expressed by means of (3.3.6), we find the von Klitzing constant or *quantum of resistivity*, related to the quantum

Hall effect

$$R_K = \frac{V_e}{I_e} = \frac{h}{e} \cdot \frac{c}{2\pi r_e} \cdot \frac{2\pi r_e}{ec} = \frac{h}{e^2} = \frac{2\pi\hbar}{e^2} \approx 25812.807\ \Omega \tag{3.3.26}$$

An alternative expression of the von Klitzing constant can be derived from the electrostatic potential φ_e and the current I_e

$$R_K = \frac{\varphi_e}{I_e} = \frac{1}{4\pi\epsilon_0} \cdot \frac{e}{r_0} \cdot \frac{2\pi r_e}{ec} = \frac{1}{2c\epsilon_0} \cdot \frac{r_e}{r_0} = \frac{\mu_0 c}{2\alpha} \approx 25812.807\ \Omega \tag{3.3.27}$$

Finally, it is possible to deduce the values of two interesting electrical parameters, namely the inductance L_e, the capacitance C_e of the electron, and the frequency f_e. In fact

$$L_e = \frac{\phi_e}{I_e} = 4\pi^2 \frac{\hbar r_e}{e^2 c} \approx 2.089\,108 \cdot 10^{-16}\ \Omega \cdot \mathrm{s} \tag{3.3.28}$$

$$C_e = \frac{e}{\varphi_e} = 4\pi\epsilon_0 r_0 \approx 3.135\,381 \cdot 10^{-25}\ \mathrm{F} \tag{3.3.29}$$

and

$$f_e = \frac{1}{\sqrt{L_e C_e}} \approx 1.235\,590 \cdot 10^{20}\ \mathrm{Hz} \tag{3.3.30}$$

3.3.5. *Electron kinematics*

The finite dimension of the elementary charge imposes the constraint that all points of the surface of the spinning charged sphere must have the same instantaneous speed of light c (see Equation (3.2.16)) and the same angular speed. In a frame rotating with the Zitterbewegung frequency, the spinning charged sphere rotates around its center with opposite speed with respect to the Zitterbewegung angular frequency:

$$\omega_0 = -\omega_e \tag{3.3.31}$$

The new (not point-like) electron model and the speed diagrams are shown in Figure 3.1(a). Here the charge rotates with angular speed ω_0 around the axis passing through the center of the sphere and, therefore, all points of the sphere have the same absolute speed c.

During the revolution around the origin C, the charge describes a torus whose cross-section is equal to πr_0^2 and having a volume equal to $2\pi^2 r_e r_0^2$. In Figure 3.1(b), the elementary charge is represented as a charged sphere.

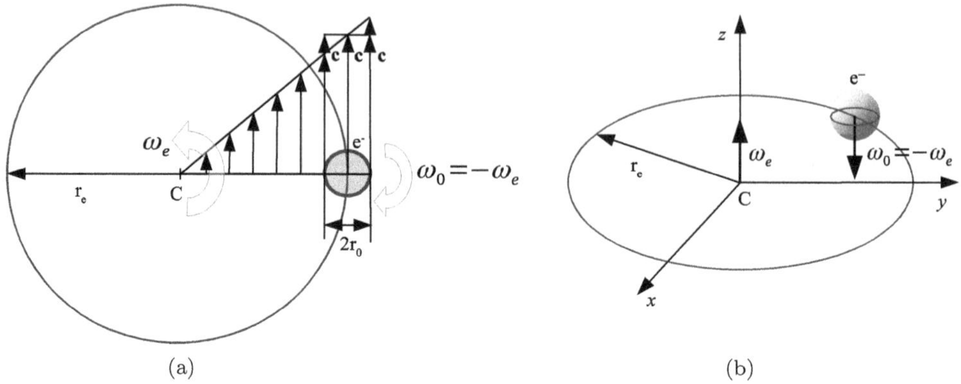

Figure 3.1. (a) Zitterbewegung model and speed diagrams of the electron charge (e^-). All points of the sphere have an absolute speed equal to c. (b) 3D representation. The charged sphere is rotating with the relative angular speed $\omega_0 = -\omega_e$ on the trajectory having radius r_e around the vertical axis passing through the center of the sphere.

3.3.6. *Electron and electromagnetic Lagrangian density*

From (3.3.9), (3.3.6) and (3.3.23), with the hypothesis that the electron is characterized by a uniform current density, we get the first term of the interaction part of the Lagrangian density

$$L_{\text{int1}} = \boldsymbol{J}_\Delta \cdot \boldsymbol{A}_\Delta = JA = \frac{I_e}{\pi r_0^2} \cdot \frac{m_e c}{e} \approx 1.352\,604 \cdot 10^{27}\ \text{J} \cdot \text{m}^{-3} \qquad (3.3.32)$$

By integration over the volume described by the electron toroidal trajectory, it is possible to recompute its rest energy:

$$\begin{aligned} W_{\text{tot}} &= \iiint_V JA\,dV = \frac{I_e}{\pi r_0^2} \cdot \frac{m_e c}{e} \cdot 2\pi^2 r_e r_0^2 \approx 8.187\,106 \cdot 10^{-14}\ J \\ &= 510.998\,946\ \text{keV} \end{aligned} \qquad (3.3.33)$$

which gives the same result calculated by means of (3.3.20). The same results are obtained, apart from a sign -, by the following relations

$$\begin{aligned} L_{\text{int2}} &= -\rho\varphi_e = \frac{e}{2\pi^2 r_e r_0^2} \cdot \frac{e}{4\pi\epsilon_0 r_e} = -\frac{e^2}{8\pi^3 \epsilon_0 (r_e r_0)^2} \\ &\approx -1.352\,604 \cdot 10^{27}\ \text{J} \cdot \text{m}^{-3} \end{aligned} \qquad (3.3.34)$$

$$\begin{aligned} W_{\text{tot}} &= \iiint_V |-\rho\varphi_e|\,dV = \frac{e^2 \cdot 2\pi^2 r_e r_0^2}{8\pi^3 \epsilon_0 (r_e r_0)^2} = \frac{e^2}{4\pi\epsilon_0 r_e} \approx 8.187\,106 \cdot 10^{-14}\ J \\ &= 510.998\,946\ \text{keV} \end{aligned} \qquad (3.3.35)$$

Table 3.2. Parameters of the Zitterbewegung model.

Item	Symbol	Value	Unit
Charge	e	$1.602\,176\,565 \cdot 10^{-19}$	$\mathrm{C = A \cdot s}$
Zitterbewegung orbit radius	$r_e = \frac{\lambda_c}{2\pi}$	$3.861\,593 \cdot 10^{-13}$	m
Intrinsic angular momentum	$\Omega = \hbar = \frac{h}{2\pi}$	$1.054\,571\,726 \cdot 10^{-34}$	$\mathrm{J \cdot s}$
Spin*	$\frac{\hbar}{2}$	$0.527\,285\,863 \cdot 10^{-34}$	$\mathrm{J \cdot s}$
Angular speed	ω_e	$7.763\,440 \cdot 10^{20}$	$\mathrm{rad \cdot s^{-1}}$
Mass	m_e	$9.109\,384 \cdot 10^{-31}$	kg
Current	I_e	19.796 331	A
Magn. moment (Bohr magn.)	μ_B	$9.274\,010 \cdot 10^{-24}$	$\mathrm{A \cdot m^2}$
Vector potential	A	$1.704\,509 \cdot 10^{-3}$	$\mathrm{V \cdot s \cdot m^{-1}}$
Magnetic flux density	B_e	$4.414\,004 \cdot 10^{9}$	$\mathrm{V \cdot s \cdot m^{-2}}$
Magnetic flux	$\phi_e = \frac{h}{e}$	$4.135\,667 \cdot 10^{-15}$	$\mathrm{V \cdot s}$
Magnetic energy	W_m	$4.093\,553 \cdot 10^{-14}$	J
Electrostatic energy	W_e	$4.093\,553 \cdot 10^{-14}$	J
Electron energy at rest	$W_{\text{tot}} = m_e c^2$	$8.187\,106 \cdot 10^{-14}$	J
Charge radius	r_0	$2.817\,940 \cdot 10^{-15}$	m
Inverse of the FSC	$\alpha^{-1} = \frac{r_e}{r_0}$	137.035 999	1
Von Klitzing constant	$R_K = \frac{h}{e^2} = \frac{\mu_0 c}{2\alpha}$	25812.807	Ω
Inductance	$L_e = \frac{4\pi^2 \hbar r_e}{e^2 c}$	$2.089\,108 \cdot 10^{-16}$	$\Omega \cdot \mathrm{s}$
Capacitance	$C_e = 4\pi\epsilon_0 r_0$	$3.135\,381 \cdot 10^{-25}$	F
1st part of L_{int}	JA	$1.352\,604 \cdot 10^{27}$	$\mathrm{J \cdot m^{-3}}$
Electron energy at rest	$\iiint_V JA\, dV$	510.998 946	keV
Electron energy at rest	$\iiint_V \rho\varphi_e\, dV$	510.998 946	keV
Memristance	$\frac{\phi_e}{e} = \frac{h}{e^2}$	25812.807	Ω

Note: The spin is the component of the angular momentum along the external magnetic field B_E (see (3.3.11)).

All parameters that can be deduced by the application of the present Zitterbewegung model are resumed in Table 3.2, where the first three rows refer to the model's input parameters.

3.3.7. *Zitterbewegung and a simple derivation of the relativistic mass*

With the Zitterbewegung model, it is possible to show a simple, original, and intuitive explanation of the relativistic mass concept. For an electron moving at constant speed v_z along the z axis orthogonal to the rotation plane, calling

$v_\perp$ the component of the velocity of the rotating charge in the $\gamma_x\gamma_y$ plane we can find the value of the Zitterbewegung radius r of the moving electron. In fact, assuming a constant value of ω_e, we have

$$v_z^2 + v_\perp^2 = c^2 \tag{3.3.36}$$

that can be written as

$$v_z^2 + \omega_e^2 r^2 = \omega_e^2 r_e^2 = c^2$$

Therefore,

$$\frac{r^2}{r_e^2} = 1 - \frac{v_z^2}{c^2}$$

or

$$r = r_e\sqrt{1 - \frac{v_z^2}{c^2}} \tag{3.3.37}$$

Finally, by considering that the mass is inversely proportional to r, it is possible to write the relativistic expression of the mass as

$$m = \frac{m_e}{\sqrt{1 - \frac{v_z^2}{c^2}}} \tag{3.3.38}$$

where m_e is the electron mass at rest. Figure 3.2 represents the Zitterbewegung trajectory of the spinning charge of an electron subjected to

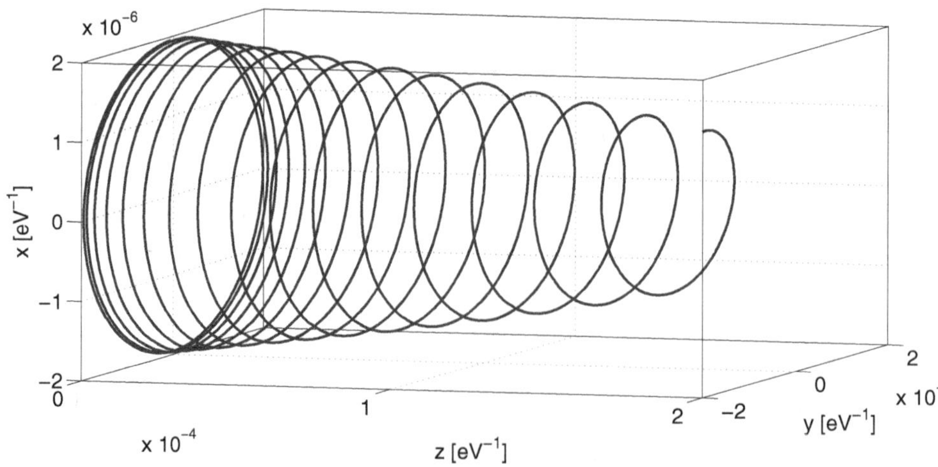

Figure 3.2. Zitterbewegung trajectory during an acceleration of the electron in the z direction.

an acceleration directed along the positive z axis. Due to the acceleration, the radius reduces itself according to (3.3.37).

3.4. Electromagnetism, Mechanics, and Lorentz Force

The "pure electromagnetic" vector $e\boldsymbol{A}_\square$ may be interpreted as the momentum–energy $\boldsymbol{P}_\square$ of a particle with electric charge e, momentum $\boldsymbol{P}_\triangle$, and energy $U = P_t c$:

$$\boldsymbol{P}_\square = e\boldsymbol{A}_\square \tag{3.4.1}$$

$$\boldsymbol{P}_\square = \gamma_x P_x + \gamma_y P_y + \gamma_z P_z + \gamma_t \tfrac{U}{c} = \boldsymbol{P}_\triangle + \gamma_t \tfrac{U}{c} \tag{3.4.2}$$

For a particle that moves with speed $\boldsymbol{v}$ along a direction z orthogonal to the Zitterbewegung rotation plane, the momentum $\boldsymbol{P}_\triangle$ can be decomposed in two vectors, one parallel and one orthogonal to $\boldsymbol{v}$. The orthogonal component is a rotating vector, that indicates the component of the momentum due to the angular frequency ω_e in the spatial plane xy orthogonal to z

$$\boldsymbol{P}_\triangle = \boldsymbol{P}_\parallel + \boldsymbol{P}_\perp \tag{3.4.3}$$

where

$$|\boldsymbol{P}_\perp| = \frac{\hbar\omega_e}{c} = m_e\omega_e r_e = m_e c$$

$$\left[|\boldsymbol{P}_\perp| = \frac{1}{r_e} = \omega_e = m_e\right]_{NU}$$

The $\boldsymbol{P}_\parallel$ component can be seen as the usual three-components momentum of a particle with mass at rest m_e. For simplicity of notation, from now, we will call $\boldsymbol{P}$ this component, so that

$$P_\square^2 = e^2 A_\square^2 = P_\triangle^2 - \frac{U^2}{c^2} = P^2 + m_e^2 c^2 - \frac{U^2}{c^2}$$

The relativistic mass m can be derived directly by the application of the Pythagorean theorem

$$m^2c^2 = m_e^2c^2 + P^2 = P_\perp^2 + P^2 = m_e^2c^2 + m^2v^2$$

Consequently, this electromagnetic four-momentum $\boldsymbol{P}_\square$, for electrons moving with uniform velocity, is a light-like vector:

$$P_\square^2 = m^2c^2 - \frac{U^2}{c^2} = 0$$

An electron that moves with velocity $v \ll c$ has an approximate momentum P given by

$$P = eA_{\|} \simeq P_{\perp}\frac{v}{c} = m_e v$$

and a variation of speed $a = \frac{dv}{dt}$ implies a force $f = \frac{dP}{dt} = e \cdot \frac{dA_{\|}}{dt} = m \cdot \frac{dv}{dt}$.

Now recalling that the bivector part of (3.2.5) is

$$\boldsymbol{\partial} \wedge \boldsymbol{A}_{\square} = \boldsymbol{F}$$

after multiplying both sides by the charge e, it becomes

$$e\boldsymbol{\partial} \wedge \boldsymbol{A}_{\square} = \boldsymbol{\partial} \wedge e\boldsymbol{A}_{\square} = e\boldsymbol{F} \tag{3.4.4}$$

By considering (3.4.1) and (3.4.2), this equation can be rewritten as

$$\boldsymbol{\partial} \wedge \left(\boldsymbol{P}_{\triangle} + \gamma_t \frac{U}{c}\right) = e\boldsymbol{F} \tag{3.4.5}$$

or, by means of (3.4.3), as

$$\boldsymbol{\partial} \wedge \left[(\boldsymbol{P} + \boldsymbol{P}_{\perp}) + \gamma_t \frac{U}{c}\right] = e\boldsymbol{F} \tag{3.4.6}$$

The term $\boldsymbol{\partial} \wedge \boldsymbol{P}_{\perp}$ can be carried out because the average value of $\boldsymbol{P}_{\perp}$, in a scale time much larger than the Zitterbewegung period, is zero:

$$\boldsymbol{\partial} \wedge \left(\boldsymbol{P} + \gamma_t \frac{U}{c}\right) = e\boldsymbol{F} \tag{3.4.7}$$

$$\boldsymbol{\partial} \wedge \left(\boldsymbol{P} + \gamma_t \frac{U}{c}\right) = \frac{e}{c}\boldsymbol{E}\gamma_t + eI\boldsymbol{B}\gamma_t$$

Equating only the components that contain bivectors with γ_t terms, we obtain

$$\left(\frac{\partial \boldsymbol{P}}{\partial t}\right)_{EU} \gamma_t + \nabla U \gamma_t = e\boldsymbol{E}\gamma_t$$

or

$$\left(\frac{\partial \boldsymbol{P}}{\partial t}\right)_{EU} = e\boldsymbol{E} - \nabla U \tag{3.4.8}$$

In (3.4.8), $\left(\frac{\partial \boldsymbol{P}}{\partial t}\right)_{EU}$ is the force acting on the charge e due both to the electric field $\boldsymbol{E}$ (Coulomb force) and to the gradient of the "potential energy" U. Instead, by equating only the components that contain pure spatial bivectors, we get

$$\nabla \wedge \boldsymbol{P} = eI\boldsymbol{B}\gamma_t = -eI\gamma_t\boldsymbol{B} = eI_\Delta \boldsymbol{B} \tag{3.4.9}$$

where the term $-I\gamma_t = \gamma_x\gamma_y\gamma_z = I_\Delta$ is the unitary volume of the three-dimensional space. Left-multiplying both sides of (3.4.9) by I_Δ gives

$$I_\Delta \nabla \wedge \boldsymbol{P} = -e\boldsymbol{B} \tag{3.4.10}$$

which is equivalent to the two following equations in ordinary algebra:

$$\nabla \times \boldsymbol{P} = e\boldsymbol{B} \tag{3.4.11}$$

As an example, the component of the above equation along the x axis is

$$\gamma_x \left(\frac{\partial P_z}{\partial y} - \frac{\partial P_y}{\partial z} \right) = \gamma_x e B_x$$

Now, by applying the cross-product of the velocity $\boldsymbol{v}$ of charge e to both terms in (3.4.11), we obtain

$$\boldsymbol{v} \times (\nabla \times \boldsymbol{P} - e\boldsymbol{B}) = 0 \tag{3.4.12}$$

The components of (3.4.12) are represented in Table 3.3 considering that

$$v_i \frac{\partial P_j}{\partial i} = \frac{\partial i}{\partial t}\frac{\partial P_j}{\partial i} = \frac{\partial P_j}{\partial t}$$

$$v_j \frac{\partial P_j}{\partial i} = \frac{\partial j}{\partial t}\frac{\partial P_j}{\partial i} = \frac{\partial j}{\partial i}\frac{\partial P_j}{\partial t} = 0 \quad \text{for } i \neq j$$

where $i, j \in \{x, y, z\}$.

For these reasons, (3.4.12) leads to the usual form of the force contribution due to the magnetic flux density field $\boldsymbol{B}$

$$\left(\frac{\partial \boldsymbol{P}}{\partial t} \right)_B = e\boldsymbol{v} \times \boldsymbol{B} \tag{3.4.13}$$

Finally, we get the whole force contribution by summing up the forces $\frac{d\boldsymbol{P}}{dt} = \left(\frac{\partial \boldsymbol{P}}{\partial t}\right)_{EU} + \left(\frac{\partial \boldsymbol{P}}{\partial t}\right)_B$ given, respectively, by (3.4.8) and (3.4.13)

$$\frac{d\boldsymbol{P}}{dt} = e\left(\boldsymbol{E} + \boldsymbol{v} \times \boldsymbol{B}\right) - \nabla U \tag{3.4.14}$$

Table 3.3. Products $\boldsymbol{v} \times (\nabla \times \boldsymbol{P} - e\boldsymbol{B})$.

$\boldsymbol{v} \times (\nabla \times \boldsymbol{P} - e\boldsymbol{B})$	$\gamma_x \left(\frac{\partial P_z}{\partial y} - \frac{\partial P_y}{\partial z} - eB_x\right)$	$\gamma_y \left(\frac{\partial P_x}{\partial z} - \frac{\partial P_z}{\partial x} - eB_y\right)$	$\gamma_z \left(\frac{\partial P_y}{\partial x} - \frac{\partial P_x}{\partial y} - eB_z\right)$
$\gamma_x v_x$	0	$\gamma_z \left(- \frac{\partial P_z}{\partial t}\Big\vert_{xy} - ev_x B_y\right)$	$\gamma_y \left(\frac{\partial P_y}{\partial t}\Big\vert_{xz} - ev_x B_z\right)$
$\gamma_y v_y$	$-\gamma_z \left(\frac{\partial P_z}{\partial t}\Big\vert_{yx} - ev_y B_x\right)$	0	$\gamma_x \left(- \frac{\partial P_x}{\partial t}\Big\vert_{yz} - ev_y B_z\right)$
$\gamma_z v_z$	$-\gamma_y \left(- \frac{\partial P_y}{\partial t}\Big\vert_{zx} - ev_z B_x\right)$	$-\gamma_x \left(\frac{\partial P_x}{\partial t}\Big\vert_{zy} - ev_z B_y\right)$	0

3.5. Energy, Momentum, and Quanta Current

In Chapter 2, we *considered the quantum of action $\hbar$ as a consequence of a "noise floor" in space–time*. The Planck relation $W_\varphi = \hbar\omega = \frac{2\pi\hbar}{T_\varphi}$ tells us that photons with energy W_φ can transmit quanta of actions with a quanta current Q_c (number of quanta of actions per time unit)

$$Q_c = n \cdot \frac{\hbar}{T_\varphi} = n \cdot \frac{W_\varphi}{2\pi} \tag{3.5.1}$$

The same quanta current can be obtained in a time unit by a large number of low-energy photons or by a small number of high-energy photons.

Calling W_H the energy of high-energy photons and W_L the energy of low-energy ones, and n_H and n_L their number, we observe that the same Q_c can be obtained with the same total energy following two different ways if $n_H W_H = n_L W_L$:

$$\frac{n\hbar}{T_\varphi} = \frac{n_H W_H}{2\pi} \tag{3.5.2}$$

$$\frac{n\hbar}{T_\varphi} = \frac{n_L W_L}{2\pi} \tag{3.5.3}$$

In the first case, we have few high-energy photons that guarantee the required current. In the latter case, the same result is obtained by many low-energy photons. The information per unit time is the same in both cases, but the entropy is different. Now, following the above considerations the equation $P_\varphi = \hbar k = \frac{2\pi\hbar}{\lambda_\varphi}$, where $k = \frac{2\pi}{\lambda_\varphi}$ is the wave number, should be viewed as the direction of quanta current in space. For photons, the four-momentum $\boldsymbol{P}_\square$ is a light-like vector

$$P_\square^2 = 0 \tag{3.5.4}$$

In this case, the momentum is a vector that gives the direction of quantum of action in space and the norm of momentum P_φ is the energy W_φ:

$$\left[P_\varphi^2 - W_\varphi^2 = 0\right]_{NU}$$

3.5.1. *Zitterbewegung and Heisenberg's uncertainty principle*

In natural units, the quantum of action is a dimensionless (i.e., scalar) value, as it is always the ratio (measure) of two values of the same nature. For this reason, the concepts of "energy," "momentum," "space," and "time" cannot be separated and space and time can be measured, using natural units, in eV^{-1}.

The product of the momentum P of the rotating charge and the radius r of the orbit in the proposed Zitterbewegung model is always equal to $\hbar$:

$$Pr = \hbar \tag{3.5.5}$$

This expressions points out that the angular momentum remains $\hbar$ under any Lorentz transformation.

3.6. Electromagnetic Composite at Compton Scale

If the electron is a current loop whose radius is equal to the reduced electron Compton wavelength, it is reasonable to assume the possibility of existence of "super chemical" structures of pico-metric (1 pm $= 10^{-12}$ m) dimensions. These dimensions are intermediate between nuclear (1 fm $= 10^{-15}$ m) and atomic scale (1 Å $= 10^{-10}$ m).

A simple Zitterbewegung model of the proton consists in a current loop generated by an elementary positive charge that rotates at the speed of light along a circumference with a length equal to the proton Compton wavelength ($\approx 1.32141 \cdot 10^{-15}$ m) [22]. According to this model, the proton is much smaller than the electron ($\frac{r_e}{r_p} = \frac{m_p}{m_e} \approx 1836.153$).

The hypothesis of existence of Compton-scale composites (CSCs) has been experimentally confirmed by Holmlid [2, 3, 23]. The inter-nuclear distance in ultra-dense deuterium (UDD) of ≈2.3 pm, found by Holmlid [2], seems compatible with deuteron–electron (or proton–electron in ultra-dense hydrogen (UDH)) structures where the Zitterbewegung phases of adjacent electrons are correlated. Such distance may be obtained imposing, as a first step, the condition that the space–time distance $d_\square$ between adjacent electrons rotating charges is a light-like vector:

$$d_\square^2 = d_\triangle^2 - c^2\delta t^2 = 0$$
$$\left[d_\square^2 = d_\triangle^2 - \delta t^2 = 0\right]_{NU}$$

where $d_\triangle$ is the ordinary euclidean distance in space. This condition is satisfied if $d_\triangle$ is equal to electron Compton wavelength ($d_\triangle = \lambda_c$), $\delta t = T$ is the Zitterbewegung period, and the phase difference between adjacent electrons is equal to π because of Coulomb repulsion. Considering the Compton wavelength distance between neighboring electrons, by a direct application of the Pythagorean theorem we can find the internuclear deuteron distance d_i as shown in Figure 3.3.

$$d_i = \frac{\lambda_c}{\pi}\sqrt{\pi^2 - 1} \simeq 2.3 \cdot 10^{-12}\ \text{m} \tag{3.6.1}$$

Figure 3.4 shows a hypothetical chain of these deuteron–electron pairs. We must remark that the hypothesis of existence of exotic forms of hydrogen

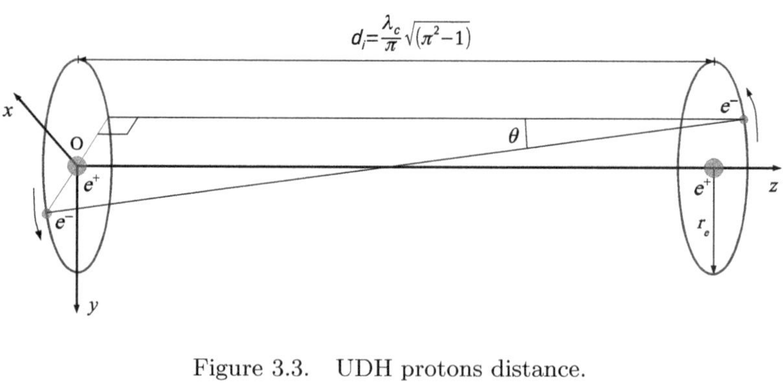

Figure 3.3. UDH protons distance.

Figure 3.4. UDH model.

is not new and has been proposed in different ways by many authors ([8, 24], [25] and many others).

Indirect support for these hypotheses comes also from the numerous claims of observation of anomalous heat generation in metal-hydrogen systems. We must remark that these hypothetical "CSCs" should be electrical neutral or negatively charged objects that cannot be stopped by the Coulomb barrier. For this reason, they may generate unusual nuclear reactions and transmutations, considering the different sizes, time-scales, and energies of this composites with respect to the particles (such as neutrons) normally used in nuclear experiments.

The terms "CSC," "UDH," and "UDD" all refer to the same experimentally observed state of hydrogen isotopes, characterized by Compton-scale radius. We prefer the CSC terminology over UDH/UDD because it expresses that the electron and proton form such a composite structure which is qualitatively different from an ordinary electron orbital. We will return to the physics of CSCs in the following chapters.

3.7. Some Other Spinning Charge Models

In 1915, Alfred Lauck Parson published *A Magneton Theory of the Structure of the Atom* [28], where he proposed a spinning ring model of the electron. Various forms of the spinning charge model of electrons have been rediscovered by many authors. However, the incompatibility with the most widely accepted interpretations of quantum mechanics prevented them from receiving proper attention.

According to Randell L. Mills, the free electron *is a spinning two-dimensional disk of charge. The mass and current density increase towards the center, but the angular velocity is constant. It produces an angular momentum vector perpendicular to the plane of the disk* [29]. As in our proposed model, the intrinsic angular momentum of a free electron is $\hbar$, but there is an important difference in charge distribution shape and speed. A constant angular velocity for a flat charge distribution implies that the charge speed is not always equal to the speed of light [30], as strictly demanded by our model. Mills's theory [31] *assumes physical laws apply on all scales including the atomic scale*, in agreement with Occam's razor principle.

Using geometric algebra and starting from Dirac theory, David Hestenes has proposed a Zitterbewegung model according to which *the electron is a massless point particle executing circular motion in the rest system* and *with an intrinsic orbital angular momentum or spin of fixed magnitude*

$s = \frac{\hbar}{2}$ [1]. The phase of the probability amplitude wave function is related to the Zitterbewegung rotation phase, a concept usually hidden in the traditional mathematical formalism used in quantum mechanics based on complex numbers and matrices. However, we should remark that the concept of point-like charges in quantum mechanics should be considered unrealistic. It violates Occam's razor principle and may be used only as a first approximation. We note also that in our model the value of intrinsic angular momentum for a free electron is $\hbar$ and that the "spin" is interpreted as the component of the angular momentum along an external magnetic field, as in the Stern–Gerlach experiment. Another interesting electron model has been proposed by David L. Bergman [13, 14]. According to this model, the electron is a very thin, torus-shaped, rotating charge distribution with intrinsic angular momentum of the electron equal to its spin value $s = \frac{\hbar}{2}$. The torus radius has a length $R = \frac{\hbar}{mc}$ and half thickness $r = 8\ \mathrm{R}e^{-\frac{\pi}{\alpha}}$, where α is the fine structure constant.

3.8. Conclusions

We wish to underline that simplicity is an important and concrete value in scientific research. Connections between very different concepts in physics can be evidenced if we use the language of geometric algebra, recognizing also the fundamental role of the electromagnetic four-vector potential in physics.

The application of Occam's razor principle to Maxwell's equations suggests, as a natural choice, a Zitterbewegung interpretation of quantum mechanics, similar but not identical to the one proposed by D. Hestenes. According to this framework, the electron structure consists of a massless charge distribution that rotates at the speed of light along a circumference with a length equal to electron Compton wavelength. Following this interpretation the electron mass-energy, expressed in natural units, is equal to the angular speed of the Zitterbewegung rotation and to the inverse of the orbit radius. Inertia has a pure electromagnetic origin related to the vector potential generated by the Zitterbewegung current. The proposed model supports the ideas of some authors [3, 8] that the Zitterbewegung may explain the existence of "super-chemical structures," such as UDD, at pico-metric scale. A preliminary hypothesis on the structure of Holmlid's UDD, in which the Zitterbewegung phase of adjacent electrons are synchronized, has been presented demonstrating good agreement with Holmlid's experimental results.

It is a delusion to think of electrons and fields as two physically different, independent entities. Since neither can exist without the other, there is only one reality to be described, which happens to have two different aspects; and the theory ought to recognize this from the outset instead of doing things twice! — A. Einstein, cited in [26]

In atomic theory, we have fields and we have particles. The fields and the particles are not two different things. They are two ways of describing the same thing, two different points of view. — P. A. M. Dirac, cited in [27]

References

[1] D. Hestenes, Quantum mechanics from self-interaction. *Foundations of Physics*, **15**(1) (1985) 63–87.

[2] S. Badiei, P.U. Andersson, and L. Holmlid, High-energy Coulomb explosions in ultra-dense deuterium: Time-of-flight-mass spectrometry with variable energy and flight length. *International Journal of Mass Spectrometry*, **282**(1–2) (2009) 70–76.

[3] L. Holmlid and S. Olafsson, Spontaneous ejection of high-energy particles from ultra-dense deuterium D(0). *International Journal of Hydrogen Energy*, **40**(33) (2015) 10559–10567.

[4] K. Huang, On the zitterbewegung of the dirac electron. *American Journal of Physics*, **20**(8) (1952), 479–484.

[5] B.G. Sidharth, Revisiting zitterbewegung. *International Journal of Theoretical Physics*, **48** (2009) 497–506.

[6] D. Hestenes, *Mysteries and insights of dirac theory. In Annales de la Fondation Louis de Broglie*, volume 28, Fondation Louis de Broglie, 2003, p. 3.

[7] D. Hestenes, Hunting for snarks in quantum mechanics. In P.M. Goggans and C.-Y. Chan (eds.), *American Institute of Physics Conference Series*, volume 1193 of American Institute of Physics Conference Series, December 2009, pp. 115–131.

[8] F.J. Mayer and J.R. Reitz, Electromagnetic composites at the Compton scale. *International Journal of Theoretical Physics*, **51**(1) (2012) 322–330.

[9] Y. Aharonov and D. Bohm, Significance of electromagnetic potentials in the quantum theory. *Physical Review*, **115** (1959) 485–491.

[10] D. Hestenes, Oersted Medal Lecture 2002: Reforming the mathematical language of physics. *American Journal of Physics*, **71** (2003) 104–121.

[11] D. Hestenes, Spacetime physics with geometric algebra. *American Journal of Physics*, **71**(3) (2003) 691–714.

[12] D. Hestenes, Zitterbewegung modeling. *Foundations of Physics*, **23**(3) (1993) 365–387.

[13] D.L. Bergman and J.P. Wesley, Spinning charged ring model of electron yielding anomalous magnetic moment. *Galilean Electrodynamics*, **1**(5) (1990) 63–67.

[14] D. Bergman and C.W. Lucas Jr, Credibility of common sense science. *Foundations of Science*, **6**(2) (2003) 1–10.

[15] F. Santandrea and P. Cirilli, Unificazione elettromagnetica, concezione elettronica dello spazio, dell'energia e della materia (2006). Atlante di numeri e lettere, http://www.atlantedinumerielettere.it/energia2006/labor.htm

[16] R.A. Treumann, W. Baumjohann, and A. Balogh, The strongest magnetic fields in the universe: how strong can they become? *Frontiers in Physics*, **2**(59) (2014) 1–4.

[17] Q. Peng and H. Tong, The physics of strong magnetic fields in neutron stars. *Monthly Notices of the Royal Astronomical Society*, **378** (2007) 159–162.
[18] D. Dew-Hughes, The critical current of superconductors: An historical review. *Low Temperature Physics*, **27**(9) (2001) 713–722.
[19] J. Dolbeault, M.J. Esteban, and M. Loss, Relativistic hydrogenic atoms in strong magnetic fields. *Annales Henri Poincaré*, **8** (2007) 749–779.
[20] J. Dolbeault, M.J. Esteban, and M. Loss, Characterization of the critical magnetic field in the Dirac-Coulomb equation. (2007), arxiv e-prints.
[21] R.P. Feynman, R.B. Leighton, and M. Sands, The Feynman Lectures on Physics: Mainly electromagnetism and matter. Volume 2. Basic Books. Basic Books, 2011.
[22] D.L. Bergman, The real proton. *Foundations of Science*, **3**(4) (2004) 1–13.
[23] L. Holmlid, Excitation levels in ultra-dense hydrogen p(-1) and d(-1) clusters: Structure of spin-based Rydberg matter. *International Journal of Mass Spectrometry*, **352** (2013) 1–8.
[24] F.J. Mayer and J.R. Reitz, Thermal energy generation in the earth. *Nonlinear Processes in Geophysics*, **21**(2) (2014), 367–378. http://www.nonlin-processes-geophys.net/21/367/2014/.
[25] V. Krasnoholovets, Y. Zabulonov, and I. Zolkin, On the nuclear coupling of proton and electron. *Universal Journal of Physics and Application*, **10**(3) (2016) 90–103.
[26] C. Mead, *Collective Electrodynamics: Quantum Foundations of Electromagnetism.* The MIT Press, Cambridge, Massachusetts, London, England (2000).
[27] A. Babin and A. Figotin, Neoclassical theory of electromagnetic interactions: A single theory for macroscopic and microscopic scales. *Theoretical and Mathematical Physics.* Springer London, 2016.
[28] A.L. Parson, *A Magneton Theory of the Structure of the Atom.* Smithsonian Miscellaneous Collection, Pub 2371, 1915.
[29] http://www.millsian.com/images/theory/Periodic-Table-Poster-light.pdf.
[30] A. Rathke, A critical analysis of the hydrino model. *New Journal of Physics*, **7**(127) (2005) 1–9.
[31] R.L. Mills, J.J. Farrell, and W.R. Good, *Unification of Spacetime, the Forces, Matter, and Energy.* Science Press, Ephrata, PA 17522, 1992.

https://doi.org/10.1142/9781800611306_0004

Chapter 4

The Aharonov–Bohm Effect, Proca Fields, and Flux Quantization

Giorgio Vassallo* and Antonino Oscar Di Tommaso†
Engineering Department, University of Palermo
viale delle Scienze, 90128 Palermo, Italy
**giorgio.vassallo@unipa.it*
†antoninooscar.ditommaso@unipa.it

Nomenclature

Symbol, name, SI units, natural units (NU).

$\boldsymbol{A}_{\square}$: electromagnetic four-potential $(\mathrm{V \cdot s \cdot m^{-1}})$, (eV);
$\boldsymbol{A}_{\triangle}$: electromagnetic vector potential $(\mathrm{V \cdot s \cdot m^{-1}})$, (eV);
A: the norm of the electromagnetic vector potential $(\mathrm{V \cdot s \cdot m^{-1}})$, (eV);
A_t: time component of electromagnetic four potential $(\mathrm{V \cdot s \cdot m^{-1}})$, (eV);
$\boldsymbol{a}_{\square}$: the four-vector Proca field $(\mathrm{V \cdot s \cdot m^{-1}})$, (eV);
$\boldsymbol{a}_{\triangle}$: spatial components of the Proca field $(\mathrm{V \cdot s \cdot m^{-1}})$, (eV);
a: the norm of the Proca field, not to be confused with acceleration $(\mathrm{V \cdot s \cdot m^{-1}})$, (eV);
a_t: time component of the Proca field $(\mathrm{V \cdot s \cdot m^{-1}})$, (eV);
m: mass (kg), (eV);
$\boldsymbol{F}$: electromagnetic field bivector derived from $\boldsymbol{A}_{\square}$ $(\mathrm{V \cdot s \cdot m^{-2}})$, $(\mathrm{eV^2})$;
$\boldsymbol{f}$: field bivector derived from $\boldsymbol{a}_{\square}$ $(\mathrm{V \cdot s \cdot m^{-2}})$, $(\mathrm{eV^2})$;
$\boldsymbol{B}$: flux density field $(\mathrm{V \cdot s \cdot m^{-2}}) = (\mathrm{T})$, $(\mathrm{eV^2})$;
$\boldsymbol{E}$: electric field $(\mathrm{V \cdot m^{-1}})$, $(\mathrm{eV^2})$;
V: potential energy $(\mathrm{J = kg \cdot m^2 \cdot s^{-2}})$, (eV);
$\boldsymbol{J}_{\square}$: four-current density field, $(\mathrm{A \cdot m^{-2}})$, $(\mathrm{eV^3})$;

$\boldsymbol{J}_\triangle$: current density field, $(\mathrm{A}\cdot\mathrm{m}^{-2})$, (eV^3);
ρ: charge density $(\mathrm{A}\cdot\mathrm{s}\cdot\mathrm{m}^{-3} = \mathrm{C}\cdot\mathrm{m}^{-3})$, (eV^3);
x, y, z: space coordinates (m), (eV^{-1}), $(1.9732705\cdot 10^{-7}\ \mathrm{m} \simeq 1\ \mathrm{eV}^{-1})$;
t: time variable (s), (eV^{-1}), $(6.5821220\cdot 10^{-16}\ \mathrm{s} \simeq 1\ \mathrm{eV}^{-1})$;
c: light speed in vacuum $(2.997\,924\,58\cdot 10^8\ \mathrm{m}\cdot\mathrm{s}^{-1})$, (1);
$\hbar$: reduced Planck constant $(\hbar = \frac{h}{2\pi})$ $(1.054\,571\,726\cdot 10^{-34}\ \mathrm{J}\cdot\mathrm{s})$, (1);
μ_0: permeability of vacuum $(4\pi\cdot 10^{-7}\ \mathrm{V}\cdot\mathrm{s}\cdot\mathrm{A}^{-1}\cdot\mathrm{m}^{-1})$, (4π);
ϵ_0: dielectric constant of vacuum $(8.854\,187\,817\cdot 10^{-12}\ \mathrm{A}\cdot\mathrm{s}\cdot\mathrm{V}^{-1}\cdot\mathrm{m}^{-1})$, $(\frac{1}{4\pi})$;
e: electron charge $(1.602\,176\,565\cdot 10^{-19}\ \mathrm{A}\cdot\mathrm{s})$, $(0.085\,424\,546)$;
α: fine structure constant $(7.2973525664\cdot 10^{-3})$, $(7.2973525664\cdot 10^{-3})$;
m_e: electron rest mass $(9.10938356\cdot 10^{-31}\mathrm{kg})$, $(0.5109989461\cdot 10^6\mathrm{eV})$;
λ_c: electron Compton wavelength $(2.426\,310\,2389\cdot 10^{-12}\ \mathrm{m})$, $(1.229\,588\,259\cdot 10^{-5}\ \mathrm{eV}^{-1})$;
K_J: Josephson constant $(0.4835978525\cdot 10^{15}\ \mathrm{Hz\,V}^{-1})$, $(2.71914766\cdot 10^{-2})$
r_e: reduced Compton electron wavelength (Compton radius) $r_e = \frac{\lambda_c}{2\pi}$;
r_c: electron charge radius $r_c = \alpha r_e$;
T_e: Zitterbewegung period $T_e = \frac{2\pi r_e}{c}$;
ω_e: Zitterbewegung angular frequency $\omega_e = \frac{2\pi}{T_e}$;
γ_i: $\gamma_x^2 = \gamma_y^2 = \gamma_z^2 = -\gamma_t^2 = 1$, where $\{\gamma_x, \gamma_y, \gamma_z, \gamma_t\}$ are the four basis vectors of $Cl_{3,1}(\mathbb{R})$ Clifford algebra, isomorphic to Majorana matrices algebra;
$\boldsymbol{\partial}$: $\boldsymbol{\partial} = \gamma_x\frac{\partial}{\partial x} + \gamma_y\frac{\partial}{\partial y} + \gamma_z\frac{\partial}{\partial z} + \gamma_t\frac{1}{c}\frac{\partial}{\partial t}$;
I: $I = \gamma_x\gamma_y\gamma_z\gamma_t$;
$I_\triangle$: $I_\triangle = \gamma_x\gamma_y\gamma_z$.

4.1. Introduction

According to Carver Mead, mainstream physics literature has a long history of hindering fundamental conceptual reasoning, often *involving assumptions that are not clearly stated* [1]. One of these is the unrealistic assumption of point-like elementary particles with *intrinsic* properties as mass, charge, angular momentum, magnetic moment, and spin. According to the laws of mechanics and electromagnetism, a point-like particle cannot have an "intrinsic angular momentum." A magnetic moment must necessarily be generated by a current loop, which cannot exist in a point-like particle. Furthermore, the electric field generated by a point-like charged particle should have an infinite energy. Therefore, an alternative realistic approach that fully addresses these *very basic* problems is indispensable. We developed a *Zitterbewegung* solution of the electron structure in Chapter 3, according to which charged elementary particles can be modeled by a current ring generated by a massless charge distribution rotating at light speed along a circumference whose length is equal to the particle's Compton wavelength. As a consequence, every elementary charge is always associated with a magnetic flux quantum and every charge is coupled to all other charges on its light cone by time-symmetric interactions [1, 9]. The aim of this chapter is to further develop the electron Zitterbewegung model of Chapter 3.

The present chapter is structured in the following way. In Section 4.2, the deep connection between some basic concepts such as space, time, energy, mass, frequency, and information is exposed. Section 4.3 presents a geometric-electromagnetic interpretation of Proca fields and Aharonov–Bohm equations. In Section 4.4, the relation of electronic spin resonance (ESR) frequency with Larmor precession frequency of the Zitterbewegung orbit is presented. Finally, in Section 4.5, some hypotheses on the structure of ultra-dense hydrogen are formulated, whereas Section 4.6 deals with the possible role of ultra-dense hydrogen in low-energy nuclear reactions.

N.B. all equations enclosed in square brackets with subscript NU have dimensions expressed in natural units.

4.2. Energy, Mass, Frequency, and Information

The concept of measurement plays a fundamental role in all scientific disciplines based on experimental evidence. The most used measurement units (such as the international system, SI) are based mainly on human conventions not directly related to fundamental constants. To simplify the conceptual understanding of certain physical quantities, it is convenient to adopt in some cases a measurement system based on universal constants, such as the speed of light c and Planck's quantum $\hbar$.

Considering that a measure is an event localized in space and time, the quantum of action can be seen, in some cases, as an objective entity in some respects analogous to a bit of information located in the space–time continuum. In accordance with Heisenberg's uncertainty principle, the result of the measurement of some values (such as angular momentum) cannot have an accuracy less than half a single Planck's quantum. Therefore, to simplify the interpretation of physical quantities, it may be useful to adopt a system in which both the speed of light and the quantum of action are dimensionless quantities (pure numbers) having a unit value, i.e., $c = 1$ and $\hbar = 1$. In this system, the constancy of light speed makes possible to use a single measurement unit for space and time, simplifying, in many cases, the conceptual interpretation of physical quantities. The energy of a photon, a "particle of light," is equal to Planck's quantum multiplied by the photon angular frequency. By using the symbol T to indicate the period of a single complete oscillation and λ the relative wavelength, it is, therefore, possible to write

$$E = \hbar\omega = \frac{2\pi\hbar}{T} = \frac{2\pi\hbar c}{\lambda} \tag{4.2.1}$$

By using natural units, period and wavelength coincide and the above expression is simplified in

$$\left[E = \omega = \frac{2\pi}{T} = \frac{2\pi}{\lambda}\right]_{\mathrm{NU}} \tag{4.2.2}$$

The NU subscript highlights the use of natural units for expressions contained within square brackets. This equation indissolubly links some fundamental concepts, as space, time, energy, and mass, giving the possibility to express an energy value simply as a frequency or as the inverse of a time, or even as the inverse of a length. *Vice versa*, it allows to use as a measurement unit of both space and time a value equal to the inverse of a particular energy value as the electron-volt. Therefore, to compute a photon wavelength in vacuum with natural units, it is sufficient to divide the constant 2π by its energy. This value will correspond exactly to the period of a complete oscillation. Hence, in natural units the inverse of an eV can be used as a measurement unit for space and time:

$$L_{(1\mathrm{eV})} = 1\ \mathrm{eV}^{-1} \approx 1.9732705 \cdot 10^{-7}\ \mathrm{m} \approx 0.2\ \mu\mathrm{m}$$
$$T_{(1eV)} = 1\ \mathrm{eV}^{-1} \approx 6.582122 \cdot 10^{-16}\ \mathrm{s} \approx 0.66\ \mathrm{fs}$$

Consequently, an angular frequency can be measured in electron volts:

$$1\ \mathrm{eV} \approx 1.519268 \cdot 10^{15}\ \mathrm{rad\,s^{-1}}$$

Following these concepts, it is possible to define a link between fundamental concepts of information, space, time, frequency, and energy. A "quantum of information" carried by a single photon will have a "necessary reading time" and a "spatial dimension" inversely proportional to its energy. A simple example is given by radio antennas (dipoles), whose length is proportional to the received (or transmitted) "radio photons" wavelength and inversely proportional to their frequency and to the number of bits that can be received in a unit of time. From this perspective, the concept of energy is closely linked to the "density" of measurable information in space and in time.

4.3. Magnetic Flux, Phase Shift, Proca Field, and Charge Quantization

Einstein's famous formula $E = mc^2$ becomes particularly explanatory if expressed in natural units:

$$[E = m]_{\text{NU}}$$

Mass is energy and it is, therefore, possible to associate a precise amount of energy to a particle having a given mass. Taking up the considerations made on the deep bond existing between the concepts of space, time, frequency, and energy, it is interesting trying to associate the electron rest mass m_e to an angular frequency ω_e, a length r_e, and a time T_e. In fact, Einstein's formula can be expressed as

$$E_e = m_e c^2 = \hbar\omega_e = \frac{\hbar c}{r_e} = \frac{h}{T_e} \tag{4.3.1}$$

or adopting natural units

$$\left[E_e = m_e = \omega_e = \frac{1}{r_e} = \frac{2\pi}{T_e}\right]_{\text{NU}} \tag{4.3.2}$$

These constants have a simple and clear interpretation under our electron model consisting of a current ring generated by a massless charge rotating at the speed of light along a circumference whose radius is equal to the electron-reduced Compton wavelength, defined as $r_e = \frac{\lambda_c}{2\pi} \approx 0.38616 \cdot 10^{-12}$ m. According to the model described in the preceding chapters, the charge is not a point-like entity, but it is distributed on a spherical surface whose radius is equal to the electron classical radius $r_c \approx 2.8179 \cdot 10^{-15}$ m. In Equation (4.3.2), ω_e is the angular frequency of the rotating charge, r_e is its orbit radius, and T_e its period. The current loop is associated with a quantized magnetic flux Φ_M equal to Planck's constant ($h = 2\pi\hbar$) divided by the elementary charge e (see Equation (3.3.18) in Chapter 3)

$$\Phi_M = h/e$$

or in natural units

$$[\Phi_M = 2\pi/e]_{\text{NU}}$$

The rotation is caused by the centripetal Lorentz force due to the magnetic field associated with the current loop generated by the elementary rotating charge. The value of this elementary charge, in natural units, is a pure number and is equal to the square root of the ratio between the charge radius r_c and the orbit radius r_e (see Equations (3.3.23) and (3.3.24) in Chapter 3):

$$\left[e = \sqrt{\frac{r_c}{r_e}} = \sqrt{\alpha} \approx 0.0854245 \right]_{\text{NU}} \tag{4.3.3}$$

Similar models, based on the concept of "current loop," have been proposed by many authors, but have often been ignored for their incompatibility with the most widespread interpretations of quantum mechanics [2–7]. It is interesting to remember how, already in his Nobel lecture of 1933, P.A.M. Dirac referred to an internal high-frequency oscillation of the electron: *It is found that an electron which seems to us to be moving slowly, must actually have a very high frequency oscillatory motion of small amplitude superposed on the regular motion which appears to us. As a result of this oscillatory motion, the velocity of the electron at any time equals the velocity of light. This is a prediction which cannot be directly verified by experiment, since the frequency of the oscillatory motion is so high and its amplitude is so small.* In the scientific literature, the German word *Zitterbewegung* (ZBW) is often used to indicate this rapid oscillation/rotation of the electron charge. The rotating charge is characterized by a momentum p_c of purely electromagnetic nature:

$$p_c = eA = e\frac{\Phi_M}{2\pi r_e} = \frac{\hbar\omega_e}{c} = \frac{\hbar}{r_e} = m_e c$$

As defined in Equation (3.3.9) of Chapter 3, in this formula the variable $A = \frac{\hbar}{er_e}$ indicates the norm of the vector potential seen by the rotating charge.

Multiplying the above-defined charge momentum p_c by the radius r_e, we obtain the "intrinsic" angular momentum $\hbar$ of the electron:

$$p_c r_e = \hbar \tag{4.3.4}$$

Using natural units, the momentum p_c has the dimension of energy and it is exactly equal to the electron mass-energy at rest m_e:

$$\left[p_c = eA = E_e = \frac{1}{r_e} = m_e = \omega_e \right]_{\text{NU}}$$

4.3.1. *Aharonov–Bohm equations and Zitterbewegung model*

The magnetic Aharonov–Bohm effect is described by a quantum law that gives the phase variation φ of the "electron wave function" starting from the integral of the vector potential $\boldsymbol{A}_\Delta$ along a path [8], i.e.,

$$\varphi = \frac{e}{\hbar}\int \boldsymbol{A}_\Delta \cdot \boldsymbol{dl} \tag{4.3.5}$$

In the proposed Zitterbewegung model, the "wave function phase" of the electron has a precise geometric meaning and indicates the charge rotation phase. By using (4.3.5), a possible counter-test consists in verifying that the phase shift φ along the circumference of the Zitterbewegung orbit is equal exactly to 2π radians. In fact,

$$\varphi = \frac{e}{\hbar}\oint \boldsymbol{A}_\Delta \cdot \boldsymbol{dl} = \frac{e}{\hbar}\int_0^{2\pi r_e} A dl = \frac{e}{\hbar}\int_0^{2\pi r_e} \frac{\hbar}{e r_e} dl = \frac{e}{\hbar}\frac{\hbar}{e r_e} 2\pi r_e = 2\pi \tag{4.3.6}$$

because vectors $\boldsymbol{A}_\Delta$ and $\boldsymbol{dl}$ have the same direction tangent to the elementary charge trajectory.

This result is also consistent with the prediction of the *electric* Aharonov–Bohm effect, a quantum phenomenon that establishes the variation of phase φ as a function of the integral of electric potential V in a time interval T, i.e.,

$$\varphi = \frac{e}{\hbar}\int_T V dt \tag{4.3.7}$$

Applying the electric Aharonov–Bohm effect formula to compute the phase shift φ within a time interval $T_e = \frac{2\pi}{\omega_e}$ equal to a Zitterbewegung period, we obtain the expected result, i.e., $\varphi = 2\pi$. In fact, the electric potential of the electron rotating charge can be expressed as

$$V = \frac{e}{4\pi\varepsilon_0 r_c} = \left[\frac{e}{r_c}\right]_{\text{NU}}$$

and its period as

$$T_e = \frac{2\pi r_e}{c} = [2\pi r_e]_{\text{NU}}$$

A simple calculation, applying (4.3.7) and (4.3.3), yields the same results:

$$\varphi = \frac{e}{\hbar}\int_0^{T_e} V dt = \frac{e}{\hbar} V T_e = \frac{e}{\hbar} V \frac{2\pi r_e}{c} = \left[\frac{e^2}{r_c} 2\pi r_e\right]_{\text{NU}} = 2\pi \tag{4.3.8}$$

Now, by equating the $\frac{e}{\hbar}\frac{\hbar}{er_e}2\pi r_e$ term of (4.3.6) and the $\frac{e}{\hbar}V\frac{2\pi r_e}{c}$ term of (4.3.8) it is possible to demonstrate that

$$A_t = \frac{V}{c} = A = |\boldsymbol{A}_\triangle|$$

$$[A_t = V = A = |\boldsymbol{A}_\triangle|]_{\mathrm{NU}}$$

$$\boldsymbol{A}_\square^2 = (\boldsymbol{A}_\triangle + \gamma_t A_t)^2 = \boldsymbol{A}_\triangle^2 - A_t^2 = 0 \tag{4.3.9}$$

We emphasize that the above relationships pertain to the electromagnetic potential field seen by the electron charge at the $S \neq 0$ space–time locations.

By introducing the differential form of (4.3.7), we obtain

$$d\varphi = \frac{e}{\hbar}V\,dt$$

and this yields the phase speed

$$\frac{d\varphi}{dt} = \omega_e = \frac{e}{\hbar}V = \frac{e^2}{4\pi\varepsilon_0\hbar r_c} = \frac{c\alpha}{r_c} = \frac{c}{r_e} = \frac{m_e c^2}{\hbar} = \frac{ce}{\hbar}A$$

In natural units, the above phase speed is expressed as

$$\left[\frac{d\varphi}{dt} = \omega_e = m_e = eV = eA\right]_{\mathrm{NU}} \tag{4.3.10}$$

4.3.2. *Proca equation and Zitterbewegung electron model*

A deep connection of Maxwell's equations (see Equation (1.3.26) of Chapter 1)

$$\boldsymbol{\partial}\left(\boldsymbol{\partial} \wedge \boldsymbol{A}_\square\right) + \mu_0 \boldsymbol{J}_\square = 0 \tag{4.3.11}$$

with Proca equation for a particle of mass m

$$\boldsymbol{\partial}\left(\boldsymbol{\partial} \wedge \boldsymbol{a}_\square\right) + \left(\frac{mc}{\hbar}\right)^2 \boldsymbol{a}_\square = \boldsymbol{\partial}\boldsymbol{f} + \left(\frac{mc}{\hbar}\right)^2 \boldsymbol{a}_\square = 0 \tag{4.3.12}$$

$$\left[\boldsymbol{\partial}\left(\boldsymbol{\partial} \wedge \boldsymbol{a}_\square\right) + m^2\boldsymbol{a}_\square = \boldsymbol{\partial}\boldsymbol{f} + m^2\boldsymbol{a}_\square = 0\right]_{\mathrm{NU}} \tag{4.3.13}$$

emerges if we prove that equation $\left[\mu_0\boldsymbol{J}_\square = m^2\boldsymbol{a}_\square\right]_{\mathrm{NU}}$ can be applied to the electron Zitterbewegung model introduced in Chapter 3. The $\boldsymbol{\partial}$ operator of $Cl_{3,1}(\mathbb{R})$ algebra is used in the same way as in Chapter 1.

In the Proca equation, $\boldsymbol{a}_\square$ denotes a field which is analogous to the electromagnetic four-potential, and $\boldsymbol{f} = \boldsymbol{\partial} \wedge \boldsymbol{a}_\square$ is the corresponding Proca field. But we don't know yet the physical interpretation of $\boldsymbol{a}_\square$.

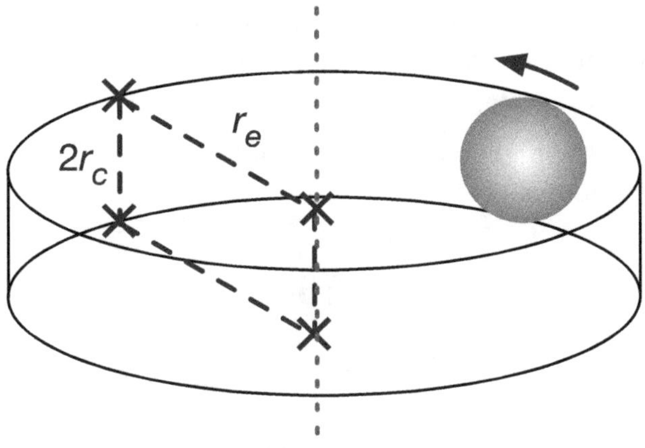

Figure 4.1. Illustration of the "quantum current conductor" method of the Proca equation's current density averaging. The conductor is disc-shaped with radius r_e and height $2r_c$. In this figure, the disc is punctured in the middle, represented by the vertical line. The hole is not real; it is used only to visualize the averaged current flow inside this disc-shaped volume. The area enclosed by the dashed lines is the cross-section of the current conductor.

As shown in Figure 4.1, the electron's Zitterbewegung charge orbit delimits a disc-shaped volume with radius r_e and height $2r_c$.

Inside this volume, the *average* Zitterbewegung current density $\bar{\boldsymbol{J}}_e$ can be computed dividing the Zitterbewegung current by the $\mathcal{A}_{\text{conductor}}$ cross-section area of the "quantum mechanical current conductor" illustrated in Figure 4.1:

$$\bar{\boldsymbol{J}}_e = \frac{\boldsymbol{I}_e}{\mathcal{A}_{\text{conductor}}}$$

where

$$\mathcal{A}_{\text{conductor}} = 2r_e r_c = 2\alpha r_e^2$$

and $\boldsymbol{I}_e$ is the Zitterbewegung component of the electron current, which remains orthogonal to the electron's kinetic velocity vector.

Combining the two equations above, we get:

$$\bar{\boldsymbol{J}}_e = \frac{\boldsymbol{I}_e}{\mathcal{A}_{\text{conductor}}} = \frac{\boldsymbol{I}_e}{2\alpha r_e^2} \tag{4.3.14}$$

From Equation (3.3.9), we have the following relationship:

$$\left[\boldsymbol{I}_e = \frac{\alpha \boldsymbol{A}_\triangle}{2\pi}\right]_{\text{NU}}$$

In the above equation, $\boldsymbol{A}_\triangle$ denotes the electromagnetic vector potential seen by the spinning electron charge. Substituting the above expression into (4.3.14), we get $\left[\bar{\boldsymbol{J}}_e = \frac{\boldsymbol{A}_\triangle}{4\pi r_e^2}\right]_{\text{NU}}$.

Now we choose to define $\boldsymbol{a}_\triangle$ in terms of the *averaged* current density $\bar{\boldsymbol{J}}_e$ at all space–time locations:

$$\left[\bar{\boldsymbol{J}}_e = \frac{\boldsymbol{a}_\triangle}{4\pi r_e^2}\right]_{\text{NU}}$$

$$\left[\mu_0 \bar{\boldsymbol{J}}_e = 4\pi \bar{\boldsymbol{J}}_e = \frac{\boldsymbol{a}_\triangle}{r_e^2} = \omega_e^2 \boldsymbol{a}_\triangle = m_e^2 \boldsymbol{a}_\triangle\right]_{\text{NU}}$$

Thus, $\boldsymbol{a}_\triangle$ is the electromagnetic vector potential seen by the spinning electron charge. This definition tells us how to translate between treating the electron as a particle and treating it as a wave: i.e., it is the expression of wave–particle duality. The detailed explanation of wave–particle duality will be focus of Chapter 5. By treating the electron as a particle, $\bar{\boldsymbol{J}}_e$ is seen to be non-zero only in the region of space shown in Figure 4.1. By treating the electron as a wave, $\bar{\boldsymbol{J}}_e$ reflects the electron's position uncertainty: it becomes defined at all space–time locations, but its magnitude reflects the probabilistic amount of "electron charge" at a given space–time location. As will be shown in what follows, these definitions of $\bar{\boldsymbol{J}}_e$ are not only compatible with the internal electron structure, but lead to a relativistically co-variant wave equation.

Remembering that the electron's electromagnetic four potential $\boldsymbol{A}_\square = \boldsymbol{A}_\triangle + \gamma_t A_t$ associated to the rotating charge is a light-like vector (i.e., $\boldsymbol{A}_\square^2 = 0$ and $A_t = |\boldsymbol{A}_\triangle|$, see Equation (4.3.9)) we can write $\left[\mu_0 J_{et} = \frac{A_t}{r_e^2} = \omega_e^2 A_t = m_e^2 A_t\right]_{\text{NU}}$. Therefore, we define a_t as follows: $a_t = |\boldsymbol{a}_\triangle|$. This definition yields the following relation:

$$\left[\mu_0 J_{et} = \frac{a_t}{r_e^2} = \omega_e^2 a_t = m_e^2 a_t\right]_{\text{NU}}$$

Using the $\boldsymbol{a}_\square = \boldsymbol{a}_\triangle + \gamma_t a_t$ notation, we can write the overall relationship between $\boldsymbol{a}_\square$ and $\bar{\boldsymbol{J}}_{e\square}$ as follows:

$$\left[\mu_0 \bar{\boldsymbol{J}}_{e\square} = \mu_0 \left(\bar{\boldsymbol{J}}_e + \gamma_t J_{et}\right) = m_e^2 \left(\boldsymbol{a}_\triangle + \gamma_t a_t\right) = m_e^2 \boldsymbol{a}_\square\right]_{\text{NU}}$$

and consequently (QED):

$$\left[\mu_0 \bar{\boldsymbol{J}}_{e\square} = m_e^2 \boldsymbol{a}_\square\right]_{\text{NU}} \tag{4.3.15}$$

In summary, defining $\boldsymbol{a}_{\square}$ according to Equation (4.3.15) establishes a connection between the Proca and Maxwell equations. We note that since both $\bar{\boldsymbol{J}}_{e\square}$ and $\boldsymbol{a}_{\square}$ are four vectors, m_e^2 is a scalar. In the point-like macroscopic limit of $\bar{\boldsymbol{J}}_{e\square} \to \boldsymbol{J}_{e\square}$, the Proca equation becomes equivalent to the Maxwell equation. Let's investigate the above-defined Proca equation's meaning in quantum mechanics.

4.3.3. *An equivalence between the electromagnetic Proca and the Klein–Gordon equations*

In the next two sections, we will use only natural units, omitting the subscript NU. Our aim is to show the equivalence between the Proca equation and the "electromagnetic version" of the Klein–Gordon equation. By applying the operator $\boldsymbol{\partial}\wedge$ to the Proca equation

$$\begin{aligned} \boldsymbol{\partial}(\boldsymbol{\partial} \wedge \boldsymbol{a}_{\square}) + m^2 \boldsymbol{a}_{\square} &= 0 \\ \boldsymbol{\partial} \boldsymbol{f} + m^2 \boldsymbol{a}_{\square} &= 0 \end{aligned} \tag{4.3.16}$$

we get

$$\begin{aligned} \boldsymbol{\partial} \wedge \boldsymbol{\partial} \boldsymbol{f} + m^2 \boldsymbol{\partial} \wedge \boldsymbol{a}_{\square} &= 0 \\ \boldsymbol{\partial} \wedge \boldsymbol{\partial} \boldsymbol{f} + m^2 \boldsymbol{f} &= 0 \end{aligned}$$

Now, by writing the Proca equation with the averaged four-current vector density

$$\boldsymbol{\partial} \boldsymbol{f} = -4\pi \bar{\boldsymbol{J}}_{\square} \tag{4.3.17}$$

and by applying to both members the operator $\boldsymbol{\partial}\cdot$, we obtain the following expression:

$$\boldsymbol{\partial} \cdot \boldsymbol{\partial} \boldsymbol{f} = -4\pi\, \boldsymbol{\partial} \cdot \bar{\boldsymbol{J}}_{\square} = 0$$

that is equal to zero as a consequence of the charge-current conservation law. The charge-current conservation law remains still valid for the averaged $\bar{\boldsymbol{J}}_{\square}$. For this reason, the term $\boldsymbol{\partial} \wedge \boldsymbol{\partial} \boldsymbol{f}$ can be safely substituted by the term $\boldsymbol{\partial}^2 \boldsymbol{f}$:

$$\boldsymbol{\partial} \wedge \boldsymbol{\partial} \boldsymbol{f} = \boldsymbol{\partial}^2 \boldsymbol{f} - \boldsymbol{\partial} \cdot \boldsymbol{\partial} \boldsymbol{f} = \boldsymbol{\partial}^2 \boldsymbol{f}$$

As a result we obtain a Klein–Gordon-like equation where the Clifford bivector $\boldsymbol{f}$ plays the "wavefunction" role:

$$\boldsymbol{\partial}^2 \boldsymbol{f} + m^2 \boldsymbol{f} = 0 \tag{4.3.18}$$

A similar equation for the electromagnetic potential-like field $\boldsymbol{a}_{\square}$ can be obtained simply by applying the Lorenz gauge condition $\boldsymbol{\partial a}_{\square} = \boldsymbol{\partial} \wedge \boldsymbol{a}_{\square}$ to the Proca equation:

$$\boldsymbol{\partial}^2 \boldsymbol{a}_{\square} + m^2 a_{\square} = 0 \tag{4.3.19}$$

or

$$\boldsymbol{\partial}^2 \boldsymbol{a}_{\square} + \omega^2 a_{\square} = 0 \tag{4.3.20}$$

It is important to note that the Lorenz gauge condition can be applied only to the $\boldsymbol{a}_{\square}$ field, which was defined from *an averaged* four-current density vector field.

4.3.4. *The electromagnetic Dirac equation*

As explained in Chapter 2, the solutions of Klein–Gordon equation are also solutions of a corresponding Dirac equation. We want to find the Dirac type equations whose solutions are also solutions of Equations (4.3.19) and (4.3.18). We try the following form of the sought Dirac–type equations:

$$\boldsymbol{\partial a}_{\square} = -\boldsymbol{m a}_{\square} \tag{4.3.21}$$

$$\boldsymbol{\partial f} = \boldsymbol{m f} \tag{4.3.22}$$

where $\boldsymbol{m}$ is a space-like vector with norm m. Let's see whether these equations lead to the Klein–Gordon equations (4.3.19) and (4.3.18). Upon taking $\boldsymbol{m}^2$, we find that the requirement of m^2 being a scalar is satisfied.

Since $\boldsymbol{\partial a}_{\square} = \boldsymbol{f}$, we can write Equation (4.3.21) as

$$\boldsymbol{f} = -\boldsymbol{m a}_{\square} \tag{4.3.23}$$

Upon left-multiplying the above equation by $\boldsymbol{m}$, we get

$$\boldsymbol{m f} = -\boldsymbol{m}^2 \boldsymbol{a}_{\square} = -m^2 \boldsymbol{a}_{\square} \tag{4.3.24}$$

where m^2 is a scalar. We evaluate $\boldsymbol{\partial}^2 \boldsymbol{f} = \boldsymbol{\partial}(\boldsymbol{\partial f})$, and substitute in Equations (4.3.22) and (4.3.24):

$$\boldsymbol{\partial}(\boldsymbol{\partial f}) = \boldsymbol{\partial}(\boldsymbol{m f}) = -m^2 \boldsymbol{\partial a}_{\square} = -m^2 \boldsymbol{f} \tag{4.3.25}$$

The above result is equivalent to the Klein–Gordon equation (4.3.18). In Section 4.3.3, we proved that Equations (4.3.19) and (4.3.18) follow from each other. Therefore, we proved that solutions of the Dirac equations (4.3.21) and (4.3.22) are also solutions of the Klein–Gordon equations (4.3.19) and (4.3.18).

Our candidate for $\boldsymbol{m}$ is a rotating space-like vector that points from the center of Zitterbewegung circle to the electron charge, and has a norm $m = \frac{1}{r} = \omega$. Calling $\boldsymbol{r}_u$ a unit vector in the same direction as $\boldsymbol{m}$, Equation (4.3.22) becomes

$$\boldsymbol{\partial f} - \omega \boldsymbol{r}_u \boldsymbol{f} = 0 \tag{4.3.26}$$

where the Zitterbewegung angular frequency ω substitutes the electron mass, and the bivector $\boldsymbol{f}$ is identified with the Proca field. The unit vector $\boldsymbol{r}_u$ is always orthogonal to the vector potential $\boldsymbol{a}_\square$ and therefore:

$$\begin{aligned} \boldsymbol{r}_u^2 &= 1 \\ \omega \boldsymbol{r} &= \boldsymbol{r}_u \\ \boldsymbol{r} \cdot \boldsymbol{a}_\square &= 0 \end{aligned}$$

By applying (4.3.17) to (4.3.26), we can write

$$4\pi \bar{\boldsymbol{J}}_\square + \omega \boldsymbol{r}_u \boldsymbol{f} = 0 \tag{4.3.27}$$

whereas, by applying (4.3.15) to (4.3.27) and remembering that $\boldsymbol{f} = \boldsymbol{\partial a}_\square$, we obtain:

$$\omega^2 \boldsymbol{a}_\square + \omega \boldsymbol{r}_u \boldsymbol{\partial a}_\square = 0$$

that can be written as

$$\boldsymbol{r}_u \boldsymbol{\partial a}_\square + \omega \boldsymbol{a}_\square = 0$$

Now, by left-multiplying the last equation for the unit vector $\boldsymbol{r}_u$ we obtain a Dirac-like equation for the electromagnetic four-potential

$$\boldsymbol{\partial a}_\square + \omega \boldsymbol{r}_u \boldsymbol{a}_\square = 0 \tag{4.3.28}$$

Multiplying for the elementary charge e, Equation (4.3.28) becomes

$$e\boldsymbol{\partial a}_\square + e\omega \boldsymbol{r}_u \boldsymbol{a}_\square = 0 \tag{4.3.29}$$

Moreover, upon multiplying the electromagnetic four-potential by the ratio $\frac{e}{\omega}$, we obtain a light-like vector that can be interpreted as the charge four-velocity $\boldsymbol{c} - \gamma_t$ (see Equation (1.2.52) of Chapter 1 and Equation (4.3.10)):

$$\frac{e}{\omega}\boldsymbol{a}_{\square} = \boldsymbol{c} - \gamma_t \tag{4.3.30}$$

In the above equation, $\boldsymbol{c}$ is a unit vector pointing in the direction of charge movement, i.e., $\boldsymbol{c}^2 = 1$. Upon left-multiplying by $\boldsymbol{r}_u$, the above equation becomes

$$\frac{e}{\omega}\boldsymbol{r}_u\boldsymbol{a}_{\square} = \boldsymbol{r}_u\boldsymbol{c} - \boldsymbol{r}_u\gamma_t \tag{4.3.31}$$

Now, by applying (4.3.31) to (4.3.29) and remembering that $\boldsymbol{\partial a}_{\square} = \boldsymbol{f}$, (4.3.29) becomes

$$e\boldsymbol{f} = -\omega^2\left(\boldsymbol{r}_u\boldsymbol{c} - \boldsymbol{r}_u\gamma_t\right) \tag{4.3.32}$$

Given that the Proca field $\boldsymbol{f}$ can be interpreted as the electromagnetic bi-vector field seen/generated by the rotating charge, we make the $\boldsymbol{f} = (\boldsymbol{E}_P + I\boldsymbol{B}_P)\gamma_t$ split into electric and magnetic components, according to Equations (1.3.2) and (4.3.32). Applying the $\boldsymbol{f} = (\boldsymbol{E}_P + I\boldsymbol{B}_P)\gamma_t$ definition, the above equation becomes

$$e\boldsymbol{f} = -\omega^2\left(\boldsymbol{r}_u\boldsymbol{c} - \boldsymbol{r}_u\gamma_t\right) \tag{4.3.33}$$

This last equation can be split into two equations. The first one deals with the electric Proca field $\boldsymbol{E}_P$:

$$e\boldsymbol{E}_P\gamma_t = \omega^2\boldsymbol{r}_u\gamma_t$$

Applying the identity $ea = \omega$, the square ω^2 can be written as $ea\omega$, which is a term that is equal to the norm of the force generated on an elementary electric charge by the time derivative of a rotating vector potential:

$$e\boldsymbol{E}_P = ea\omega\boldsymbol{r_u} = -e\frac{d\boldsymbol{a}_{\triangle}}{dt} \tag{4.3.34}$$

This electric force has the same value of the centrifugal force acting on a mass m rotating with angular frequency ω at distance r from its orbit center:

$$\boldsymbol{r}_u = \omega\boldsymbol{r} = m\boldsymbol{r}$$

Combining the above equations, the rotating charge becomes subjected to a centrifugal electric force $\omega^2\boldsymbol{r}_u$:

$$e\boldsymbol{E}_P = m\omega^2\boldsymbol{r} = \omega^2\boldsymbol{r}_u \tag{4.3.35}$$

The second part of (4.3.33) deals with the magnetic Proca field $\boldsymbol{B}_P$:

$$eI\boldsymbol{B}_P\gamma_t = -\omega^2\boldsymbol{r}_u\boldsymbol{c}$$
$$eI_\Delta\boldsymbol{B}_P = -\omega^2\boldsymbol{r}_u\boldsymbol{c}$$

Since $\boldsymbol{B}$ and $\boldsymbol{c}$ are orthogonal vectors in the Zitterbewegung model, and since $\boldsymbol{c}^2 = 1$, upon right-multiplying the above equation by $\boldsymbol{c}$ we get

$$eI_\Delta\boldsymbol{B}_P \wedge \boldsymbol{c} = -\omega^2\boldsymbol{r}_u$$

Using ordinary vector algebra, the above equation becomes:

$$e\boldsymbol{c} \times \boldsymbol{B}_P = -\omega^2\boldsymbol{r}_u \tag{4.3.36}$$

Finally, by merging (4.3.34) with (4.3.36) we obtain an equation which tells us that the rotating charge, which has Zitterbewegung momentum $\boldsymbol{p} = e\boldsymbol{a}_\Delta$, is subjected to a centripetal magnetic force $-\omega^2\boldsymbol{r}_u$:

$$e\boldsymbol{c} \times \boldsymbol{B}_P = e\frac{d\boldsymbol{a}_\Delta}{dt} = \frac{d\boldsymbol{p}}{dt} \tag{4.3.37}$$

$$e\boldsymbol{c} \times \boldsymbol{B}_P = -m\omega^2\boldsymbol{r} = -\omega^2\boldsymbol{r}_u \tag{4.3.38}$$

These easy to interpret equations confirm the correctness of the original choice of $\omega\boldsymbol{r}_u$ for the vector $\boldsymbol{m}$ in Equation (4.3.22).

Now we may finally see the meaning of Proca equation (4.3.12): the derived Equations (4.3.35) and (4.3.38) represent the balance of electric and magnetic forces acting upon the electron's charge. Therefore, the Proca equation conveys that the electron remains a stable particle due to the balance of electric and magnetic forces acting upon its spherical charge surface. As we demonstrated in Section 4.3.2, the Proca equation's macroscopic limit converges to the Maxwell equation, and thus the Proca equation remains valid at all distance scales.

Treating the electron as a localized particle, Chapter 3 results demonstrated how Zitterbewegung dynamics corresponds to the mutual induction between electric and magnetic fields within the electron. Treating the electron as a delocalized wave, the results of this chapter demonstrate how Zitterbewegung dynamics correspond to the balance of forces acting upon the electron. In this context, the eigenvalue solution of the Dirac equation represents the force equilibrium state. As explained in Chapter 6.4 of [16], the Proca equation corresponds to a Lorentz-invariant Lagrangian. Therefore, the derived equilibrium of forces is invariant under any reference frame transformation. Theoretical physicists consider the Proca equation to describe the dynamics of a massive particle having $\hbar$ angular momentum: this view is certainly compatible with our electron model.

So far, we have encountered two versions of the Dirac equation. In Chapter 2, the Dirac equation was formulated over a scalar valued field. In the above discussion, the Dirac equation was formulated over a bivector valued field. These perspectives shall be reconciled in Chapter 7, where we show that these two perspectives are just factorizations of one single Dirac equation, acting on a "spinor valued" field.

4.3.5. *Proca equation, electric charge quantization, and Josephson constant*

An interesting consequence of Equation (4.3.20) is the magnetic flux and electric charge quantization. In this paragraph, we call "wave amplitude" a the quantity derived from the vector potential $\boldsymbol{a}_\triangle$ in Equation (4.3.20):

$$\boldsymbol{a}_\Box = \boldsymbol{a}_\triangle + \gamma_t a_t$$

$$a = |\boldsymbol{a}_\triangle| = a_t$$

Substituting ω with ea in Equation (4.3.20), we obtain *a nonlinear wave equation for the electromagnetic four-potential, where the wave angular frequency is proportional to the wave amplitude and the proportionality coefficient is the "electric charge quantum," i.e., the elementary charge e*

$$\left[\boldsymbol{\partial}^2 \boldsymbol{a}_\Box + e^2 a^2 \boldsymbol{a}_\Box = 0\right]_{\text{NU}} \tag{4.3.39}$$

$$\left[\boldsymbol{\partial}^2 \boldsymbol{a}_\Box + \alpha a^2 \boldsymbol{a}_\Box = 0\right]_{\text{NU}} \tag{4.3.40}$$

In this equation, the frequency/amplitude ratio, $\frac{\nu}{a}$, expressed in natural units is a pure number equal to half the value of Josephson constant $K_J = \Phi_o^{-1}$:

$$\left[\frac{v}{a} = \frac{1}{2} K_J\right]_{\text{NU}}$$

The product of the wave amplitude a and the wave period T is constant, and exactly equal to a magnetic flux Φ_M, a value two times the magnetic flux quantum Φ_o. This relationship is illustrated on Figure 4.2. It's a reasonable conjecture to consider Equation (4.3.39) to be also valid for other charged elementary particles. Elementary particles' charge quantization is

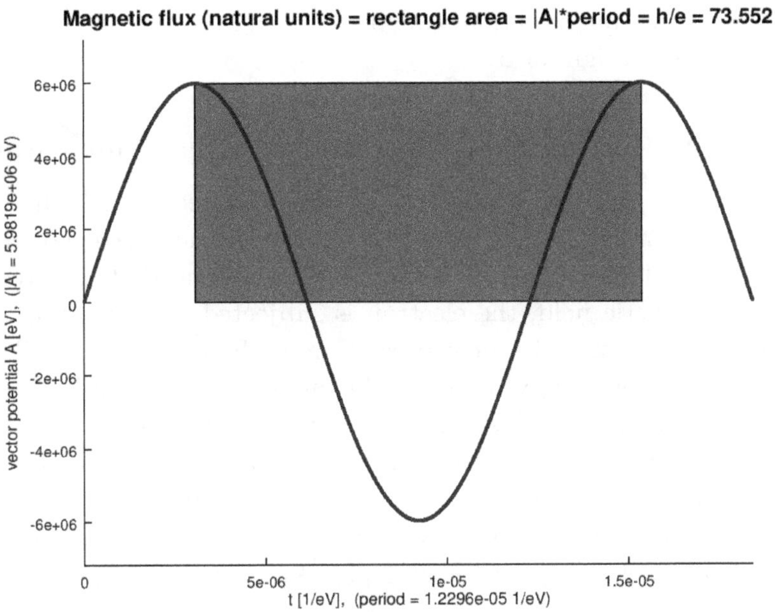

Figure 4.2. Illustration of the relationship between magnetic flux and electric charge quantization. In the electromagnetic Klein–Gordon/Proca equation, the vector potential amplitude times the wave period is a constant: $\Phi_m = \frac{h}{e}$.

therefore related to their magnetic flux quantization, which was derived in Equation (3.3.18). In natural units, we have

$$\left[aT = \frac{\omega T}{e} = \frac{h}{e} = \Phi_M = 2K_J^{-1} \right]_{\mathrm{NU}}$$

where

$$\Phi_M = 2\Phi_o = 4.13566766 \cdot 10^{-15}\, V \cdot s$$

$$[\Phi_M = 73.55246018]_{\mathrm{NU}}$$

From the perspective of macroscopic electrodynamics, charge quantization was a mystery: how do quantized electric and magnetic charges emerge from the scale-free electromagnetic wave equation? We showed that charge quantization is a consequence of the nonlinear equation (4.3.39), and that it implies magnetic flux quantization. Φ_M is observed in various areas of physics, such as superconducting electrons, and therefore the usefulness of knowing the correct electron structure appears to be far-reaching. While the value of α is not yet derived from fundamental constants, discovering the magnetic flux quantization constraint is a meaningful step towards understanding the origin of α.

4.4. ESR, NMR, Spin, and "Intrinsic" Angular Momentum

As shown in the previous paragraph, in the proposed model, the electron has an angular momentum $\boldsymbol{\hbar}$ and a magnetic moment $\boldsymbol{\mu}_B$, equal to the Bohr magneton. It is, therefore, reasonable to assume that, in presence of an external magnetic field, the electron is subjected, as a small gyroscope, to a torque $\boldsymbol{\tau}$ and to a Larmor precession with frequency ω_p. The only difference with a classical gyroscope is the quantization of the $\hbar_{\|}$ component of the angular momentum $\boldsymbol{\hbar}$ along the external flux density field $\boldsymbol{B}_E$. This component can take only two possible spin values, namely $\hbar_{\|} = \pm\frac{1}{2}\hbar$ (see Chapter 3, Section 3.3.2). The two spin values will correspond to two possible values for the angle θ formed between the angular momentum vector and the external magnetic field vector: $\theta \in \left\{\frac{\pi}{3}, \frac{2\pi}{3}\right\}$:

$$\hbar_{\|}^2 + \hbar_{\perp}^2 = \hbar^2$$

$$\hbar_{\|} = \pm\tfrac{1}{2}\hbar$$

The torque exerted by the external flux density field $\boldsymbol{B}_E$ is

$$\tau = |\boldsymbol{\mu}_B \times \boldsymbol{B}_E| = \mu_B B_E \sin(\theta)$$

and the related Larmor precession angular frequency is

$$\omega_p = \frac{B_E \mu_B}{\hbar} \tag{4.4.1}$$

The precession angular frequency will correspond to two possible energy levels:

$$E_H = \hbar\omega_p \quad \text{if } \theta = \frac{2\pi}{3}$$

and

$$E_L = -\hbar\omega_p \quad \text{if } \theta = \frac{\pi}{3}$$

The difference of energy levels corresponds to the spin electronic resonance (ESR) frequency ν_{ESR}:

$$\Delta E = E_H - E_L = 2\hbar\omega_p = \hbar\omega_{\text{ESR}} = h\nu_{\text{ESR}} \tag{4.4.2}$$

From (4.4.2) and (4.4.1), it is possible to determine the ESR frequency as

$$\nu_{\text{ESR}} = 2\frac{B_E \mu_B}{h} \tag{4.4.3}$$

For instance, an external magnetic flux density field equal to $B_E = 1.5$ T yields a frequency $\nu_{\text{ESR}} \approx 42$ GHz. The ESR method is already used for various applications, such as quantum computing or chemical analysis, even though its theoretical foundations have been lacking. With a proper understanding of the ESR theory, scientists will have a solid foundation to guide their progress.

By calling s the spin value and μ the nuclear magnetic moment, we can also generalize (4.4.3) for particles other than the electron. In this case, the term used is nuclear magnetic resonance (NMR) frequency, which is equal to

$$\nu_{\text{NMR}} \approx \frac{B_E \mu}{hs} \tag{4.4.4}$$

For instance, for isotope ${}^7_3Li$, with $s = \frac{3}{2}$, $\mu \approx 1.645 \cdot 10^{-26}$, and $B_E = 1.5$ T, the NMR frequency is $\nu_{\text{NMR}} \approx 24.8$ MHz, whereas for isotope ${}^{11}_5B$ we have $s = \frac{3}{2}$, $\mu \approx 1.36 \cdot 10^{-26}$ J $\cdot$ T^{-1}, and NMR frequency is $\nu_{\text{NMR}} \approx 20.5$ MHz. Another example deals with isotope ${}^{87}_{38}Sr$ with $s = \frac{9}{2}$ and $\mu \approx 5.52 \cdot 10^{-27}$ J $\cdot$ T^{-1}. In this case, NMR frequency is $\nu_{\text{NMR}} \approx 278$ kHz for $B_E = 0.15$ T with a Larmor frequency $\frac{\omega_p}{2\pi} = \frac{1}{2}\nu_{\text{NMR}} \approx 139$ kHz.

The electron, in the presence of an external magnetic field, is subjected to Larmor precession and its spin value $\pm\frac{\hbar}{2}$ is interpreted as the intrinsic angular momentum component parallel to the magnetic field. It is interesting to note that a hypothetical technology, able to align the intrinsic angular momentum of a sufficient number of electrons, could favor the formation of a coherent superconducting and super-fluid condensate state. In this state, the electrons would behave as particles with whole spin $\hbar$ and would no longer be subject to the Fermi–Dirac statistic. The compression effect (pinch) of an electrical discharge, accurately localized in a very small "capillary" volume, inside which a very rapid and uniform variation of the electric potential occurs, could favor the formation of a superconducting plasma. This conjecture is based on the possibility that, as a consequence of Aharonov-Bohm effect, a rapid, collective, and simultaneous variation of the Zitterbewegung phase catalyzes the creation of coherent systems like those described by Shoulders and Puthoff [10]: *Laboratory observation of high-density filamentation or clustering of electronic charge suggests that under certain conditions strong coulomb repulsion can be overcome by cohesive forces as yet imprecisely defined.*

4.5. Hypotheses on the Structure of Ultra-Dense Hydrogen

In relativistic quantum mechanics, the Klein–Gordon equation describes a charge density distribution in space and time. In this equation, a term $\frac{m^2c^2}{\hbar^2}$ appears, whose interpretation becomes simple and intuitive if one uses natural units and the principle of mass-energy-frequency equivalence. In particular, it is possible to recognize this term as the square of the Zitterbewegung angular frequency ω:

$$\left[\frac{m^2c^2}{\hbar^2} = m^2 = \omega^2\right]_{\text{NU}}$$

In Chapter 3, we introduced the idea of a special hydrogen state, where the average nucleus–electron distance is

$$r_0 \approx \frac{\hbar}{m_e c} \approx 0.39 \cdot 10^{-12}\ \text{m}$$

A series of experiments conducted by Leif Holmlid of the University of Gothenburg, recently replicated by Sindre Zeiner-Gundersen [11], demonstrates the existence of a very compact form of hydrogen or deuterium [12–14]. Based on the kinetic energy (about 630 eV) of the nuclei emitted in some experiments, achieved by irradiating this particular form of ultra-dense deuterium with a small laser, a distance between deuterium nuclei of about $2.3 \cdot 10^{-12}$ m has been computed, a value much smaller than the distance of about $74 \cdot 10^{-12}$ m that separates the nuclei of a normal H_2 or D_2 molecule. Therefore, it is possible to advance a hypothesis on the structure of ultra-dense hydrogen (UDH) starting from the electron Zitterbewegung model. The proton is considerably smaller than Zitterbewegung orbit radius r_e, consequently a hypothetical structure formed by a "Zitterbewegung-orbit" electron with a proton (or a deuteron) in its center would have a potential energy of $\left[\frac{-e^2}{r_e} \approx -3.7\ \text{keV}\right]_{\text{NU}}$.

The distance between the deuterium nuclei in the Holmlid experiment can be explained by an ordered linear sequence of ultra-dense particles in which the rotation planes of the electron charges are parallel and equidistant. In these hypothetical aggregates, the Zitterbewegung phases of two neighboring electrons differ by π radians due to electrostatic repulsion, and the distance d_c between the charges of the two electrons is equal to

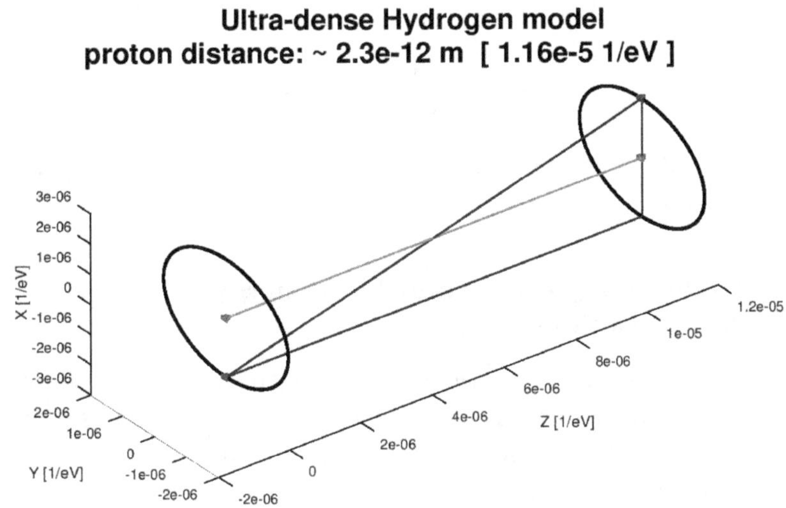

Figure 4.3. Ultra-dense hydrogen model.

the distance traveled by light in a time equal to a rotation period T. This distance amounts to $d_c = cT = \lambda_c \approx 2.42 \cdot 10^{-12}$ m.

This UDH model is also compatible with the third assumption of Carver Mead "Alternate World View": *every element of matter is coupled to all other charges on its light cone by time-symmetric interactions* [1].

The distance between the nuclei d_i can be obtained by applying the Pythagorean theorem, as shown in Figure 4.3, yielding exactly the experimentally measured value:

$$d_i = \sqrt{\lambda_c^2 - \left(\frac{\lambda_c}{\pi}\right)^2} \approx 2.3 \cdot 10^{-12} \text{ m}$$

We consider the consequence of the $\alpha m_e c^2 \approx 3.7$ keV potential energy gain on the Zitterbewegung orbit radius reduction. By defining m_{eu} and r_{eu} the mass and the radius, respectively, in this new state we have:

$$m_{eu}c^2 = m_e c^2 + \alpha m_e c^2 \approx 514.728 \text{ keV} \tag{4.5.1}$$

The mass increase implies a Zitterbewegung radius reduction. In fact,

$$m_e c^2 = \hbar\omega_e = \frac{\hbar c}{r_e}$$
$$m_{eu}c^2 = m_e (1 + \alpha) c^2 = \hbar\omega_{eu} = \frac{\hbar c}{r_{eu}}$$

and therefore,

$$r_{eu} = \frac{\hbar}{m_e (1+\alpha) c} = \frac{r_e}{1+\alpha}$$

In Chapter 12, we shall further investigate the physics of such ultra-dense state of hydrogen.

4.6. Ultra-Dense Hydrogen and Low-Energy Nuclear Reactions

In the proposed model, the particles of hydrogen or ultra-dense deuterium are electrically neutral but have a magnetic moment almost equal to that of the electron. This is a value 960 times higher than the neutron magnetic moment. A particle with magnetic moment $\boldsymbol{\mu}$ is subjected, in presence of a magnetic field $\boldsymbol{B}$, to a force $\boldsymbol{\mathcal{F}}$ proportional to the gradient of $\boldsymbol{B}$

$$\boldsymbol{\mathcal{F}} = \nabla (\boldsymbol{B} \cdot \boldsymbol{\mu})$$

Therefore, the magnetic field $\boldsymbol{B}$ generated by a nucleus could exert a considerable "remote action" on the particles of ultra-dense hydrogen.

Hence, it is possible that, according to this scenario, electrons would have a fundamental dual role as catalysts of low-energy nuclear reactions (LENRs): firstly by creating a pseudo-neutron effect by masking the positive charge of hydrogen or deuterium nuclei, a necessary condition to overcome the Coulomb barrier, and secondly as the source of a relatively long-range magnetic force.

By using the Holmlid notation "H(0)" to indicate ultra-dense hydrogen particles, it is possible to hypothesize a LENR reaction involving the ${}^7_3Li$, an isotope that constitutes more than 92% of the natural Lithium

$${}^7_3Li + H(0) \rightarrow 2{}^4_2He + e \tag{4.6.1}$$

This reaction would produce an energy of about 17.34 MeV mainly in the form of kinetic energy of the emitted electron and helium nuclei, without emission of neutrons or penetrating gamma rays. A similar reaction, able to release about 8.67 MeV, could be hypothesized for the isotope ${}^{11}_5B$

$${}^{11}_5B + H(0) \rightarrow 3{}^4_2He + e \tag{4.6.2}$$

Emissions in the X-ray range would still be present in the form of braking radiation (*Bremsstrahlung*) generated by the deceleration of the reaction products.

The kinetic energy of the reaction products could be converted with a reasonable yield directly into electrical energy or into usable mechanical energy [15], avoiding the need for an intermediate conversion into thermal energy.

4.7. Conclusions

In this chapter, a simple Zitterbewegung electron model has been introduced, where the concepts of mass-energy, momentum, magnetic momentum, and spin naturally emerge from its geometric and electromagnetic parameters, thus avoiding the obscure concept of "intrinsic property" of a "point-like" particle. Using only electromagnetic and geometric concepts, an interpretation of Proca, Klein–Gordon, Dirac, and Aharonov–Bohm equations based on this particular electron model has been presented. The results of this chapter also demonstrate that the phase of a quantum mechanical wavefunction is derived from the Zitterbewegung phase; these two phases are related by Lorentz transformations, which will be further investigated in Section 5.4. A nonlinear equation for the electromagnetic four-potential has been introduced, which directly implies elementary particlcs' clcctric charge and magnetic flux quantization.

ESR frequency has been computed starting from a spin model based on the Larmor precession frequency of Zitterbewegung rotation plane. A very simple model for ultra-dense hydrogen, where electron has only spin angular momentum, has been proposed.

Acknowledgments

Authors wish to thank Francesco Celani and Giuliano Bettini for helpful discussions and suggestions.

References

[1] C. Mead, The nature of light: What are photons? *Proceedings of the SPIE*, **8832** (2013) 8832–8837.

[2] D. Hestenes, Quantum mechanics from self-interaction. *Foundations of Physics*, **15**(1) (1985) 63–87.

[3] A.L. Parson and Smithsonian Institution, *A magneton Theory of the Structure of the Atom (with two plates)*, vol. 65. Publication of Smithsonian Institution. Smithsonian Institution, 1915.

[4] J.P. Wesley and D.L. Bergman, Spinning charged ring model of electron yielding anomalous magnetic moment. *Galilean Electrodynamics*, **1** (1990) 63–67.

[5] D.L. Bergman and C.W. Lucas, Credibility of common sense science. *Foundations of Science*, (2003) 1–17.
[6] D. Hestenes, The zitterbewegung interpretation of quantum mechanics. *Foundations of Physics*, **20**(10) (1990) 1213–1232.
[7] R. Gauthier, The electron is a charged photon. Volume 60 of APS April Meeting 2015, 2015. http://meetings.aps.org/link/BAPS.2015.APR.Y16.4.
[8] Y. Aharonov, and D. Bohm, Significance of electromagnetic potentials in the quantum theory. *Physical Review*, **115** (1959) 485–491.
[9] R.P. Feynman, *QED: The Strange Theory of Light and Matter*. Penguin Books. Penguin, 1990.
[10] H.E. Puthoff and M.A. Piestrup, Charge confinement by casimir forces (2004) arXiv:physics/0408114.
[11] L. Holmlid and S. Zeiner-Gundersen, Ultradense protium p(0) and deuterium D(0) and their relation to ordinary Rydberg matter: A review. *Physica Scripta*, **94** (2019) 7.
[12] S. Badiei and P. U. Andersson, and L. Holmlid, High-energy Coulomb explosions in ultra-dense deuterium: Time-of-flight-mass spectrometry with variable energy and flight length. *International Journal of Mass Spectrometry*, **282**(1–2) (2009) 70–76.
[13] L. Holmlid, Excitation levels in ultra-dense hydrogen p(-1) and d(-1) clusters: Structure of spin-based Rydberg matter. *International Journal of Mass Spectrometry*, **352** (2013) 1–8.
[14] L. Holmlid and S. Olafsson, Spontaneous Ejection of High-energy Particles from Ultra-dense Deuterium D(0). *International Journal of Hydrogen Energy*, **40**(33) (2015) 10559–10567.
[15] A.G. Tarditi, Aneutronic Fusion Spacecraft Architecture, NASA-NIAC 2001 Phase I Research Grant — Final Research Activity Report (2012).
[16] J. Schwichtenberg, *Physics from Symmetry*, 2nd edition, Springer Berlin, 2018.

https://doi.org/10.1142/9781800611306_0005

Chapter 5

Wave–Particle Duality

Paul O'Hara[*,§], András Kovács[†], and Giorgio Vassallo[‡,¶]
Sophia University Institute, Loppiano, Italy
†BroadBit Energy Technologies, Metallimiehenkuja 8 C 02150 Espoo, Finland
‡Engineering Department, University of Palermo viale delle Scienze, 90128 Palermo, Italy
§paul.ohara@sophiauniversity.org
¶giorgio.vassallo@unipa.it

5.1. Introduction

In this chapter, we focus on the wave-like description of a moving particle, especially of an electron. We look at wave–particle duality from three different perspectives, which turn out to be complementary. However, it should be noted that by definition of wave–particle duality, there are no particles in a classical sense. When we refer to an electron as a particle, in reality we are referring to the "particle properties" associated with the center-of-mass.

Firstly, in Section 5.2 we treat the subject from the perspective of general relativity. The mathematics of Section 5.2 is more complex than the rest of the book, and assumes that the reader is familiar with tensors and differential geometry.

We find that there is a one-to-one correspondence between the metrics of general relativity and wave equations, which will permit us to associate the particle momentum with its corresponding quantum mechanical operator. In particular, this approach will allow us to discuss Zitterbewegung from a new perspective. Before entering into the specifics, we first give an overview of some differential geometry, Clifford algebra, and the relationship between them, emphasizing the role of the Principle of Equivalence.

Secondly, in Section 5.3 we build upon the results of Chapter 3 to develop a geometric interpretation of wave–particle duality: we find that the approach of Chapter 3 matches the results of Section 5.2.

Finally, in Section 5.4 we look at the subject from the perspective of electromagnetic waves' Lorentz transformation, and establish the correspondence between Lorentz transformation rules and the previously derived results. Moreover, a natural interpretation of the quantum mechanical uncertainty relation, $\triangle x \triangle p \geq \frac{\hbar}{2}$, is suggested by the electromagnetic vacuum fluctuations.

5.2. Metrics and the Dirac Equation

What is the role of space–time metrics in the description of fundamental forces and interactions? As mentioned in the book's introduction, the motion of planets was first described and predicted by epicycloid formulas. Newton discovered that planetary motion is caused by gravitational forces, given by the $F_{\text{gravity}} = G\frac{mM}{r^2}$ formula. While Newton's formula has the same accuracy as epicycloids for predicting planetary motion, it represents a deeper understanding of elementary interactions and gives a unified method to calculate the motion of planets, falling apples, and rockets. A next level of understanding gravitational forces was given by Einstein's general relativity theory. As illustrated in Figure 5.1, general relativity allows us to identify mass with space–time curvature or more precisely to identify the stress–energy–momentum tensor with the Ricci curvature tensor. In this theory, a planet moves along a straight line (geodesic) in its locally flat space–time, which corresponds to a trajectory of a closed ellipsoid loop in the globally curved space–time. Moreover, although one may calculate approximately the same planetary trajectory via Newton's gravitational force over a flat space–time or via Einstein's space–time gravitational field equations, they are different from both an ontological and physics perspective. The new insights gained through the concepts and equations of general relativity involve a paradigm shift. For example, as illustrated in Figure 5.1, general relativity gives a unified method to calculate the motion of planets and the gravitational deflection of electromagnetic waves. Thus, we gained new knowledge with respect to Newton's theory, which failed to say anything about light deflection.

The Reissner–Nordström metric of general relativity describes how electromagnetic field energy curves space–time. This suggests the following thought experiment: consider two electrically charged cannon balls orbiting

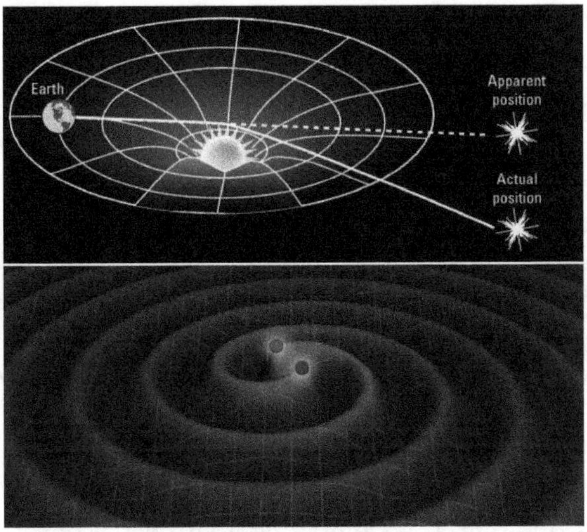

Figure 5.1. An illustration of gravitational space–time curvature (top) and electromagnetic space–time curvature (bottom).

around their center of mass, such that the gravitational forces are negligible in comparison with electric forces. Applying Maxwell's equations over a flat space–time, we may calculate the trajectory of these balls spiraling into each other, as well as the radiated electromagnetic wave during this process. Alternatively, we can calculate the space–time curvature, which is not constant in this case, but rather a rotating spiral as shown in Figure 5.1. Upon lengthy calculations, we would arrive at the same geodesic trajectory result as obtained by applying Maxwell's equations over a flat space–time metric. So far, there is nothing controversial, and the involved relativistic equations are well known. We might now ask: what happens when the balls are scaled down to a microscopic level so that the mass-to-charge ratio is very small? As the ball radius shrinks, electromagnetic fields get more intense, which generates more curved space–time. Intuitively, the role of space–time curvature should be immense in quantum mechanics. However, this subject is still unexplored: quantum theorists work exclusively over flat space–time metrics. We try to remedy this shortcoming, and explore what new insights can be gained by calculating particle geodesics in curved space–time.

Another dramatic consequence of the free movement of charged particles along space–time geodesics is worked out in Section 8.9: it explains the single-frequency light emission and absorption during quantum mechanical state transitions. In other words, it is at the heart of clarifying over 100 years of misconceptions about the so-called "photon particles."

5.2.1. *Dual equations*

We begin with an intuitive and non-rigorous approach to our methodology by indicating two ways in which quantum mechanical wave equations can be obtained from the metrics of general relativity, without any explicit recourse to Lagrangians or Hamiltonians. We will then combine the results of the two approaches into a mathematical theorem. In this section, we will impose more rigorous constraints which will enable us to identify the spinor formulation given here with the usual Hilbert Space formulation of quantum mechanics. In general, Einsteinian notation will be used for index summations throughout this chapter, although at times it will be necessary to distinguish the time component from the space ones, in which case we will do so using the $\sum$ notation.

In his book entitled *Differential Geometry and Its Applications,* Oprea loosely describes differential geometry as a subject "concerned with understanding shapes and their properties in terms of calculus" [6]. These shapes are usually expressed in terms of curves, areas, and volumes embedded in locally flat surfaces which are given the technical name of "manifolds." The General Theory of Relativity associates gravitational fields and also electromagnetic fields with four-dimensional differential manifolds. Consequently, the language of differential geometry is at the heart of the subject. We denote a manifold and its local metric tensor as $(\mathcal{M}, g)$. Particles move on curves, which in turn can be defined in terms of metrics $ds^2 = g_{\mu\nu}dx^\mu dx^\nu$ where (dx^μ) represents a tangent vector along the curve, expressed in terms of differentials defined with respect to a basis. At each point along the curve, we can also assign a vector field, $(\partial/\partial x^\mu)$, whose operation on a function ψ can be associated with a gradient $\frac{\partial\psi}{\partial x^\mu}$ defined over the manifold. In effect, there is a canonical 1-1 correspondence between vectors defined on the tangent space, associated with the directional derivative along a curve, and those which can be identified as gradients defined over the dual space associated with the same curve:

$$T_0^1(M) \leftrightarrow T_1^0(M) \tag{5.2.1}$$

$$v^\mu \partial_\mu \leftrightarrow \tilde{v}_\mu dx^\mu \tag{5.2.2}$$

Note that this is true for general coordinate systems defined over the manifold. The reader is advised to review Sections 0.3–0.4, which introduce the mathematics of exterior algebra and tensors.

Since a manifold is a locally flat surface, we can erect an orthogonal tetrad (vierbien) at each point. In terms of general relativity and the principle of equivalence, this means that there exists such a (local) tetrad at every point of an electromagnetic field that for each $v \in T_0^1$ and the corresponding $\tilde{v} \in T_1^0$ we can write, respectively,

$$v = v^a \partial_a \quad \text{and} \quad \tilde{v} = v_a dx^a$$

The use of tetrads is essential when we work with spinors.

We will now exploit this one-to-one correspondence to identify the quantum-mechanical wave equations as the dual of the Dirac "square-root" of the metric (see also [7] and [8, 9]). Let

$$ds^2 = g_{\mu\nu} dy^\mu dy^\nu = \eta_{ab} dx^a dx^b \tag{5.2.3}$$

where a and b refer to local tetrad coordinates and η to a rigid (local) Minkowski metric. We used (dy^μ) and (dx^a) symbols for the respective tangent vectors in order to emphasize that these are two different coordinate systems. Associated with this metric η are the scalar ds and a vector $\tilde{ds} \equiv \gamma_a dx^a$ such that $\tilde{ds}^2 = ds^2$. Here, γ_a is a Clifford basis, which transforms as a covariant vector under coordinate transformations. Geometrically, $\gamma_a dx^a$ is a vector and ds is its norm. We will work with a matrix representation of $\tilde{ds}$; such a representation was introduced in Section 0.9. There are many possible choices for the matrix representation of γ_a. Regardless of our choice of basis, it is important to define $\{\gamma_a, \gamma_b\} \equiv \gamma_a\gamma_b + \gamma_b\gamma_a = 2\eta_{ab}$ in accordance with the commutation rules of a Clifford basis.

Note also that the relationship between the two metric tensors is given by $g_{\mu\nu}(x) = \eta_{ab} e^a_\mu(x) e^b_\nu(x)$ with $e^a_\mu(x)$ forming local tetrads at x, which are the tensor relationships between the two metric tensors.

Furthermore, ds can be considered as an "eigenvalue" of the linear operator $\tilde{ds}$ by forming the spinor eigenvector equation $\tilde{ds}\xi = ds\xi$. We introduced such an eigenvalue equation already in Section 0.9.3, where the matrix X was an equivalent of $\tilde{ds}$.

Based on the above definitions, we associate the quadratic polynomial of the metric

$$ds^2 = g_{\mu\nu} dy^\mu dy^\nu = \eta_{ab} dx^a dx^b \tag{5.2.4}$$

with the spinor equation:

$$ds\xi = \gamma_a dx^a \xi \tag{5.2.5}$$

This equation, in a natural way, associates spinors directly with the metrics of general relativity. As a next step, we investigate how the spinor ξ of Equation (5.2.5) relates to a quantum-mechanical wave function.

Each vector $\frac{\partial}{\partial x^a}$ can be mapped to a dual one-form dx^a. In a similar way, the $\tilde{ds}$ matrix above can be seen as the dual of the expression $\tilde{\partial}_s \equiv \gamma^a \frac{\partial}{\partial x^a}$, where γ^a is defined by the relationship $\{\gamma^a, \gamma_b\} = 2\delta^a_b$, or equivalently $\eta_{ab}\gamma^a = \gamma_b$.

If we let s describe the length of a particle's trajectory along a curve defined by $(x^0(s), x^1(s), x^2(s), x^3(s)) \in (\mathcal{M}, g)$, then s can be regarded as an independent parameter with an associated 1-form ds, which is the dual of the tangent vector ∂_s. Note that in terms of the basis vectors for $T_p(\mathcal{M})$ and $T_p^*(\mathcal{M})$, we can write $\partial_s = \frac{\partial x^a}{\partial s}\partial_a$ and $ds = \frac{\partial s}{\partial x^a}dx^a$. It also follows that this dual map is given by $ds\partial_s \equiv \frac{\partial s}{\partial x^i}\frac{\partial x^i}{\partial s} = 1$. Putting these two results together allows us to consider Equation (5.2.5) as the dual of the equation:

$$\frac{\partial \psi}{\partial s} = \gamma^a \frac{\partial \psi}{\partial x^a} \tag{5.2.6}$$

where $\frac{\partial}{\partial s}$ refers to differentiation along a curve parametrized by s, and ψ denotes a quantum-mechanical wave function. Geometrically, Equation (5.2.6) is the dual of Equation (5.2.5).

We shall refer to Equation (5.2.6) as a (generalized) Dirac equation and later show how it relates to the usual form of this equation. At times, we shall also refer loosely to it as a "dual wave-equation." **This one-to-one map between Equations (5.2.5) and (5.2.6) formally defines what we mean by wave–particle duality.** They both go together and one cannot exist without the other. The metric corresponds to the particle property and the wave equation to its wave property. Given the metric, we can write down the wave, and conversely, given the wave we can write down the metric. It is only a matter of convention whether we begin first with the wave and then the metric or *vice versa*. It remains to better understand the origin and properties of this wave from a physical point of view. The first step in this understanding is to show the relationship between mass, null metrics, and the classical wave equation. This is formulated more precisely in the following theorem and lemmas.

Theorem 5. *Let $ds^2 = g_{\mu\nu}dy^\mu dy^\nu = \eta_{ab}dx^a dx^b$ define a metric along a curve defined over T_1^0. Then*

$$ds^2 = \eta_{ab}dx^a dx^b \iff 0 = v^2dt^2 - dx_1^2 - dx_2^2 - dx_3^2 \quad \textit{where } v_i = \frac{dx_i}{dt},\ v = |v_i|$$

Proof. Note that $ds^2 = c^2 d\tau^2$, where τ is called proper time by definition, from which it follows that:

$$c^2 d\tau^2 = c^2 dt^2 - \sum_{i=1}^{3} dx_i^2 = c^2 dt^2 \left(1 - \frac{v^2}{c^2}\right) \tag{5.2.7}$$

$$0 = c^2 dt^2 - c^2 d\tau^2 - \sum_i dx_i^2 \tag{5.2.8}$$

$$\iff 0 = c^2 dt^2 \left(1 - \left(\frac{d\tau}{dt}\right)^2\right) - dx_1^2 - dx_2^2 - dx_3^2 \tag{5.2.9}$$

$$\iff 0 = v^2 dt^2 - dx_1^2 - dx_2^2 - dx_3^2 \tag{5.2.10}$$

□

The novelty of the theorem is not that any freely moving particle with or without mass can be identified with motion along a null metric, although this too is important and follows directly from the velocity equation $v = \frac{dx^i}{dt}$ as defined in the laboratory frame. Instead, the novelty rests on the fact that a necessary and sufficient condition for motion along a null metric to be transformed into a non-null metric in Minkowski space is that there should bc an isotropic velocity c valid for all frames of reference such that $\frac{d\tau^2}{dt^2} = 1 - \frac{v^2}{c^2}$ and $v \neq c$. In other words, the relationship between dt and $d\tau$ is defined by a Lorentz transformation. With this said, we now seek a spinor representation of the above theorem.

Lemma 6. *If* $ds^2 = \eta_{ab} dx_a dx_b$ *then*

$$ds\xi = \left(c\gamma_0 dt + \sum_{i=1}^{3} \gamma_i dx^i\right) \xi \iff 0 = vdt + \sum_{i=1}^{3} \gamma_i' dx^i$$

where $\gamma_i' = \frac{c}{v}\left(\gamma_0 + \frac{d\tau}{dt}\right)\gamma_i$, $(\gamma_i')^2 = 1$, $\{\gamma_i', \gamma_j'\} = 0$, *and* $ds = cd\tau$.

Proof. Since $ds^2 = \eta_{ab} dx_a dx_b$ the spinor square root exists such that

$$ds\xi = \left(c\gamma_0 dt + \sum \gamma_i dx^i\right) \xi \quad \text{this is the Dirac square root} \tag{5.2.11}$$

$$0 = \left[(c\gamma_0 dt - ds) + \sum \gamma_i dx^i\right] \xi \quad \text{subtract } ds\xi \text{ from both sides} \tag{5.2.12}$$

$$0 = cdt \left(\gamma_0 - \frac{d\tau}{dt}\right) \xi + \sum \gamma_i dx^i \xi \quad \text{using } ds = cd\tau,\ \text{factoring } cdt \tag{5.2.13}$$

$$0 = \frac{c}{v}\left(\gamma_0 + \frac{d\tau}{dt}\right)\left[cdt\left(\gamma_0 - \frac{d\tau}{dt}\right)\xi + \sum \gamma_i dx^i \xi\right] \quad \text{multiply by } \gamma' \tag{5.2.14}$$

$$0 = vdt\xi + \sum \gamma_i' dx^i \xi \quad \text{note}: \left(\gamma_0 + \frac{d\tau}{dt}\right)\left(\gamma_0 - \frac{d\tau}{dt}\right) = \frac{v^2}{c^2} \tag{5.2.15}$$

It is easy to check that for $1 \le i \le 3$, we get $(\gamma_i')^2 = -\gamma_i^2 = 1$, $\{\gamma_i', \gamma_j'\} = 0$ and $v^2 dt^2 \xi = \sum dx_i^2 \xi$.

It should be clear in the above lemma that γ_i and γ_i' are two different representations of the Dirac spin matrices which are dependent upon each other through the relationship $\gamma_i' = \frac{c}{v}\left(\gamma_0 + \frac{d\tau}{dt}\right)\gamma_i$. Once we have chosen a representation γ_i to linearize the metric of Theorem 5, then the linearization of the null metric will depend on this choice of γ_i or vice versa, as shown in the above lemma. We say more about the choice of γ matrices in the next section.

Corollary 7. *Given the relationship* (5.2.2), *there exists a wave equation given by*

$$\left(\frac{\partial\psi}{v\partial t} = -\sum(\gamma')^i \frac{\partial\psi}{\partial x^i}\right), \quad \text{where } \gamma_i' \textit{ is as in Lemma 6} \tag{5.2.16}$$

Proof. Take the dual of (5.2.15). □

Corollary 8. *We can associate the spinor equation given by* $vdt\xi = \sum_{i=1}^{3} \alpha_i dx^i \xi$ *with the metric* $0 = v^2dt^2 - dx_1^2 - dx_2^2 - dx_3^2$ *where* α_i *is a two-dimensional representation of the spinor matrices. Moreover, this can also be expressed in a four-dimensional representation as* $vdt = \sum\left(\gamma_i' dx^i\right)\xi$, *where* γ_0' *is defined in Lemma 6.*

Proof.

$$0 = v^2dt^2 - dx_1^2 - dx_2^2 - dx_3^2 \tag{5.2.17}$$

$$\Longleftrightarrow v^2dt^2 = dx_1^2 + dx_2^2 + dx_3^2 \tag{5.2.18}$$

Since there are only three independent random variables, we can define a square root over a two-dimensional spinor space with respect to a Pauli spinor basis σ_i where $\sigma_i^2 = 1$ and $\{\sigma_i, \sigma_j\} = 0$ such that

$$vdt\xi = \sum(\sigma_i dx^i)\xi \tag{5.2.19}$$

The σ_i can be embedded in a four-dimensional representation by defining

$$\gamma_i' = \frac{c}{v}\begin{pmatrix} 0 & \sigma_i + \sigma_i\frac{d\tau}{dt} \\ \sigma_i - \sigma_i\frac{d\tau}{dt} & 0 \end{pmatrix}$$

In which case,

$$vdt\xi = \sum(\gamma_i' dx^i)\xi \tag{5.2.20}$$

□

Remark. We chose γ_i' in the above corollary to be consistent with Lemma (6). In fact, if we define

$$\gamma_i = \begin{pmatrix} 0 & -\sigma_i \\ \sigma_i & 0 \end{pmatrix} \qquad \gamma_0 = \begin{pmatrix} I & 0 \\ 0 & -I \end{pmatrix}$$

then we see that $\gamma_i' = \frac{c}{v}\left(\gamma_0 + \frac{d\tau}{dt}\right)\gamma_i$, as in the lemma. In practice, the boundary or initial conditions will dictate the proper choice of γ_i, and if no specific choice is required, the important thing will be to maintain consistency throughout.

Corollary 9. *It follows from Corollary 7 that if v is constant, then*

$$0 = \frac{1}{v^2}\frac{\partial^2\psi}{\partial t^2} - \sum\frac{\partial^2\psi}{\partial x_i^2} \tag{5.2.21}$$

Proof. It is sufficient to multiply Equation (5.2.16) by $-\sum(\gamma')^i\frac{\partial\psi}{\partial x^i}$ or equivalently note that

$$0 = \left(\gamma^0\frac{\partial}{v\partial t} + \sum\gamma^i\frac{\partial}{\partial x^i}\right)\left(\gamma^0\frac{\partial\psi}{v\partial t} + \sum\gamma^i\frac{\partial\psi}{\partial x^i}\right) = \frac{1}{v^2}\frac{\partial^2\psi}{\partial t^2} - \sum\frac{\partial^2\psi}{\partial x_i^2} \tag{5.2.22}$$

□

Both Corollaries 7 and 9 express deep results. It means that the Dirac and the Klein–Gordon equations, respectively, can be reformulated in terms of classical wave equations defined locally over null geodesics, where the velocity of the particle is v. This v is mass (and presumably charge)-dependent. Indeed, from the perspective of special relativity, we can write $v^2 = c^2\left(1 - \left(\frac{m_0}{m}\right)^2\right)$. In other words, all wave equations for quantum particles are distinguished only by their velocity, which in turn will be dependent upon their mass and charge. The wave equation for electromagnetic radiation

Table 5.1. The relationship between space–time metrics and Dirac's original assignments

	Space–time metrics	Dual of the metric	Dirac's assignment
Length	ds	$\frac{\partial\psi}{\partial s}$	mc^2
Time difference	dt	$\frac{\partial\psi}{\partial t}$	ϵ
Spatial distance	dx	$\frac{\partial\psi}{\partial x}$	pc
Eigenvalue eq.	$ds\xi = \gamma_a dx^a \xi$	$\frac{\partial\psi}{\partial s} = \gamma^a \frac{\partial\psi}{\partial x^a}$	$m\psi = i\hbar\gamma^a \frac{\partial\psi}{\partial x^a}$

is a special case that occurs when $v = c$ or equivalently $m_0 = 0$. In the case of general relativity, this mass-charge dependence is expressed in the Reissner–Nordström metric. This result is the physical meaning of the geometric analogy between a quantum mechanical wavefunction and an electromagnetic wave, which we pointed out in Section 2.5.

We can now compare our approach with the traditional Dirac theory by referring to Table 5.1. Column 3 is associated with Dirac's approach and can be derived from Column 2, which is associated with the "dual of the metric," by means of Theorem 17 in Chapter 8. We restate it without proof, for the convenience of reading of this section:

Theorem 10. *Let $p^0 = \epsilon$. Given the relativistic Hamiltonian $\epsilon^2 - (pc)^2 = (mc^2)^2$, we get the $\gamma^a p_a \xi = mc\xi$ equation iff $\psi(p) \equiv \psi^i(\int^x p^a dx_a)e_i$ is a solution of*

$$\frac{\partial}{\partial s}\psi(p) = \gamma^a \partial_a \psi(p)$$

where $p^a dx_a$ is Lorentz invariant and integration is taken along the curve with tangent vector $p^a = m\frac{dx^a}{d\tau}$, and where τ is proper time. Regarding the meaning of $\psi(p) = \psi^i(\int^x p^a dx_a)e_i$ and $\partial_a\psi(p)$, it is important to note that there are many different p^as that obey the above Hamiltonian. Since the integration is taken along the curve with tangent vector p^a, the wave function with integral of form $\int_{x_1}^{x_2}$ will be a function of p^a and therefore we use the $\psi(p)$ notation. We also note that the Lorentz invariance of $p^a dx_a$ and the fact that $x_a = x_a(\tau)$ imply

$$\int^x p^a dx_a = \int^x p^a \frac{dx^a}{d\tau} d\tau = mc^2 \int d\tau = mc^2\tau$$

and therefore

$$\partial_a \psi(p) = \frac{\partial \psi}{\partial \tau} \frac{\partial \tau}{\partial x_a} = \psi' \partial_a \tau = \psi' p_a$$

by the chain-rule of differentiation.

We noted in Section 7.2 and also in Theorem 10 that the traditional Dirac ansatz is to write down an eigenvalue equation corresponding to the linearized $\epsilon^2 - (pc)^2 = (mc^2)^2$ energy–momentum relationship. Essentially, Dirac emulated the Schrödinger approach of first taking a Hamiltonian and substituting a differential operator for the momentum. Schrödinger seems to have historically made a "judicious guess" of the quantum momentum operator by introducing a "purely formal procedure" (his words, [12]) to replace the $\frac{\partial W}{\partial x^a}$ terms of the classical Hamilton–Jacobi equation with the quantum operators $\pm i\hbar \frac{\partial}{\partial x^a}$. Indeed, as we shall see in the next section, our generalized Dirac equation is essentially the Hamilton–Jacobi equation constrained by the quantum mechanics requirement that the wave function ψ be an element of a Hilbert space and $mc^2 = h\nu$, where m is the mass of the particle, ν its wave frequency, and h is Planck's constant. Moreover, the advantage of our approach is that it begins with metrics and not Hamiltonians, and consequently, the wave equation is not some type of hybrid as it is in the Dirac (Schrödinger) case but is the natural dual of the metric. The traditional hybridization has two other consequences: (i) the other terms get multiplied by $i\hbar$ in order to compensate the operator-to-eigenvalue substitution, and (ii) the resulting Dirac equation is no longer a local equation due to the presence of the m coefficient, which represents electromagnetic energy integration over the whole space. While mainstream theorists consider the traditional Dirac equation to be their fundamental starting point, we observed already in Chapter 2 that it is not a proper field equation because m is neither a physical constant nor a local parameter. We gave the example of electron–positron annihilation process, during which the non-local particle masses disappear completely.

In contrast, we work on a more fundamental level with the metrics of space–time. Our generalized Dirac equation is a local equation, and it is therefore the correct foundation for building a relativistically consistent field theory. The spinor appearing in our equation is nothing else than the Zitterbewegung of electromagnetic field, which we discussed in Chapter 3. Under this approach, the wave–particle duality is no longer a mystery, but a mathematical consequence of duality transformations. By avoiding the above-mentioned hybridized Dirac equation, we avoid all the wavefunction interpretation difficulties which have plagued quantum mechanics over the past 100 years. Instead of using Schrödinger's purely formal procedure to guess the eigenvalues shown in the last column of Table 5.1, we will use Hamilton–Jacobi theory to derive them.

There are also algebraic properties related to the duality structure that we explore in the following section. These will help us better understand the physics.

5.2.2. *Clifford algebra properties*

Recall the Clifford multiplication rule given by equation (0.2.5). Using the symmetry property of $\boldsymbol{u} \cdot \boldsymbol{v}$ and the antisymmetry property of $\boldsymbol{u} \wedge \boldsymbol{v}$, the Clifford product, of $\boldsymbol{u}$ and $\boldsymbol{v}$ vectors can be written as

$$2\boldsymbol{uv} = \{\boldsymbol{u}, \boldsymbol{v}\} + [\boldsymbol{u}, \boldsymbol{v}] \tag{5.2.23}$$

Consider the two operators $\tilde{ds}$ and $\tilde{\partial}_s$, which are matrix representations of Clifford vectors. On taking their Clifford product, we find that

$$2\tilde{ds}\,\tilde{\partial}_s\psi = \{\tilde{ds}, \tilde{\partial}_s\psi\} + [\tilde{ds}, \tilde{\partial}_s\psi] \tag{5.2.24}$$

$$= 2\frac{\partial\psi}{\partial s}ds + [\tilde{ds}\psi, \tilde{\partial}_s\psi] \tag{5.2.25}$$

In contrast, if we multiply Equations (5.2.6) and (5.2.5) together, we obtain

$$\tilde{ds}\,\tilde{\partial}_s\psi = \frac{\partial\psi}{\partial s}ds \tag{5.2.26}$$

Equations (5.2.25) and (5.2.26) are compatible if and only if $[\tilde{ds}, \tilde{\partial}_s\psi] = 0$. This is important for many reasons:

(1) In means that both $\tilde{ds}$ and $\tilde{\partial}_s\psi$ share a complete set of common eigenvectors, and therefore, from the perspective of quantum mechanics, both are compatible and simultaneously measurable.
(2) Geometrically, it means that $\tilde{ds}$ and $\tilde{\partial}_s\psi$ are parallel. The curve described (locally) by $\tilde{ds}$ is a geodesic. Indeed, this assumption is implicit in the use of local tetrad coordinates (vierbien).
(3) $d\psi = \frac{\partial\psi}{\partial s}ds$ is an exact differential and obeys the Hamilton–Jacobi equation. The precise meaning of this point will be discussed in what follows.
(4) Historically, Schrödinger work was motivated by the Hamilton–Jacobi equation of classical mechanics [11, 12].

For much of what follows, we will assume that $[\tilde{ds}, \tilde{\partial}_s\psi] = 0$. This is not such a restrictive assumption as we shall see from Lemma 12 in the next section.

Essentially, one can always decompose ψ into a parallel and perpendicular component given by $\psi(x) = \psi_{||} + \psi_{\perp}$. Combined with the requirement of differentiability, this condition means that $\psi_{\perp}$ is purely linear and consequently $[\tilde{ds}, \tilde{\partial}_s \psi_{\perp}] = 0$. In other words, in the case of a real function, the requirement that $\psi_{\perp}$ be an exact differential corresponds to tangential motion with an intrinsic constant angular momentum along the curve.

Now consider the motion of a test particle of mass m along a timelike geodesic. Let $p^a = m(dx^a/d\tau)$, where τ is the proper time (i.e., $ds = cd\tau$). Then

$$ds^2 = \eta_{ab} dx^a dx^b \quad \text{is equivalent to } (mc)^2 = \eta_{ab} p^a p^b \tag{5.2.27}$$

This can be expressed in spinor notation by

$$ds\xi = \gamma_a dx^a \xi \tag{5.2.28}$$

which is equivalent to Dirac's Hamiltonian: $\gamma^a p_a \xi(p) = mc\xi(p)$.

If Equation (5.2.6) is subjected to the constraints of Equation (5.2.28), as it should be for motion along the timelike geodesic, we find that $\psi = \psi^i(\int_{x_0}^{x(s)} p_\mu dx^\mu) e_i$ is a solution of Equation (5.2.6) in a general coordinate system, provided the integration is taken along the curve $s = c\tau$ and $\xi(p) = \frac{d\psi^i(p)}{d\tau} e_i$. This method of solving the Dirac equation will be proven by Theorem 17 of Section 8.2.2.

It is also worth noting that if all of ψ^i are equal, then $\psi = \psi^i(x)u$, where u is a spinor independent of x. In this particular case, the Dirac equation takes on the form

$$(\tilde{\partial}_s \psi^i)u = \frac{\partial \psi^i}{\partial s} u \tag{5.2.29}$$

Finally, we need to say something about the choice of γ matrices. Essentially, we have defined them in terms of the linearized metric $ds\xi = \gamma_a dx^a \xi$. In doing so, this requires that $\{\gamma_a, \gamma_b\} = 2\eta_{ab}$. Note that there are an infinite number of choices for these matrices. In Minkowski space, it is conventional to choose

$$\gamma^0 = \begin{pmatrix} I & 0 \\ 0 & -I \end{pmatrix} \quad \text{and} \quad \gamma^a = \begin{pmatrix} 0 & \sigma^a \\ -\sigma^a & 0 \end{pmatrix} \tag{5.2.30}$$

where the σ^as are constant and correspond to the Pauli spin matrices. However, if we switch to curvilinear coordinates, then the γ matrices may be no longer constant. For example, if

$$ds^2 = dx^2 + dy^2$$

then

$$\tilde{ds} = \alpha_x dx + \alpha_y dy$$

where we choose

$$\alpha_x = \begin{pmatrix} 0 & 1 \\ 1 & 0 \end{pmatrix} \quad \text{and} \quad \alpha_y = \begin{pmatrix} 1 & 0 \\ 0 & -1 \end{pmatrix} \tag{5.2.31}$$

On switching to polar coordinates, $x = r\cos\theta$ and $y = r\sin\theta$, such that

$$\tilde{ds} = \alpha_r dr + \alpha_\theta r d\theta$$

and

$$\alpha_r = \begin{pmatrix} \sin\theta & \cos\theta \\ \cos\theta & -\sin\theta \end{pmatrix} \quad \text{and} \quad \alpha_\theta = \begin{pmatrix} \cos\theta & -\sin\theta \\ -\sin\theta & -\cos\theta \end{pmatrix} \tag{5.2.32}$$

Clearly, α_r and α_θ are now variables. In the Heisenberg representation of quantum mechanics, operators are variables that usually evolve in time. Specifically, when discussing the Zitterbewegung problem, the α operators are variables such that $i\dot{\alpha} = [\alpha, H_o]$, where $H_o = \alpha.\mathbf{p} + \beta m$. We will discuss this later on.

5.2.3. *Hamilton–Jacobi functions and the Dirac equation*

If we return to Equations (5.2.25) and (5.2.26), there is another way of viewing the relationship between both expressions in terms of Hamilton–Jacobi functions and the notion of exact differentials. In fact, (5.2.26) is an exact differential in that [1]

$$\tilde{ds}\,\tilde{\partial}_s\psi = \frac{\partial\psi}{\partial s}ds = \frac{\partial\psi}{\partial x^a}dx^a \tag{5.2.33}$$

and as we shall see, can be associated with a coherent set of natural motions [13]. Indeed, to remove any ambiguity, we begin with the following definitions:

Definition 11. A function $W = \int_{\sigma(\lambda)} \left(\mathbf{p}\frac{\mathbf{dx}}{d\lambda} - H\frac{dt}{d\lambda}\right) d\lambda$ is called a Hamilton–Jacobi function if the integral is path independent for all curves in $\{\sigma(\lambda) \in (M, g)\}$ and $\frac{\partial W}{\partial t} = -H(x_1, x_2, x_3, t, \frac{\partial W}{\partial x_1}, \frac{\partial W}{\partial x_2}, \frac{\partial W}{\partial x_3})$ defined with respect to a local tetrad. Equivalently, we can say that $dW = \mathbf{pdx} - Hdt$ is an exact differential.

As we saw in Chapter 3, the electron's spherical charge surface is at a constant potential. **The Hamilton–Jacobi function W defines a potential, and it is related to the electromagnetic vector potential field A.** The relationship between the electron momentum vector and the vector potential field A is discussed in Section 5.3.3:

$$\mathbf{p} = m_e v = eA_{\|} \tag{5.2.34}$$

where $A_{\|}$ is the component of the vector potential seen by the electron's charge parallel to velocity vector $\boldsymbol{v}$.

In relation to Equations (5.2.25) and (5.2.26), **we state (and prove) the following important theorem**:

Lemma 12. *Let $\psi(W(\mathbf{x},t))$ be a differentiable function and $\{\sigma(\lambda)\}$ a family of curves on the manifold with unit tangent vectors $\frac{\tilde{ds}}{ds}$ with respect to a local tetrad, then $\tilde{ds}.\tilde{\partial}_s\psi(W)$ is an exact differential iff $\psi(W)$ is a Hamilton–Jacobi function such that $p_a^* = \frac{d\psi}{ds}\frac{dx_a}{dt} = \frac{\partial\psi_{\|}(W)}{\partial x^a}$, where $\psi(W) = \psi_{\|}(W) + \psi_{\perp}(W)$ and $\psi_{\perp}(W) = c_0 t + c_1 x_1 + c_2 x_2 + c_3 x_3$, with c_0, c_1, c_2, c_3 being constants.*

Proof. See Appendix, Lemma 16. □

Corollary 13. *Let $\tilde{ds}.\tilde{\partial}_s\psi(W)$ be an exact differential, then $2\tilde{ds}.\tilde{\partial}_s\psi(W) = 2\frac{\partial\psi}{\partial x^a}dx^a = \{\tilde{ds}, \tilde{\partial}_s\psi(W)\}$.*

Proof. This follows both from Equation (5.2.23) and as noted in proof of Lemma (12), and from the exact differentiability of $\tilde{ds}.\tilde{\partial}_s\psi(W)$, which means $[\tilde{ds}, \tilde{\partial}_s\psi(W)] = 0$. □

In effect, Lemma 12 establishes a relationship between Hamilton–Jacobi functions and the commutator relationship, $[\tilde{ds}, \tilde{\partial}_s\psi(W)]$. We now use the same commutator relationship to establish another important property relating the dual operator $\tilde{\partial}_s$ and the metric operator $\tilde{ds}$ associated with the increments along a curve. Intuitively, we could think of $\tilde{\partial}\psi(W(s))$ as a wave associated with the vibration of a curve $\sigma(s)$ in space–time, whose tangent is $\tilde{ds}$, with respect to a local tetrad coordinate system. This leads to the following lemma:

Lemma 14. *If $\psi(W)$ is a Hamilton–Jacobi function such that $[\tilde{\partial}_s W, \tilde{ds}] = 0$, then there exists a simultaneous eigenfunction ξ such that*

$$(\tilde{\partial}_s\psi)\xi(p) = \partial_s\psi\xi(p) \tag{5.2.35}$$

where $\partial\psi_s = \frac{\partial\psi}{\partial s}$, *which in the case of geodesic motion reduces to*

$$\tilde{\partial}_s \Psi = \frac{d\Psi}{ds}, \quad \textit{where } \Psi = \psi\xi \tag{5.2.36}$$

Remark. $(\tilde{\partial}_s\psi)\xi = \tilde{\partial}_s\Psi$ in general, since x is independent of p in phase space. However, $\frac{d\Psi}{ds} \neq \psi'(p)\xi$ unless motion is along a geodesic.

Proof. First note that $[\tilde{\partial}_s W, \tilde{ds}] = 0$ implies $[\tilde{\partial}_s\psi, \tilde{ds}] = [\psi'(W)\tilde{\partial}_s W, \tilde{ds}] = 0$. Therefore, there exist simultaneous eigenvectors $\xi = \xi(p)$ such that $\tilde{ds}\xi = ds\xi$ and $(\tilde{\partial}_s\psi)\xi = \gamma^a p_a^* = \gamma^a p_a \psi'\xi(p) = mc\psi'(p)\xi(p) = (\partial_s\psi)\xi(p)$. Also, $\xi(p)$ is constant along a geodesic and therefore

$$\tilde{\partial}_s \Psi = \frac{d\Psi}{ds}, \quad \text{where } \Psi = \psi\xi$$

The result follows. □

Remark: We refer to Equation (5.2.35) as a generalized Dirac equation associated with a curve, and (5.2.36) as a generalized Dirac equation associated with geodesics. Both of these are subsumed by Equation (5.2.6) which we have also referred to as a generalized Dirac equation. It reduces to the usual form of the Dirac equation if we let $\psi = Ae^{\Lambda W}$, where A is an arbitrary constant and $\Lambda = \frac{i}{\hbar}$.

5.2.4. *Exact differentials and metrics*

In this section, we investigate the relationship between the Hamilton–Jacobi function W considered as an exact differential and its relationship to a family of metrics, ds. Specifically, since dW is exact, then $p_a = \frac{\partial W}{\partial x^a}$ depends explicitly only on the coordinates and not on the parametrization *per se*. Therefore, to be consistent we are required to choose a λ in Definition 11, so that p_a is invariant with respect to some agreed (universal) standard parameter, which emerges naturally from the geometrical and dynamical requirements of the system. For example, following the usual rule of dynamics, we require that in the instantaneous rest frame

$$p_a = m(s)c\frac{dx^a}{ds}$$

where $m(s)$ is the instantaneous mass on a specific curve, parametrized with respect to its proper time ds. Note that $m(s)$ is not necessarily a constant except along a geodesic. Moreover, if we were to change the parameter from s to some other s', then for p_a to be invariant at each point with respect to its coordinate system would require that

$$p_a = m(s)c\frac{dx^a}{ds} = m(s')c\frac{dx^a}{ds'} \text{ with } \frac{ds'}{m(s')} \equiv \frac{ds}{m(s)} \tag{5.2.37}$$

This suggests that in a gravitational field we should define $d\lambda = \frac{ds}{m(s)}$ as a universal parameter with respect to some standard particle. Note in the laboratory frame, we would have $\frac{ds}{m(s)} = \frac{dt}{m(t)}$ [10]. Other references to this particular parametrization can also be found in [2, 4].

With that said, we now address the important question of which comes first, the Hamilton–Jacobi function or the metric associated with the dynamics of the particle. In reality it does not matter provided one or the other is an exact differential, although there are two points to keep in mind:

(1) Given an arbitrary metric $ds^2 = dx_a dx^a$ does not necessarily mean that we can construct a Hamilton–Jacobi function, W, from it. More precisely, given a metric ds there does not necessarily exist an integrating factor $\rho = m(x^a)c$ such that $dW = \rho ds$ is an exact differential. For example, if

$$ds = -ydx + xdy + kdz$$

and we require $dW = \rho(x, y, z)ds$, then on solving we find that $\rho = 0$. On the other hand, if we let

$$ds = \frac{yz}{\rho}dx + \frac{xz}{\rho}dy + \frac{xy}{\rho}dz$$

then for all $\rho(x, y, z) \neq \rho(xyz)$, ds is not exact although $dW = \rho ds$ is always exact.

(2) Given an exact differential dW can we associate a metric with it? The answer is yes.

Theorem 15. *Let dW be an exact differential such that $dW = \frac{\partial W}{\partial x^a}dx^a$, then a family of curves with metric $ds^2 = dx_a dx^a$ exists such that $dW = \rho ds$, where ρ is an integrating factor such that $\rho^2 = \frac{\partial W}{\partial x^a}\frac{\partial W}{\partial x_a}$.*

Proof. Note that if there exists a metric ds such that $dW = \rho ds$, then one can also define $ds_1 = \Lambda ds$ and $\rho_1 = \frac{\rho}{\Lambda}$ such that $dW = \rho_1 ds_1$, indeed, one can choose $\rho_1 = 1$. This presents us with the question of choosing a standard (universal) parameter for the system. For what follows, we choose ds such that dx^a/ds is the unit tangent vector along a family of regular curves (meaning $ds \neq 0$) on some interval of s. Since dW is exact, the value of $W(x_a)$ is independent of the path, and therefore the choice of ρ will depend on the parametrized coordinate system (see what follows). In fact,

$$\begin{aligned} dW &= \rho ds && (5.2.38)\\ &= \rho\left(\frac{dx_a}{ds}dx^a\right) && (5.2.39)\\ &= \left(\rho\frac{dx_a}{ds}\right)dx^a && (5.2.40)\\ &= \frac{\partial W}{\partial x^a}dx^a \text{ since } dW \text{ is an exact differential} && (5.2.41)\end{aligned}$$

Therefore,

$$\frac{\partial W}{\partial x^a} = \rho\frac{dx_a}{ds}$$

But by definition of ds^2, we have that $\frac{dx_a}{ds}\frac{dx^a}{ds} = 1$ and consequently $\rho^2 = \frac{\partial W}{\partial x_a}\frac{\partial W}{\partial x^a}$ and invariant under Lorentz transformations. □

Remark 1. In the case of a Hamilton–Jacobi function W, we note that the integrating factor ρ has the units of momentum. Indeed, if we let $d\lambda = ds/m(x^a)$ be the universal parameter as suggested above, then $\rho = m(x^a)c$ along the curve.

Remark 2. If $\rho = \frac{dW(s)}{ds}$, then by the chain rule both dW and ds are exact. To see this it is sufficient to note that $dW = \frac{dW}{ds}ds$ and $ds = \frac{ds}{dW}dW$.

Remark 3. In the case of thermodynamics, W would be entropy, s the total heat energy and $\rho = 1/T$ where T is temperature.

5.2.5. *Some examples*

The above theorem is subtle. It does not claim that every metric can be associated with an exact differential but the opposite, namely that every exact differential can be associated with a metric and consequently a specific family of curves. We give three examples:

(1) Consider a metric given locally by $ds^2 = dx_a dx^a$ such that

$$s = \cos(\theta_o)t - i\cos(\theta_j)x^j,\ j \in \{1, 2, 3\}$$

and $\cos(\theta_o)$ a constant. s is clearly exact. Consequently, we may take $W = m_o cs$ and $\rho = m_o c$. Or applying the theorem directly, we obtain

$$\rho^2 = \frac{\partial W}{\partial x_a}\frac{\partial W}{\partial x^a} = (m_o c)^2 \cos(\theta^a)\cos(\theta^a) = (m_o c)^2$$

Note $\frac{\partial s}{\partial x^a} = \frac{dx^a}{ds} = \cos(\theta^a)$ by definition of directed cosines on a local Minkowski space. It is easy to check that $\nabla(s).\vec{dx} = ds$. In this case, W is a Hamilton–Jacobi function along a geodesic.

(2) As a second example consider $W = xy$, then we have

$$dW = \frac{\partial W}{\partial x}dx + \frac{\partial W}{\partial y}dy \tag{5.2.42}$$

$$= ydx + xdy \tag{5.2.43}$$

From the above theorem, we obtain,

$$\rho^2 = \frac{\partial W}{\partial x_a}\frac{\partial W}{\partial x^a} = y^2 + x^2$$

$$\frac{\partial W}{\partial x} = y = \rho\frac{dx}{ds} \tag{5.2.44}$$

and

$$\frac{\partial W}{\partial y} = x = \rho\frac{dy}{ds} \tag{5.2.45}$$

This means that

$$\frac{dy}{dx} = \frac{x}{y} \quad \text{and} \quad y^2 - x^2 = \kappa^2 \tag{5.2.46}$$

which defines a family of hyperbolae. Therefore, the standard metric associated with W is given by

$$ds = \frac{y}{\sqrt{x^2+y^2}}dx + \frac{x}{\sqrt{x^2+y^2}}dy$$

Note that ds is not an exact differential but $\sqrt{(x^2+y^2)}ds$ is exact.

(3) As a third example, we define $W = x^2 + y$, then $dW = 2xdx + dy$. From the above theorem, we obtain,

$$\frac{\partial W}{\partial x} = 2x = \rho\frac{dx}{ds} \tag{5.2.47}$$

and

$$\frac{\partial W}{\partial y} = 1 = \rho \frac{dy}{ds} \tag{5.2.48}$$

This means that

$$\frac{dy}{dx} = \frac{1}{2x} \iff y = \ln \sqrt{x} + \kappa, \text{ and } \rho^2 = \left(\frac{\partial W}{\partial x}\right)^2 + \left(\frac{\partial W}{\partial x}\right)^2 = 4x^2 + 1 \tag{5.2.49}$$

which defines an exponential family of curves given by $x = A\exp(2y)$.

5.2.6. *Wave–particle duality and the Zitterbewegung phenomenon*

We now analyze the motion of a quantum particle keeping in mind that, unlike the classical case, the quantum particle is subjected to quantization conditions or equivalently the Heisenberg uncertainty relations. Indeed, regular quantum mechanics presupposes the wave function to be an element of a Hilbert space that obeys continuity conditions at the boundaries, which corresponds to standing wave solutions for atomic oscillators. The energy of these standing waves are related by the Planck–Einstein formula given by $E = h\nu$, where h is Planck's constant and ν is a frequency of an atomic oscillator or equivalently using the de Broglie relation $p = h/\lambda$, where λ corresponds to the particle wavelength and h is a constant of proportionality that in effect behaves as a scaling factor for measuring energy and momentum. h takes on different values according to the system of units that is being used. In the preceding chapters, natural units were frequently used. However, it is also important to note that h can never be 0.

The second point is that in the case of a Hilbert space, or more precisely L^2 functions, the inner product $\langle\psi|\psi\rangle$ not only has a statistical interpretation but is gauge invariant with respect to phase factors. This means that if $|\psi\rangle = e^{iW}|\xi\rangle$, then $\langle\psi|\psi\rangle = \langle\xi|\xi\rangle$ is invariant for all W. Note that if the i were dropped, then $\langle\psi|\psi\rangle$ would not necessarily be invariant.

This brings us to a third point. In conventional quantum mechanics, the momentum operator is usually defined by $p\psi = -i\hbar\frac{\partial\psi}{\partial x^\mu}$. This effectively changes nothing from the perspective of the generalized Dirac equation in that

$$\frac{\partial \psi}{\partial s} = \gamma^a \frac{\partial \psi}{\partial x^a} \iff i\hbar\frac{\partial \psi}{\partial s} = i\hbar\gamma^a \frac{\partial \psi}{\partial x^a}$$

To fully exploit the gauge invariance mentioned above, when we refer to the original Dirac equation, convention has it that $i\hbar\frac{\partial\psi}{\partial s} = mc^2\psi$ where $\psi = \exp(-(i/\hbar)mc^2\xi)$. In other words, the requirement of gauge invariance and that the eigenvalues be real suggests redefining the Hamilton–Jacobi operators by multiplying them by an $i\hbar$ term. The important thing is to be consistent throughout. Therefore, we define $\mathbf{p_a} = i\hbar\partial_a$ with the understanding that $p_a = i\hbar\partial_a W$. Note that this implies $\mathbf{p^a} = \eta^{ab}\mathbf{p_b}$, which means that $\mathbf{p}^0 = \mathbf{p}_0$ but $\mathbf{p}^1 = -i\hbar\partial_1$, $\mathbf{p}^2 = -i\hbar\partial_2$ and $\mathbf{p}^3 = -i\hbar\partial_3$. Recall, in terms of Hamilton–Jacobi theory, we have $p = \frac{\partial\psi(W)}{\partial x^\mu}$. Indeed, if $W = mc^2\tau$, then from Theorem 10 we can write

$$H_0 = \frac{\partial W}{\partial \tau} = mc^2 \quad \text{and} \quad H_0^* \equiv \frac{\partial e^W}{\partial \tau} = mc^2 e^W$$

However, if we replace W with $(-i/\hbar)W$ and write the Hamilton–Jacobi operator as $i\hbar\frac{\partial}{\partial\tau}$, then we find that once again,

$$H_0 = i\hbar\frac{\partial(-i/\hbar)W}{\partial \tau} = \frac{\partial W}{\partial \tau} = mc^2 \quad \text{and} \quad H_0^* \equiv i\hbar\frac{\partial e^{(-i/\hbar)W}}{\partial \tau} = mc^2 e^{(-i/\hbar)W}$$

To conclude, the generalized Dirac equation is a Hamilton–Jacobi equation both for classical (relativistic) mechanics and quantum mechanics. In the classical case, the line increment from which the Hamilton–Jacobi function is derived gives rise to a dual non-quantum "wave-function," which is a point mass. In contrast, in the case of quantum mechanics the same line increment is dual to a family of L^2 functions in such a way that the initial boundary conditions coming from the physics are statistical, non-deterministic, and incorporate quantization by seeking standing wave solutions. In other words, the mechanics of a strictly classical particle can be determined (in principle) from the initial conditions applied directly to the properties of the line increments, with the non-quantum "wave-equation" representing a point-mass and not contributing any additional information. In the case, of a quantum particle, because of the Heisenberg uncertainty principle, the opposite appears to be true. It is precisely the "wave-equation" that encapsulates the kinematics of the particle, although the solution to the "wave-equation" is dependent upon the line increments associated with the classical particle. This also gives a new insight into the Principle of Complementarity, in that the general solution to the Dirac wave equation is composed of a point mass solution (δ function) plus a wave solution (an L^2 function). This in turn, enables us to have a more complete understanding of the Zitterbewegung phenomenon.

Consider a particle of rest mass m_o moving with uniform velocity u with respect to proper time along the x-axis in Minkowski space. This also means that the motion of the particle with respect to two different frames in uniform motion relative to each other are related by Lorentz transformations. The Hamilton–Jacobi function of the particle is given by $W = m_o u x - m_o c^2 \dot{t} t = -m_o c^2 \tau$ such that $H = m_o c^2 \dot{t}$. In the rest frame, when $t = \tau$, we have $\dot{t} = 1$ and consequently $H_0 = m_0 c^2$. Indeed, as a classical (relativistic) particle with $x = 0$ when $\tau = 0$, then $x = u\tau$, where $u = \frac{dx}{d\tau}$. In terms of the coordinate system (x, t) of the laboratory frame, this can be written as $x = vt$, where $v = (u\frac{d\tau}{dt})$ is constant and obeys the wave equation

$$\frac{1}{v^2}\frac{\partial^2 \psi}{\partial t^2} - \frac{\partial^2 \psi}{\partial x^2} = 0$$

associated with the null metric $0 = v^2 dt^2 - dx^2$ (cf. Corollary 9). The general solution of this equation is given by $\psi(t, x) = Af(x - vt) + Bg(x + vt)$. However, from the initial conditions, and the fact that we are describing a point "quantum" particle, we obtain a specific solution of the form $\psi(x, t) = \delta_k(x - vt)$, k a constant [3].

This information can also be derived by noting that the Hamilton–Jacobi function for a null metric is given by $W_0 = m_0 ck$, an unknown constant where W_0 refers to the Hamilton–Jacobi function associated with the null metric. Therefore, for a point particle the general solution is of the form $\psi(W_0) = \delta_k$. But $x - vt = k$ for a particle moving with uniform velocity v which means $\psi(W_0) = \delta_k(x - vt)$. It might be worth noting that W and W_0 are related by the following identities:

$$\frac{\partial W_0}{\partial t} = \frac{v^2}{c^2}\frac{\partial W}{\partial t} \quad \frac{\partial W_0}{\partial x} = \frac{\partial W}{\partial x}$$

or in other words, W_0 is derived from W by rescaling the local time variable.

As discussed in Chapter 2, in physical terms this means that the particle's center of mass moves in a noise-free environment. The wave function *per se* adds no new information. For elementary particles like electrons, such initial conditions are unknown in principle because of the uncertainty relations. Indeed, precisely because the initial position is unknown, at best we can write $\psi(x, t) = \delta_k(x - vt)$, where k is some unknown constant, which also means that while motion might be deterministic in principle, in practice (and in principle) it is unknowable. We can only hope to glean more information by way of probabilistic methods which are not a cloak for ignorance but rather capture the ontological reality of vacuum–matter interactions.

The above equations strongly indicate that an elementary particle is not a point particle in a classical sense and appears to have some structure, implicitly suggested by combining Equations (5.2.10) and (5.2.21) with the Zitterbewegung. They suggest that a free falling particle obeys the same equation as a transversal electromagnetic wave, but with a different velocity such that $v = v(m)$ with $v(0) = c$. This interesting analogy was already introduced and motivated in Section 2.5. In terms of quantum mechanics, it is as if wave–particle duality corresponds to an entrapped electromagnetic wave with energy given by $E = mc^2 = nh\nu$, with each elementary particle being characterized by its own distinct ν. As we shall now see, this latter characteristic offers one explanation of the Zitterbewegung phenomenon.

In contrast to the classical particle problem, the Heisenberg uncertainty relations, which are given by $\Delta x \Delta p \geq \hbar/2$ and $\Delta E \Delta t \geq \hbar/2$, change things radically and suggest that there is an underlying interference due to the presence of electromagnetic fluctuations interacting with the particle structure. Indeed, the whole study of decoherence related to entanglement is another indication that the background noise is a determining factor in particle motion. In that regard, the act of measurement is just another form of interference which is also subjected to the uncertainty relations. And these relations, combined with de Broglie's formula, lead to quantum mechanics.

To better understand this, and also based on the results of Section 8.9, let us restrict energy exchanges between particles to a multiple of nh and write $W + \Delta W = W + nh$ ([13, p. 461]). This leads to the simple formula:

$$(W + \Delta W - W) = \Delta W = nh \tag{5.2.50}$$

from which it follows that in the rest frame of a free particle, we have

$$\frac{dW}{d\tau} = H_0 \approx \frac{\Delta W}{\Delta \tau} = \frac{nh}{\Delta \tau} = nh\nu_o, \quad \text{where } \nu_o = 1/\Delta\tau \tag{5.2.51}$$

This means that every elementary particle with rest mass m_0 can be characterized by a standard frequency (wavelength) given by the equation $m_0c^2 = h\nu_o$. This is de Broglie's formula, which we already introduced in Chapter 2.

In this case, because of the uncertainty principle, the particle characteristics are embedded within the wave function and not *vice versa*. This can be especially seen in the Zitterbewegung effect. As previously noted, the position of a particle constrained to move on the line is unknown because of the uncertainty relations. The best we can do is describe the position by means of a uniform density function $f(x, t) = 1/\xi$ for $x \in [0, \xi]$ and introduce a wave function on a Hilbert space whose inner product gives the probability distribution. We associate the Hamilton–Jacobi equation not with the probability f but with the wave function ψ and also encapsulate Planck's constant into ψ. In other words, $f = \langle\psi,\psi\rangle$ and ψ is called the quantum wave function. This suggests writing the wave function for a free particle (constant p^a) in spinor notation as a self-adjoint eigenfunction $\psi(W) = \psi(\int p^a dx_a)$. Moreover, for our notation to be consistent with the traditional Heisenberg approach of solving the $i\dot{\alpha} = [\alpha, H]$ equation to explain the Zitterbewegung problem, we note by Corollary 13 that we can use inner product notation, analogous to Heisenberg's outer product notation, to write $2\int p^a dx_a = \left\{\tilde{\partial}_s\psi(W), \tilde{ds}\right\}$. It follows that:

$$\psi(W) = \psi^i(2\int^x p^a dx_a)e_i \tag{5.2.52}$$

$$= \frac{\exp(-2i\hbar^{-1}H_o s)}{\sqrt{\xi}}\psi_o \tag{5.2.53}$$

$$= \frac{\exp(-2i\hbar^{-1}m_0 c^2\tau)}{\sqrt{\xi}}\psi_o \tag{5.2.54}$$

$$= \frac{\exp(-2i\hbar^{-1}m_0 c\lambda)}{\sqrt{\xi}}\psi_o \tag{5.2.55}$$

This Zitterbewegung structure can be associated with a periodic (simple harmonic) isotropic vibration of a particle with rest energy mc^2 and with de Broglie wavelength λ. Regarding Zitterbewegung frequency, we arrive at the same result as Chapter 2: the frequency of Zitterbewegung vibration is the de Broglie frequency, and the above derivation shows the origin of double frequency in Schrödinger's Zitterbewegung analysis. Many textbooks refer to the frequency of Equation (5.2.55) as their Zitterbewegung frequency, but we showed in Chapter 2 that it should be interpreted as the optical spinor frequency.

It should also be noted that the above formula contains no mention of electric charge. In the next section, we check the relationship between this result and the electromagnetic structure identified in Chapter 3 results.

Remark. In many textbooks, the Zitterbewegung problem is analyzed from the perspective of the Heisenberg equation of motion $i\dot{\alpha} = [\alpha, H_0]$, for the "velocity operator" α and the free particle Hamiltonian, $H_0 = \alpha\mathbf{p} + \beta m$. This equation of motion has the formal solution [5]:

$$x(t) = k + vt - ke^{-2iH_o t/\hbar}$$

This contains the linear part that corresponds to the particle motion and expressed by $x(t) = k + vt$ and the wave part (the Zitterbewegung effect) expressed by the $e^{-2iH_o t/\hbar}$. However, the presentation given above using a Dirac (Schrödinger) approach is more intuitive and helps us better understand the wave–particle properties implicit in the Heisenberg approach.

5.2.7. *Summary of this section*

We summarize the key equations and results of Section 5.2. From our perspective, wave–particle duality is associated with two spinor equations pertaining to a quantum mechanical wave equation and a metric with associated spinor ξ, respectively. Each one is the dual of the other and we have used this fact to give a more intuitive understanding of the Zitterbewegung frequency.

In this regard, we make some final observations before discussing the electron:

- In general, for any Hamilton–Jacobi function $\psi(W)$ it is possible to define $\tilde{\partial}_s\psi = \tilde{\partial}_s\psi_{\parallel} + \tilde{\partial}_s\psi_{\perp}$ such that $[\tilde{\partial}_s\psi_{\parallel}, \tilde{d}s] = 0$ and $\{\tilde{\partial}_s\psi_{\parallel}, \tilde{d}s\} = 0$. $\tilde{\partial}_s\psi_{\parallel}$ is the projected cosine along $\tilde{d}s$ and satisfies
$$(\tilde{\partial}_s\psi)_{\parallel}\xi(p) = (\partial_s\psi)_{\parallel}\xi(p), \quad \text{with } p_a = \frac{\partial W}{\partial x^a}$$
where $\tilde{d}s\xi(p) = ds\xi(p)$. Also, $\tilde{\partial}_s\psi_{\perp}$ defines the spin along $\tilde{d}s$.
- The Hamilton–Jacobi function can be re-written in covariant form for a general coordinate system as follows:
$$dW = g^{\mu\nu}p_\mu dx_\nu$$
with the corresponding wave operator
$$\tilde{\gamma}^\mu \frac{\partial\psi}{\partial x^\mu}\xi = \tilde{\gamma}^\mu p_\mu \psi' \xi$$
associated with the action along a curve, provided $2g^{\mu\nu} = \tilde{\gamma}^\mu\tilde{\gamma}^\nu + \tilde{\gamma}^\nu\tilde{\gamma}^\mu$, where $\tilde{\gamma}^\mu = \frac{\partial x^\mu}{\partial x^a}\gamma^a$.
- The generalized Dirac equation
$$\tilde{\gamma}^\mu \frac{\partial\psi}{\partial x^\mu}\xi = \frac{\partial\psi}{\partial s}\xi$$
can be defined along an arbitrary curve and is always covariant in such a manner that when it is multiplied by its dual (metric) equation, we obtain a Hamilton–Jacobi equation.

- Gauge potentials of the form A^{μ} can be introduced into the system by defining

$$p_{\mu} = m_o \frac{dx_{\mu}}{d\tau} = \partial_{\mu} W - eA_{\mu}$$

However, in general $\oint A^{\mu} dx_{\mu} \neq 0$ and therefore is not an exact differential. In other words, $\tilde{ds}\partial_s \tilde{W} \neq \frac{d\psi}{ds} ds$ but does obey Equation (5.2.25).
- As already noted, the above approach deepens our understanding of the Principle of Complementarity. The particle properties should be directly associated with the path increments. The wave properties emerge from Equation (5.2.5).
- It should be clear that the strict form of the Dirac equation (5.2.35) pertains to the kinematics and not the dynamics of the motion. It describes the kinematics with respect to a local tetrad. The dynamics requires further work. Indeed, the restriction of the motion to geodesics also explains why we obtain distinct energy and momentum levels. Geodesic motion presupposes constant momentum and energy, which can either be discrete or form a continuum depending on the boundary conditions. Quantum mechanics associates these constants with quantum numbers.

$$ds^2 = g_{\mu\nu} dx^{\mu} dx^{\nu} = \eta_{ab} dx^a dx^b \longleftrightarrow ds\xi = \gamma_a dx^a \xi$$

$$\frac{\partial^2 \psi}{\partial s^2} = \eta_{ab} \frac{\partial^2 \psi}{\partial x^a \partial x^b} \longleftrightarrow \frac{\partial \psi}{\partial s} = \gamma^a \frac{\partial \psi}{\partial x^a}$$

$$\frac{\partial \psi}{\partial s} = \gamma^a \frac{\partial \psi}{\partial x^a}$$

$$d\psi = \frac{\partial \psi}{\partial x^a} dx^a$$

$$ds^2 = \eta_{ab} dx^a dx^b \longleftrightarrow 0 = v^2 dt^2 - dx_1^2 - dx_2^2 - dx_3^2$$

$$\frac{\partial^2 \psi}{\partial s^2} = \eta_{ab} \frac{\partial^2 \psi}{\partial x^a \partial x^b} \longleftrightarrow 0 = \frac{1}{v^2} \frac{\partial^2 \psi}{\partial t^2} - \frac{\partial^2 \psi}{\partial x_1^2} - \frac{\partial^2 \psi}{\partial x_2^2} - \frac{\partial^2 \psi}{\partial x_3^2}$$

$$ds\xi = \gamma_a dx^a \xi \longleftrightarrow 0 = v\gamma_0 dx^0 + \gamma_1 dx^1 + \gamma_2 dx^2 + \gamma_3 dx^3$$

$$\frac{\partial \psi}{\partial s} = \gamma^a \frac{\partial \psi}{\partial x^a} \longleftrightarrow 0 = \frac{1}{v} \gamma^0 \frac{\partial \psi}{\partial t} + \gamma^1 \frac{\partial \psi}{\partial x_1} + \gamma^2 \frac{\partial \psi}{\partial x_2} + \gamma^3 \frac{\partial \psi}{\partial x_3}$$

5.3. Geometric Interpretation of e^- Mass and de Broglie Wavelength

5.3.1. *Electromagnetic analysis of electron mass and Zitterbewegung radius*

We briefly recap some relevant results of Chapter 3: if an electron moves along an axis z orthogonal to its charge rotation plane, it will describe a helical trajectory whose length is $L = c\Delta t$ and whose position change along the z axis is $l = v_z\Delta t$. The electron mass is exactly equal to the inverse of the helix radius r if expressed in NU, i.e., $m = r^{-1}$. An acceleration along z implies a smaller radius and, hence, a mass increase. We showed in Chapter 3 that using the Pythagorean theorem it is possible to write the value of the radius r as a function of v_z:

$$r = r_e\sqrt{1 - \frac{v_z^2}{c^2}}$$

and the related mass variation is

$$m = \frac{\hbar\omega}{c^2} = \frac{m_e}{\sqrt{1 - \frac{v_z^2}{c^2}}}$$

The electron mass, expressed in NU units, is exactly equal to the inverse of the helix radius r, which means $m = r^{-1}$.

5.3.2. *Relativistic analysis of electron mass and Zitterbewegung radius*

Taking the relativistic perspective of Section 5.2, we assume that the wave function for a free particle with charge, moving along the z axis, is a Hamilton–Jacobi function and consequently an exact differential. Therefore, by Lemma 12, it can be decomposed into $\psi = \psi_{||} + \psi_{\perp}$, where $\psi_{\perp} = -Et + p_r r + p_\theta r\theta$ in tetrad coordinates. The angular momentum is defined by the $\psi_{\perp}$ term.

It follows that the charge momentum in cylindrical (tetrad) coordinates is given by the four-momentum

$$p_c^* = \left(-\frac{E}{c}, \frac{\partial\psi_{\perp}}{\partial r}, \frac{\partial\psi_{\perp}}{r\partial\theta}, \frac{\partial\psi_{||}}{\partial z}\right)$$

For what follows, we will consider r to be a constant and decompose p_c into two components given respectively by

$$\boldsymbol{p}_{||}^{*} = \frac{\partial \psi_{||}}{\partial z} \equiv m v_z \quad \text{and} \quad p_{\perp}^{*} = \left(-\frac{E}{c}, \frac{\partial \psi_{\perp}}{r \partial \theta}\right) \tag{5.3.1}$$

where $\psi_{||}$ is a Hamilton–Jacobi function, p^* is defined as in Lemma 12, and consequently m is defined by Equation (5.3.1). These components can be calculated from the Minkowski metric in the following manner:

$$ds^2 = c^2 dt^2 - r^2 d\theta^2 - dz^2 \tag{5.3.2}$$

$$c^2 \left(\frac{d\tau^2}{dt^2}\right) = c^2 - r^2 \dot{\theta}^2 - v_z^2 \tag{5.3.3}$$

Therefore, for the purpose of what follows we define

$$v_{\perp}^2 \equiv c^2 - v_z^2 = c^2 \left(\frac{d\tau}{dt}\right)^2 + r^2 \dot{\theta}^2 = c^2 \left[\left(\frac{d\tau}{dt}\right)^2 + \frac{r^2}{c^2} \dot{\theta}^2\right] \tag{5.3.4}$$

from which it follows upon multiplying by the mass that

$$m v_{\perp} = m_e c, \quad \text{where} \quad m_e^2 = m^2 \left[\left(\frac{d\tau}{dt}\right)^2 + \frac{r^2}{c^2} \dot{\theta}^2\right] < m^2 \tag{5.3.5}$$

In the above equation, m_e denotes the electron rest mass. Since v_z is a constant, the electron will prescribe a helical motion in the z direction of constant radius r.

We define a hypothetical Zitterbewegung radius r_e in the electron rest frame. Its Lorentz transformation rule is determined by the transversal Doppler effect, which will be derived in Section 5.4:

$$r = r_e \sqrt{1 - \frac{v_z^2}{c^2}}$$

The corresponding mass variation can be evaluated from Equation (5.2.51), and gives

$$m = \frac{\hbar \omega}{c^2} = \frac{m_e}{\sqrt{1 - \frac{v_z^2}{c^2}}} \quad \text{by the de Broglie principle}$$

The above mass and Zitterbewegung radius formulas demonstrate that the Maxwell equation-based approach of Chapter 3 is equivalent to the general relativity-based approach of this chapter: in both cases, we derive the same electron structure.

5.3.3. *Electromagnetic analysis of electron momentum*

Recalling Section 3.4 results, the charge momentum is proportional to the angular frequency and it has a direction tangent to the helical path. The relativistic momentum of charge is, then,

$$p_c = eA = \frac{\hbar\omega}{c} = \frac{\hbar}{r} \tag{5.3.6}$$

or using natural units,

$$\left[p_c = \omega = \frac{1}{r} = m\right]_{\text{NU}}$$

Now, the charge momentum vector $\boldsymbol{p}_c = e\boldsymbol{A}_{\vartriangle}$ can be decomposed into two components: $\boldsymbol{p}_\perp$, which is orthogonal to electron velocity and another one, $\boldsymbol{p}_\parallel$, parallel to the electron direction of motion, i.e., in the z direction:

$$\boldsymbol{p}_c = \boldsymbol{p}_\perp + \boldsymbol{p}_\parallel$$

The magnitude of component $\boldsymbol{p}_\perp$ is a constant, independent from velocity v_z, and is proportional to the charge angular speed ω_e in the xy plane. Therefore,

$$p_\perp = \frac{\hbar\omega_e}{c} = m_e c$$

or in natural units

$$[p_\perp = \omega_e = m_e]_{\text{NU}}$$

whereas the component $p_\parallel$ is the momentum of the electron and is proportional to the *instantaneous* angular frequency $\omega_z = \frac{v_z}{r}$

$$p_\parallel = \frac{\hbar\omega_z}{c} = \frac{\hbar v_z}{cr} = \frac{\hbar\omega}{c^2}v_z = mv_z$$

or in natural units

$$\left[p_\parallel = \omega_z = \frac{v_z}{r} = mv_z\right]_{\text{NU}}$$

Using again the Pythagorean theorem, it is possible to write the following equations:

$$\omega_e = \frac{v_\perp}{r} = \frac{\sqrt{c^2 - v_z^2}}{r} = \frac{\sqrt{c^2 - v_z^2}}{r_e\sqrt{1 - \frac{v_z^2}{c^2}}} = \frac{c}{r_e} \tag{5.3.7}$$

and, as a consequence of (5.3.7), also

$$\omega = \frac{c}{r}$$

But

$$\omega_z = \frac{v_z}{r}$$

and, therefore, the sum of squares of the angular frequencies yields the following relations:

$$\omega^2 = \omega_e^2 + \omega_z^2$$

$$p_c^2 = p_\perp^2 + p_\parallel^2$$

5.3.4. *Relativistic analysis of electron momentum*

Equations (5.3.1) and (5.3.5) already give an orthogonal decomposition of the particle momentum:

$$p_c^2 = p_\perp^2 + p_\parallel^2$$

From Equations (5.3.1) and (5.3.5), we have

$$m^2c^2 = m_e^2c^2 + m^2v_z^2, \quad m_e^2 = m^2\left[\left(\frac{d\tau}{dt}\right)^2 + \frac{r^2}{c^2}\dot{\theta}^2\right] \tag{5.3.8}$$

where $p_\parallel$ is the momentum of the particle's kinetic motion

$$p_\parallel = mv_z$$

The Maxwell equation-based electron momentum calculations of Chapter 3 are thus equivalent to the general relativity-based analysis developed in this chapter.

5.3.5. *Quantum mechanical wavelength from de Broglie principle*

According to the de Broglie principle, ω_z is the instantaneous angular frequency associated to a particle with rest mass m_e, relativistic mass m, and velocity $v_z = \omega_z r$. As a consequence,

$$p_\parallel = mv_z = \frac{\hbar\omega}{c^2}v_z = \frac{\hbar}{cr}v_z = \frac{\hbar\omega_z}{c} = \hbar\frac{2\pi}{\lambda} = \hbar k \tag{5.3.9}$$

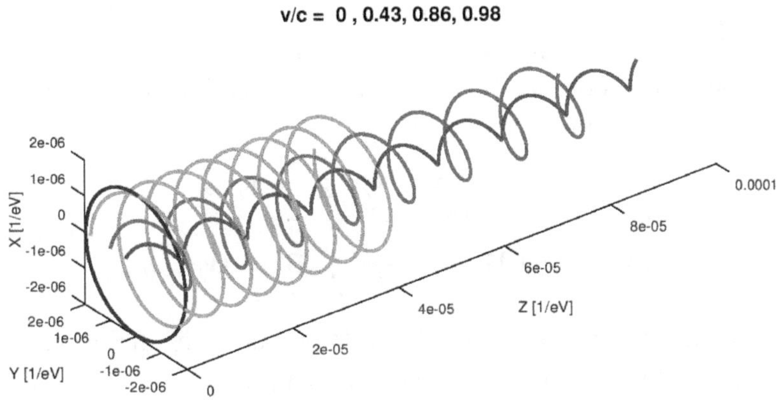

Figure 5.2. Zitterbewegung trajectories for different speeds.

or

$$\left[p_{\|} = mv_z = \omega v_z = \frac{v_z}{r} = \omega_z = \frac{2\pi}{\lambda} = k\right]_{\mathrm{NU}}$$

Equation (5.3.9) yields

$$\frac{p_{\|}}{k} = p_{\|}\frac{\lambda}{2\pi} = \hbar \tag{5.3.10}$$

where the term $k = \frac{2\pi}{\lambda}$ is the wave number of the electron and λ the related de Broglie wavelength. Of course, if we observe the electron at a spatial scale much larger than its Compton wavelength and at a time scale much higher than the very short period $T \approx 8.1 \cdot 10^{-21}$ s of the Zitterbewegung rotation period, for a constant speed v_z, the electron can be approximated to a point particle, provided with "mass" and charge, which moves with a uniform motion along the z-axis of the helix. Particularly, Figure 5.2 represents the helical trajectories of electrons moving at different speeds.

5.4. Lorentz Transformations of Electromagnetic Waves

In this section, we take the wave perspective of electromagnetism, and consider the relativistic transformation of electromagnetic fields between various reference frames. As seen in the previous section, a Lorentz boost γ_L causes the Zitterbewegung radius to contract by γ_L^{-1}. At first sight, such contraction is paradoxical since a Lorentz boost contracts only longitudinal distances along the boost, while leaving transversal distances unchanged. To understand what is happening, let's first consider a simple plane-wave solution.

In a circularly polarized electromagnetic wave $\mathbf{F}$, the electric and magnetic fields are always orthogonal to the direction of propagation. In a plane wave, upon Lorentz boost along the direction of propagation, the fields transform as follows:

$$\mathbf{E}'_{\perp} = \gamma_L \left(\mathbf{E}_{\perp} + \mathbf{v} \times \mathbf{B}\right), \mathbf{E}'_{\parallel} = \mathbf{E}_{\parallel} = 0$$

$$\mathbf{B}'_{\perp} = \gamma_L \left(\mathbf{B}_{\perp} - \frac{\mathbf{v} \times \mathbf{E}}{c^2}\right), \mathbf{B}'_{\parallel} = \mathbf{B}_{\parallel} = 0$$

Since $\mathbf{E}_{\perp}$ and $\mathbf{B}_{\perp}$ are orthogonal, $\mathbf{v} \times \mathbf{B}$ points in the direction of electric field and $\frac{\mathbf{v}\times\mathbf{E}}{c^2}$ points in the direction of the magnetic field. One may observe that the $+\mathbf{v} \times \mathbf{B}$ and $-\frac{\mathbf{v}\times\mathbf{E}}{c^2}$ cross-product terms are either both positive or both negative, depending on the direction of the boost. Thus, $\mathbf{E}'_{\perp}$ and $\mathbf{B}'_{\perp}$ remain orthogonal, and their magnitude is changed by a factor of $\gamma_L (1 \pm \beta_L)$, where $\beta_L = \frac{v}{c}$. Consequently, the local energy density changes by the square of the field strengths, i.e.,

$$\frac{\varepsilon'}{V'} = \gamma_L^2 (1 \pm \beta_L)^2 \frac{\varepsilon}{V} \tag{5.4.1}$$

where the sign depends on the boost direction relative to the wave propagation. Recognizing that $\gamma_L^{-2} = \frac{c \mp v}{c \pm v} (1 \pm \beta_L)^2$, the above expression can be simplified as follows:

$$\frac{\varepsilon'}{V'} = \frac{c \pm v}{c \mp v} \frac{\varepsilon}{V} \tag{5.4.2}$$

One may recognize in the above expression the square of the $\sqrt{\frac{c\pm v}{c\mp v}}$ factor, which is the relativistic longitudinal Doppler shift factor. Not surprisingly, the energy density of an electromagnetic wave changes according to its frequency shift.

Let's consider now longitudinal electromagnetic waves, which were introduced in Section 2.10. In that case, the electric and magnetic fields are parallel to the direction of propagation, and remain constant upon Lorentz boost along the direction of propagation. Since $S = E = B$ in the longitudinal field, the scalar part also remains constant. However, the wavelength must change according to the $\sqrt{\frac{c\pm v}{c\mp v}}$ longitudinal Doppler shift factor. Thus, even though the frequency of the longitudinal electromagnetic field Doppler shifts with Lorentz boost along the propagation direction, its local energy density remains constant: $\frac{\varepsilon'}{V'} = \frac{\varepsilon}{V}$.

We obtain the following Lorentz transformation properties of a longitudinal electromagnetic wave:

- It always moves at the speed of light along the direction of its propagation and maintains a constant energy density.
- Its frequency varies according to the $\sqrt{\frac{c\pm v}{c\mp v}}$ longitudinal Doppler shift factor, and the energy of the $\hbar\omega$ energy quantum varies accordingly. Depending on the applied Lorentz transformation, the $\hbar\omega$ energy quantum may become arbitrarily small.
- $\hbar\omega$ may be Doppler shifted to any value upon a change of reference frames.

We will prove in Chapter 7 that neutrinos are in fact longitudinal electromagnetic waves. It follows from the above analysis that a neutrino wave always propagates at the speed of light and has no rest mass.

We consider now the electromagnetic wave picture of an electron. We derived in Chapter 3 that the electron's internal electromagnetic fields rotate around its Zitterbewegung axis, and this axis point is in the direction of the electron's spatial movement. According to Heisenberg uncertainty, the more we closely approximate the electron's rest frame, the less we know its position along the Zitterbewegung axis. Let's consider the following rest frame orientation: its Zitterbewegung plane is the $x-y$ plane, and its position is being infinitely spread along the z direction. Such infinite spreading can be understood as having many copies of the electron field solution of Chapter 3 superposed along the z axis, all in the same phase. Since Maxwell's equation is linear, such superpositions along the z axis form a valid solution. As the number of copies approaches infinity, the electromagnetic wave decomposition becomes like the one illustrated in Figure 5.3: nothing changes along the z axis. In the electron's rest frame, the Fourier series of its electromagnetic waves may be described in a cylindrical coordinate system: these waves are all transversally propagating with respect to its Zitterbewegung axis. Figure 5.3 illustrates the Lorentz transformation of these electron generating waves. We observe the following consequences of Lorentz boost:

- Due to the transversal Doppler shift effect, wavelengths in the $x-y$ plane of Zitterbewegung are contracted by γ_L^{-1}.
- The Lorentz boost transformation contracts distances along the z direction of the boost by γ_L^{-1}.

The above relativistic transformation rules are the physical cause for the boost effects on the electron; they describe a contracting electron size by

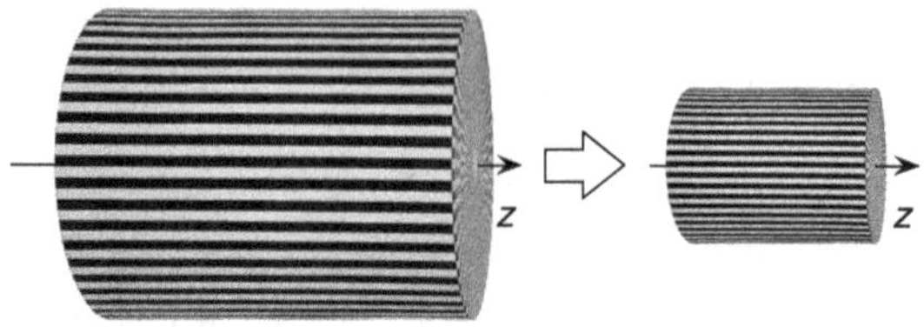

Figure 5.3. Wavelength contraction through the transversal Doppler effect. The contraction factor is γ_L^{-1}.

γ_L^{-1} along each spatial dimension as its mass increases by γ_L. This result resolves the apparent paradox between Lorentz transformation rules and the electron model which we derived in Chapter 3.

A relativistic Lorentz boost along the z direction is also mixing $z - t$ coordinates. Geometrically, this is produced by a rotor R_{zt}, that generates a hyperbolic rotation with rapidity φ in the $\gamma_z\gamma_t$ plane, in order to encode the motion at speed v_z along the γ_z direction:

$$R_{zt} = \cosh(\varphi)^{-\frac{1}{2}} e^{-\gamma_z\gamma_t\frac{\varphi}{2}}$$

where $\varphi = \tanh^{-1}\left(\frac{v_z}{c}\right)$.

A boosted space–time position vector $\mathbf{r}'_{\square}(t)$ is obtained by applying R_{zt} onto $\mathbf{r}_{\square}(t)$:

$$\mathbf{r}'_{\square}(t) = R_{zt}\mathbf{r}_{\square}\widetilde{R_{zt}}$$

The above boost transformation rotates the t and z axes into each other by angle φ. Thus, a time-wise Zitterbewegung oscillation of the rest frame acquires an oscillation component along the z direction in the boosted reference frame. The "appearance" or "disappearance" of waves along the z direction is a direct consequence of Lorentz transformations. The wavenumber of this oscillation along the z direction was derived from the canonical charge momentum vector $\boldsymbol{p}_c = e\boldsymbol{A}_{\vartriangle}$ via Equation (5.3.9), and it was found to be related to the particle momentum as $p = mv_z = \frac{\hbar\omega}{c^2}v_z = \hbar k$.

In the above case of a well-defined momentum, the generated spatial wave is homogeneous along the z direction: i.e., the electron's position is still infinitely spread-out along the z direction. The electron's position is determined by the Fourier transform of its wave frequencies along this z direction. There must be an uncertainty of the boost value, i.e., a momentum uncertainty, in order to generate a localization of the wave packet along the z direction. The applied uncertainty of the boost value can be described as an uncertainty of the wave frequency in the z direction. This property of the wave transformations is the physical basis of conjugate position-momentum variables in quantum mechanics. The e^{-lx^2} Gaussian function is of particular interest as it becomes its own Fourier transform for some choice of l. Therefore, a Gaussian probability distribution of the electron's position corresponds to a Gaussian probability distribution of its momentum. The Heisenberg uncertainty limit is the Fourier transforms' uncertainty limit.

We summarize some electron properties, which follow from the Lorentz transformation of electromagnetic waves:

- Upon Lorentz boost, the electron size is contracting by γ_L^{-1} along each spatial dimension. Along the two transversal axes to the boost, the γ_L^{-1} contraction is caused by the relativistic Doppler effect. Along the parallel axis to the boost, the γ_L^{-1} contraction is caused by the relativistic Lorentz contraction effect.
- A Lorentz boost corresponds either to a change of reference frame or to an accelerated electron. Such transformation causes the appearance or disappearance of a quantum mechanical wave frequency in the direction of the particle's velocity. This corresponds to the wave "creation" and "annihilation" operators of quantum field theory.
- The electron's position and momentum variables form a conjugate pair due to the Fourier transformation rules of an electromagnetic Zitterbewegung wave.

5.5. Conclusions

The origin and meaning of wave–particle duality have been shrouded in mystery for over 100 years. We resolved this puzzle, and our solution shows that the proper use of general relativity is indispensible for a correct understanding of quantum mechanics. These results re-establish the compatibility between different branches of physics. The obtained generalized Dirac equation is a local differential equation, and it is a foundation for building a consistent field theory.

In terms of the electron structure, we demonstrated in Section 5.3 that the Maxwell equation-based Zitterbewegung analysis of Chapter 3 leads to the same results as the general relativity-based analysis in this chapter. These complementing perspectives give confidence in our electron model. Regarding the Dirac equation, the meaning of parallel and orthogonal components to the Zitterbewegung axis will be further investigated in the following chapters.

If a quantum mechanical oscillation mode exists in a certain direction, the corresponding position and momentum values must always form a conjugate pair, linked by Fourier transformation rules. As discussed in Chapter 2, elementary particles are subject to the effect of electromagnetic vacuum fluctuations, which sets the scale of this quantum mechanical uncertainty. From this, we may understand that the presence or absence of an oscillation mode depends on the interaction details between an elementary particle and the electromagnetic vacuum noise. What do we gain by knowing the origin of Heisenberg uncertainty? Without this insight, we would not be able to tell whether the $\triangle x \triangle p \geq \frac{\hbar}{2}$ uncertainty rule always and unconditionally applies along each spatial coordinate, or whether it conditionally applies to those oscillation modes which exist. This distinction will become relevant when we discuss the physics of composite particles in later chapters.

Appendix: Hamilton–Jacobi Functions and Exact Differentials

The following lemma gives a necessary and sufficient condition for the existence of Hamilton–Jacobi functions, and was labeled as Lemma 12 in Section 5.2:

Lemma 16. *Let $\psi(W(\mathbf{x},t))$ be a differentiable function and $\{\sigma(\lambda)\}$ a family of curves on the manifold with unit tangent vectors $\frac{\tilde{ds}}{ds}$ with respect to a local tetrad, then $\tilde{ds}.\tilde{\partial}_s\psi(W)$ is an exact differential iff $\psi(W)$ is a Hamilton–Jacobi function such that $p^*_a = \frac{d\psi}{ds}\frac{dx_a}{dt} = \frac{\partial\psi_{\parallel}(W)}{\partial x^a}$, where $\psi(W) = \psi_{\parallel}(W) + \psi_{\perp}(W)$.*

Proof. From Equations (5.2.25), if $\frac{\tilde{ds}}{ds}\tilde{\partial}_s\psi(W)$ is an exact differential, then $[\frac{\tilde{ds}}{ds}, \frac{\partial\tilde{\psi}(W)}{\partial s}]$ is also an exact differential. This means that $\psi(W) = \psi_{\parallel}(W) + \psi_{\perp}(W)$ where $[\frac{\tilde{ds}}{ds}, \frac{\partial\tilde{\psi_{\parallel}}(W)}{\partial s}] = 0$ and $\psi_{\perp}(W) = c_o t + c_1x_1 + c_2x_2 + c_3x_3$, with c_o, c_1, c_2, c_3 being constants.

Also, $[\frac{\tilde{ds}}{ds}, \frac{\partial\tilde{\psi_{\parallel}}(W)}{\partial s}] = 0$ means $\frac{dx^a}{d\lambda}$ is parallel (for any parameter λ including the curve length s) to $\frac{\partial\tilde{\psi}_{\parallel}(W)}{\partial s}$, and $\exists m(\lambda)$ such that $p^*_{\parallel a} \equiv m(\lambda)c\frac{dx_a}{d\lambda} = \frac{\partial\psi_{\parallel}(W)}{\partial x^a}$ (where c, the velocity of light, is a scaling factor).

Also, given $\frac{\tilde{ds}}{ds}\tilde{\partial}_s\psi_{\parallel}(W)$ is an exact differential and denoting $x_0 = t$, we obtain

$$\begin{aligned}\frac{\tilde{ds}}{ds}\tilde{\partial}_s\psi_{\parallel}(W) &= \frac{\partial\psi_{\parallel}(W)}{\partial x^1}\frac{dx^1}{ds} + \frac{\partial\psi_{\parallel}(W)}{\partial x^2}\frac{dx^2}{ds} + \frac{\partial\psi_{\parallel}(W)}{\partial x^3}\frac{dx^3}{ds} + \frac{\partial\psi_{\parallel}(W)}{\partial t}\frac{dt}{ds}\\ &= \frac{d\psi_{\parallel}(W)}{ds}\end{aligned}$$

On substituting $m(s)c\frac{dx_a}{ds} = \frac{\partial\psi_{\|}(W)}{\partial x^a}$ and noting that $\frac{dx^a}{ds}\frac{dx_a}{ds} = 1$, we obtain $m(s)c = \frac{d\psi_{\|}(W)}{ds}$. It follows that

$$\left(\frac{\partial\psi_{\|}(W)}{\partial t}\right)^2 = (p_{\|}^{*1})^2 + (p_{\|}^{*2})^2 + (p_{\|}^{*3})^2 + \left(\frac{d\psi_{\|}(W)}{ds}\right)^2$$

$$= \left(H_o^*\left(x^a, p_{\|}^{*1}, p_{\|}^{*2}, p_{\|}^{*3}\right)\right)^2 \tag{5.5.1}$$

Therefore, $\psi_{\|}(W)$ is a Hamilton–Jacobi function, as also is $\psi(W) = \psi_{\|}(W) + \psi_{\perp}(W)$ with $p_a^* = p_{\|a}^* + c_a$.

Conversely, given $p_a^* = \frac{d\psi}{ds}\frac{dx_a}{dt} + c_a = \frac{\partial\psi_{\|}(W)}{\partial x^a} + \frac{\partial\psi_{\perp}(W)}{\partial x^a} = \frac{\partial\psi(W)}{\partial x^a}$, then $\left[\tilde{ds}, \tilde{\partial}_s\psi_{\|}\right] = 0$ and since $\psi(W)$ is a Hamilton–Jacobi function, it follows from the definition of $\psi_{\perp}$ that

$$\tilde{ds}\,\tilde{\partial}_s\psi_{\|}(W) = d\psi_{\|}(W) \quad \text{and} \quad \left[\frac{\tilde{ds}}{ds}, \frac{\partial\psi_{\perp}(W)}{\partial s}\right]$$

are both integrable. Therefore, $\tilde{ds}\,\tilde{\partial}_s\psi(W)$ is an exact differential. The result follows.

Remark. The duality relating line increments and the Dirac equation brings to the foreground philosophical issues regarding the difference between quantum and classical mechanics, and in particular how the two might be related.

Appendix: Clifford Algebra and Directional Derivatives

Lemma 16 is in its own way surprising. It requires that a Hamilton–Jacobi nonlinear wave function can only be decomposed into a parallel component. In other words, if $\psi(W) = \psi_{\|}(W) + \psi_{\perp}(W)$ where $\left[\frac{\tilde{ds}}{ds}, \frac{\tilde{\partial}\psi_{\|}(W)}{\partial s}\right] = 0$, then $\psi_{\perp}(W) = c_0 t + c_1 x_1 + c_2 x_2 + c_3 x_3$, with c_0, c_1, c_2, c_3 being constants.
One might have expected that $\psi_{\perp}(W)$ would be arbitrary and not simply linear. So why is this so? First of all, by definition of $\psi_{\perp}(W)$ we expect that

$$\left\{\frac{\tilde{ds}}{ds}, \frac{\tilde{\partial}\psi_{\perp}(W)}{\partial s}\right\} = 0$$

which means

$$\frac{d\psi_{\perp}(w)}{ds} = \frac{\partial\psi_{\perp}(W)}{\partial x^i}dx^i = 0$$

This implies that $\psi_\perp(W) = constant$ and this is the surprise. In contrast,

$$\left[\frac{\tilde{ds}}{ds}, \frac{\tilde{\partial}\psi_\parallel(W)}{\partial s}\right] = 0$$

does not imply $\psi_\parallel(W) = constant$. Usually in geometry, parallel and perpendicular decompositions are relative to a basis. For example, if $\mathbf{x} = \mathbf{x_1 e_1} + \mathbf{x_2 e_2}$, then the component $x_1\mathbf{e_1}$ is parallel to $\mathbf{e_1}$ and perpendicular to $\mathbf{e_2}$, while $x_2\mathbf{e_2}$ is perpendicular to $\mathbf{e_1}$ and parallel to $\mathbf{e_2}$ and *vice versa*. The choice of x_1 and x_2 are arbitrary and independent. It might be instructive to take a concrete example and see what happens with respect to the Clifford product. Consider

$$\tilde{ds}\tilde{\partial}_s\psi(W) = \left(\frac{\partial\psi}{\partial x}\gamma_x + \frac{\partial\psi}{\partial y}\gamma_y\right)(dx\gamma_x + dy\gamma_y) \tag{5.5.2}$$

$$= \frac{\partial\psi}{\partial x}dx + \frac{\partial\psi}{\partial y}dy + \gamma_x\gamma_y\left(\frac{\partial\psi}{\partial x}dy - \frac{\partial\psi}{\partial y}dx\right) \tag{5.5.3}$$

On the one hand, if $\left(\frac{\partial\psi}{\partial x}, \frac{\partial\psi}{\partial y}\right)$ is parallel to $(\dot{x}, \dot{y})$, then on letting $\frac{\partial\psi}{\partial x} = g(x, y)$ and $\frac{\partial\psi}{\partial y} = g(x, y)\frac{dy}{dx}$, we obtain

$$\left[\frac{\tilde{ds}}{ds}, \frac{\tilde{\partial}\psi_\parallel(W)}{\partial s}\right] = 0$$

and

$$\left\{\frac{\tilde{ds}}{ds}, \frac{\tilde{\partial}\psi_\parallel(W)}{\partial s}\right\} = \left(\frac{\partial\psi}{\partial x} + \frac{\partial\psi}{\partial y}\frac{dy}{dx}\right)\dot{x} \tag{5.5.4}$$

$$= \left[1 + \left(\frac{dy}{dx}\right)^2\right]g(x, y)\dot{x} \tag{5.5.5}$$

On the other hand, if $\left(\frac{\partial\psi}{\partial x}, \frac{\partial\psi}{\partial y}\right)$ is perpendicular to $(\dot{x}, \dot{y})$ and we let $\frac{\partial\psi}{\partial x} = g(x, y)\frac{dy}{dx}$ and $\frac{\partial\psi}{\partial y} = -g(x, y)$, we obtain

$$\left\{\frac{\tilde{ds}}{ds}, \frac{\tilde{\partial}\psi_\perp(W)}{\partial s}\right\} = 0$$

and

$$\left[\frac{\tilde{ds}}{ds}, \frac{\tilde{\partial}\psi_{\perp}(W)}{\partial s}\right] = \left(\frac{\partial\psi}{\partial x}\frac{dy}{dx} - \frac{\partial\psi}{\partial y}\right)\dot{x} \tag{5.5.6}$$

$$= \left[\frac{dy}{dx}^2 + 1\right] g(x,y)\dot{x} \tag{5.5.7}$$

This demonstrates the point that there is a one-to-one correspondence between the set of projections onto the γ_x and the γ_y axes and yet when we imposed the property that ψ should be an **exact differential**, this one-to-one correspondence breaks down and we are forced to chose $g(x,y) = 0$. This follows because a total derivative, although as an inner product it is rotationally invariant, as a directional derivative defined relative to a basis vector its value varies according to the direction.

Appendix: Clifford Algebra and Harmonic Functions

There is a class of functions called the harmonic functions where the derivative is the same in all directions. In two dimensions they are obtained from the Cauchy–Riemann (CR) equations which can be generalized to the Cauchy–Riemann–Fueter equations in four dimensions. Specifically, in two dimensions if ψ is a complex differentiable function, then the Cauchy–Riemann condition requires that

$$\frac{\partial\psi}{\partial x} + i\frac{\partial\psi}{\partial y} = 0 \quad \text{and} \quad \frac{d\psi}{dz} = \frac{\partial\psi}{\partial x}$$

In this case, the Clifford product becomes

$$\tilde{\partial}_s\psi.\tilde{ds} = \left(\frac{\partial\psi}{\partial x}\gamma_x + \frac{\partial\psi}{\partial y}\gamma_y\right)(dx\gamma_x + dy\gamma_y) \tag{5.5.8}$$

$$= \frac{\partial\psi}{\partial x}dx + \frac{\partial\psi}{\partial y}dy + \gamma_x\gamma_y\left(\frac{\partial\psi}{\partial x}dy - \frac{\partial\psi}{\partial y}dx\right) \tag{5.5.9}$$

$$= \frac{\partial\psi}{\partial x}dx + i\frac{\partial\psi}{\partial x}dy + \gamma_x\gamma_y\left(\frac{\partial\psi}{\partial x}dy - i\frac{\partial\psi}{\partial x}dx\right) \quad \text{by CR} \tag{5.5.10}$$

$$= \frac{\partial\psi}{\partial x}(dx + idy) - i\gamma_x\gamma_y\left(\frac{\partial\psi}{\partial x}dx + i\frac{\partial\psi}{\partial x}dy\right) \tag{5.5.11}$$

$$= \frac{d\psi}{dz}(1 - i\gamma_x\gamma_y)dz \tag{5.5.12}$$

It also follows from the Cauchy–Riemann equations that

$$\frac{\partial^2\psi}{\partial x^2} + \frac{\partial^2\psi}{\partial y^2} = 0 \tag{5.5.13}$$

References

[1] É. Cartan, *The Theory of Spinors.* Dover, New York, 1981, p. 134.

[2] J. Costella, B. McKellar, and A. Rawlinson, Classical anti-particles. *American Journal of Physics*, **65** (1997) 835–841.

[3] G. Duff and D. Naylor, *Differerential Equations of Applied Mathematics.* Wiley, New York, 1966, p. 71.

[4] L. Horwitz and R. Arshansky, Relativistic entanglement. *Physics Letters A*, **382**(26) (2018) 1701–1708.

[5] P. Milonni, *The Quantum Vacuum.* Academic Press, New York, 1994, pp. 322–323.

[6] J. Oprea, *Differential Geometry and its Applications.* Pearson/Prentice Hall, 1997, pp. xv.

[7] P. O'Hara, Wave-particle duality in general relativity. *Il Nuovo Cimento B*, **111**(7) (1996) 799–810.

[8] P. O'Hara, Quantum mechanics and the metrics of general relativity. *Foundations of Physics*, **35**(9) (2005) 1563–1584.

[9] P. O'Hara, Equations of motion in general relativity and quantum mechanics. *Journal of Physics: Conference Series*, **330** (2011) 012013.

[10] P. O'Hara, Constants of the motion, universal time and the Hamilton-Jacobi function in general relativity. *Journal of Physics: Conference Series*, **437** (2013) 012007.

[11] L. O'Raifeartaigh, *The Dawning of Gauge Theory.* Princeton University Press, Princeton, 1997, p. 112.

[12] E. Schrodinger, Quantization as an eigenvalue problem. *Annalen der Physik*, **81** (1926) 162.

[13] J. Synge and B. Griffith, *Principles of Mechanics*, McGraw-Hill, New York, 1959, pp. 440–463.

https://doi.org/10.1142/9781800611306_0006

Chapter 6

Battle of Theories: Magnetic Moment and Lamb Shift Calculations

András Kovács
BroadBit Energy Technologies, Metallimiehenkuja 8 C
02150 Espoo, Finland
andras.kovacs@broadbit.com

6.1. Introduction

Electron orbitals' wavefunction and energy eigenvalues can be calculated from the Dirac equation. The methodology of such calculations may be found for example in [1]. The match between calculated and experimentally measured electron energy eigenvalues historically validated that the Dirac equation gives a correct description of electron dynamics. However, some unexpected experimental data emerged already during the 1940s, which remains unexplained in the context of traditional quantum mechanics. Specifically, these unexpected new data were the anomalous magnetic moment of the electron and the Lamb shift in the relative energies of s and p electron orbitals. To make matters worse, experimenters discovered that these minor magnetic moment deviations of a weakly bound electron become very strong magnetic moment deviations for a proton, a neutron, or a strongly bound electron.

Today, these experimental data are considered to be successfully explained by quantum field theory (QFT), which has become the mainstream theory of electrodynamics. Some scientists expect that any new theory must first of all account for those experimental data which are already successfully explained by the prevailing theory. Therefore, we calculate the electron's anomalous magnetic moment and Lamb shift in this chapter.

Up to now, it was assumed to be impossible to calculate the magnetic moments of the electron, proton, and neutron from one universally applicable equation. We prove this assumption to be wrong. The obtained result also provides an insight into the origin of fine structure constant α.

The final step of our magnetic moment investigation is to consider the magnetic moment of electrons bound to highly charged nuclei, which significantly deviates from the free electron values. We further validate our electron model by comparing the calculated magnetic moment of a bound electron against experimental data.

As we shall see in this chapter, the correct explanation of these deviations is much more than just an accounting of exotic and inconsequential effects: it provides deep insight into elementary particles' internal geometry and vacuum–matter interactions.

6.2. The Electron's Anomalous Magnetic Moment

In the preceding chapters, we showed that the electron's apparent magnetic moment is generated by a charge precession around the externally applied magnetic field lines. The angle between the electron's angular momentum vector and external magnetic field lines is $\frac{\pi}{3}$. From the analysis of fundamental electromagnetic relationships, we derived that the electron's intrinsic angular momentum is $\hbar$ in its rest frame.

The magnetic moment is traditionally given by the $\mu = \frac{e\hbar}{2m_e}$ formula, derived in Equation (3.3.7), where the only non-constant parameter is the electron mass. Therefore, the measurement of the electron's magnetic moment can be interpreted as a measurement of its mass, according to this formula. The presence of the small anomalous part means that the formula is not exactly correct, and the implicit assumptions going into this formula must be checked. This magnetic moment formula was derived in Equation (3.3.7) by calculating the area enclosed by the electron current. It has the implicit assumption that the dipole magnetic field is created by the electric charge, upon exactly one full rotation. Let's consider whether this assumption is valid for a non-zero electron charge radius.

Starting from the fundamentals, the $\partial G = 0$ Maxwell equation describes how electric and magnetic fields induce each other. Let's consider a point on the electron's spherical charge surface. Although we can't do a Lorentz transformation to move along with this point at the speed of light, we can write Maxwell equation at consecutive spacetime points corresponding to a constant scalar field value S at this selected point. As seen from Equations (2.8.3) and (2.8.4), the electromagnetic energy density has symmetric contributions from the electric and magnetic fields, even in the presence of a non-zero scalar field. In Chapter 3, we derived that the electron indeed comprises equal amounts of electric and magnetic field energy. In terms of local coupling, we showed in Chapter 3 that the internal electron structure is defined by Maxwell's equation: electric and magnetic fields of equal strength are induced at the non-zero scalar field locations. Since the electron's energy is accounted by the electric and magnetic field energies, the correct physical coupling model is the following: while the non-zero scalar field induces a certain coupling between the electric and magnetic fields, it is finally just an intermediary. The incorrect physical model is that the circulation of the non-zero scalar field "creates" the magnetic dipole field. Field creation is different from the intermediation of field induction between existing electric and magnetic fields. This distinction may initially sound like semantics, because in the case of a point charge the induced magnetic field can be precisely calculated from the motion of the point-like non-zero scalar field.

But let's consider the circular motion of an extended electron, as illustrated in Figure 6.1. As it makes one circle at radius R_e, it is influenced by electromagnetic signals propagating at the speed of light from within a sphere of $2\pi R_e$ radius. As derived in Equation (3.3.22), the total electric field energy of the electron is

$$W_e = \frac{e^2}{32\pi^2\epsilon_0}\int_{r_0}^{\infty}\frac{1}{r^4}\cdot 4\pi r^2 dr = \frac{e^2}{8\pi\epsilon_0}\int_{r_0}^{\infty}\frac{1}{r^2}dr = -\frac{e^2}{8\pi\epsilon_0}\left.\frac{1}{r}\right|_{r_0}^{\infty} = \frac{e^2}{8\pi\epsilon_0 r_0} \tag{6.2.1}$$

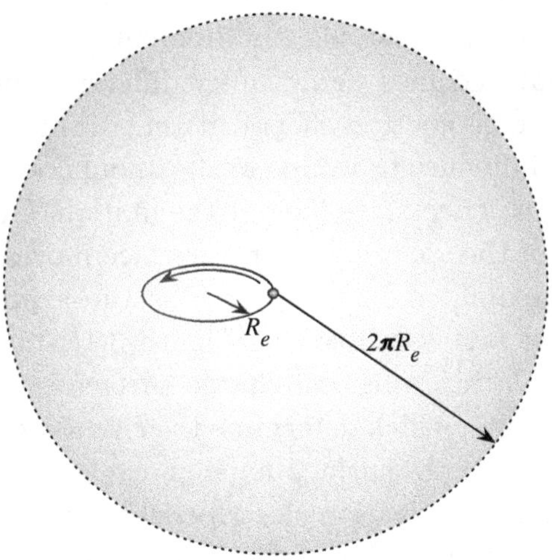

Figure 6.1. The spherical volume element within the reach of electromagnetic signals during one full Zitterbewegung circulation.

where r_0 is its spherical charge radius. However, the electric field energy within a sphere of $2\pi R_e$ radius is

$$W_{e,2\pi R} = \frac{e^2}{8\pi\epsilon_0}\int_{r_0}^{2\pi R_e}\frac{1}{r^2}dr = -\frac{e^2}{8\pi\epsilon_0}\frac{1}{r}\bigg|_{r_0}^{2\pi R_e} = \frac{e^2}{8\pi\epsilon_0}\left(\frac{1}{r_0}-\frac{1}{2\pi R_e}\right) \tag{6.2.2}$$

The above equation means that only a part of the total electric field could exert its magnetic field-inducing effect during a Zitterbewegung rotation by 2π phase. When the charge is point-like, $\frac{1}{r_0}\to\infty$, and therefore the unaccounted electric field energy goes asymptotically to zero. However, in case of a real electron, the accounted ratio of electric field energy is

$$\frac{W_{e,2\pi R}}{W_e} = 1 - \frac{r_0}{2\pi R_e} \tag{6.2.3}$$

In order to calculate the correct magnetic dipole field strength, we must account for the induction of the total electric field energy. Therefore, the anomalous magnetic moment is the inverse of the above already accounted electric field energy ratio:

$$g = \left(1 - \frac{r_0}{2\pi R_e}\right)^{-1} \approx 1 + \frac{\alpha}{2\pi} \tag{6.2.4}$$

The approximation part in the above formula corresponds to the Schwinger factor, which is the first-order QFT estimation of the anomalous magnetic moment. It can be observed from the above result that the correct $g = (1-\frac{r_0}{2\pi R_e})^{-1}$ formula is just slightly different from the Schwinger factor when $\frac{r_0}{R_e}$ is small, but becomes significantly different when $\frac{r_0}{R_e} > 1$.

Are there further geometric effects arising from the non-zero charge radius, which might influence the value of the anomalous magnetic moment? To consider order-of-α^2 effects, we look at the motion of the spherical charge surface. Each point of the spherical charge surface moves at light-speed, and this effect was approximated in Chapter 3 as the azimuthal displacement of a counter-rotating ball as shown in Figure 3.1(b). However, the exact geometry of the circulating spherical charge surface is shown in Figure 6.2. There are two constraints, which determine the circulation geometry: (i) each point of the spherical charge surface must move at the speed of light and (ii) the magnetic flux quantization described by Equation (3.3.18) implies that the average circulation radius of the charge surface is exactly R_e. These constraints are met when each point of the charged sphere is circulating around a point on a virtual sphere located in the center of Zitterbewegung.

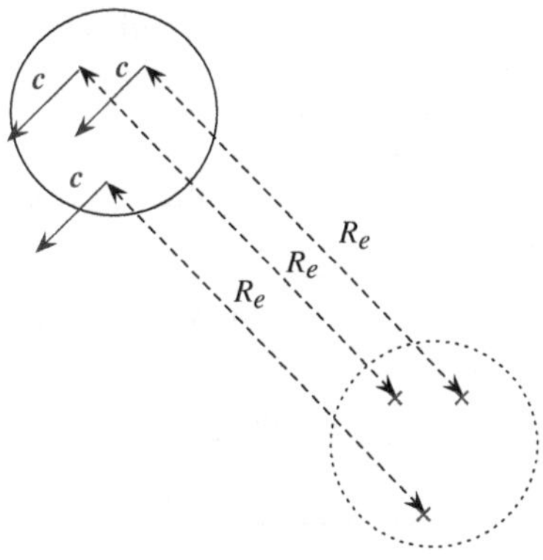

Figure 6.2. The geometry of a circulating spherical charge surface, where each point is moving at the speed of light.

The arrows of Figure 6.2 illustrate the mapping between each point on the charge sphere and the corresponding rotation center-point on the virtual sphere. Since each point of the charged sphere has exactly R_e Zitterbewegung radius, there is no order-of-α^2 contribution to the anomalous magnetic moment.

In summary, the electron's anomalous magnetic moment is estimated by the following formula:

$$g = \left(1 - \frac{r_0}{2\pi R_e}\right)^{-1} = \left(1 - \frac{\alpha}{2\pi}\right)^{-1} \tag{6.2.5}$$

Substituting in the $\alpha^{-1} = 137.035999$ value of the fine structure constant, we obtain $g = 1.00116276$ value of the electron's anomalous magnetic moment, which is in good agreement with the experimental value of $g_e = 1.00115965$.

Table 6.1 compares the calculated and experimental anomalous magnetic moment values for the electron and muon. Since this value is very similar for both particles, we obtain an important result: the muon is approximately a scaled-down version of the electron, having the same $\frac{r_0}{R_e}$ ratio between their charge radius and Zitterbewegung radius. The analysis of the difference between calculated and experimental values reveals that the deviation is $\frac{8\pi}{3}\alpha^3$. This deviation is clearly related to the volume of a sphere, and has opposite signs depending on the particle type: $-\frac{8\pi}{3}\alpha^3$ for the electron, and $+\frac{8\pi}{3}\alpha^3$ for the muon.

A logical next step in the development of our theory is to investigate the origin of the $\frac{8\pi}{3}\alpha^3$ term, which would improve the accuracy of our

Table 6.1. The anomalous magnetic moment of the electron and muon: Calculated values, experimental values, and the analysis of the remaining deviations.

Theoretical prediction	Experimental value	Order-of-α^3 deviation
$g = 1.00116276$	$g_e = 1.00115965$	$g - \frac{8\pi}{3}\alpha^3$ $= 1.00115951$
$g = 1.00116276$	$g_\mu = 1.00116592$	$g + \frac{8\pi}{3}\alpha^3$ $= 1.00116602$

calculation by nearly two orders of magnitude. Since it has opposite signs for the electron and muon, its understanding would be an important starting point for revealing the structural difference between these two particle types. Since the electric field is zero inside the electron's spherical charge, this effect reduces the effective volume of the large sphere shown in Figure 6.1. Thus, the electron's $-\frac{8\pi}{3}\alpha^3$ term appears to be a consequence of having a constant charge radius. The muon's $+\frac{8\pi}{3}\alpha^3$ term will be further discussed in Section 16.4.

The QFT methodology of representing g as a Taylor series expansion occults the fact there is no order-of-α^2 term and prevents the recognition of the order-of-α^3 term's spherical geometry. Given that the QFT calculation of the electron's g factor is a gradually developed post-experimental explanation of measurement results, and given that it requires different so-called "correction terms" for each different particle type, one is lead to conclude that the QFT methodology has no predictive power. In contrast, the above results show that our methodology is applicable in the same way to the electron and to the muon, and yields same type of order-of-α^3 terms for both particles. In Section 6.4, we will apply Equation (6.2.5) for calculating the proton's anomalous magnetic moment, without any particle-specific "correction terms." In a later chapter, we will use the same equation for calculating the neutron's anomalous magnetic moment.

6.3. A Possible Connection Between α, Φ_M, and the Feigenbaum Constant

In Chapter 4, we showed that charge quantization is in fact a quantization of magnetic flux. However, we could not yet derive a formula for α. In the preceding section, we saw that the induction between the more localized

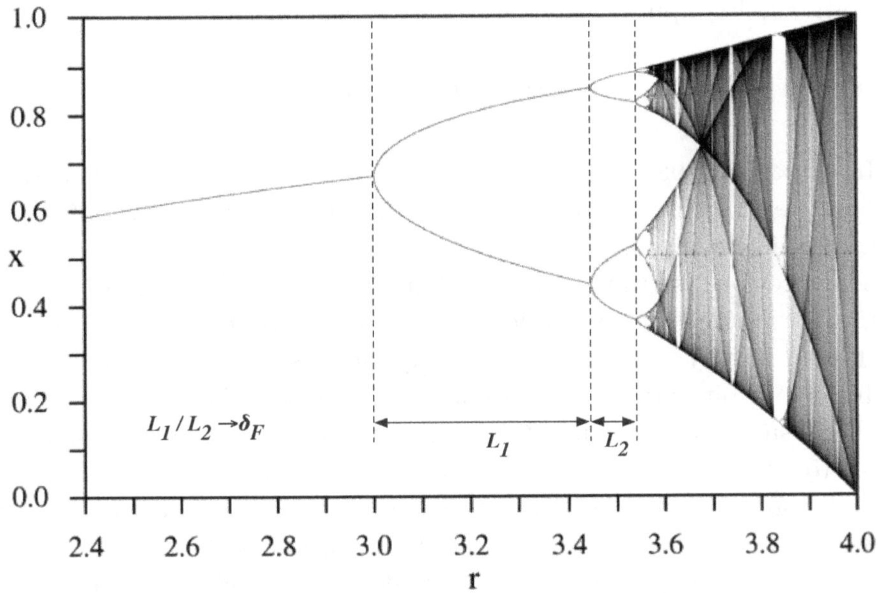

Figure 6.3. An exemplary bifurcation map. For any one-parameter map, the $\frac{L_n}{L_{n+1}}$ ratio of period-doubling distances converges to the Feigenbaum constant δ_F as $n \to \infty$.

magnetic field and the less localized electric field of the electron is an iterative process: each Zitterbewegung turn is a process step.

The Feigenbaum constant $\delta_F \approx 4.6692016$ is a fundamental mathematical constant of iterative processes, analogous to π in geometry or to e in calculus. This δ_F constant corresponds to the control parameter's distance ratio between successive period-doubling bifurcations, as shown in Figure 6.3. It is a universal constant for any single-parameter quadratic map. Up to now, no connection was suspected between α and δ_F, since the electron was assumed to be just a structureless point-particle, without any relationship to iterative processes. However, at least one publication already pointed out a phenomenological relationship [10].

We observe that $\frac{\alpha^{-1}}{2\pi} \approx \delta_F^2$. This simple relationship is 99.96% accurate with respect to the value of δ_F, and is therefore unlikely to be a coincidence. The phenomenological fine structure formula can be expressed as follows:

$$\alpha = \frac{\delta_F^{-2}}{2\pi} \tag{6.3.1}$$

In Chapter 4, we showed that the electron's charge and magnetic flux are both quantized. Recalling from Section 4.3 the $\left[\Phi_M = \frac{2\pi}{\sqrt{\alpha}}\right]_{\text{NU}}$ magnetic flux formula, Φ_M appears to be directly proportional to δ_F:

$$\left[(2\pi)^{-\frac{3}{2}}\,\Phi_M = \delta_F\right]_{NU} \tag{6.3.2}$$

In case the 99.98% accuracy of Equation (6.3.2) is not coincidental, it implies a very direct relationship between the Feigenbaum constant and the electron's magnetic flux.

The previous section demonstrated that the induction between the electron's electric and magnetic fields is an iterative process. Thus, the electron's stability requires the stability of this iterative process. When an iterative orbit converges to a stable value faster than exponentially, it is called "superstable orbit." Interestingly, the Feigenbaum constant also relates to superstable orbits of any iterative process. It turns out that δ_F yields not only the bifurcation point distance ratio shown in Figure (6.3), but also the distance ratio of superstable n-period orbits as $n \to \infty$. The mystery of α's origin is perhaps reduced to the challenge of deriving Equation (6.3.2).

6.4. The Proton's Anomalous Magnetic Moment

As we saw in Section 6.2, the charge radius and Zitterbewegung radius are the real control parameters of the anomalous magnetic moment calculation. According to Equation (3.3.5) of Chapter 3, the Zitterbewegung radius is just the inverse of particle mass in natural units: $\left[R_p = \frac{1}{m}\right]_{NU}$. Thus, we know the value of R_p exactly from mass measurements: R_p=0.2103 fm. However, the measurement of the proton charge radius has been more controversial. For a long time, a so-called "proton radius puzzle" persisted: proton–electron system-based measurements yielded 0.875 fm proton radius value, proton–muon system-based measurements yielded 0.8414 fm proton radius value, but electron–proton scattering measurements again indicated a radius value in the 0.84–0.85 fm range [11]. Recently, a new technique allows to measure the proton radius at improved accuracy [12]; this more precise radius value is $r_p = 0.8482\,\text{fm}$. Applying Equation (6.2.5) to the proton, its "anomalous" magnetic moment is calculated from its $r_p = 0.8482$ fm charge radius and its $R_p = 0.2103$ fm Zitterbewegung radius parameters: $g_p = (1 - \frac{r_p}{2\pi R_p})^{-1} = 2.7927$. Our result is 99.99% accurate with respect to the experimentally measured "anomalous" proton magnetic moment of 2.7929, which was considered to be a surprisingly large value at the time of its discovery. This calculation demonstrates an important universality of our elementary particle model.

Table 6.2. A phenomenological relationship between elementary particles' charge radius (r) and Zitterbewegung radius (R).

Particle type	$\frac{r}{R}$	$2\pi\sqrt{\frac{R}{r}} = [\Phi_M]_{NU}$
Electron	α	$(2\pi)^{\frac{3}{2}}\,\delta_F$
Muon	α	$(2\pi)^{\frac{3}{2}}\,\delta_F$
Proton	4.03	$\approx \pi$
Pion	$\frac{4.03}{9}$	$\approx 3\pi$

We also note that the $\frac{r_p}{R_p}$ ratio is 4.03. Possibly, this ratio encodes some information about the proton's geometry.

The charged pion is another particle for which the charge radius and Zitterbewegung radius data are available. The charged pion has 139.57 MeV mass, and the corresponding Zitterbewegung radius is $R_\pi = 1.4141$ fm. The pion charge radius is measured to be 0.657 fm [13], but this measurement is carried out with the same technique as the 0.875 fm proton radius measurements. While both measurements are subject to the same systemic error, we still get the accurate ratio of these charge radii: $r_\pi/r_p = 0.657/0.875 = \frac{3}{4}$. We therefore estimate the pion charge radius to be $r_\pi = \frac{3}{4}r_p = 0.636$ fm.

Table 6.2 summarizes elementary particles' charge radius to Zitterbewegung radius ratios. At this point, we don't know yet whether the phenomenological values shown in Table 6.2 are coincidental, or have a deeper meaning. It would be very significant if Table 6.2 values were not coincidental: in that case, elementary particles' charge quantization is related to mathematical constants such as π and δ_F, and it should be possible to mathematically derive the elementary charge value.

6.5. Electromagnetic Vacuum Fluctuations

In Chapter 2, we interpreted the Heisenberg uncertainty effect to be caused by the presence of an electromagnetic vacuum fluctuation. In Chapter 5, we showed that the presence of vacuum noise floor leads to the $\triangle x\,\triangle p \geq \frac{\hbar}{2}$ uncertainty of conjugate variables.

One way to characterize electromagnetic vacuum fluctuation is to describe its average electric and magnetic field intensity. In a harmonic wave, the average field intensities are related to the energy density as

$\frac{1}{2}\left(\varepsilon_0 E^2 + \frac{1}{\mu_0}B^2\right) = \varepsilon_0 \bar{E}^2 = \frac{\bar{B}^2}{\mu_0}$. The average vacuum field intensity is given by the following formula [4]:

$$\varepsilon_0 \bar{E}^2 = \frac{\bar{B}^2}{\mu_0} = \frac{\hbar c}{4\pi^2} \int_0^\infty k^3 dk \tag{6.5.1}$$

where k is the wavenumber of the given wave. There is no special wavelength; each electromagnetic wavelength contributes uniformly to the field intensity of the vacuum fluctuation. Such spectral distribution is the only Lorentz invariant possibility — under non-accelerating frame transformations. If there was any other type of spectral distribution in the vacuum, it would be subject to Doppler shift in various reference frames, and our perception of the vacuum would then depend on our frame of reference. We emphasize that Equation (6.5.1) is not just a theory for explaining the Casimir force measurements, but was also directly measured in [4].

We rewrite the above formula as follows:

$$\begin{aligned} \varepsilon_0 \bar{E}^2 &= \frac{\bar{B}^2}{\mu_0} = \frac{\hbar c}{4\pi^2} \int_0^\infty k^3 dk = \frac{\hbar c}{16\pi^3} 4\pi \int_0^\infty k k^2 dk \\ &= \frac{\hbar c}{16\pi^3} \iiint_0^\infty k\, d^3k = \frac{1}{8\pi^3} \iiint_0^\infty \left(\frac{1}{2}\hbar\omega\right) d^3k = \frac{1}{V}\sum_k \frac{1}{2}\hbar\omega \end{aligned} \tag{6.5.2}$$

where V is a unit volume element. This result implies that each possible vacuum oscillation mode has $\frac{1}{2}\hbar\omega$ energy, which is generally referred to as the "zero-point energy" of the electromagnetic noise floor. The above expression is the field intensity integration over the abstract space of all possible wavenumbers, and this form will be used in the Lamb shift calculation. The overall electric field intensity can also be written as an integration over its spectral components:

$$\varepsilon_0 \bar{E}^2 = \frac{V}{8\pi^3} \iiint_0^\infty \varepsilon_0 \bar{E}_k^2\, d^3k \tag{6.5.3}$$

From the above equations, we get the relationship between $\bar{E}_k$ and k:

$$V \bar{E}_k^2 = \frac{\hbar c}{2\varepsilon_0} k \tag{6.5.4}$$

6.6. Lamb Shift

The s and p electron orbital wavefunctions are calculated from the Dirac equation, according to the well-known methodology explained in [1]. As derived from the Dirac equation, the following formulas describe the radial part of the wavefunctions which are relevant to this section:

$$\psi_{2s}(r) = \frac{1}{2\sqrt{2}}\left(\frac{Z}{a_0}\right)^{\frac{3}{2}}\left(2-\frac{Z}{a_0}r\right)e^{-\frac{Z}{2a_0}r} \tag{6.6.1}$$

$$\psi_{2p}(r) = \frac{1}{2\sqrt{6}}\left(\frac{Z}{a_0}\right)^{\frac{3}{2}}\frac{Z}{a_0}re^{-\frac{Z}{2a_0}r} \tag{6.6.2}$$

$$\psi_{3s}(r) = \frac{2}{81\sqrt{3}}\left(\frac{Z}{a_0}\right)^{\frac{3}{2}}\left(27-18\frac{Z}{a_0}r+2\left(\frac{Z}{a_0}r\right)^2\right)e^{-\frac{Z}{3a_0}r} \tag{6.6.3}$$

$$\psi_{3p}(r) = \frac{4}{81\sqrt{6}}\left(\frac{Z}{a_0}\right)^{\frac{3}{2}}\left(\frac{6Z}{a_0}r-\left(\frac{Z}{a_0}r\right)^2\right)e^{-\frac{Z}{3a_0}r} \tag{6.6.4}$$

where r is the radial coordinate, Z is the nuclear charge, and a_0 is the Bohr radius parameter. The electrostatic potential energy of the ψ_{2s} and ψ_{2p} functions is exactly the same. Therefore, according to the virial theorem, the kinetic energy and binding energy should also be the same in both cases. Nonetheless measurements reveal a small difference between the energy of these two states; this difference is referred to as the Lamb shift. Likewise, the electrostatic potential energy of the ψ_{3s} and ψ_{3p} functions is exactly the same, yet there is also a measured Lamb shift. The Lamb shift is a tiny effect; the binding energy of the above-mentioned two electron states differs by about one part in a million.

In the following paragraphs, we calculate the interaction between the electromagnetic vacuum fluctuations and a bound electron. The calculation methodology is based on the idea introduced by Welton in [2], which we develop into a simpler and assumption-free calculation.

The ground state electron orbital is characterized by the $R_0 = \frac{a_0}{Z}$ mean radius. As the electric field of a harmonic noise component pushes the electron back-and-forth, it "blurs out" the electron's position. To characterize this blurring, we calculate the mean distance δr_k into which the noise component with wavenumber k pushes the electron.

The maximum and mean electric field intensities of a harmonic noise component are related by the $E_{\max}^2 = 2\bar{E}^2$ relationship. The harmonic noise component accelerates the electron according to the $a = \frac{eE_{\max}}{m}\sin(\omega t)$ relationship. Consequently, the displacement generating the electron blur is $x = \frac{eE_{\max}}{m}\omega^{-2}\sin(\omega t) = \frac{eE_{\max}}{m}(kc)^{-2}\sin(\omega t)$. In a harmonic oscillation, the maximum and mean displacements are related by the $x_{\max}^2 = 2\bar{x}^2$ relationship.

As the above relationships hold independently for each spatial direction, we are now able to estimate the mean displacement δr_k as follows:

$$(\delta r_k)^2 = \left(\frac{e\bar{E}_k}{m}\right)^2 (kc)^{-4} \tag{6.6.5}$$

The effect of nuclear charge Z is not directly evident in the above equation.

How to add up the contributions of various wavenumbered noise components? Since the various noise wavelengths are independent of each other, their contributions are adding up in quadrature:

$$(\delta r)^2 = \sum_k (\delta r_k)^2 \tag{6.6.6}$$

We now rewrite the above summation as an integration over the possible wave numbers, replacing $\sum_k$ by $\frac{V}{8\pi^3}\iiint d^3k$. The integral formula thus takes the following form:

$$\begin{aligned}(\delta r)^2 &= \frac{V}{8\pi^3}\iiint \left(\frac{e\bar{E}_k}{m}\right)^2 (kc)^{-4}\, d^3k = \frac{1}{16\pi^3}\frac{e^2\hbar}{\varepsilon_0 m^2 c^3}\iiint kk^{-4}\, d^3k \\ &= \frac{e^2\hbar}{4\pi^2\varepsilon_0 m^2 c^3}\int_{k_{\min}}^{k_{\max}} \frac{1}{k}\, dk = \left(\frac{e}{mc^2}\right)^2 \frac{\hbar c}{4\pi^2\varepsilon_0} \ln\left(\frac{k_{\max}}{k_{\min}}\right)\end{aligned} \tag{6.6.7}$$

In the above equation, we substituted in the average field intensity formula according to Equation (6.5.4). The $\left(\frac{e}{mc^2}\right)^2 \frac{\hbar c}{4\pi^2\varepsilon_0}$ factor evaluates to $(18.608 \text{ fm})^2$.

The above displacement formula yields infinity if either $k_{\min}$ goes to zero or $k_{\max}$ goes to infinity. Regarding $k_{\min}$, the above calculation considers the electron in isolation. The isolated electron model is certainly not valid for small k values, where the displacing action of vacuum fluctuations is counterbalanced by the dynamic interactions between the electron and the nucleus.

In the $k \to 0$ case, the corresponding E_k field points in similar direction all across the electron wavefunction and nucleus sites, causing opposing displacement of the oppositely charged electron and nucleus. It is interesting to note that the ground state wavefunction's energy and radius can be well described by calculating the low-frequency electromagnetic vacuum noise effect on the electron, using the vacuum noise formulas presented in Section 6.5. Such calculation is presented on introductory level in [7] and is shown in detail in [8].

The k values which may induce quantum mechanical state transitions are quantized values at low quantum numbers, but approach a continuum as the main quantum number n goes to infinity. The process of such light-induced quantum mechanical state transitions will be investigated in Section 8.9. The region of $n \to \infty$ is the weakly bound region, where the electron dynamics gradually becomes similar to that of a free particle. As quantum mechanical transitions from the ground state approach toward this continuum, the limiting value of their transition energy is characterized by the Rydberg constant R_H, which denotes the lowest k value capable of generating a free electron from its ground state. Thus, we may conclude that in the $k < R_H$ region the electron–nucleus interactions counteract the effect of electromagnetic vacuum fluctuations, while in the $k > R_H$ region the electron responds as a free particle.

Based on the above reasoning, we set $k_{\min}$ to the Rydberg constant R_H:

$$k_{\min} = R_H = \frac{\alpha}{4\pi a_0} \frac{m_p}{m_e + m_p} \tag{6.6.8}$$

We note that the above formula does not depend on the main quantum number n; i.e., $k_{\min}$ must be the same value for any electron orbital around a given nucleus.

What is the appropriate choice for $k_{\max}$? Beyond a certain k value, the electron cannot respond by oscillating, and becomes transparent to higher frequencies just like metallic conduction electrons become transparent to radiation frequencies beyond their plasma frequency. Welton and Bethe also recognized that this issue represents yet another divergence of the abstract point-particle theory, and they chose $k = \frac{m_e c}{\hbar} = \frac{\omega_e}{c}$ as their largest wavenumber. We need to critically consider the highest k value where the $a = \frac{e\bar{E}}{m}$ expression is valid. The $k = \frac{\omega_e}{c}$ value corresponds to the Zitterbewegung wavenumber, but we are looking for the largest wavenumber in the orthogonal direction to the Zitterbewegung plane. As discussed in Chapter 5, the quantum mechanical wave is generated through a Lorentz transformation of the Zitterbewegung wave. A high frequency oscillation generates close to light-speed at the oscillation midpoint, and at that point a displacement Δx represents the Lorentz transformation of a $c\Delta\tau$ passage of the electron's internal time. In a harmonic oscillation, the $\frac{1}{2\pi}$ normalization coefficient of one-dimensional Fourier transformation represents the ratio between unit displacement Δx and unit passage of $2\pi\Delta t$ time per one complete phase. Since $\Delta x \approx c\Delta\tau$, the $\frac{1}{2\pi}$ normalization coefficient of one-dimensional Fourier transformation now represents the ratio between Δt and $\Delta\tau$. For a three-dimensional Fourier transformation, the normalization coefficient is $\frac{1}{8\pi^3}$, but we evaluate the Fourier integral by using the $\frac{1}{8\pi^3}\iiint d^3k = \frac{1}{2\pi^2}\int k^2 dk$ identity. The high frequency oscillation limit requires a $\frac{1}{2\pi^2}$ ratio between Δt and $\Delta\tau$, and therefore we must set $k_{\max}$ to the following value:

$$k_{\max} = \frac{\omega_e}{2\pi^2 c} \tag{6.6.9}$$

Figure 6.4 illustrates the electron–noise interaction characteristics in various frequency ranges. We note that according to [2, 3] Welton and Bethe chose $k_{\min} = 17.8R_H$ and $k_{\max} = \frac{m_e c}{\hbar} = \frac{\omega_e}{c}$. However, according to our derivation, the correct choice is $k_{\min} = R_H$ and $k_{\max} = \frac{1}{2\pi^2}\frac{\omega_e}{c}$. The $\ln(\frac{k_{\max}}{k_{\min}})$ factor comes out to a very similar value in either case, so this discrepancy does not directly show up in the electron Lamb shift calculation results. For the electron orbitals around a proton the $\ln(\frac{k_{\max}}{k_{\min}})$ factor evaluates to $\ln(\frac{1}{2\pi^2}\frac{\omega_e}{R_H c}) = 2.482$. This relatively small number means that the spectral range, where the bound electron responds in the same way as free particle, is actually quite narrow. From Equation (6.6.7), the δr parameter evaluates to 29.316 fm.

In order to evaluate the effect of the δr parameter on the average potential, we calculate the first- and second-order Taylor expansion terms of the potential at $\vec{r} + \vec{\xi}$, where $\vec{\xi}$ is a small random displacement from $\vec{r}$:

$$U\left(\vec{r} + \vec{\xi}\right) = U\left(\vec{r}\right) + \vec{\xi}\cdot\nabla U\left(\vec{r}\right) + \frac{1}{2}\sum_{ij}\xi_i\xi_j\partial_i\partial_j U\left(\vec{r}\right) \tag{6.6.10}$$

The timewise averages of $\vec{\xi}$ are $\bar{\xi} = 0$ and $\overline{\xi_i\xi_j} = \frac{1}{3}\vec{\xi}^{\,2}\delta_{ij} = \frac{1}{3}\left(\delta r\right)^2\delta_{ij}$, where δ_{ij} is the Kronecker delta symbol. The $\frac{1}{3}$ factor appears since the

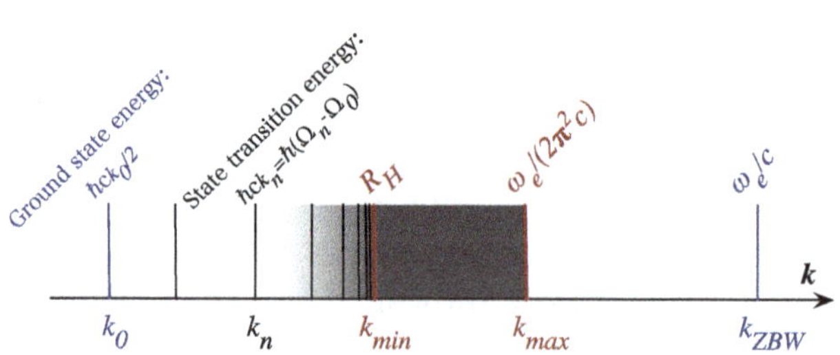

Figure 6.4. A schematic illustration of electron – light interaction in various frequency regimes, classified by the light's wavenumber k. The lower frequency range is characterized by quantized interactions, corresponding to transitions between Dirac spinor eigenstates. Ω_n is the spinor frequency of an eigenstate. These quantized wavenumbers approach continuum in the Rydberg constant limit, and the electron responds like a free particle in the $k_{\min} - k_{\max}$ range.

variance of a random displacement is equally split among the three spatial dimensions. We evaluate the time-averaged potential value at some location:

$$\overline{U\left(\vec{r}+\vec{\xi}\right)} = U\left(\vec{r}\right) + \frac{1}{6}\left(\delta r\right)^2 \nabla^2 U\left(\vec{r}\right) \tag{6.6.11}$$

The second term of the above equation gives the Lamb shift effect. $\nabla^2 U\left(\vec{r}\right)$ is non-zero only at the nucleus, and we use Gauss's law to evaluate it:

$$\nabla^2 U\left(\vec{r}\right) = -\nabla^2 \frac{Ze}{4\pi\epsilon_0 r} = \frac{\rho\left(\vec{r}\right)}{\epsilon_0} \tag{6.6.12}$$

In the above expression, the $\rho\left(\vec{r}\right)$ symbol denotes the charge density at a given coordinate $\vec{r}$. Now we can evaluate the electron's potential energy by integrating over the radial coordinate:

$$U_p = e\int \psi^2\left(r\right)\overline{U}4\pi r^2 dr = e\int \psi^2\left(r\right)U\left(r\right)4\pi r^2 dr + \frac{Ze^2}{6\epsilon_0}\left(\delta r\right)^2 \psi^2(0) \tag{6.6.13}$$

The first term of the above formula is the usual Coulomb potential energy, and the second term is the Lamb shift of this potential. The relationship between the electron's potential and kinetic energies is given by the virial theorem. In the non-relativistic limit, the electron's kinetic energy is half of its electrostatic potential, and we may write the Lamb shift of the electron energy as follows:

$$E_{\text{Lamb}} = \frac{Ze^2}{12\epsilon_0}\left(\delta r\right)^2 \psi^2(0) \tag{6.6.14}$$

Equation (6.6.14) is remarkable because we already have all the information for evaluating it. The p-type wavefunction has a vanishing probability at the origin, while the s-type wavefunction has a non-zero probability at the origin. Therefore, the Lamb shift influences the potential energy of s-type wavefunctions only.

We use the experimental $2s$–$2p$ and $3s$–$3p$ Lamb shift values of a hydrogen atom for validating our calculation. The results are shown in Table 6.3; our straightforward derivation indeed gives the precise Lamb shift values. This means that the Lamb shift effect may also be understood as a consequence of the Maxwell and Dirac equations, without making any new postulates.

Recent measurements [9] also reveal the Lamb shift values of an electron around a $Z = 2$ nucleus. Regarding this $Z > 1$ case, Equation (6.6.14)

Table 6.3. The calculated and experimental $2s\frac{1}{2}$ - $2p\frac{1}{2}$ and $3s\frac{1}{2}$ - $3p\frac{1}{2}$ Lamb shift values for atomic hydrogen.

Orbital transition	Calculated $s-p$ Lamb shift	Experimental $s-p$ Lamb shift
$2s\frac{1}{2}$ to $2p\frac{1}{2}$	1058 MHz	1058 MHz
$3s\frac{1}{2}$ to $3p\frac{1}{2}$	314 MHz	315 MHz

Note: The calculation is based on Equation (6.6.14).

implies approximately Z^4 dependence: there is a Z dependence of the first term, a Z^3 dependence of the $\psi^2(0)$ term, and as a first approximation we neglect the Z dependence of the logarithmic term. In the $Z = 2$ case, the $2s\frac{1}{2}$ to $2p\frac{1}{2}$ Lamb shift estimation with this simple Z^4 dependence yields 16.9 GHz, which compares reasonably well with the measured value of 14.04 GHz. This result further validates that our calculation yields the correct electron dynamics. Experimentally, the 14.04 GHz value of $2s\frac{1}{2}$ to $2p\frac{1}{2}$ Lamb shift at $Z = 2$ implies $Z^{3.7}$ dependence of the Lamb shift. The $2s\frac{1}{2}$ to $2p\frac{1}{2}$ Lamb shift measurement on heavier nuclei, such as oxygen [15], demonstrates that this $Z^{3.7}$ dependence remains fairly constant.

The above-calculated δr parameter is an order of magnitude smaller than the Zitterbewegung radius, which raises major questions. Should not Zitterbewegung cause then an even stronger blurring of the quantum mechanical wavefunction? Why is the effect of Zitterbewegung not making the $2s\frac{1}{2}$ to $2p\frac{1}{2}$ transition energy even larger? To answer this question, let's take a closer look at the Dirac equation's solutions. Upon solving for the Dirac eigenfunctions, one finds the following binding energy contributions:

- The non-relativistic binding energy contribution, which shows up in the Schrödinger equation: $\frac{1}{2}(Z\alpha)^2 mc^2$.
- The so-called "Darwin term," which is $-\frac{Ze^2\hbar^2}{8\epsilon_0(mc)^2}\psi^2(0)$.
- Relativistic corrections to the binding energy: $(Z\alpha)^4 mc^2\left(\frac{1}{(2j+1)n^3} - \frac{3}{8n^4}\right)$.

Comparing the Darwin term against Equation (6.6.14), one obtains $\delta r = \sqrt{\frac{3}{2}}\frac{\hbar}{mc}$. This is of similar magnitude as the $R_e = \frac{\hbar}{mc}$ Zitterbewegung radius, but not quite the same. This similarity led quantum theorists to interpret the Darwin term as a wavefunction blurring, caused by Zitterbewegung. Plugging in the $\psi(0)$ formula for quantum number n, the Darwin term can be written as $(Z\alpha)^4 mc^2 \frac{1}{2n^3}$, but only for s orbitals. For p orbitals, $\psi(0)$ evaluates to 0, and the Darwin term vanishes.

The first part of the relativistic corrections is $(Z\alpha)^4 mc^2 \frac{1}{(2j+1)n^3}$, where j describes the total angular momentum. The $j = \frac{1}{2}, \frac{3}{2}, \ldots$ values are applicable to the p state, i.e., $j = 0$ for the s state. The effect of the j term is interpreted as the "spin–orbit coupling" energy contribution. Curiously, upon summing up the Darwin term and $(Z\alpha)^4 mc^2 \frac{1}{(2j+1)n^3}$ for an s state, we get

$$(Z\alpha)^4 mc^2 \frac{1}{n^3}\left(\frac{1}{1} - \frac{1}{2}\right) = (Z\alpha)^4 mc^2 \frac{1}{n^3}\frac{1}{2} = (Z\alpha)^4 mc^2 \frac{1}{n^3}\frac{1}{2\frac{1}{2}+1} \tag{6.6.15}$$

In other words, an $s\frac{1}{2}$ orbital's Darwin term reduces the binding energy by exactly the same amount as a $p\frac{1}{2}$ orbital's spin–orbit coupling effect, for any n. Quantum theorists consider this exact match to be a coincidence, and use a somewhat mismatching explanation to ascribe the Darwin term to a Zitterbewegung effect. It creates further confusion that QFT practitioners deny the existence of Zitterbewegung altogether, but provide no interpretation to the Darwin term. We postpone resolving this puzzle till Section 6.8.

6.7. Which Microscopic Vacuum Model is the Correct One?

Historically, two different Lamb shift calculations emerged in the 1940s. Bethe modeled the vacuum noise as random charge fluctuations appearing and disappearing [3]. Since his thinking was bound by the electromagnetic gauge invariance hypothesis, he could not talk about scalar field fluctuations. Instead, he considered that electron–positron pairs would emerge spontaneously from the vacuum, and these emerging "virtual positrons" have the magical property of not producing any 511 keV gamma radiation, which actual positrons always produce upon annihilation. In Bethe's model, such "virtual positrons" behave just like real positrons, except for their magical ability of disappearing without any trace. Bethe's model was inspired by considering positrons as "negative energy solutions" to the Dirac equation, but we showed this assumption to be a fallacy in Chapter 2. Bethe's model gives rise to a "vacuum polarization" effect near the nucleus, as the hypothetical virtual electron–positron pairs partially shield the electric field in close nuclear proximity. His calculations apparently matched the observed Lamb shift effect. In contrast, Welton modeled the vacuum noise as random electromagnetic field fluctuations, which are Lorentz invariant

[2]. This model gives rise to a "blurring" effect on the electron's position, which is present everywhere, but only consequential near the nucleus. His calculations also apparently matched the observed Lamb shift effect. These two models describe distinct microscopic processes. If both are correct, both calculations must be simultaneously applied. However, they cannot be both correct, because that would over-estimate the Lamb shift by a factor of two. Recent QFT textbooks choose to refer to the Lamb shift as Bethe's "vacuum polarization" effect caused by virtual electron–positron pairs [5], and ignore Welton's model.

In contrast, our Lamb shift calculation simply accounts for the effect of Lorentz invariant electromagnetic vacuum field fluctuations. These electromagnetic vacuum field fluctuations have been experimentally measured, as presented, for example, in [4], and also in Casimir–Polder force measurements. Therefore, experiments clearly prove that Welton's approach is the one which uses the correct physical model of the vacuum. The inescapable conclusion is that QFT textbooks use an unrealistic vacuum model for discussing the Lamb shift effect.

For the purpose of anomalous magnetic moment calculation, some QFT practitioners endow the above-discussed unrealistic vacuum model also with the ability to temporarily violate energy–momentum conservation, leading to so-called off-mass–shell photons discussed in [5], while other scientists claim that such energy-momentum conservation violation is not necessary [17]. Such uncertainty about basic principles is not the hallmark of a mature theory. The QFT calculation of these hypothetical vacuum processes gets very complicated already for the second-order correction term: the overall process is modeled through numerous Feynman diagrams. Each Feynman diagram represents lengthy calculations, and for some diagrams the claimed result is not accompanied by any published calculation [18]. In contrast, our simple calculation reveals that the anomalous magnetic moment is unrelated to electromagnetic vacuum fluctuations, which is the presumed causation in the QFT model. We showed that the $\frac{\alpha}{2\pi}$ factor should be more appropriately written as $\frac{r_0}{2\pi R_e}$. In this chapter, we investigated the magnetic moments of the electron, muon, and proton, and in a later chapter we will use the same methodology to calculate the anomalous magnetic moment of the neutron. Thus, our methodology is universally applicable, without any major particle-specific "correction factors." We proved that the second-order correction is zero, and regarding third-order corrections we identified the $\frac{8\pi}{3}\alpha^3$ factor; its simple structure suggests the existence of a straightforward geometric interpretation. In summary, the anomalous magnetic moment calculation also does not require any complicated vacuum model.

Our results reveal that the simple, Lorentz invariant, and experimentally validated vacuum model of Section 6.5 is the correct one.

6.8. The Magnetic Moment of a Bound-State Electron

In Equation (3.3.7) of Chapter 3 we derived that — in the point-particle approximation — the electron's magnetic moment is described by the simple $\mu_B = \frac{e\hbar}{2m_e}$ formula. As we saw in Section 6.2, the precise expression of the magnetic moment becomes $\mu = g_e \frac{e\hbar}{2m_e}$. The only non-constant term in this formula is the electron mass. The relativistic electron mass is $m_e = \gamma_L m_0$, where m_0 is the electron's rest mass and γ_L is the Lorentz boost factor. Therefore, the electron's magnetic moment must decrease as its kinetic energy increases, and any measurement of its magnetic moment thus becomes a measurement of its kinetic energy via the γ_L factor.

Equation (3.3.7) is partially recognized in mainstream textbooks, which refer to $\frac{e\hbar}{2m_e}$ and $\frac{e\hbar}{2m_p}$ as the Bohr magneton and nuclear magneton, respectively. However, traditional quantum theory considers the magnetic moment value to be an inherent "spin" property of the electron, and treats μ_B as a constant, without any relativistic mass adjustment.

As the kinetic energy of inner-shell electrons around highly charged nuclei becomes relativistic, we expect their magnetic moment to be significantly different from the free electron value. Recent experiments succeeded to measure the magnetic moment of such tightly bound electrons [14]. These measurements were performed on $Z - 1$ charged ions, where only a single electron is orbiting the Z-charged nucleus. As the reader may expect, the magnetic moment of a tightly bound electron indeed significantly differs from the magnetic moment of a free electron. Despite such significant deviation, quantum theorists cling to their assumption that the magnetic moment of a relativistic electron should maintain a constant value, and refer an explanation of this deviation to be in the scope of QFT. In turn, QFT practitioners take their problematic g-factor calculation as a starting point, add many new assumptions on top of such a foundation, and fill dozens of pages with tedious equations. A summary of such "explanation" can be found for example in [6].

In contrast, we just calculate the kinetic energy of a given electron orbital, and then apply the $\mu = g_e \frac{e\hbar}{2\gamma_L m_0}$ formula to find the bound electron's magnetic moment. Since this formula is derived directly from Maxwell's

Table 6.4. The binding energy of a bound electron around highly charged nuclei.

Nucleus	$\frac{1}{2}Z^2U_0$ (eV)	$\frac{1}{8}\alpha^2Z^4U_0$ (eV)	Lamb shift (eV)
H ($Z = 1$)	13.6057	0.0001811	−0.000035
C ($Z = 6$)	489.81	0.235	
O ($Z = 8$)	870.78	0.74	
Si ($Z = 14$)	2666.8	6.96	
Ca ($Z = 20$)	5442.4	29.0	
Xe ($Z = 54$)	39675	1540.2	
Pb ($Z = 82$)	91486	8189.4	
U ($Z = 92$)	115160	12976	

Note: U_0 is the calculated potential energy of the hydrogen ground state, $\frac{1}{2}Z^2U_0$ is the non-relativistic binding energy, $\frac{1}{8}\alpha^2Z^4U_0$ is the relativistic correction, and the last column is the 1s Lamb shift.

equation, it must be universally valid for all other particle types as well, without any additional "correction factors."

As discussed in Chapter 2, the Dirac spinor eigenstate of an orbital essentially corresponds to the relativistic $\epsilon^2 - (pc)^2 = \left(mc^2\right)^2$ energy–momentum relation. While being conceptually straightforward, finding the eigenstate belonging to 1s orbitals is not an easy problem. Nevertheless, this problem is already solved; the binding energy formula of the 1s orbital is derived for example in [1]. Table 6.4 lists the calculated binding energy values for some nuclei of interest. The non-relativistic 1s binding energy is $\frac{1}{2}Z^2U_0$, where $U_0 = \alpha^2m_0c^2$ is the potential energy of the ψ_{1s} eigenfunction for $Z = 1$. The relativistic correction, which we described at the end of Section 6.6, is $\frac{1}{8}\alpha^2Z^4U_0$. The last column of Table 6.4 shows the 1s Lamb shift for $Z = 1$, calculated according to Equation (6.6.14). The Lamb shift is five times smaller than the relativistic correction in the $Z = 1$ case. While the relativistic correction term grows as Z^4, the Lamb shift energy grows only as $Z^{3.7}$, i.e., it is 646 eV for $Z = 92$. The Lamb shift contribution will therefore relatively remain small, and can be neglected.

Knowing the binding energy values, we need to determine the electron's kinetic energy. We use the virial theorem to find out how the binding energy splits into kinetic energy and potential energy parts. The virial theorem is a very general mathematical theorem which is valid not only in classical mechanics, but also for quantum mechanical wavefunctions. The relativistic form of the virial equation is

$$T = E_p \frac{\gamma_L}{\gamma_L + 1} \tag{6.8.1}$$

where $T \equiv (\gamma_L - 1)\, m_0 c^2$ is the kinetic energy and E_p is the potential energy gain with respect to a free electron state. By rearranging the above virial equation, E_p can be written as $E_p = \frac{\gamma_L^2 - 1}{\gamma_L} m_0 c^2$. The binding energy is the difference of potential and kinetic energies:

$$E_b = E_p - T = \left(\frac{\gamma_L^2 - 1}{\gamma_L} - (\gamma_L - 1) \right) m_0 c^2 = \frac{\gamma_L - 1}{\gamma_L} m_0 c^2 \tag{6.8.2}$$

Now we can solve for γ_L:

$$\gamma_L = \frac{m_0 c^2}{m_0 c^2 - E_b} \tag{6.8.3}$$

Finally, we need to consider the electron's frame rotation. At this point, we revisit the curiosity of Equation (6.6.15) an $s\frac{1}{2}$ orbital's Darwin term reduces the binding energy by exactly the same amount as a $p\frac{1}{2}$ orbital's spin–orbit coupling effect. This coincidence motivates us to consider whether the Darwin term in fact corresponds to spin–orbit coupling. Spin–orbit coupling implies that the electron is in some kind of rotation around the nucleus. Although an s orbital has no net angular momentum, the electron is nevertheless in a singlet state with non-zero orbital angular momentum, which was discussed in Section 0.10.

We make a simple thought experiment to demonstrate that the electron indeed must have orbital rotation even for an s orbital. Let the electron be in the $2p\frac{1}{2}$ state, which has an orbital angular momentum. This orbital angular momentum can be measured under an applied magnetic field. Without an applied external field, the orbital angular momentum has no preferred orientation, and the time-averaged angular momentum of the $2p\frac{1}{2}$ state is 0. The electron transitions into the $2s\frac{1}{2}$ state by absorbing one quantum from 1058 MHz radiation. As the frequency and energy of the absorbed radiation quantum is negligibly small, the corresponding angular momentum is also negligibly small. There is no physical process which might have made the electron's orbital angular momentum disappear during the $2p\frac{1}{2} \rightarrow 2s\frac{1}{2}$ transition. All forces acting on the electron are centrally oriented toward the nucleus. The inevitable conclusion follows from the conservation of angular momentum: the electron still has the same amount of orbital angular momentum in the $2s\frac{1}{2}$ state as it had in the $2p\frac{1}{2}$ state. We emphasize that this statement is not a hypothesis, but a direct consequence of angular momentum conservation. Thus, we corrected the long-held mistaken assumption that s states have no orbital angular momentum. A further mathematical discussion of s states' orbital angular momentum will be given in Section 8.8.

We note that Hestenes came to similar conclusion already in 1979, by comparing the Schrödinger and Pauli theories of quantum mechanics [16]: "As shown in every text in quantum mechanics, the probability distribution for the ground state of a hydrogen-like atom is spherically symmetric, and it is assumed to be static, because the Schrödinger current vanishes. However, ... the Pauli current assigns to the hydrogen ground state an average angular momentum with magnitude $\hbar$. Is it a mere coincidence that this assignment agrees exactly with the one made in the Bohr theory and that, as the textbooks show, the ground state radial distribution function in the Schrödinger theory has its maximum at the Bohr radius?"

Having clarified that the electron has non-zero orbital angular momentum in any hydrogen-like state, our calculation must take into account the Thomas precession effect which arises due to the electron's rotating reference frame. The relativistic Thomas precession effect reduces the apparent lab-frame speed of a circularly orbiting object in proportion to the γ_L factor:

$$\beta_{\text{lab}} = \frac{\beta'}{\gamma_{\text{lab}}} \tag{6.8.4}$$

where $c\beta'$ is the electron speed without the Thomas precession effect, $c\beta_{lab}$ is the apparent electron speed in the lab frame, and $\gamma_{\text{lab}}=\sqrt{1-\beta_{lab}^2}^{-1}$.

Now we are ready to calculate the magnetic moment values from the binding energy values. Firstly, we calculate the γ_L values from Equation (6.8.3), then apply Equation (6.8.4) to find the γ_L values in the lab-frame, and finally we use the $\mu = g_e \frac{e\hbar}{2\gamma_L m_0}$ equation to find the magnetic moment values. Table 6.5 shows the calculation result in tabular format, and Figure 6.5 charts the calculated versus measured magnetic moment values.

The match between the calculated and measured values in Figure 6.5 validates our theory. As seen from Table 6.5, the Thomas precession effect is negligible for light nuclei, but indeed becomes a significant factor for heavy nuclei. In summary, a bound particle's magnetic moment is described by the following equation:

$$\mu = g\frac{e\hbar}{2\gamma_L m_0} \tag{6.8.5}$$

Table 6.5. The calculated and measured magnetic moment values for a single electron bound to various nuclei.

Nucleus	E_b (eV)	γ_L without Thomas precession	γ_L in the lab frame	μ calculated	μ experimental
H ($Z = 1$)	13.606	1.000027	1.000027	$1.001133\mu_B$	$1.001142\mu_B$
C ($Z = 6$)	490.1	1.000960	1.000958	$1.000202\mu_B$	$1.000521\mu_B$
O ($Z = 8$)	871.5	1.001708	1.001702	$0.9995\mu_B$	$1.000024\mu_B$
Si ($Z = 14$)	2673.7	1.00526	1.00520	$0.996\mu_B$	$0.99768\mu_B$
Ca ($Z = 20$)	5471.4	1.01082	1.01059	$0.991\mu_B$	$0.99403\mu_B$
Xe ($Z = 54$)	41215	1.088	1.0748	$0.932\mu_B$	$0.9473\mu_B$
Pb ($Z = 82$)	99675	1.242	1.163	$0.861\mu_B$	$0.869\mu_B$
U ($Z = 92$)	128136	1.335	1.199	$0.835\mu_B$	$0.830\mu_B$

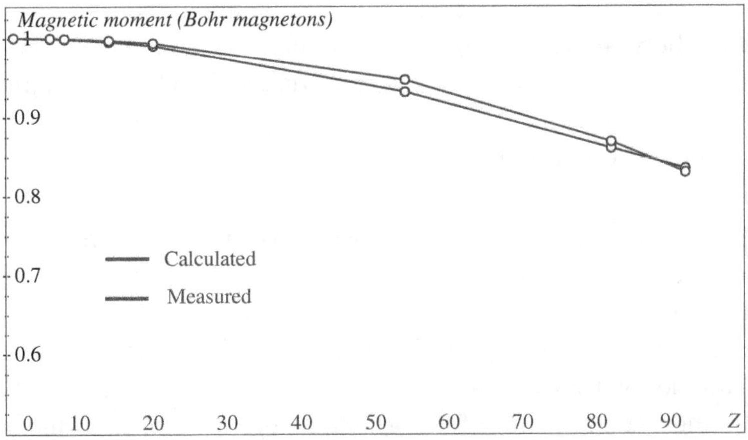

Figure 6.5. A graph of the calculated and measured magnetic moment values for a single electron bound to various nuclei. The horizontal axis shows the nuclear charge, and the vertical axis shows the magnetic moment value.

where g is the anomalous magnetic moment factor given by Equation (6.2.5), m_0 is the particle's rest mass, and γ_L is the Lorentz boost factor in the lab frame.

6.9. Orbital Angular Momentum Entanglement

In Section 0.10, we discussed the mathematics of an s-orbital's $L_o \wedge L_c$ singlet state, where L_o represents the electron's orbital angular momentum, and L_c was a yet undetermined companion angular momentum. What could

L_c stand for? Classically, the electron is orbiting not around the nucleus, but around the center-of-mass in the electron–nucleus system. In a classical Kepler-orbit, the electron and nucleus have the same angular momentum around their center-of-mass.

In quantum mechanics, it is still true that the electron orbits around the center-of-mass in the electron-nucleus system. The center-of-mass effect is given by the Rydberg constant, which we discussed in Section 6.6. How do the angular momenta of an electron and nucleus relate to each other?

Mathematically, there are two possibilities for the s-orbital's $L_o \wedge L_c$ singlet state:

(1) Coupling between the orbital angular momenta of the electron and nucleus. In this case, the L_c term of $L_o \wedge L_c$ stands for the nuclear orbital angular moment. Regarding spin angular momenta, the electron spin and nuclear spin can analogously be a singlet state.
(2) Coupling between the orbital and spin angular momenta of the electron. In this case, the L_c term of $L_o \wedge L_c$ stands for the electron's spin angular moment. Analogously, the orbital and spin angular momenta of the nucleus can also entangle into a singlet state.

We now try to identify which of the above two options is physically realized. A quantum mechanical s-orbital is isotropic, which differs from the elliptical Kepler-orbit of classical mechanics. Classically, if an electron is in a Kepler-orbit, and we measure its angular momentum to be, e.g., in a $|+\rangle$ state, we automatically know that the nuclear angular momentum is also $|+\rangle$. In other words, the angular momenta of an orbiting electron and nucleus are always entangled. This classical starting point suggests option 1 for the quantum mechanical orbital, i.e., classical and quantum mechanical states may differ just in the geometry of their angular momentum entanglement. The actual proof for option 1 will be given in Chapter 9, where the hyperfine splitting of the hydrogen atom's s-orbital energy will be discussed. The hyperfine splitting is an experimental signature for the singlet-state entanglement between the electron spin and nuclear spin. It will be proven in Chapter 9 that no more than two angular moments can be entangled into a singlet state. Since the s-orbital electron's spin is already entangled with the nuclear spin, it cannot be simultaneously entangled with the orbital angular moment. This reasoning proves that option 1 is the correct choice.

Based on the above discussion, a quantum mechanical angular momentum measurement is in fact simultaneously measuring the entangled angular momenta of the electron and nucleus. The isotropic entanglement is mathematically described by the $L_o \wedge L_c$ singlet state, and thus we can finally identify L_c with the nuclear angular momentum.

6.10. Conclusions

We successfully calculated elementary particles' anomalous magnetic moment, the Lamb shift effect, and the magnetic moment deviation of bound elementary particles. Our calculations conclude in three simple formulas: Equations (6.2.5), (6.6.14), and (6.8.5). Why did we bother working on these seemingly exotic problems? Because from the perspective of many theorists, a correct calculation of these effects is a required "rite of passage" for a new theory which differs from QFT. Thus, we demonstrated that our theory replicates all QFT results, but without any untenable postulates, such as "renormalization," "virtual particles," "off-shell photons," "Furry picture," or particle specific "correction factors," Even after applying all these controversial postulates, QFT calculations remain extremely complex [5, 17] and practically impossible to peer-review. In contrast, we succeeded solving all three problems with just a few lines of calculations.

In the course of this work, we obtained some surprising results, beyond the scope of the above-mentioned specific problems. The $\pm\frac{8\pi}{3}\alpha^3$ difference between the electron's and muon's anomalous magnetic moment conveys some important information about these elementary particles' internal structure. There might be a connection between α and the Feigenbaum constant. A simple vacuum model characterized by the Lorentz invariant noise equation of Section 6.5 is the only realistic vacuum model.

We realized that the Darwin term accounts for spin–orbit coupling, and is therefore unrelated to Zitterbewegung. This leads us to conclude that Zitterbewegung does not cause any blurring of the quantum mechanical wavefunction. Such a conclusion is not surprising in the light of preceding chapters.

When Pauli was searching for an explanation to the exclusion principle, which is named after him, he was working under the assumption that the electron's internal magnetic moment has a constant value, and that this constant value is generated by $\frac{\hbar}{2}$ internal angular momentum. The results of Chapters 2–3 and Section 6.8 prove all these assumptions to be wrong. Had Pauli known these results, he would have probably proposed a different explanation to the exclusion principle. This leads to the uncomfortable conclusion that we don't know yet why no more than two electrons may occupy any quantum mechanical state, i.e., we don't know what makes chemistry work! In order to remedy this important problem, a mathematically sound derivation of the exclusion principle will be given in Chapter 9.

References

[1] W.R. Johnson, K.T. Cheng, and M.H. Chen, Accurate relativistic calculations including QED contributions for few-electron systems, *Relativistic Electronic Structure Theory*, Vol. 14, Chapter 3, Elsevier, Amsterdam, The Netherlands, 2004, pp. 120–187.
[2] T.A. Welton, Some observable effects of the quantum-mechanical fluctuations of the electromagnetic field. *Physical Review*, **74**(9) (1948) 1157–1167.
[3] H.A. Bethe, The electromagnetic shift of energy levels. *Physical Review*, **72**(4) (1947) 339–341.
[4] C. Riek *et al.*, Direct sampling of electric-field vacuum fluctuations. *Science*, **350** (2015) DOI: 10.1126/science.aac9788.
[5] T. Lancaster and S.J. Blundell, Quantum Field Theory for the Gifted Amateur. Oxford University Press, Oxford Scholarship Online 2014. DOI:10.1093/acprof:oso/9780199699322.001.0001.
[6] A. Czarnecki *et al.*, Two-loop binding corrections to the electron gyromagnetic factor, arXiv:1711.00190v2.
[7] T.H. Boyer, Any classical description of nature requires classical electromagnetic zero-point radiation. arXiv:1107.3455v1.
[8] H.E. Puthoff, Quantum ground states as equilibrium particle–vacuum interaction states. *Quantum Studies: Mathematics and Foundations*, **3** (2016) 5–10.
[9] A. van Wijngaarden *et al.*, Lamb shift in He^+. *Physical Review A*, **63**(1) (2001) 12501–12505. https://scholar.uwindsor.ca/physicspub/128.
[10] M. Hieb, Feigenbaum's constant and the Sommerfeld fine-structure constant, *Journal of Modern Physics*, **8**(8) (2017). Unpublished.
[11] W. Xiong *et al.*, A small proton charge radius from an electron–proton scattering experiment. *Nature*, **575** (2019) 7781.
[12] A. Grinin *et al.*, Two-photon frequency comb spectroscopy of atomic hydrogen. *Science*, **370** (2020) 6520.
[13] B. Ananthanarayan *et al.*, Electromagnetic charge radius of the pion at high precision. arXiv:1706.04020v2.
[14] W. Quint and M. Vogel, Magnetic moment of the bound electron. *Fundamental Physics in Particle Traps*. Springer Tracts in Modern Physics, vol. 256, Chapter 3. Springer, Berlin, Heidelberg, 2014, pp. 73–135.
[15] G.P. Lawrence *et al.*, Measurement of the Lamb shift in the $^{16}O^{7+}$ ion. *Physical Review Letters*, **28** (1972) 1612–1615.
[16] D. Hestenes, Spin and uncertainty in the interpretation of quantum mechanics. *American Journal of Physics*, **47**(5) (1979) 399–415.
[17] A.O. Barut, Brief History and Recent Developments in Electron Theory and Quantumelectrodynamics. *The Electron — New Theory and Experiment*, vol. 45, Chapter 6. Springer, Dordrecht, 1991, pp. 105–148.
[18] O. Consa, The Unpublished Feynman Diagram IIc. *Progress in Physics*, **16**(2) (2020) arXiv:2010.10345.

https://doi.org/10.1142/9781800611306_0007

Chapter 7

Spinor Fields

András Kovács*,‡ and Giorgio Vassallo†,§
*BroadBit Energy Technologies, Metallimiehenkuja 8 C
02150 Espoo, Finland
†Engineering Department, University of Palermo
viale delle Scienze, 90128 Palermo, Italy
‡andras.kovacs@broadbit.com
§giorgio.vassallo@unipa.it

7.1. Introduction

The Dirac equation is generally known to be "spinor field" valued, but the physical and mathematical meaning of this statement remains unclear in most textbooks. We clarified the mathematical definition of spinors in Section 0.9, and here we focus on the physical meaning of Dirac spinors.

Important progress was made in this field by David Hestenes and William Baylis, who recognized that a spinor field represents Lorentz boost and spatial rotations of the particle orientation. With their approach, the spinor value factorizes into a scalar valued probability density, and representations of a Lorentz boost and a spatial rotation. All of these factors have a well-understood physical meaning. The corresponding formulations of the Dirac equation are referred to as the Dirac–Hestenes and Dirac–Baylis equations. These approaches are discussed in Refs. [1, 2], along with applications that demonstrate the utility of these new insights.

We introduced a novel interpretation of the spinor field in Chapter 5. A Dirac spinor was found to be the eigenvector of a relativistic equation, which defines the motion of a steady-state particle along a space–time geodesic. As space–time curvature is generated by electromagnetic energy density, the Dirac equation can be seen as an energy eigenstate equation.

The usefulness of this approach was demonstrated through numerous examples in Chapter 5.

We explore a new formulation of the Dirac equation in this chapter, where the particle mass is represented as a vector quantity. In chapter 4, we already worked with a vectorial mass representation. In that context, the Dirac equation was found to correspond to the balance of electric and magnetic forces acting on the electron charge. Here, we extend that approach to show how the $\epsilon^2 - (pc)^2 = (mc^2)^2$ energy–momentum relationship leads to understanding the Dirac spinor field as a representation of energy and momentum density.

Finally, we discuss the factorization of electron state parameters. By the end of this chapter, the reader will gain a deeper understanding of Dirac spinors, quantum mechanical state representation, isospin, and neutrinos.

7.2. What is the Dirac Spinor Field?

Macroscopically, the acceleration of charges produces transversal electromagnetic waves. This macroscopic situation changes when the energy of a harmonically oscillating charge coincides with the noise floor, which is given by the Heisenberg uncertainty relation. We outlined this phenomenon in Section 6.6. Under such quantum mechanical scenario, a radiationless equilibrium state is obtained. In the 1920s, Dirac discovered that these radiationless equilibrium states can be found by solving the equation named after him, which is a differential equation over the so-called "four-component spinor field."

In the Dirac equation context, a spinor wavefunction is denoted by ψ. The spinor field is defined as the quantum mechanical wavefunction whose timewise derivative yields the particle energy, and whose spatial derivative in a certain direction yields canonical momentum along that direction. Since ψ is a quantum mechanical field, the quantity $\psi\bar{\psi}$ gives the particle's probability density distribution.

Consider a spinor field undergoing a harmonic oscillation. Its spinor wavefunction then takes the $\psi = \psi_0 e^{i(\Omega t - \mathbf{K}\cdot\mathbf{r})}$ form, and with reference to Chapter 2, we denote the corresponding quantum mechanical wavefunction as ψ_+ field type. The spinor field's differentiation along the time coordinate gives $\hbar\partial_t\psi = i\hbar\Omega\psi_0 e^{i(\Omega t - \mathbf{K}\cdot\mathbf{r})} = i\epsilon\psi$ which produces an imaginary energy eigenvalue of the wavefunction. Similarly, differentiation along a space coordinate gives the imaginary momentum eigenvalue. Again with reference to Chapter 2, the ψ_- type quantum mechanical field corresponds to an exponentially decaying spinor wavefunction defined by $\psi = \psi_0 e^{-(\Omega t - \mathbf{K}\cdot\mathbf{r})}$. Its time differentiation gives $\hbar\partial_t\psi = -\hbar\Omega\psi_0 e^{-(\Omega t-\mathbf{K}\cdot\mathbf{r})} = -\epsilon\psi$ which produces a negative real energy eigenvalue. As we discussed in Section 2.6, it would be a mistake to equate such a negative energy eigenvalue with a paradoxical "negative energy state," because the transient ψ_- field represents negative energy only with respect to a positive valued and static initial condition.

For either wavefunction type, the $\epsilon^2 - (pc)^2 = (mc^2)^2$ energy–momentum relation turns into the Klein–Gordon equation. Using the above generic formulas, we first consider the Klein–Gordon equation for sinusoidal wavefunctions:

$$i^2\hbar^2(\partial_t^2 - \partial_x^2 - \partial_y^2 - \partial_z^2)\psi = -m^2\psi \tag{7.2.1}$$

We aim to find a linear equation for ψ, whose solution is also a solution of the above Klein–Gordon equation. We write this linear equation using the following matrix notation:

$$i\hbar(\gamma_0\partial_t - \gamma_1\partial_x - \gamma_2\partial_y - \gamma_3\partial_z)\psi = m\psi \tag{7.2.2}$$

The above equation is the Dirac equation. It implies the Klein–Gordon equation if the matrices satisfy the following properties:

$$\gamma_0^2 = \mathbf{I}, \gamma_1^2 = \gamma_2^2 = \gamma_3^2 = -\mathbf{I} \tag{7.2.3}$$

$$\gamma_i\gamma_y = -\gamma_j\gamma_i(i \neq j) \tag{7.2.4}$$

It is recognized that the above relations are the same as the Clifford commutation relations defined by Equations (0.1.1) and (0.1.2). The spinor geometry of the ψ field is equivalent to the Clifford geometry. There are several possible choices to write the Dirac γ_μ matrices. Dirac originally discovered the following matrix representation of Clifford algebra:

$$\gamma_0 = \begin{pmatrix} \mathbf{I} & 0 \\ 0 & \mathbf{I} \end{pmatrix}, \gamma_1 = \begin{pmatrix} 0 & -\sigma_x \\ \sigma_x & 0 \end{pmatrix}, \gamma_2 = \begin{pmatrix} 0 & -\sigma_y \\ \sigma_y & 0 \end{pmatrix}, \gamma_3 = \begin{pmatrix} 0 & -\sigma_z \\ \sigma_z & 0 \end{pmatrix} \tag{7.2.5}$$

where $\mathbf{I}$ is the identity matrix and σ_i are the Pauli spin-matrices.

Instead of working with matrix representation of base elements, we can choose to write the Dirac equation using Clifford bases, with the differentiation operator d defined as follows:

- $d_l = \partial_t\mathbf{e}_t - \partial_x\mathbf{e}_x - \partial_y\mathbf{e}_y - \partial_z\mathbf{e}_z$,
- $d_r = \partial_r\mathbf{e}_t + \partial_x\mathbf{e}_x + \partial_y\mathbf{e}_y + \partial_z\mathbf{e}_z$.

So we can now write the Dirac equation using the above-defined operator, for either left- or right-hand polarized sinusoidal spinor wavefunctions, and identify i with the Clifford-pseudoscalar:

$$i\hbar d_l \psi_{l+} = \dddot{m}_l \psi_{l+} \Leftrightarrow \hbar d_l \psi_{l+} = \mathbf{m}_l \psi_{l+} \tag{7.2.6}$$

$$i\hbar d_r \psi_{r+} = \dddot{m}_r \psi_{r+} \Leftrightarrow \hbar d_r \psi_{r+} = \mathbf{m}_r \psi_{r+} \tag{7.2.7}$$

where $i\mathbf{m} = \dddot{m}$ is the tri-vector representation of mass, while $\mathbf{m}$ is the vector representation of mass. It follows from the dimensional analysis of the above equations that as long as ψ is scalar, mass must be represented as either a vector or a tri-vector quantity, but certainly can no longer be a scalar quantity. The above equations highlight the equivalence between these two mass representations. This might strike the reader as odd: all textbook discussions of the Dirac equation consider mass to be a scalar quantity, even geometric algebra oriented studies. Due to the dimensional considerations of Equations (7.2.6) and (7.2.7), a scalar mass representation implies that ψ can no longer be scalar. Since ψ appears on both sides of the Dirac equation, it is not evident what its dimensionality should be. Moreover, mainstream theorists claim that we are not supposed to write down Equation (7.2.6) or (7.2.7) individually, but should always use a coupled system of four Dirac equations. Given that the sinusoidal wavefunction of a free electron remains a sinusoidal wavefunction under any Lorentz boost, we find that the solution type of the quantum mechanical wavefunction ψ is Lorentz-invariant. In other words, the Lorentz-invariance of ψ_+ versus ψ_- solution types appears to support our choice of individually discussing these eigensolutions.

Let's recall that our starting point was the wish to study solutions of the scalar type $\psi = \psi_0 e^{i(\Omega t - \mathbf{K}\cdot\mathbf{r})}$ quantum mechanical wavefunction. We found that working with the Dirac equation implies that either m is not scalar or that ψ is not scalar. Historically, scientists made the "ψ is not scalar" choice, and ψ was understood as a spinor-valued function. Based on the above-presented dimensional analysis, the "m is not scalar" choice has also a strong justification. Therefore, the vectorial representation of m will be pursued further in this chapter.

What Equations (7.2.6) and (7.2.7) really tell us is that applying the $\hbar d$ operator to ψ is equivalent to multiplying it by the $\mathbf{m}$ vector. There is no need to use complex numbers anywhere. Since d_l and d_r are related via a reversal of spatial directions, the corresponding mass vectors are also related via a reversal of spatial orientation. Into which direction is $\mathbf{m}$ pointing? As a particle moves through the four-dimensional spacetime,

Equations (7.2.6) and (7.2.7) imply that $\mathbf{m}$ is pointing along the particle's spacetime trajectory. In the particle's rest frame, the $\mathbf{m}$ vector is pointing into the $\mathbf{e}_t$ direction.

For an exponential spinor wave function with $\psi_- = \psi_0 e^{-\Omega t}$ type time dependency, ψ_- becomes pseudoscalar valued. The corresponding Dirac equation nevertheless retains the same form as Equations (7.2.6) and (7.2.7):

$$i\hbar d_l \psi_{l-} = \dddot{m}_l \psi_{l-} \Leftrightarrow \hbar d_l \psi_{l-} = \mathbf{m}_l \psi_{l-} \tag{7.2.8}$$

$$i\hbar d_r \psi_{r-} = \dddot{m}_r \psi_{r-} \Leftrightarrow \hbar d_r \psi_{r-} = \mathbf{m}_r \psi_{r-} \tag{7.2.9}$$

We emphasize that the above spinor terms, which have exponential time dependence, are misleadingly referred to as "negative energy" spinors in textbooks: the correct interpretation of the apparently negative spinor energy follows from the analysis of Section 2.6.

In the above discussion, the spinor field ψ represents the total energy ϵ of a particle. This spinor field ψ moves through spacetime in the direction of its $\mathbf{m}$ vector. Instead of the total energy, one may consider just the $T = \epsilon - m_0 c^2$ kinetic energy of an electron: the spatial propagation of an electron is described by the wavefunction Ψ, which we derived in Section 5.4 via a Lorentz transformation of its spinor field ψ. We emphasize that the frequency of $\Psi(t)$ is very different from the spinor frequency.

This analysis of the Dirac equation also sheds light on why the imaginary unit i appears in the Dirac equation; it is the Clifford pseudo-scalar $\gamma_t\gamma_x\gamma_y\gamma_z$. The traditional way of treating i just as an imaginary number occults its geometric meaning of being the unit of space–time volume.

7.3. Optical Spinor Representation of Electromagnetic Fields

It is possible to write the $\partial G = 0$ Maxwell equation in a semantically different form. Note that the Maxwell and Dirac equations use the same differentiation operator: $\partial \equiv \partial_t \mathbf{e}_t + \partial_x \mathbf{e}_x + \partial_y \mathbf{e}_y + \partial_z \mathbf{e}_z$.

Since the right-hand side of Maxwell's equation is zero, we can just multiply it by $i\hbar$ to get $i\hbar\partial G = 0$. This is just semantics so far. Some scientists refer to such semantics as their "optical Dirac equation" [3], arguing that the zero on the right-hand side can be equated to the zero field mass. Such semantically adjusted Maxwell equation is however still just Maxwell's equation, and the zero on the right-hand side has nothing to do with the value of a mass term. As we saw in the previous section, the actual meaning of the spinor field is that its differentiation yields energy–momentum eigenvalues, and thus the appropriate definition of the optical Dirac equation is the following:

$$i\hbar\partial\psi_{\mathrm{EM}} = \frac{1}{2}G\mathbf{e}_t\widetilde{G} \tag{7.3.1}$$

Solutions to the above equation are also solutions to the optical Klein–Gordon equation:

$$i^2\hbar^2(\partial_t^2 - \partial_x^2 - \partial_y^2 - \partial_z^2)\psi_{\mathrm{EM}} = N^2 \tag{7.3.2}$$

$N^2 = 0$ for a scalar-free transversal wave; above equation then simply states that the $\epsilon^2 - (pc)^2 = 0$ energy–momentum relationship is valid at each space–time point for the scalar-free photonic wave. The above equation of course has four variants corresponding to left/right polarization and sinusoidal/exponential solution types. It is clearly seen that the $i\hbar\partial\psi_{\mathrm{EM}} = \frac{1}{2}G\mathbf{e}_t\widetilde{G}$ equation is not the same as the $i\hbar\partial G = 0$ form of Maxwell's equation.

How does the azimuthal frequency of optical spinors relate to the frequency of electromagnetic fields? A sinusoidal optical spinor has $\psi_{\mathrm{EM}} = \psi_0 e^{i(\Omega t - \mathbf{k}\cdot\mathbf{r})}$ form. In Equations (2.8.3) and (2.8.4), we showed that each term of the electromagnetic energy–momentum vector comprises a multiplication between two selections from S, E, or B field. When the electromagnetic field oscillates at a frequency ω, we have $S \sim e^{i\omega t}$, $E \sim e^{i\omega t}$, and $B \sim e^{i\omega t}$. Therefore, multiplications between any two selections from S, E, or B yield $\partial\psi_{\mathrm{EM}} \sim e^{i2\omega t}$, and since a differentiation operator does not change the frequency, we get $\partial\psi_{\mathrm{EM}} \sim e^{i2\omega t}$. Thereby we obtain an important relationship between the optical spinor field's and the transversal electromagnetic field's angular frequencies: $\Omega = 2\omega$. This means that the frequency of the optical spinor field is twice the electromagnetic frequency.

How does the energy–momentum of ψ_{EM} relate to a particle-like energy–momentum representation? Since $\Omega = 2\omega$, we have $\hbar\partial_t\psi_{\mathrm{EM}} = i2\hbar\omega\psi_0 e^{i2(\omega t - \mathbf{k}\cdot\mathbf{r})} = i\epsilon\psi_{\mathrm{EM}}$, where ϵ is the energy eigenvalue of a particle. One may notice that the $2\hbar\omega$ eigenvalue is twice the energy value given by the $\hbar\omega = mc^2$ relationship. This result means that if one naively considers the particle-internal electromagnetic energy–momentum to represent the kinetic motion of a particle, the $\hbar\partial_t$ operator yields the eigenvalues of a mass $2m$ particle. A "normalization" into mass m energy–momentum representation is in the background of declaring the electron's internal angular momentum to be $\frac{\hbar}{2}$, which is only half of its actual electromagnetic angular momentum; we already derived in Chapter 3 that an electron's internal angular momentum is $\hbar$.

We are now ready to clarify several major misconceptions about the Dirac equation, which all persisted for nearly 100 years:

- The Dirac spinor field is considered to describe a "half-spin" particle. Under an applied magnetic field, the electron's angular momentum measurement yields $\frac{\hbar}{2}$. This measured value was interpreted as the inherent angular momentum of the electron, despite the obvious contradiction with angular momentum conservation in electron–positron annihilations. This interpretation was then cited as evidence that the Dirac spinor field "describes the electron." Regarding measurements, one must realize that under an applied magnetic field the electron's angular momentum is precessing around magnetic field lines, and the measured $\frac{\hbar}{2}$ value is only its projection onto the direction of the applied magnetic field. The equations of this precession are described in Chapter 3. Our interpretation is in line with the operating principle of NMR devices, while the traditional interpretation of the measured $\frac{\hbar}{2}$ value is contradictory also to the operating principle of commercial NMR technology. Regarding theory, the above analysis reveals that the apparent $\frac{\hbar}{2}$ angular momentum value results from counting only half of the electromagnetic energy–momentum in the optical spinor field. This misunderstanding went unnoticed for such a long time because the spinor interpretation error and the measurement interpretation error hide each other.
- Schrödinger suggested in 1930 to interpret the above-discussed optical spinor frequency as a real oscillation frequency of an elementary particle [4]. Subsequently, Hestenes and also other researchers of Zitterbewegung treated the complex phase of the spinor wavefunction field as the phase of Zitterbewegung rotation. However, the above presented analysis reveals that this optical spinor frequency is twice the frequency of the particle's electromagnetic field oscillation. The correct Zitterbewegung frequency is the electromagnetic frequency.
- Dirac was perplexed by the existence of apparently negative energy solutions to the Dirac equation. He suggested equating anti-particles with the negative energy solutions of his equation, but the relation between particles and their anti-particles has been shrouded in controversy. The initial model of anti-particles involved a confusing discussion about "holes in the Dirac sea of filled-up electron states, everywhere in the vacuum." Such a model implies infinite energy density and infinite charge density of the vacuum, which is very contrary to experience. As discussed in Chapter 1, anti-particles were later thought of as particles moving

backward in time. However, the arrow of time defines the direction of entropy increase. If anti-particles would be moving backward in time, they would be evolving toward lower entropy configurations, which would mean an unphysical asymmetry between particles and anti-particles. Our proposed anti-particle model is straightforward: anti-particles differ from particles in the opposite charge value of the scalar field S. During electron–positron annihilation, the oppositely charged scalar components of the electromagnetic fields induce transversal electromagnetic fields while canceling each other. Section 2.6 outlined how this process may be compared to the discharge of a capacitor and a solenoid. As was shown in Chapter 3, the electron can be characterized by the following capacitance and inductance values: $C = 3.135381 \times 10^{-25}$ F and $L = 2.089108 \times 10^{-16}\Omega$s. The corresponding frequency of this quantum LC circuit is $f = \frac{1}{\sqrt{LC}} = 1.23559 \times 10^{-20}$ Hz, which is the frequency of the produced electromagnetic radiation. In other words, the electron-positron annihilation process converts massive particles' electromagnetic fields into fully transversal configuration. The energy of the resulting photons is certainly a positive value; i.e., the negative energy solutions of the Dirac equation cannot be equated to anti-particles. What are then the negative energy solutions? As we discussed in Section 2.6, they are transient wavefunctions of an annihilating electron, and their energy is negative only with respect to the static energy level of a stable electron.

- Mainstream quantum theorists insist that solutions of Dirac and Klein–Gordon equations are not the same. They talk about a "difference of spin" and about supposedly not positive-definite probability density of Klein–Gordon solutions. In light of the above discussion, such an opinion is mathematically untenable. Over the past century, theorists used a single Dirac equation to describe the dynamics of the electron. The spinor representations of two fields, namely ψ_{QM} and ψ_{EM}, were mixed together into one linear Dirac equation. One must carefully study the origin of these two fields, in order to understand how to work with them. The conceptually proper calculation of atomic orbitals is further addressed in Section 8.8, where we indeed find that the Dirac and Klein–Gordon equations yield the same solutions.

As a summary of this section, Table 7.1 writes electromagnetic expressions in a way that is normally used in quantum mechanics.

Table 7.1. A summary of mathematical expressions describing electromagnetism, and their quantum mechanics-related formulation.

Mathematical expression	Physical interpretation	Q.M.-like formulation
A	Electromagnetic vector potential	
$G = \partial A$	Electromagnetic field: $G = (S, \mathbf{0}) + (0, F)$	$F = E + iB$ when $S = 0$
$N = \frac{1}{2} G\mathbf{e}_t\widetilde{G}$	Electromagnetic energy–momentum density	Optical spinor field ψ: $\boldsymbol{\partial}\psi = N$
$L = \frac{1}{2} G\widetilde{G}$	Electromagnetic Lagrangian	Probability density: $G\widetilde{G} = F\bar{F}$ when $S = 0$

7.4. One Particle — Two Fields

With reference to the preceding chapters, we established that a massive particle comprises two apparently distinct fields: the quantum mechanical spinor field ψ and the particle's electromagnetic vector potential field A. The two fields are obviously not independent, as they relate to the same particle. Thus, both fields must give compatible results about the particle's energy, momentum, and position.

By Equations (7.2.6)–(7.2.9), the space–time differential of ψ yields the energy–momentum eigenvalues of a particle. It was shown in Section 3.4 that the energy–momentum $\boldsymbol{P}_\square$ of a particle with electric charge e is given by the $\boldsymbol{P}_\square = e\boldsymbol{A}_\square$ relationship, where $\boldsymbol{A}_\square$ is the vector potential seen by the electric charge. $\boldsymbol{P}_\square$ refers to the total energy–momentum of the moving charge, which is the sum of its internal Zitterbewegung momentum and its overall particle-like momentum. Since $\boldsymbol{P}_\square$ means here the vector potential seen by the electric charge, an externally applied vector potential adds onto this vector potential seen by the electric charge, and thus we directly obtain the effect of external electromagnetic fields on the Dirac equation:

$$i\hbar d\psi = \dddot{m}\psi + ieA_{\text{applied}} \tag{7.4.1}$$

The above equation was historically validated by comparing the calculated energy levels of electron orbitals against experimental transition energy measurements.

Let us look at the dimension of the terms appearing in Equation (7.4.1). ψ is a scalar, $d\psi$ becomes Clifford vector, and thus we get a tri-vector upon further multiplication by the pseudoscalar i. On the right hand side

Table 7.2. The dimension of terms appearing in the Dirac equation.

Term	Dimension
ψ	Scalar
$id\psi$	Tri-vector
eA	Vector
ieA	Tri-vector
$\dddot{m}$	Tri-vector

eA_{applied} is a Clifford vector, and thus ieA_{applied} also becomes tri-vector. These relationships are summarized in Table 7.2, and directly support our original choice of $\dddot{m}$ being a tri-vector. In this way, Equation (7.4.1) is yet again telling us that the particle mass $\dddot{m}$ is the result of an integration over a volume element, and not just some "inherent scalar property of an abstract point particle." It is not a new idea that particle mass may be equated to electromagnetic field energy, i.e., the integral of the electromagnetic energy density over the particle volume. However, according to Feynman [6], past attempts at developing this idea into a valid physical model remained unsuccessful: "Many attempts have been made, and some of the theories were even able to arrange things so that all the electron mass was electromagnetic. But all of these theories have died." Nevertheless, the electron–positron interaction example given at the end of Section 2.8 directly demonstrates the electromagnetic origin of elementary particle mass. Equations (2.8.3) and (2.8.4), imply the following integration formula for the particle mass:

$$\iiint_v \frac{1}{2}(S^2 + E^2 + B^2)\, dV = m_0 \tag{7.4.2}$$

In Chapter 3, we indeed successfully identified the electron mass with electric and magnetic field energy. In order to fully understand the relationship between electromagnetism and quantum mechanics, we must understand what the $\dddot{m}\psi$ term really means. At first glance, it seems to imply that the electromagnetic field at any local point is interacting with the electromagnetic field everywhere else, as expressed through the above integral. That may feel like a paradox since a field theory is supposed to be a local theory. This paradox was however resolved in Chapter 5, where we discovered the more fundamental way of writing the Dirac equation, as a local equation.

All these considerations lead up to the question: can we factorize Dirac spinors into the scalar ψ and vectorial A fields? In other words, are these two fields really just two projections of a single unified field? In the following, we develop such a factorization.

7.5. A Factorization of the Electron State

7.5.1. *The Dirac–Hestenes and Dirac–Baylis factorization*

Suppose that we exactly know the electron's state. We saw in Chapter 5 that any perception of an electron is related via a Lorentz transformation to its rest frame perception. One might try to model the electron state as a function of Lorentz transformations at each point in space; i.e., at each point the local perception of the electron state is obtained via some rotation plus Lorentz boost with respect to rest frame state. To each point in space we assign then a $\sqrt{\rho}\mathcal{S}$ value, where ρ is the scalar valued probability density and $\mathcal{S}$ is an element of the $SU(2) \times SU(2)$ Lorentz group, corresponding to a rotation of axial orientation and a boost into a certain direction. Therefore, $\mathcal{S}$ factorizes into UB terms, where U represents a rotation and B represents a boost value. In general, we are interested in eigenvalue problems where $\rho(\boldsymbol{x},t)\,\mathcal{S}(\boldsymbol{x},t)$ does not have any time dependence, i.e., the electron is in a steady state. The first and foremost thing which we learn about an electron eigenstate, e.g., about an electron orbital, is that it has a characteristic energy and momentum. Therefore, not only does $\mathcal{S}(\boldsymbol{x},t)$ not have time dependence, but the electron boost part is just a constant parameter of the eigenstate.

As explained in [1, 2], the Dirac–Hestenes and Dirac–Baylis formulation of the Dirac equation maps the well-known four-component Dirac spinors into a matrix form. Equation (20) of [2] presents the following factorization of the resulting Dirac–Hestenes spinor:

$$\Psi = \sqrt{\rho}\exp\left(\mathbf{i}\frac{\beta}{2}\right)UB \tag{7.5.1}$$

where ρ is the scalar valued probability density, U represents a spatial rotation, B represents a boost value, β is known in the literature as the Yvon–Takabayashi angle, and $\mathbf{i}$ is defined from the Dirac gamma matrices as $\mathbf{i} \equiv \gamma_0\gamma_1\gamma_2\gamma_3$. This factorization of Ψ corresponds to the above-formulated Lorentz transformation of a particle's rest frame perception, and also includes an additional phase factor involving the so-called Yvon–Takabayashi angle. Dirac spinors gain now a well-understood physical meaning.

Importantly for our present discussion, we recognize the $\mathbf{i} \equiv \gamma_0\gamma_1\gamma_2\gamma_3$ term as the Clifford pseudo-scalar I. Therefore, the question arises how the $\exp\left(\mathbf{i}\frac{\beta}{2}\right)$ term is related to the $\gamma_t \to e^{I\theta}\gamma_t$ transformation, which we introduced in Chapter 2. This question will be addressed in Section 7.9.

Solving the Dirac equation historically meant finding the complex-valued four-component functions which comprise the Dirac-spinor eigenstate. Here, the eigenstate means that Ψ has no time dependence, i.e., the eigenstate has $\Psi(\boldsymbol{x})$ form. The Dirac-spinor eigenstate solutions were successfully found, e.g., for atomic orbitals discussed in Section 6.6.

Several advantages of the Dirac–Hestenes approach are described in [1, 2]. It allows calculating certain non-stationary problems, such as the dynamics of electron beams. However, it is not helpful with understanding what the internal structure of the "rest frame state" really is. We now turn to another factorization, which is directly related to the electromagnetic vector potential.

7.5.2. *A factorization with vectorial mass representation*

In the preceding chapters, we developed a new insight into the quantum mechanical electron state. We learned in Chapter 2 that the scalar valued spinor field ψ is coupled to a mass vector field $\boldsymbol{m}_{\|}$, which points along the particle's space–time trajectory. For eigenvalue problems, the spatial part of $\boldsymbol{m}_{\|}$ points in the direction of orbital movement. We learned in Chapter 4 that the vector valued Proca field $\boldsymbol{a}_{\square}$ is coupled to a rotating mass vector field $\boldsymbol{m}_{\perp}$, which points from the center of Zitterbewegung to the location of the charged electron sphere. The $\boldsymbol{m}_{\|}$ and $\boldsymbol{m}_{\perp}$ notation emphasizes that these mass vectors are always orthogonal to each other. Figure 7.1 depicts these mass vectors, and illustrates the effect of a reference transformation $\mathcal{S}$.

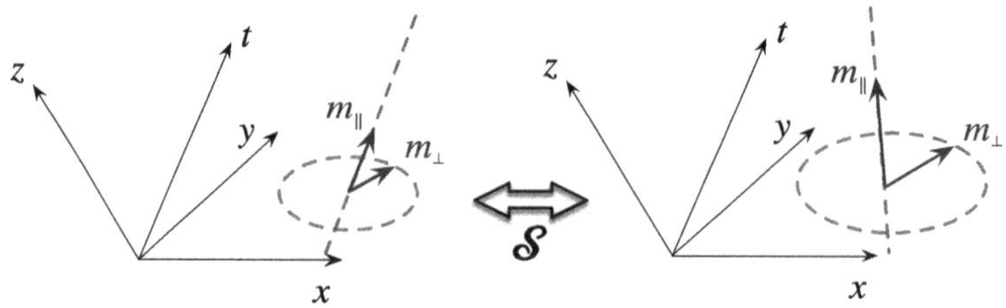

Figure 7.1. An illustration of the two mass vectors characterizing the electron's state. The four axes are all orthogonal to each other, and the two mass vectors are always orthogonal as well. The dashed straight line represents the electron's space–time trajectory, while the dashed circle represents the Zitterbewegung rotation of each mass vector. The left and right sides are linked via a reference frame transformation $\mathcal{S}$.

Having understood what the ψ and $\boldsymbol{a}_\square$ fields represent, our preferred strategy is to carefully search for eigenstate solutions of the corresponding Dirac equations, without overlooking any solution type. Our approach is more fundamental than traditional ones because we can directly work with the physical fields comprising the electron.

In the following, we use the same Clifford algebra notation as in Chapter 4.

We found that the $\epsilon^2 - (pc)^2 = (mc^2)^2$ energy–momentum relationship leads to the Dirac equation, i.e., Equation (7.4.1), which we write in the following form:

$$\hbar \boldsymbol{\partial} \psi = \boldsymbol{m}_{||} \psi + e A_{\text{applied}||} \tag{7.5.2}$$

where ψ is scalar, and $\boldsymbol{m}_{||}$ is a vector. We found that the ψ field represents the energy–momentum distribution of the electron. To solve practical problems, the impact of external electromagnetic fields is evaluated via the $eA_{\text{applied}||}$ term of the above equation. Here, the $A_{\text{applied}||}$ notation means that the external electromagnetic field couples to the $\boldsymbol{m}_{||}$ vector, and the electron gains momentum with respect to the external field source. Therefore: **when $A_{\text{applied}||}$ originates from another particle, the eigenstate solution of Equation (7.5.2) is an electron orbit whose kinetic and potential energy distribution is in accordance with the virial theorem.**

In Chapter 4, we worked with the Proca field-valued Dirac equation. We obtained the electromagnetic Dirac equation $\hbar \boldsymbol{\partial} \boldsymbol{a}_\square = -\boldsymbol{m}_\perp \boldsymbol{a}_\square$, where $\boldsymbol{a}_\square$ is the vector valued Proca field. We found that the vector potential $\boldsymbol{a}_\square$, which is generated by the Zitterbewegung charge rotation, is at the very origin of the electron mass concept. Since the Proca field was obtained via a spatial averaging of the Zitterbewegung-generated electromagnetic field, the presence of externally applied electromagnetic fields simply adds onto the Proca field, and therefore this Dirac-like equation can be written as follows:

$$\hbar \boldsymbol{\partial} \left(\boldsymbol{a}_\square + A_{\text{applied}\perp}\right) = -\boldsymbol{m}_\perp \left(\boldsymbol{a}_\square + A_{\text{applied}\perp}\right) \tag{7.5.3}$$

In practical eigenvalue problems, $A_{\text{applied}\perp}$ has no time dependence. Moreover, in many problems $A_{\text{applied}\perp}$ can be considered uniform on the electron's dimension scale, and then we may simplify the above equation into

$$\hbar \boldsymbol{\partial} \boldsymbol{a}_\square = -\boldsymbol{m}_\perp \left(\boldsymbol{a}_\square + A_{\text{applied}\perp}\right) \tag{7.5.4}$$

Here, the $A_{\text{applied}\perp}$ notation means that the external electromagnetic field couples to the $\boldsymbol{m}_\perp$ vector, and the electron gains or loses mass via this external field interaction. It is important to keep in mind that the electron does not gain kinetic momentum through $A_{\text{applied}\perp}$ interaction with respect to the external field source. Therefore: **when $A_{\text{applied}\perp}$ originates from another particle, the eigenstate solution of Equation (7.5.4) is an adjusted electron mass, in accordance with the electric Aharonov–Bohm Equation (4.3.10).**

We come to a puzzle: does the external field of another particle interact with the electron via Equation (7.5.2) or (7.5.4) or both? In a hypothetical case where the external field interacts only via Equation (7.5.2), the result is an electron orbital without any change in the electron mass. The mean value of $eA_{\text{applied}\|}$ corresponds to the sum of the binding energy plus kinetic energy. In a hypothetical case where the external field interacts only via Equation (7.5.4), the result is a composite particle where the electron does not have any kinetic momentum with respect to the other interacting particle. The mean value of $eA_{\text{applied}\perp}$ corresponds to the change in the electron mass. In the general case, two particles interact both via Equation (7.5.2) and (7.5.4): Equation (7.5.2) determines the resulting orbital energy–momentum eigenstate, while Equation (7.5.4) determines the resulting mass change and spin correlations.

To understand this issue more clearly, let's consider the following interaction types and ask pertinent questions about the origin of binding energy:

- **Bohr orbit of the electron:** The $1s$ electron orbital around a proton has 13.6 eV binding energy, and 13.6 eV kinetic energy. Let's take an electron–proton pair, initially at rest. Where does the 27.2 eV energy come from that is produced upon the formation of the $1s$ electron orbital? As the electron and proton move toward each other, the conservation of momentum dictates that they each have the same magnitude of momentum, with opposing direction. The equal amount of momenta implies that the electron gains most of the kinetic energy, specifically $\frac{m_p}{m_e+m_p}$ ratio. Since the electron cannot accelerate itself, its kinetic energy originates from the proton's electromagnetic field. In the same way, the proton's much smaller kinetic energy originates from the electron's electromagnetic field. This implies $A_{\text{applied}\|} = \frac{m_p}{m_e+m_p} A_{\text{applied}}$ and $A_{\text{applied}\perp} = \frac{m_e}{m_e+m_p} A_{\text{applied}}$ for the electron, while the two particles accelerate toward each other. Does the same principle also apply upon the formation of $1s$ electron orbital? As mentioned in Chapter 6, the $\frac{m_p}{m_e+m_p}$ ratio indeed shows up in the energy of quantum mechanical state transitions, and it is therefore incorporated into the Rydberg constant. As an additional check, we can ascertain the electron mass through a measurement of its magnetic moment, which is related to the electron mass via Equation (3.3.7). For a free electron, the magnetic moment value is $1.00115965\mu_B$, where μ_B is the Bohr magneton. Considering the values in the last table of Section 6.8, which are calculated by assuming that electron mass only changes due to the Lorentz boost, the calculated $1.001133\,\mu_B$ magnetic moment value in the 1s state well matches the measured $1.001142\,\mu_B$ value. If the 27.2 eV energy originated from the electron, its magnetic moment would be $1.001186\,\mu_B$, which is much further from the measured value. Therefore, the 13.6 eV binding energy plus 13.6 eV kinetic energy originate from the proton's change of mass. The electron and proton spins may be correlated either parallelly or anti-parallelly, and the choice of this orientation generates the so-called hyperfine splitting of binding energy levels.

- **Positronium state of an electron–positron pair:** Let's take a separated electron–positron pair, initially at rest. They start moving toward each other due to electric attraction, which generates momentum according to Equation (7.5.2), and before annihilating they form a so-called positronium structure, which has $7\,\mu$s lifetime. Experimental scientists measure the binding energy of positronium to be 6.8 eV, which is half of the electron Bohr orbit case. Because of the electron–positron symmetry, each particle contributes equally to the binding energy and generates the kinetic energy of the other particle. The virial theorem implies that both particles have 3.4 eV kinetic energy. Therefore, each constituent particle of the positronium looses 6.8 eV mass with respect to their rest mass. This implies $A_{\text{applied}||} = \frac{m_{p+}}{m_{e-}+m_{p+}} A_{\text{applied}}$ and $A_{\text{applied}\perp} = \frac{m_{e-}}{m_{e-}+m_{p+}} A_{\text{applied}}$ for the electron, which is an analogous formula to the Bohr orbit case. The electron and positron spins may be correlated either parallelly or anti-parallelly, and the choice of this orientation generates the so-called para-positronium and ortho-positronium variants.
- **Ultra-dense hydrogen:** As shown in Chapter 4, ultra-dense hydrogen forms structures where the electron has no kinetic momentum with respect to protons. This must be the case also for a single electron–proton pair in ultra-dense hydrogen state. The implication is that the electron forms a composite particle with the proton by interacting solely via Equation (7.5.4), i.e., $A_{\text{applied}||} = 0$ and $A_{\text{applied}\perp} = A_{\text{applied}}$. This type of interaction will be examined in Chapter 12.
- **Nuclear bonds:** It is experimentally recognized that nuclear binding energy corresponds to mass change. In some reactions, this mass change is radiated away as a single-frequency electromagnetic radiation, which is analogous to the single-frequency electromagnetic radiation emitted

during chemical bond formation. A basic example is the deuteron formation from a proton and neutron: the 2.22 MeV mass change is radiated away as a single-frequency gamma radiation. Assuming an electromagnetic origin of nuclear bonds, the $A_{\text{applied}\|} = \frac{m_1}{m_2+m_1} A_{\text{applied}}$ and $A_{\text{applied}\perp} = \frac{m_2}{m_2+m_1} A_{\text{applied}}$ relations must hold for an interaction between particles of mass m_1 and m_2. This assumption will be investigated in Chapter 14.

One may conclude from the above examples that the rest mass of elementary particles in an atom or molecule is not exactly the same as a free particle mass. In other words, elementary particle mass is not strictly an invariant, for example, the electron rest mass within the positronium is 0.0027% different from the free electron mass.

Let's consider an electron interacting with a particle of mass M, and forming a steady state. We introduce the parameter $\kappa = \frac{M}{m_e+M}$, where m_e is the rest mass of a free electron. The resulting steady state is an eigenstate solution of the Dirac equations, which we now write with the κ parameter:

$$\hbar \boldsymbol{\partial} \psi = \boldsymbol{m}_{\|} \psi + \kappa e A_{\text{applied}} \tag{7.5.5}$$

$$\hbar \boldsymbol{\partial} \boldsymbol{a}_{\square} = -\boldsymbol{m}_{\perp} \left(\boldsymbol{a}_{\square} + (1-\kappa) A_{\text{applied}} \right) \tag{7.5.6}$$

where A_{applied} is the electromagnetic vector potential generated by the mass M particle. According to the above examples, there are two choices for κ:

(1) $\kappa = \frac{M}{m_e+M}$, which is an orbital state,
(2) $\kappa = 0$, which is a composite particle state.

The $\kappa = \frac{M}{m_e+M}$ value follows from the conservation of momentum. The $\kappa = 0$ value is also compatible with the conservation of momentum because the constituents of the composite particle have no relative kinetic momentum.

One may wonder whether it is possible to write Equations (7.5.5) and (7.5.6) as a single equation. Upon introducing the $\boldsymbol{\phi} = \psi \boldsymbol{a}_{\square}$ field, we evaluate its space–time gradient for a free particle:

$$\hbar \boldsymbol{\partial} \boldsymbol{\phi} = \hbar \psi \left(\boldsymbol{\partial} \boldsymbol{a}_{\square} \right) + \hbar \left(\boldsymbol{\partial} \psi \right) \boldsymbol{a}_{\square} = \left(\boldsymbol{m}_{\|} - \boldsymbol{m}_{\perp} \right) \psi \boldsymbol{a}_{\square} \tag{7.5.7}$$

In the above equation, we made use of ψ being scalar, which commutes with the vectorial fields. Consequently, we may write a unified Dirac equation of a free particle:

$$\boxed{\hbar \boldsymbol{\partial} \boldsymbol{\phi} = \left(\boldsymbol{m}_{\|} - \boldsymbol{m}_{\perp} \right) \boldsymbol{\phi}} \tag{7.5.8}$$

We recognize that the quantum mechanical electron state is represented by a single field, which is the $\boldsymbol{\phi} = \psi \boldsymbol{a}_{\square}$ field. This $\boldsymbol{\phi}$ field is therefore a representation of the spinor field, and its dynamics is given by the Dirac Equation (7.5.8).

When the electron interacts with another particle, the unified Dirac equation includes an electromagnetic vector potential A_{applied}, which is generated by the other particle:

$$\begin{aligned} \hbar \boldsymbol{\partial} \boldsymbol{\phi} &= \hbar \psi \left(\boldsymbol{\partial} \boldsymbol{a}_{\square} \right) + \hbar \left(\boldsymbol{\partial} \psi \right) \boldsymbol{a}_{\square} \\ &= -\boldsymbol{m}_{\perp} \psi \left(\left(1 - \kappa \right) A_{\text{applied}} + \boldsymbol{a}_{\square} \right) + \left(\kappa e A_{\text{applied}} + \boldsymbol{m}_{\parallel} \psi \right) \boldsymbol{a}_{\square} \end{aligned} \tag{7.5.9}$$

We simplify the above into the following Dirac equation of an interacting electron:

$$\hbar \boldsymbol{\partial} \boldsymbol{\phi} = \left(\boldsymbol{m}_{\parallel} - \boldsymbol{m}_{\perp} \right) \boldsymbol{\phi} + \kappa e A_{\text{applied}} \boldsymbol{a}_{\square} - \boldsymbol{m}_{\perp} \psi \left(1 - \kappa \right) A_{\text{applied}} \tag{7.5.10}$$

When $\kappa \approx 1$, which is the case for the electron orbitals, the above equation may be approximated as follows:

$$\hbar \boldsymbol{\partial} \boldsymbol{\phi} = \left(\boldsymbol{m}_{\parallel} - \boldsymbol{m}_{\perp} \right) \boldsymbol{\phi} + \kappa e A_{\text{applied}} \boldsymbol{a}_{\square} \tag{7.5.11}$$

The resulting electron eigenstate is determined by the above equation, but it is no longer an equation of just $\boldsymbol{\phi}$.

We make some concluding observations on the remarkably simple free electron Equation (7.5.8) and orbital electron Equation (7.5.11):

- Since the free electron field $\boldsymbol{\phi}$ factorizes into two fields, the spinless ψ field and the spin-carrying $\boldsymbol{a}_{\square}$ field, the quantum mechanical state of an electron can be approximated by independent state coordinates $|q, s\rangle$, where q labels the electron's orbital state while s labels its spin state. This result will be used in a later chapter on the Pauli exclusion principle. It follows from Equation (7.5.11) that the $|q, s\rangle$ factorization gives almost stationary $|q, s\rangle$ eigenstates in the $eA_{\text{applied}} \ll mc^2$ regime, and gives a strong oscillatory dynamics among $|q, s\rangle$ states in the $eA_{\text{applied}} \sim mc^2$ regime.
- The above introduced $\boldsymbol{\phi} = \psi \boldsymbol{a}_{\square}$ field has a well-defined physical meaning, without any imaginary-valued component. The origin of the relativistic $\epsilon^2 - (pc)^2 = \left(mc^2 \right)^2$ energy–momentum relationship will be further discussed in Chapter 8, where ψ will be shown to originate from the space–time curvature. Thus, Chapter 8 shall complete the reconciliation of quantum mechanics with general relativity.

- Based on Chapter 2 results, $\boldsymbol{\phi}$ may have four variants, corresponding to left-/right-handed and sinusoidal/exponential eigenstate types of ψ. These four variants carry the proper meaning of the four components in the well-known Dirac spinor structure.
- A free electron stays in an eigenstate of $\boldsymbol{\phi}$, and theorists say that it is in some particular state of their four-component spinor structure. A bound electron however is no longer in the eigenstate of $\boldsymbol{\phi}$: it has an oscillatory component corresponding given by $\kappa e A_{\text{applied}} \boldsymbol{a}_{\square}$, which cannot be described by a single eigenstate of $\boldsymbol{\phi}$. In the $eA_{\text{applied}} mc^2$ regime, mainstream textbooks the $\kappa e A_{\text{applied}} \boldsymbol{a}_{\square}$ oscillation as a sinusoidal oscillation between two eigenstates of $\boldsymbol{\phi}$. These two eigenstates are the left-handed and right-handed spinors. Therefore, textbooks claim that a bound electron is oscillating between left-handed and right-handed spinor states. Such a textbook model creates confusion by obscuring the physical meaning of the $\kappa e A_{\text{applied}} \boldsymbol{a}_{\square}$ term. In the $eA_{\text{applied}} \sim mc^2$ regime, textbooks model the $\kappa e A_{\text{applied}} \boldsymbol{a}_{\square}$ oscillation as a sinusoidal oscillation involving all four spinor types.
- With our approach, the calculation of the positronium eigenstate is not any more a difficult problem than the calculation of an ordinary Bohr orbit eigenstate. Once we know the A_{applied} field, the corresponding Dirac equations become clearly defined.

In summary, the Dirac spinor field becomes rather intuitive and physically meaningful under the proposed vectorial representation of the electron mass.

7.6. An Electron Wave at Potential Steps

In optics, a basic problem to study is the transmission and reflection of an electromagnetic plane wave at material interfaces, e.g., at vacuum–glass or at vacuum–metal interface. The equivalent basic problem of quantum mechanics is to study the plane electron wave behavior at an interface represented by a voltage step function. There are no electromagnetic fields on either side of the step function, and therefore, the spatial solution will be given everywhere by Equation (7.5.8). So the basic problem is to find a consistent solution of Equation (7.5.8) in a voltage step scenario. A valid solution must respect energy-, momentum-, and charge-conservation.

In 1929, Oskar Klein obtained a strange result by applying the Dirac equation to this voltage step scenario. According to his calculations, when the barrier-crossing energy is similar to the electron mass, i.e., $V \sim \frac{mc^2}{e}$, the barrier is nearly transparent. As the voltage is increased even higher, the reflection diminishes and the electron is always transmitted. This nonsensical result has become known as the Klein paradox. A paradox is always a red flag, signaling that a mathematical mistake was made. To this day, mainstream theorists have not been able to resolve this basic problem. Instead of correcting mathematical mistakes, quantum field theorists add on more structures and assumptions till the problem seems to disappear. Ref. [6] represents such a mainstream patchwork, which handwaves the Klein paradox away.

To resolve the Klein paradox, we now discuss Equation (7.5.8) in a voltage step scenario. A voltage step represents an infinite electric field at the step location: according to Chapter 5, this generates infinite space–time curvature. Thus, the idealized voltage step function scenario is ill-posed, and we use a voltage ramp function instead of the step function. The voltage ramp converges to the step function as the ramp width becomes small.

One must recognize that the electron spin is not arbitrarily oriented at the potential step. When a plane wave propagates perpendicularly onto the voltage step, the electron spin is oriented along its direction of propagation. To illustrate this issue through a practical example, consider a cube-shaped metal. This metal has a voltage step at each surface, which keeps the delocalized electron wavefunction bounded within the metal. In the interior, the electron wavefunction is a superposition of three orthogonal plane waves. The electron spin can be measured in x, y, or z directions: a corresponding plane wave component will give analogous spin measurement result along any direction in the interior. Thus, the electron spin is isotropic in the interior of the metal. However, this isotropy is no longer true at the interface: only one plane wave component propagates and gets reflected at each surface. In other words, the electron spin points along the direction of wave propagation in the plane wave scenario.

First, we consider a small potential step, where the wavefunction remains harmonic. The considered scenario is shown in Figure 7.2. Let the zero potential denote the potential level where the electron is in its rest frame: this is illustrated in the middle of Figure 7.2. On the left and right sides of this potential barrier, we can view the electron either from a wave perspective or a particle perspective:

- **Wave perspective:** As the wavefunction enters into a lower potential region, it gains $\epsilon = -eV'$ amount of kinetic energy according to the principle of energy conservation. In this lower potential region, the quantum mechanical wavefunction becomes a harmonic wave in space, and its direction of propagation through space–time is given by the $\boldsymbol{m}'_{\|}$ vector.

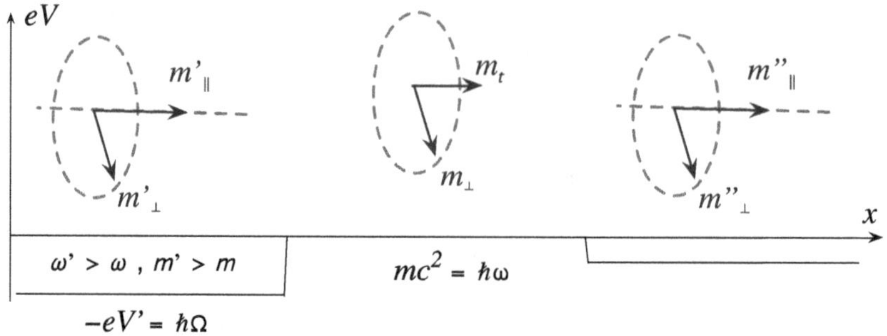

Figure 7.2. An electron moving across a potential barrier. The electron is in its rest frame in the middle section, and its parallel mass vector is illustrated pointing along the time axis.

The difference between the de Broglie frequencies corresponds to the potential energy gain: $\hbar\,(\omega' - \omega) = -eV' = \hbar\Omega$, where Ω is the angular frequency of the quantum mechanical wavefunction.

- **Particle perspective:** The electron moves along a geodesic, and its charge surface remains at a constant potential. A slow ramp would correspond to constant acceleration, and we could calculate the electromagnetic radiation emitted by an accelerating charge. However, a step-like fast ramp corresponds to an adiabatic process, where $\kappa = 0$. The relativistic change of electron mass in fact corresponds to a constant potential of its charge surface, given by the $e\,(\boldsymbol{A}'_\square - V') - e\boldsymbol{A}_\square = 0$ relationship, where $\boldsymbol{A}'_\square$ and $\boldsymbol{A}_\square$ denote the electron-internal electromagnetic potential in the two regions. The lower external potential is counter-balanced by an increased electron-internal potential, and the corresponding relativistic electron mass increase is $(\gamma_L - 1)\,mc^2 = -eV$, where γ_L is the Lorentz boost factor and m is the rest mass.

In Chapter 5, we discussed the equivalence between these wave and particle perspectives. We also derived the electron mass by using Hamilton–Jacobi theory, and that calculation also leads to the above-discussed adiabatic mass adjustment process.

Seemingly, the above harmonic wave scenario represents an interesting analogy with a free particle, considering a Lorentz boost effect or the emission/absorption of electromagnetic energy. It is nicely illustrated in Figure 7.2 that this analogy is based on the potential step acting as a sink or source of the quantum mechanical wave. However, with respect to the electron rest frame, an electron can only gain positive Lorentz boost and can

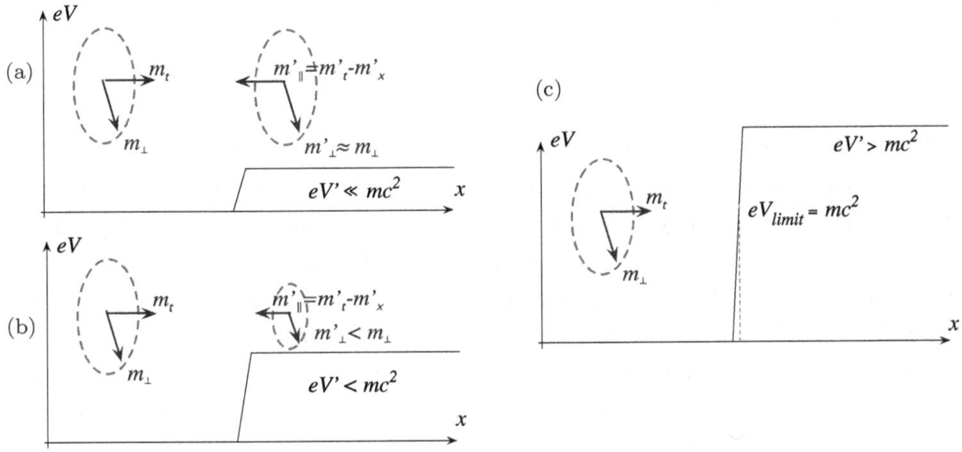

Figure 7.3. An electron encountering a potential step. The incident electron wave is nearly at rest, and its parallel mass vector is illustrated pointing along the time axis. (a) $eV \ll mc^2$, (b) $eV < mc^2$, (c) $eV > mc^2$.

only absorb electromagnetic wave energy. Therefore, increasing the potential step into the positive region would be analogous either to an unphysical Lorentz boost or to emitting electromagnetic energy from a rest-frame state. A high potential barrier transforms the spinor field into a state which is not observed for a free particle.

We discuss the high-potential scenarios of Figure 7.3 using exactly the same wave or particle perspective as we used in the low-potential case:

- **Wave perspective:** As the wavefunction enters into a potential barrier region, it looses $\epsilon = eV'$ amount of kinetic energy according to the principle of energy conservation. In this higher potential region, the quantum mechanical wavefunction becomes exponential in space. Its trajectory through space–time is given by the $\boldsymbol{m}'_\parallel$ vector, and the x component of this vector points toward the edge of the potential barrier because the exponential tail represents a negative momentum. The difference between the de Broglie frequencies corresponds to the potential energy gain: $\hbar\,(\omega - \omega') = eV'$.
- **Particle perspective:** The electron moves along a geodesic, and its charge surface remains at a constant potential. A step-like fast ramp corresponds to an adiabatic process, where $\kappa = 0$. The relativistic change of electron mass corresponds to a constant potential of its charge surface, given by the $e\,(\boldsymbol{A}'_\square + V') - e\boldsymbol{A}_\square = 0$ relationship, where $\boldsymbol{A}'_\square$ and $\boldsymbol{A}_\square$

denote the electron-internal electromagnetic potential in the two regions. The higher external potential is counter-balanced by a decreased electron-internal potential.

A key insight of our analysis is that the electron's spherical charge surface remains at the same potential during any potential step encounter. The above calculated changes to the de Broglie frequency also directly follow from Equation (4.3.10), which was derived in Chapter 4. On both sides of the potential step, the spinor field evolves according to the simple looking Equation (7.5.8). The above examples demonstrate that the solution of Equation (7.5.8) is in fact determined by the boundary conditions. Only the rest mass term of the electron is a constant value, but the mass term appearing in Equation (7.5.8) is set by the boundary conditions.

We are now ready to address the Klein paradox. Figure 7.3 distinguishes three scenarios. In the $eV \ll mc^2$ scenario, the electron mass of the exponential tail is nearly the same as the rest mass. In the second $eV < mc^2$ scenario, the electron mass of the exponential tail is noticeably smaller than the rest mass. Regarding the $eV > mc^2$ scenario, one must recognize that the electron wavefunction cannot extend beyond the indicated $eV = mc^2$ boundary: at this boundary, the mass term goes to zero while the Zitterbewegung radius goes to infinity. One may consider this boundary as an event horizon of the wavefunction. This resolves the Klein paradox.

The above analysis relates to the single-particle electron mass term appearing in Equation (7.5.8). In many-particle solid-state systems, the electron wavefunction is calculated using an effective mass term, which takes into account multi-particle interactions. Such multi-particle effect is unrelated to the present discussion, i.e., the physical mass variation is a relativistic single-particle process.

7.7. Rotor Representation of Zitterbewegung Motion

Continuing with the vectorial mass representation concept, we are interested to describe the evolution of $\boldsymbol{m}_{\parallel}$ and $\boldsymbol{m}_{\perp}$ vectors, which represent the electron's space–time trajectory and Zitterbewegung rotation. In natural units, the $\boldsymbol{m}_{\perp}$ vector's norm is the inverse of Zitterbewegung radius, and it rotates along with the electron charge. Furthermore, the $\boldsymbol{m}_{\parallel}$ vector can be identified with the space–time trajectory of the electron's center-point. Therefore, there is a direct mapping between the evolution of $\boldsymbol{m}_{\parallel}$ and $\boldsymbol{m}_{\perp}$

vectors and the helical space–time trajectory of the electron's charge. In this section, we focus on geometrically describing the space–time trajectory of the electron's spherical charge.

A rotor is a mathematical object that in space–time algebra is simply a multivector with only even grade components. As discussed in the Mathematical Preliminaries chapter, the geometric product of an even number of vectors is always a rotor. A rotor that is the geometric product of two unitary vectors is a unitary rotor. The movement of a point charge that rotates in the plane $\gamma_x\gamma_y$ and at the same time moves up along the γ_z axis can be seen as the composition of an ordinary rotation in the plane $\gamma_x\gamma_y$ followed by a scaled hyperbolic rotation in the plane $\gamma_z\gamma_t$. The composition of these two rotations can be encoded with a single rotor of $Cl_{3,1}$ algebra. Let $\mathbf{r}_{e\square 0}$ represent the coordinate of the center of the charged electron sphere at $t = t_0$, and we have

$$\mathbf{r}_{e\square 0} = \gamma_x r_e + \gamma_t c t_0$$

$$[\mathbf{r}_{e\square 0} = \gamma_x r_e + \gamma_t t_0]_{\mathrm{NU}}$$

where r_e is the Compton radius. By introducing the rotor R_{xy}, that generates ordinary rotation in the $\gamma_x\gamma_y$ plane, we get

$$R_{xy}(t) = \cos\left(\frac{\omega_e t}{2}\right) + \gamma_x\gamma_y \sin\left(\frac{\omega_e t}{2}\right) = \exp\left(\gamma_x\gamma_y\frac{\omega_e t}{2}\right)$$

We then obtain the instantaneous position of the center of the charged sphere:

$$\mathbf{r}_{e\square}(t) = R_{xy}\left(\mathbf{r}_{e\square 0} + \gamma_t c t\right)\widetilde{R_{xy}}$$

We introduce now the rotor R_{zt}, which generates a hyperbolic rotation with rapidity φ in the $\gamma_z\gamma_t$ plane, in order to encode the motion at speed v_z along the γ_z direction

$$\begin{aligned} R_{zt} &= \cosh(\varphi)^{-\frac{1}{2}}\left[\cosh\left(\frac{\varphi}{2}\right) - \gamma_z\gamma_t \sinh\left(\frac{\varphi}{2}\right)\right] \\ &= \cosh(\varphi)^{-\frac{1}{2}} \exp\left(-\gamma_z\gamma_t\frac{\varphi}{2}\right) \end{aligned}$$

where

$$\varphi = \tanh^{-1}\left(\frac{v_z}{c}\right)$$

The instantaneous position $\mathbf{r}'_{e\square}(t)$ of the center of the rotating and translating sphere can be obtained by applying R_{zt}:

$$\mathbf{r}'_{e\square}(t) = R_{zt}\mathbf{r}_{e\square}\widetilde{R_{zt}}$$

Upon defining rotor R as

$$R = R_{zt}R_{xy} = \cosh(\varphi)^{-\frac{1}{2}} \exp\left(-\gamma_z\gamma_t\frac{\varphi}{2}\right) \exp\left(\gamma_x\gamma_y\frac{\omega_e t}{2}\right)$$

we can rewrite the instantaneous position in a compact form as

$$\mathbf{r}'_{e\square}(t) = R\left(\mathbf{r}_{e\square 0} + \gamma_t ct\right)\widetilde{R}$$

The instantaneous coordinate of a generic point on the surface of the charged sphere can be obtained adding a specific fixed vector $\mathbf{r}_{o\triangle}$

$$\mathbf{r}_{surf\square}(t) = \mathbf{r}'_{e\square}(t) + \mathbf{r}_{o\triangle}$$

The norm of vector $\mathbf{r}_{o\triangle}$ is equal to $r_{o\triangle} = \alpha r'_{e\triangle}$, which evaluates to the classical radius of the electron for non-relativistic speeds. For relativistic speeds, the $\mathbf{r}_{ox}$ and $\mathbf{r}_{oy}$ components are contracted by the rotor R, while $\mathbf{r}_{oz}$ is contracted at the same rate by the usual relativistic Lorentz contraction along z. Thus, the electron's charge always maintains its spherical shape.

As discussed in Chapter 5, the physical mechanism behind the contraction of $\mathbf{r}_{ox}$ and $\mathbf{r}_{oy}$ components is the transversal Doppler shift effect. Therefore, the rotor R encodes this Doppler shift.

7.8. From Rotors to Spinors

One may establish a mapping between a rotor R defined in the preceding section and the Weyl spinor ξ defined in Section 0.9. It is evident from Equation (0.9.15) that the role of a spinor ξ is analogous to the role of a rotor R. To construct R, we used two parameters: ω_e and φ. We need two more angular parameters to describe the ordinary rotation of spatial coordinate axes into a desired orientation.

Also, a mapping between an electromagnetic field bivector F and a Weyl spinor ξ was recently discovered. The mathematics of such mapping is discussed in Section 11.3 of [5]. This allows one to study the electromagnetic field configuration F, which generates the rotor R of Section 7.7.

7.9. Neutrino Waves and Isospin

Neutrinos are emitted in nuclear reactions, for example, during neutron decay. Up to now, all measurements of neutrino speed indicate that neutrinos travel exactly at the speed of light. If one assumes that a neutrino is a particle, it appears logical to try representing neutrinos as spinors under the $m \to 0$ condition of the Dirac equation. Mainstream neutrino researchers therefore pursue this direction and study lightspeed traveling Dirac spinor solutions.

Under the $m \to 0$ constraint, the Dirac Equation (7.5.2) indeed tells us that neutrinos must travel at the speed of light. As discussed in Chapter 4, under the $m \to 0$ condition the Proca equation converges to the Maxwell equation and thereby the Dirac Equation (7.5.4) tells us that neutrinos are solutions to the Maxwell equation. In summary, we are looking for lightspeed traveling solutions to the Maxwell equation.[a]

The neutrino solution cannot be a transversal electromagnetic wave, otherwise it would have been detected as such. The transversal electromagnetic wave is a trivial solution to the Maxwell equation: any quantum mechanical de-excitation of an electron orbit and also any quantum mechanical de-excitation of a nucleus emits this trivial solution. A neutrino is emitted in all nuclear electron capture events, and also in all muon-to-electron decays. It is therefore reasonable to expect a neutrino wave to be also a trivial solution to the Maxwell equation. Based on Chapter 2 analysis of Maxwell equation, the only remaining possibility is for neutrinos to be longitudinal electromagnetic waves: we described such waves in Section 2.10. As discussed in Chapter 5, a longitudinal electromagnetic wave does not have any rest mass, which matches the $m \to 0$ constraint. Do experimentally measured neutrino properties match the properties of a longitudinal electromagnetic wave? The most intense neutrino production experiment is a nuclear detonation: the explosion of a uranium-fueled bomb releases over 4% of its energy in the form of neutrinos. The lightspeed traveling component of a nuclear detonation comprises gamma rays, X-rays, and a so-called "electromagnetic pulse." The "electromagnetic pulse" penetrates matter just like gamma rays,

[a] Mainstream theorists consider neutrinos to be fermions, i.e., particles which "obey Fermi-Dirac statistics due to their half-spin angular momentum value." A group of mainstream theorists communicated to one author that they can't consider neutrinos obeying Maxwell equation because "Maxwell equation solutions integer-spin angular valued bosons." Such opinion stems from the mistaken assumption that Fermi–Dirac statistics originates from some particular angular momentum value. This long-held assumption will be proven wrong in the chapter on Pauli exclusion.

but has distinct characteristics from gamma rays: it induces a surge of electron current in electric conductors. This property of the "electromagnetic pulse" has been well documented upon numerous experiments. Such current surge induction matches the characteristics of a low-frequency longitudinal electromagnetic wave, which has its electric and magnetic fields aligned in the same direction. While low-frequency longitudinal waves cause a displacement of conducting electrons, high-frequency longitudinal waves pass through matter and rarely interact. This distinction is rather similar to the low-frequency radio waves versus high-frequency gamma rays classification of ordinary transversal waves. The "electromagnetic pulse" is therefore just the low-frequency component in a wide-spectrum emission of longitudinal electromagnetic waves. Neutrino scientists build elaborate instruments in large slabs of ice to detect high-frequency longitudinal waves.

Thus, both theoretical considerations and experimental observations suggest that a neutrino is in fact a longitudinal electromagnetic wave: it exists as pure wave phenomenon, and does not exist as a real particle. We identified the exact neutrino wave solution already in Equation (2.10.1) of Chapter 2. Textbook statements about the supposedly "weakly interacting" properties of neutrinos relate to the observed properties of high-frequency longitudinal waves. One must avoid *ad hoc* generalizations about the entire electromagnetic spectrum. Since low-frequency longitudinal electromagnetic waves cause electric current surges, one may look for signatures of such current surges in nature. The most well-known examples are solar storm induced current surges in satellites, which imply that there is a significant amount of nuclear reactions occurring during solar storms. Fortunately, the Earth's ionosphere protects terrestrial electronics from such exposure.

Our results correct three mistaken assumptions, which are currently held by mainstream neutrino researchers: (i) the assumption that neutrinos have nothing to do with electromagnetism, (ii) the assumption that the neutrino is a particle, and (iii) the assumption that the neutrino has a rest-mass despite no observable deviation from lightspeed propagation.

As discussed in Chapter 2, the conserved current which is associated with longitudinal waves is the $\gamma_t \to e^{I\theta}\gamma_t$ rotational symmetry of electromagnetic energy density, where I is the Clifford pseudo-scalar. In the context of a charged elementary particle, the $\gamma_t \to e^{I\theta}\gamma_t$ rotation may be generated for example by electric-magnetic charge oscillation of its scalar field.

There is in fact a rotation-like property which physicists assign to nuclear particles: the so-called "isospin" property. The isospin of a neutron and a proton indeed differ from each other, and thus there is a strong reason to think that isospin is the conserved current associated with the $\gamma_t \to e^{I\theta}\gamma_t$ rotational symmetry of electromagnetic energy. It is the isospin conservation which prompts neutrino emission during a neutron decay.

The concept of neutrinos was introduced by Pauli in 1930 to explain energy and momentum conservation during neutron decay, and the isospin concept was introduced by Heisenberg and Wigner in the 1930s. After nearly a century of confusion, our results bring some clarity to these concepts. With this new understanding of the isospin concept, we can easily extend the theory of spinors to also incorporate the isospin quantity.

In Section 7.7, we used a rotor R to generate the geometric coordinates of the electron's charge surface $\mathbf{r}'_{e\Box}(t)$ at a given time t:

$$\mathbf{r}'_{e\Box}(t) = R\left(\mathbf{r}_{e\Box 0} + \gamma_t ct\right)\widetilde{R}$$

The rotor R comprises a spatial rotation U and a boost B, which were used in Section 7.5 to build the Dirac–Hestenes spinor representation. The above expression carries no isospin rotation. When our particle carries isospin rotation, the above reasoning dictates that its charge surface coordinate $\mathbf{r}'_{e\Box}(t)$ is given by the following expression:

$$\mathbf{r}'_{e\Box}(t) = R\left(\mathbf{r}_{e\Box 0} + e^{I\omega_i t}\gamma_t ct\right)\widetilde{R}$$

where ω_i is the angular speed of isospin rotation and R is the usual rotor. Let's compare the above expression to the factorization of a Dirac–Hestenes spinor, which was given by Equation (7.5.1). We recognize that a non-zero ω_i implies a rotation of the Yvon–Takabayashi angle β. It is quite remarkable that Dirac–Hestenes spinors already have the correct structure to incorporate isospin.

When the Dirac–Hestenes equation is used to calculate the motion or orbit of a particle with zero isospin, the solution must have a constant β angle in Equation (7.5.1). When the Dirac–Hestenes equation is used to calculate the motion or orbit of a particle with non-zero isospin, the solution must have a rotating β angle in Equation (7.5.1).

7.10. Conclusions

We identified three interpretations of the Dirac equation. Each of these interpretations has its own distinct advantage, and the Dirac spinor field gains a well-understood physical meaning in each case. Also, each of these interpretations is consistent with the electron model that we developed in the preceding chapters.

The Dirac–Hestenes and Dirac–Baylis formulations are summarized in [1, 2], along with the applications and advantages of this interpretation. We introduced a deeper understanding of the Dirac equation in Chapter 5, where the Dirac spinor field was seen as an eigenvector along a space–time geodesic, and was calculated using the Hamilton–Jacobi equation. Some applications of that approach were discussed in Chapter 5, and it will be further explored in the following chapter. In this chapter, we introduced an electromagnetic formulation of the Dirac equation. Our approach has the following advantages:

- The Dirac spinor field gains a physical meaning which is directly related to the electromagnetic vector potential.
- It allows us to study both quantum mechanical orbitals and composite particle formations. The process of composite particle formation becomes calculable for the first time.
- It demonstrates that the Dirac, Klein–Gordon, and Proca equations are equivalent, and each of them applies to any elementary particle. This result is in stark contrast to the present-day understanding of these three equations, and also implies that neutral particles must be composite particles. This topic will be further studied in Chapter 14, where the internal structures of the neutron and neutral pion will be identified.
- We resolved some long-standing paradoxes, such as the Klein paradox or the issue of "negative energy" states.

Up to now, neutrinos were thought to be described by spinor fields that are $m \to 0$ solutions to the Dirac equation, without any connection to electromagnetism. We recognized that neutrinos are in fact electromagnetic waves, and identified the corresponding electromagnetic wave equation. The emission of neutrinos is prompted by a change in the isospin value of a particle, and we recognized the isospin to be also an electromagnetic property. We extended the theory of spinors to also incorporate isospin. These results indicate further opportunities to unify the theory of electromagnetism with the theory of nuclear forces. We will revisit this topic in Chapter 14.

References

[1] A.G. Campos *et al.*, Analytic solutions to coherent control of the dirac equation, *Physical Review Letters*, **119**(17) (2017). DOI: 10.1103/PhysRevLett.119.173203.

[2] A.G. Campos *et al.*, Construction of dirac spinors for electron vortex beams in background electromagnetic fields, *Physical Review Research*, **3**(1) (2021). DOI: 10.1103/PhysRevResearch.3.013245.

[3] S.M. Barnett, Optical Dirac equation, *New Journal of Physics*, **16** (2014). DOI: 10.1088/1367-2630/16/9/093008.

[4] E. Schrödinger, Über die kräftefreie Bewegung in der relativistichen Quantenmechanik, *Sitz.ber. Preuss. Akad. Wiss. Phys.-Math.*, **XXIV** (1930) 418–428.

[5] D. Lundholm and L. Svensson, Clifford algebra, geometric algebra, and applications, (2016) arXiv:0907.5356.

[6] A.I. Nikishov, Barrier scattering in field theory and removal of the Klein paradox, *Nuclear Physics B*, **21** (1970) 346–358.

https://doi.org/10.1142/9781800611306_0008

Chapter 8

Electron Orbitals and Space–Time Curvature

Paul O'Hara[*,‡] and András Kovács[†,§]
[*] *Sophia University Institute, Loppiano, Italy*
[†] *BroadBit Energy Technologies, Metallimiehenkuja 8 C, 02150 Espoo, Finland*
[‡] *paul.ohara@sophiauniversity.org*
[§] *andras.kovacs@broadbit.com*

8.1. Introduction

This chapter investigates the fundamental relationship between the metrics of general relativity and quantum mechanics and, in the process, gives a deeper understanding of Einstein's $E^2 - (pc)^2 = (mc^2)^2$ formula, as it pertains to the Klein–Gordon and Dirac equations. The key role of boundary conditions in distinguishing between quantum and classical mechanics will emerge naturally from the procedure. We will find that the methodology will enable us to introduce both test charges and test masses by means of gauges, and will help us to better understand light–matter interaction processes.

In Chapter 7, we showed how the electron state is defined by a spinor field which factorizes into a scalar ψ field and the Proca field $\boldsymbol{a}_\square$. We identified that the ψ field's space–time gradient encodes the particle's energy and momentum, in accordance with the $E^2 - (pc)^2 = (mc^2)^2$ formula. In Chapter 4, we discussed the derivation of the Proca field $\boldsymbol{a}_\square$ from the electromagnetic vector potential field comprising a given particle. What remains is to understand the origin of the $E^2 - (pc)^2 = (mc^2)^2$ formula, and the process by which it generates the ψ field. Most of what follows can be found in [9] and [10].

There is a fundamental paradigm shift between general relativity and classical mechanics. In general relativity, the energy–momentum tensor is the effective cause of the space–time curvature, while in classical physics, the structure of space–time is treated as an accidental cause, serving only as a backdrop against which the laws of physics unfold. This split in turn is inherited by quantum mechanics, which is usually developed by changing classical (including special relativity) Hamiltonians into quantum wave equations. In this chapter, we will try to remedy this situation by taking the metrics of general relativity as our starting point and calculating geodesics to define electron orbital states. This new perspective gives a deeper understanding of orbital eigenstates in quantum mechanics.

This chapter builds upon the concepts and results derived in Chapter 5. In the same way as in Chapter 5, we will associate wave equations with those operators which are duals of differential one-forms, rather than with operators derived from a Hamiltonian. Thus, we enable the structure of space–time itself to determine in a natural and unique way the wave equations of quantum mechanics. Moreover, it is precisely the presence of gauge terms in the form of test masses and charge that permit new laws of physics to emerge independently within the context of the space–time structure in which they are embedded. Throughout the chapter, $(\mathcal{M}, g)$ [4] will denote a space–time pair, where "$\mathcal{M}$ is a connected four-dimensional Hausdorff manifold" and g is a metric on $\mathcal{M}$. At every point $p \in \mathcal{M}$, we can erect a local tetrad $e_0(p), e_1(p), e_2(p), e_3(p)$ and a point x will have coordinates $x = (x^0,\ x^1,\ x^2,\ x^3) = x^a e_a$ in this tetrad coordinate system [13]. We will use Roman letters a, b, c, etc. to index coordinates with respect to a tetrad. In this regard, we will refer to x^a as the coordinates of x. Also for spinors, we can write $\psi = \psi^i e_i(p)$, where ψ^i will represent the coordinates of the spinor with respect to the tetrad at p. Also at p we can establish a tangent vector space $T_p(\mathcal{M})$, with basis $\{\partial_0, \partial_1, \partial_2, \partial_3\}$ and a dual 1-form space, denoted by T_p^* with basis $\{dx_0, dx_1, dx_2, dx_3\}$ at p, defined by

$$dx^\mu \partial_\nu \equiv \partial_\nu x^\mu = \delta^\mu_\nu \tag{8.1.1}$$

We refer to the basis $\{dx^0, dx^1, dx^2, dx^3\}$ as "the basis of one forms dual to the basis $\{\partial_0, \partial_1, \partial_2, \partial_3\}$ of vectors at p" [4]. Finally, note that Greek letters will be used to represent general coordinate systems and Einsteinian notation will be used for summations.

8.2. A Method of Solving the Dirac Equation

8.2.1. *A mass gauge*

Let's start by recalling the derivation of Dirac equation in Section 5.2. Another approach to the formalism of Section 5.2 is to introduce test particles by means of a gauge term. Specifically, let $(\mathcal{M}, g)$ be a pseudo-Riemannian manifold, with metric tensor g determined by Einstein's field equations. Consider a massless test particle introduced into the field, then by the principle of equivalence, we can choose a local tetrad $\{dx_0, dx_1, dx_2, dx_3\}$, such that the massless test particle travels along a null geodesic given by $0 = dx^a dx_a$, which in terms of a spinor basis can be written as

$$ds\xi = \gamma^a dx_a \xi = 0 \tag{8.2.1}$$

Next define the wave equation corresponding to the metric by taking the dual of the 1-form space:

$$\gamma^a \partial_a \psi = 0 \tag{8.2.2}$$

which can be interpreted as the wave equation of a massless particle.

We now introduce a test particle of mass m by means of a minimal principle [13], by adopting the same technique that is usually used to introduce test charges into a field. In other words, let

$$0 = \gamma^a (\partial_a - p_a)\psi \tag{8.2.3}$$

which in turn gives the fundamental wave equation

$$\gamma^a \partial_a \psi = \gamma^a p_a \psi \tag{8.2.4}$$

This immediately suggests the particular solution

$$\psi = e^{\int^x p^a dy_a} \xi(p_0) \tag{8.2.5}$$

Moreover, if the gauge term describes a test particle of mass m moving along a timelike geodesic as defined in (5.2.28), then

$$\gamma^a \partial_a \psi = mc\psi \tag{8.2.6}$$

Once again, we have obtained a Dirac equation.

8.2.2. *A useful theorem*

To conclude this section, we summarize the results with the following theorem:

Theorem 17. *Let* $\xi(p) = [\frac{d}{ds}\psi^i(s)]e_i$, *where* $mcs = \int^x p^a dx_a$ *along a timelike geodesic, then* $\gamma^a p_a \xi = mc\xi$ *iff* $\psi(p) = \psi^i(\int^x p^a dx_a)e_i$ *is a solution of*

$$\frac{\partial}{\partial s}\psi(p) = \gamma^a \partial_a \psi(p)$$

where $\int^x p^a dx_a$ *is Lorentz invariant, and integration is taken along the curve with tangent vector* $p^a = m\frac{dx^a}{d\tau}$, *where* τ *is proper time.*

Proof. Noting that $\psi(p) = \psi^i(\int^x p^a dx_a)e_i$ and assuming $\gamma^a p_a \xi = mc\xi$, then

$$\gamma^a \frac{\partial \psi}{\partial x^a} = \gamma^a p_a \frac{\partial \psi^i}{\partial s} e_i \tag{8.2.7}$$

$$= \gamma^a p_a \xi \tag{8.2.8}$$

$$= mc\xi \quad \text{given} \tag{8.2.9}$$

$$= \frac{\partial \psi}{\partial s} \tag{8.2.10}$$

To prove the converse, it is sufficient to substitute $\psi(p) = \psi^i(\int^x p^a dx_a)e_i$ into $\tilde{\partial}_s \psi = \partial_s \psi$ to get the answer. □

Corollary 18. *In the case* $\psi^i(\int^x p^a dx_a) = e^{\int^x p^a dx_a}$, *then* $\gamma^a \partial_a \psi = mc\psi$.

Proof. Clearly, $\frac{\partial \psi}{\partial s} = mc\psi$. □

Corollary 19. *If* $\psi(\int^x p^a dx_a) = f(\int^x p^a dx_a)u$ *where* u *is a spinor independent of* x^a, *then the equation*

$$\tilde{\partial}_s f u = \frac{\partial f}{\partial s} u \tag{8.2.11}$$

has the same solutions as

$$\tilde{\partial}_s \psi = \frac{\partial}{\partial s} \psi \tag{8.2.12}$$

Proof. Substitute. □

By way of conclusion, note that if Equation (5.2.28) is relaxed to incorporate gauge terms other than mass, such as electric charge, then Equation (5.2.6) can be extended to incorporate the electromagnetic field and, in particular, the motion of the electron in a hydrogen atom. We will discuss this in more detail later. For the moment, suffice to say that if $p_a = p'_a - \frac{e}{c} A_a$ with p'_a corresponding to the mass gauge component and subject to the conditions of Equation (5.2.28) and A_a corresponding to the electromagnetic potential, then

$$\gamma^a \left(\partial_a + \frac{e}{c} A_a \right) \psi = (mc + \frac{e}{c} A_0) \psi \tag{8.2.13}$$

8.3. Covariance

Theorem 17 also enables us to write down a covariant form for the generalized Dirac equation which depends directly upon the covariance of its dual metric equation. We begin by showing that the equation $\tilde{ds}\xi = ds\xi$ is covariant under Lorentz transformations. Specifically, since ds is a scalar, if $dx^a = \frac{\partial x^a}{\partial x'^b}dx'^b = \Lambda^a_b dx'^b$, then $\tilde{ds}\xi = ds\xi$ transforms under Lorentz transformations $D(\Lambda(x)) = D(x)$ into

$$D(x)\tilde{ds}\xi(x) = dsD(x)\xi(x) \tag{8.3.1}$$

It is instructive to also use a pure spinor approach since it brings out the $SL(2,\mathbb{C})$ characteristics of the Dirac equation in relation to Lorentz transformations. For what follows, it is important to recall that if D is an element of $SL(2,\mathbb{C})$, then $\tilde{x}' = D\tilde{x}D^\dagger$ is the spinor representation of the Lorentz transformation $\mathbf{x}' = \Lambda\mathbf{x}$ where Λ corresponds to a Lorentz transformation in Minkowski space and $D^\dagger$ represents the adjoint of D. Moreover, since the elements if $SL(2,\mathbb{C})$ are not necessarily unitary, it is important to understand the relationship between $D^\dagger$ and D^{-1} for this group. This leads to the following definition and lemma.

Definition 20. Let V be an n dimensional metric space with metric tensor g and $A \in GL(n,\mathbb{C})$ be a linear transformation from $V \to V$. If $x \in V$ and $x' = Ax$ such that

$$\langle \mathbf{y}, A\mathbf{x}\rangle = \langle B\mathbf{y}, \mathbf{x}\rangle$$

where $B \in GL(n,\mathbb{C})$, then B is called the adjoint of A and denoted by $A^\dagger$.

Note: in the case of a metric space $B = \bar{A}^T$ (the transpose of the complex conjugate of A) and $B^\dagger = A$. In general, as we shall see in the next lemma, for any given A, $A^\dagger$ is not unique but depends on the choice of metric tensor.

Lemma 21. *Let V be an n dimensional metric space with metric tensor g and $A, B \in GL(n,\mathbb{C})$ be linear transformations from $V \to V$, then*

$$\langle B\mathbf{y}, A\mathbf{x}\rangle = \langle \mathbf{y}, \mathbf{x}\rangle \text{ iff } B^\dagger = gA^{-1}g^{-1}$$

Proof. Let $\mathbf{x}$, $\mathbf{y} \in V$. Then, by definition of a metric space, $\langle \mathbf{y}, \mathbf{x}\rangle \equiv \mathbf{y}^\dagger g\mathbf{x}$. It follows that

$$\langle B\mathbf{y}, A\mathbf{x}\rangle = \mathbf{y}^\dagger B^\dagger gA\mathbf{x} = \mathbf{y}^\dagger g\mathbf{x}$$

if and only if $B^\dagger gA = g$. The result follows. □

Corollary 22. *Let* $B^\dagger = gA^{-1}g^{-1}$, *then* $A \in SL(n, \mathbb{C})$ *iff* $B \in SL(n, \mathbb{C})$

Proof. $\det(B^\dagger) = \det(gA^{-1}g^{-1}) = \det(A^{-1})$. Therefore, $A \in SL(n, \mathbb{C})$ iff $\det(A^\dagger) = \det(A^{-1}) = 1$ iff $B \in SL(n, \mathbb{C})$. □

Definition 23. If V is Minkowski space and $g = \eta \equiv \text{diag}(1, -1, -1, -1)$ such that $A^T \eta A = \eta$, where A^T is the transpose of A, then A is called a Lorentz transformation and is usually denoted by Λ. Note: in this case $\Lambda^\dagger = \Lambda^T$.

Switching to spinor notation, if $\tilde{x}$ and $\tilde{y}$ are four-dimensional vectors, then, as previously noted, $\tilde{y} = D(\Lambda)\tilde{x}D^\dagger(\Lambda)$ defines a Lorentz transformation on $\tilde{x}$ if and only if $D \in SL(2, \mathbb{C})$. It then follows by Lemma 4 and the theory of group representations that $D(\Lambda^\dagger \eta \Lambda) = D(\eta)$ or equivalently by linearity

$$D(\Lambda^\dagger)D(\eta)D(\Lambda) = D(\eta) \iff D^{-1}(\Lambda^\dagger)D(\eta)D^{-1}(\Lambda) = D(\eta) \tag{8.3.2}$$

Since, the only requirement is that the metric tensor $D(\eta)$ should be symmetric, for what follows we choose $D(\eta) = \gamma_0$.

It is important to note that since the four vector $\tilde{x}$ is Hermitian, then $\tilde{y} = D(\Lambda)\tilde{x}D^\dagger(\Lambda)$ is also Hermitian and therefore a vector. In that regard, the usual generators of the Dirac (Clifford) algebra, denoted by $(\beta, \alpha_1, \alpha_2, \alpha_3)$ and defined by

$$\beta^2 = 1, \quad \alpha_i\beta + \beta\alpha_i = 0 \quad \{\alpha_i, \alpha_j\} = \delta_{ij} \tag{8.3.3}$$

are also unit vectors (see [7, pp. 306–307]). However, it is important to note that apart from $\gamma_0 \equiv \beta$, $\gamma_i \equiv \beta\alpha_i$ is not Hermitian and therefore not a vector but a bivector ($\gamma^\dagger = (\beta\alpha_i)^\dagger = \alpha_i^\dagger\beta^\dagger = \alpha_i\beta = -\gamma$). It also follows that $\tilde{ds} = \gamma_a dx^a$ is a bivector although $\tilde{ds}\gamma_0 = \gamma_a\gamma_0 dx^a$ is a vector and therefore $\tilde{ds}'\gamma_0 = D\tilde{ds}\gamma_0 D^\dagger$ defines a Lorentz transformation of $\tilde{ds}\gamma_0$. Consequently,

$$\tilde{ds}'\xi' = \tilde{ds}'\gamma_0\gamma_0\xi' \tag{8.3.4}$$

$$= D(x)\tilde{ds}\gamma_0 D^\dagger(x)\gamma_0\xi' \tag{8.3.5}$$

$$= D\tilde{ds}\gamma_0\gamma_0 D^{-1}\xi', \quad \text{using Equation (8.3.2)} \tag{8.3.6}$$

$$= D\tilde{ds}\xi \quad \text{where } D^{-1}\xi' \equiv \xi \tag{8.3.7}$$

$$= ds\xi', \quad \text{using Equation (8.3.1)} \tag{8.3.8}$$

which establishes the covariance.

To show the covariance of the corresponding Dirac equation, we rewrite Equation (5.2.6) in the form

$$(\tilde{\partial}_s \psi^i) e_i = \frac{\partial \psi^i}{\partial s} e_i \tag{8.3.9}$$

From Theorem 17, we already know that this is equivalent to the covariant metric $\tilde{ds}\xi = ds\xi$ provided $\xi = (d\psi^i/ds)e_i$, where ds indicates differentiation along timelike geodesic parametrized by s (recall $mcs = \int^x p_a dx^a$). It follows that $\tilde{ds}\xi = ds\xi$ is covariant iff $\gamma_a p^a \xi = mc\xi$ is covariant iff $\gamma_a p^a (d\psi^i/ds)e_i = mc(d\psi^i/ds)e_i$ is covariant iff Equation (8.3.9) is covariant with respect to the Lorentz transformation $D(x)$, along the geodesic. Moreover, this latter restriction of motion along a geodesic may actually be relaxed and the following more general theorem can be proven:

Theorem 24. *The Dirac equation defined over the manifold* $(\mathcal{M}, g)$ *is Lorentz covariant under the transformation* $D(x)$ *defined with respect to a tetrad* $e_i(x)$, *provided the equation is written in the form*

$$\gamma^a \frac{\partial \psi^i}{\partial x^a} e_i(x) = \frac{\partial \psi^i}{\partial s} e_i(x)$$

Proof. Let $D(x)$ be a local Lorentz transformation at x, then:

$$D(x)\tilde{\partial}_s \psi(s) = D(x)(\tilde{\partial}_s \psi^i) e_i \tag{8.3.10}$$

$$= (D(x)\tilde{\partial}_s \psi^i) D^{-1}(x) D(x) e_i \tag{8.3.11}$$

$$= (\tilde{\partial}_{s'} \psi^i) e'_i \tag{8.3.12}$$

$$= \left(\frac{\partial \psi^i}{\partial s}\right) e'_i \tag{8.3.13}$$

□

Remark. In regular Minkowski space, the covariant form of the generalized Dirac equation can be reduced to the form of Equation (5.2.6).

8.4. Wave Equations for Geodesic

At this stage, the reader may be wondering how the usual formulation of quantum mechanics emerges. Indeed, the wave equations above seem to express the wave equations of classical mechanics more than quantum mechanics, in that there is no expression for Planck's constant h, nor does the expression $i=\sqrt{-1}$ appear with the operators. With regard to the latter point, we note that i could be seen as absorbed into the γ matrices, but we postpone a full discussion of this until the next section. First, we analyze the solutions of the wave equation for a massless particle from three perspectives to help us better grasp the formal difference between classical and quantum mechanics. Later on, we will formulate the axioms of quantum mechanics as suggested by our analysis.

The linearized metric for a massless particle is given by

$$0 = \gamma^0 cdt - \gamma^1 dx_1 - \gamma^2 dx_2 - \gamma^3 dx_3 \tag{8.4.1}$$

from which it follows by the canonical correspondence established above that the associated wave equation for the particle is given by

$$0 = \gamma_0 \frac{\partial \psi}{c\partial t} - \gamma_1 \frac{\partial \psi}{\partial x_1} - \gamma_2 \frac{\partial \psi}{\partial x_2} - \gamma_3 \frac{\partial \psi}{\partial x_3} \tag{8.4.2}$$

This is the Dirac equation for a massless particle. Squaring this out we get the Klein–Gordan equation for a massless particle, as follows:

$$\frac{1}{c^2} \frac{\partial^2 \psi}{\partial t^2} = \sum_{i=1}^{3} \frac{\partial^2 \psi}{\partial x_i^2} \tag{8.4.3}$$

Technically speaking, this is a spinor equation with $\psi = \psi^i e_i$ (see Equation (8.3.9)) and e_i a constant unit vector. The equation can be rewritten as

$$\frac{1}{c^2} \frac{\partial^2 \psi^i}{\partial t^2} e_i = \sum_{a=1}^{3} \frac{\partial^2 \psi^i}{\partial x_a^2} e_i \tag{8.4.4}$$

Note that in this latter formulation each $\psi^i = Af(\mathbf{x} - ct) + Bg(\mathbf{x} + ct)$ gives the same general solution for each i and that the solutions of the massless Dirac equation are also solutions of the usual massless Klein–Gordan equation. This observation is in line with the discussion of optical spinors in Chapter 2.

Since the wave equation emerges from the structure of space–time itself, the question arises as to how to distinguish classical mechanics from quantum mechanics. We investigate this by analyzing the motion of a massless particle in a Minkowski space, subject to different sets of boundary conditions. In the first case, we consider the motion of a classical massless particle moving on the x-axis with uniform velocity c, but constrained by two mirrors placed at $x = 0$ and $x = \xi$ to move uniformly on the interval $[0, \xi]$. We will assume that perfect reflection takes place at the mirrors and that no energy is exchanged. In this case, the equation of motion for a strictly classical particle with position $x = 0$ at $t = 0$ is given by

$$x = \begin{cases} ct - 2n\xi, & \text{for } t \in \left[\dfrac{2n\xi}{c}, \dfrac{(2n+1)\xi}{c}\right] \\ 2(n+1)\xi - ct, & \text{for } t \in \left[\dfrac{(2n+1)\xi}{c}, \dfrac{(2n+2)\xi}{c}\right] \end{cases}$$

and its wave function $\psi(x,t)$ takes on the form

$$\psi(x,t) = \begin{cases} \delta[k(x-ct)], & \text{for } x - ct = -2n\xi \\ \delta[k(x+ct)], & \text{for } x + ct = 2(n+1)\xi \\ 0, & \text{otherwise} \end{cases}$$

The wave function in this case pinpoints the position of the particle with probability 1. Moreover, there is no restriction on the energy (implicit in the term k) in this case. Theoretically, it may have values ranging from 0 to ∞.

However, the classical particle is an idealized situation. In reality, the position of a massless particle constrained to move on the line is unknown and any attempt to know its exact position will be subject to Heisenberg's uncertainty relations, which we will formulate in the next section. In other words, its exact position cannot be known in principle, because any attempt to pinpoint it will scuttle the position and defeat the whole purpose of the experiment. The best we can do is to describe the position by means of a uniform probability density $f(x-ct) = 1/\xi$ for $x \in [0, \xi]$ which also suggests writing $\psi(x,t) = e^{\pm ik(x-ct)}/\sqrt{\xi}$ to preserve both the boundedness and the periodic motion of the particle, as described by the above wave equation. This does not mean that causality is violated nor that the particle does not have an exact position, at least in the above case. It simply affirms that our initial conditions have to be defined statistically, and also in such a way as to reflect the periodic motion of the particle. As a consequence, the future evolution of the wave function of the system is best interpretated in a statistical way. Finally, note that in this model the energy of the particle can once again vary from 0 to ∞ in a continuous manner.

Thirdly, the particle may be constrained to move in a potential well in such a way that the wave function is continuous ($= 0$) at the boundaries. In the case of the above problem, this means that the wave function has harmonic solutions of the form $\psi(\xi,t) = Ae^{i\nu' t}\sin(kx)$, where A is a constant. Substituting, we will find that $k = \frac{n\pi}{\xi}$ and the photon energy becomes quantized and of the form $E \equiv kc = \nu'$. It should be noted that this solution corresponds to the motion of a harmonic oscillator, and is the key to the quantization process associated with quantum field theory in general [7]. Indeed, if we were to rescale our units of energy by defining $\nu' = h\nu$, then $E = h\nu$, with h having units $J.s$, and the standard wavelength becomes $\lambda = \frac{ch}{E}$. In the next section, h will be introduced in a more formal way.

The purpose of the above three examples is to highlight the importance of the boundary conditions when distinguishing between a classical type problem and a quantum mechanical problem, a point also stressed by Lindsey and Margenau [5]. Classical and quantum laws are not in opposition to each other. There is not one set of laws on the microscopic level and another on the macroscopic. On the contrary, classical and statistical methodologies are complementary to each other and are, in principle, applicable at all levels. However, on the microscopic level, statistical fluctuations will be more pronounced, and consequently, in practice (and in principle) the effects associated with quantum physics will become more apparent.

8.5. Quantum Mechanics and Hilbert Spaces

The above analysis permits us to better understand something of the difference between quantum mechanics and classical mechanics from the perspective of general relativity. As we have noted, it suggests the difference is to be found in the boundary conditions, which in the case of quantum mechanics is subjected to statistical conditions. With this in mind, we formulate a few axioms which not only respect the manifold structure of general relativity, but also enable us to distinguish quantum mechanics from classical physics, in a formal way.

Essentially what we have noted is that the metric of general relativity forces (real) eigenvalue solutions for the free particle of the form

$$\frac{\partial\psi\left(\int p^a dx_a\right)}{\partial x_a} = p^a\psi\left(\int p^a dx_a\right)$$

However, since the choice of eigenfunction associated with the specific eigenvalue in this case is not unique, we restrict ourselves for the purpose of quantum mechanics to such eigenfunctions $\psi(t, \mathbf{x})$ that for each t, $\psi(t, \mathbf{x}) \in L^2(E^3) \times H$, where H is a four-dimensional Hilbert space. Also, we associate the dual of the 1-form dx with the self-adjoint partial differential operator $i\hbar\partial/\partial x$, where $\hbar$ is a constant. Consequently, by defining the dual in this way, we not only find that

$$dx(i\hbar\partial_x) \equiv i\hbar\partial_x x = i\hbar \tag{8.5.1}$$

but that it can also be linked to the uncertainty principle (see what follows). If we look more closely at Equation (5.2.6), we will find that to associate the operator $-i\partial_x$ with a real-valued momentum eigenvalue is already implicit in this equation, and indeed is a consequence of the signature of the metric tensor $\eta_{ab} = \{\gamma_a, \gamma_b\}$. However, it should be noted that there are numerous four-dimensional representations of the set of matrices $\{\gamma_a\}$. We have already seen one in (8.3.3). For what follows, it is convenient to introduce another such representation. Specifically, let $\gamma_a = -i\alpha_a$ for $a = 1, 2, 3$, and $\gamma_0 = \alpha_0$. The α_as are the generators of the Clifford algebra $SL(2, \mathbb{C})$ and have the advantage of being self-adjoint. Also, $\gamma_0^\dagger = \gamma_0$ and $\gamma_a^\dagger = -\gamma_a = i\alpha_a$ which is in agreement with the conventional Definition [7]. Later in the section on the hydrogen atom, we will see that by redefining γs in this way the orbital angular momentum factor in the Dirac equation will be real and not imaginary as it is usually presented in the literature [6]. Moreover, $\gamma_a p^a = -\alpha_a(ip^a)$, for $a = 1, 2, 3$ and the linearized metric (5.2.4) can be written in the explicit form

$$\tilde{ds} = \alpha_0 dx^0 - i\alpha_1 dx^1 - i\alpha_2 dx^2 - i\alpha_3 dx^3 \tag{8.5.2}$$

which, in order to maintain the invariant $\langle \tilde{ds}, \tilde{\partial}_s \rangle = 1$ relationship, gives

$$\tilde{\partial}_s = \alpha^0 \partial_0 - i\alpha^1 \partial_1 - i\alpha^2 \partial_2 - i\alpha^3 \partial_3 \tag{8.5.3}$$

In other words, by letting α_a obey the Clifford algebra, we can associate the momentum operator with $-i\partial_a$ in a natural way, i.e., in the "natural units" metric. For the ordinary metric, we rescale the momentum operator by writing $-i\hbar\partial_a$ in place of $-i\partial_a$ to obtain the usual form of quantum mechanics.

What dictates this choice of scaling? We choose $\hbar$ because it seems to be the scaling constant which nature uses, and because this choice is required by the presence of vacuum fluctuations, which cause Heisenberg uncertainty. If we multiply across by $-i\hbar\alpha^0$ and note that $\hat{\alpha}_a \equiv -i\alpha_0\alpha_a$ obeys the same Clifford algebra as α_a and remain self-adjoint, then the Dirac Equation (5.2.6) can be rewritten as follows:

$$-i\hbar\alpha^0 \frac{\partial \psi}{\partial s} = (-i\hbar\partial_0 - i\hbar\hat{\alpha}^1\partial_1 - i\hbar\hat{\alpha}^2\partial_2 - i\hbar\hat{\alpha}^3\partial_3)\psi \tag{8.5.4}$$

In particular, if we denote the eigenvalue of $i\hbar\partial_0$ associated with the eigenfunction $\exp(-i\int^x p_a dx^a)$ by $p_0 = -iE/(\hbar c)$, then Corollary 18 gives

$$\alpha^0 mc^2\psi = (E - ic\hbar\hat{\alpha}^1\partial_1 - ic\hbar\hat{\alpha}^2\partial_2 - ic\hbar\hat{\alpha}^3\partial_3)\psi \tag{8.5.5}$$

which gives us back the usual form of the Dirac equation. Note too that $\hbar$ has been absorbed into the energy terms. In the following, we show that the $\Delta\tilde{X}^a\Delta\tilde{P}^a \geq \frac{\hbar}{2}$ uncertainty of conjugate variables follows as a consequence.

Finally, based on the above discussion, we formulate a few axioms which not only respect the manifold structure of general relativity, but also enable us to distinguish quantum mechanics from classical physics, in a formal way.

Definition 25. Space–time is a four-dimensional manifold $(\mathcal{M}, g)$.

Definition 26. At every point p of $\mathcal{M}$ there is a tangent vector space $T_p(\mathcal{M})$ with tetrad basis $\{-i\hbar\partial_0, -i\hbar\partial_1, -i\hbar\partial_2, -i\hbar\partial_3\}$ and a dual 1-form space, denoted by T_p^* with basis $\{dx_0, dx_1, dx_2, dx_3\}$ at p, defined by $dx^a(i\hbar\partial_b) = i\hbar\frac{\partial x^a}{\partial x^b} = i\hbar\delta_b^a$.

Definition 27. Quantum mechanical operators are elements of $SL(2,\mathbb{C})$ Clifford Algebra, which can be viewed as a representation of the vector spaces T_p and T_p^* on $\mathcal{M}$.

Note: This definition means that if O is an operator, then it is Hermitian and $AOA^\dagger$ is also a Hermitian operator, where A is an element of a group representation of $SL(2,\mathbb{C})$.

Definition 28. Each element of $SL(2,\mathbb{C})$ algebra acts on the Hilbert Space $L^2(E^4) \times H$, where H is a four-dimensional Hilbert Space. The elements $\psi \in L^2(E^4) \times H$ are called the states of the system.

Remark. It follows from the definition of the Hilbert Space that if $\psi \in L^2(E^4) \times H$, then for each t, $\psi(t, \mathbf{x}) \equiv \psi_t(\mathbf{x}) = \psi_t^i e_i \in L^2(E^3) \times H$, where each $e_i \in H$ and an inner product exists such that $\langle\psi_t, \psi_t\rangle = \int(\psi^*)^i\psi_i d^3x$. Moreover, if $\psi_t(\mathbf{x})$ is normalized for each t, then $\psi^*\psi$ can be interpreted as a probability density function for position.

Lemma 29. *Let $d\tilde{f}$ and $\tilde{\partial}_x$ be the $SL(2,\mathbb{C})$ representation of $df \in T^*$ and $\partial_x \in T$, respectively, then $\langle d\tilde{f}, \tilde{\partial}_x\rangle\psi = [\tilde{\partial}_x, \tilde{f}]\psi$, where $\psi \in L^2(E^4)$.*

Proof. Note, $d\tilde{f} = \frac{\partial f}{\partial x^a} d\tilde{x}^a$. Therefore,

$$\langle d\tilde{f}, \tilde{\partial}_a\rangle\psi = \frac{\partial f}{\partial x^a}\psi = [\partial_a, f]\psi$$

The lemma has been proven. □

Lemma 30 (The uncertainty relationships). *Let $\tilde{X} = \int_0^{x(s)} d\tilde{X}$ and $\tilde{P} = i\hbar\tilde{\partial}_s$ be the $SL(2,\mathbb{C})$ representations of position and momentum,* *respectively, defined along a curve of length s. Also, let* $\bar{X} \equiv \int \psi^* \tilde{X} \psi ds$, $\bar{P} \equiv \int \psi^* \tilde{P} \psi ds$, $\Delta^2 \tilde{X} \equiv \int \psi^* (\tilde{X} - \bar{X})^2 \psi ds$, *and* $\Delta^2 \tilde{P} \equiv \int \psi^* (\tilde{P} - \bar{P})^2 \psi ds$, *then* $\Delta \tilde{X} \Delta \tilde{P} \geq \frac{\hbar}{2} \left| \int^s \psi^* \frac{1}{i\hbar} (\tilde{P}\tilde{X} - \tilde{Q}\tilde{X}) \psi ds \right|$. *In particular, in the case of the components* $\tilde{X}^a$, $\tilde{P}^a$, *we get* $\Delta \tilde{X}^a \Delta \tilde{P}^a \geq \frac{\hbar}{2}$.

Proof. Usual proof using Cauchy–Schwartz inequality.

We note here an important point about the $\Delta \tilde{X}^a \Delta \tilde{P}^a \geq \frac{\hbar}{2}$ uncertainty relationship: this relationship is defined in the $\tilde{X}^a$ and $\tilde{P}^a$ measurement context for the wavefunction $\psi(t, \mathbf{x})$, and it applies only to those oscillation modes of $\psi(t, \mathbf{x})$ which exist. In the context of a free particle, $\tilde{X}^a$ and $\tilde{P}^a$ are measured with respect to some reference frame. In the context of a bound electron, $\tilde{X}^a$ and $\tilde{P}^a$ are measured with respect to the nucleus. In Section 7.5, we introduced the two principal interaction pathways between an electron and a nucleus: the ordinary Bohr orbit case and the composite particle formation. The precise mathematical meaning of these interaction pathways was defined in Section 7.5: they correspond to the $\kappa = \frac{M}{m_e + M}$ or $\kappa = 0$ solutions of Equations (7.5.5) and (7.5.6). In case of electron–proton composite particle formation, i.e., in $\kappa = 0$ type Compton-scale composite structures, the $\Delta \tilde{X}^a \Delta \tilde{P}^a \geq \frac{\hbar}{2}$ uncertainty relationship cannot be applied to the electron's non-existing oscillation modes with respect to the nucleus. The physics of Compton-scale composites will be further explored in Chapter 12. For the $\kappa = \frac{M}{m_e + M}$ solution type, which is the Bohr orbit case, the ground state oscillation modes of $\tilde{X}^a$ and $\tilde{P}^a$ are generated by the electromagnetic vacuum oscillations, as discussed in Chapter 6. In this usual case, the $\Delta \tilde{X}^a \Delta \tilde{P}^a \geq \frac{\hbar}{2}$ uncertainty relationship is satisfied. We pointed out in Chapter 6 that the spectrum of electromagnetic vacuum oscillation has the unique property of remaining unchanged under any reference frame transformation: the right-hand side of the $\Delta \tilde{X}^a \Delta \tilde{P}^a \geq \frac{\hbar}{2}$ relationship therefore remains constant in any reference frame. Since the Bohr orbit case is the most common one, these other bound state solution types must have a lower binding energy in their ground state. Nevertheless, understanding these other solution modes has major technological implications, as will become clear in Chapters 12 and 14. □

8.6. Classical Mechanics

The above formulation lays the ground work for distinguishing classical from quantum mechanics. Indeed, we have seen that quantum theory is highly dependent upon $\hbar > 0$ and states $\psi \in L^2(E^4)$. Moreover, this suggests that classical mechanics can be obtained by relaxing one of these two conditions either by letting $\hbar \to 0$ or by choosing $\psi \notin L^2(E^4)$ or both. In principle, what distinguishes a quantum particle from a classical one is that in contrast to quantum mechanics, the position and momentum of a classical particle can be fully pinpointed and localized. For example, if we reconsider the case of a particle moving uniformly between two mirrors, where the initial position is unknown, then for any time t, as has already been noted in Section 8.4, the wave function is given by $\exp(ikx)/\sqrt{\xi}$. However, if this really were a classical situation, then in principle both the position and momentum of the particle could be localized and measured exactly, as the distance between the mirrors ξ shrinks to 0. Moreover, the resulting wave function at any time t would be of the form $\psi(0) = \sum_k a_k \delta^{(k)}$. This last equation follows from a well-known result in distribution theory [2], which states

Theorem 31. *A distribution T which has a support of one point (i.e., is equal to zero except at one point) is a finite linear combination of the Dirac function and its derivatives:* $T = \sum_k a_k \delta^{(k)}$.

Based on this, we can formally state that

Definition 32. A classical particle is a particle whose position operator T at any time t has support of one point.

It follows from this definition that the momentum operator $\tilde{P}$ has the support of one point. It also follows from the definition and Theorem 17 that for a classical particle with constant momentum situated at (x_0) on a geodesic, the wave function is given by $\delta^4(p^a(x - x_0)_a)$. Also, the set of operators T in Definition 32 clearly form a subspace of the solution set of the Dirac equation. Indeed, the set $\{\delta, \ldots, \delta^{(k)} \ldots\}$ is a spanning set for this subspace. Hence, the uncertainty in observing its value is 0 everywhere except at the single point, and the standard deviation $\Delta\tilde{X}$ and $\Delta\tilde{P}$ are also zero. In other words, the uncertainty principle fails for a classical particle. Moreover, in order for nature to circumvent classical solutions, it is sufficient that there be a fundamental unit of wavelength given by $\lambda = \frac{ch}{E} = \frac{2\pi c\hbar}{E}$, such that $\lambda > 0$ whenever $\hbar > 0$, which is another way of saying that if $\hbar > 0$, then classical solutions need not exist. It also strongly suggests that the process of localizing a particle for measurement is equivalent to confining (at least during the measuring process) the particle to a box. This, in turn, hints that in terms of wave–particle duality, particle properties emerge when we attempt to experimentally localize and isolate the wave, causing a discontinuity in the quantum solutions, closely approximated by delta-type functions. We have seen an example of this above, when we considered a particle moving uniformly between two mirrors with wave-function $e^{\pm ik(x-ct)}/\sqrt{\xi}$.

In concluding this section, we note that classical solutions to the generalized Dirac equation describing the motion of a particle, can be reduced to distributions corresponding to point masses as described in Theorem 31 above, and live on a larger space than the L^2 functions associated with quantum mechanics. To better understand this point, it might be useful to recall the definition of L^p spaces, and some of their properties [3]. Consider a fixed measure space $(X, \mathcal{M}, \mu)$. Let f be a measurable function on X such that

$$\|f\|_p = \left(\int |f|^p d\mu\right)^{\frac{1}{p}} \tag{8.6.1}$$

then we define

Definition 33. $L^p(X, \mathcal{M}, \mu) = \{f : X \to \mathcal{C} : \|f\|_p < \infty\}$.

Also, in general, we can define a bounded linear functional on L^p by $\phi_g(f) = \int fg$, such that $g \in L^q$ where $1/p + 1/q = 1$, and $\phi_g \in (L^p)^*$, the dual space of L^p. In the case of $p = 2$, the L^2 space is also a Hilbert space, and therefore, its own dual. In turn this allows us to formulate quantum mechanics in a very elegant and simple manner. However, in the case of classical solutions, as described above, the Dirac δ functional is usually interpreted as a functional on the set of continuous differential functions of compact support denoted by $C_c^\infty(X)$, which in turn are dense in L^p, where $1 \leq p < \infty$. Moreover, in the case of a finite (probability) measure, $L^p \subset L^1$, for $p \geq 1$. With this distinction in mind, it now follows from our formulation that general relativity while permitting a natural unification of both quantum and classical mechanics by means of the generalized Dirac equation, also permits a distinction by means of L^2 functions and distributions which are duals of L^1 functions.

8.7. One-Dimensional Potential Well

Consider a quantum particle moving in a Minkowski space such that there is zero potential between $(-a, a)$ and constant potential V for $|x| \geq a$. Then the generalized Dirac equation will in both cases reduce to

$$\alpha^0 \hbar \frac{\partial \psi}{\partial ct} - i\hbar \alpha^1 \frac{\partial \psi}{\partial x} = \hbar \frac{\partial \psi}{\partial s} \tag{8.7.1}$$

Clearly, wave function solutions of the form

$$\psi = \begin{cases} Ae^{\hbar^{-1}[Eit-px]} + Be^{-\hbar^{-1}[Eit-px]} & x \le -a \\ Ce^{i\hbar^{-1}[Et-px]} + De^{-i\hbar^{-1}[Et-px]} & |x| < a \\ Fe^{\hbar^{-1}[Eit-px]} + Ge^{-\hbar^{-1}[Eit-px]} & x \ge a \end{cases}$$

can be found, although ψ is not necessarily an eigenfunction of Equation (8.7.1). Note the s dependency follows from the Lorentz invariant relationship $mcs = Et - px$, and that in order to satisfy the boundary conditions at $\pm\infty$, $A = G = 0$. Also, smooth continuity conditions can be imposed at $\pm a$ to fully solve the system in the usual way.

On the other hand, if we begin with the conventional form of the Dirac equation as given in Corollary 19 and Equation (8.2.13), namely

$$\alpha^0 \hbar \frac{\partial \psi}{\partial ct} - i\hbar\alpha^1 \frac{\partial \psi}{\partial x} = mc\psi \tag{8.7.2}$$

then solutions of the form $\psi = C(e^{i\hbar^{-1}[Et-px]} \pm e^{-i\hbar^{-1}[Et-px]})$ cannot exist as is the case in the non-relativistic approach. In other words, Equation (8.7.2) cannot be used to describe a conventional potential well problem. It is also worth noting that if one were to square out either (8.7.1) or (8.7.2), then both of these equations would have ψ as a permissible solution. This ambiguity arises from the non uniqueness of the square root, especially of the α^1 matrix, which can be defined to be either

$$\begin{pmatrix} 0 & \sigma_1 \\ \sigma_1 & 0 \end{pmatrix} \quad \text{or} \quad \begin{pmatrix} 0 & \sigma_2 \\ \sigma_2 & 0 \end{pmatrix}$$

where

$$\sigma_1 = \begin{pmatrix} 0 & 1 \\ 1 & 0 \end{pmatrix} \quad \text{and} \quad \sigma_2 = \begin{pmatrix} 0 & i \\ -i & 0 \end{pmatrix}$$

but not both simultaneously. As we saw in Chapter 2, switching between σ_1 and σ_2 switches the spatial solution of both Maxwell and Dirac equations between harmonic and exponential types. As stated in Chapter 2, neither σ_1 or σ_2 has anything to do with anti-particles. If needed, anti-particle solutions can be independently obtained by solving the Dirac equation directly for them.

8.8. The Hydrogen Atom

We now apply the techniques of this chapter to describe the hydrogen atom. For this, we use a non-flat space–time metric dominated by the electric field energy of a static, non-rotating spherical mass with charge. The simplest such solution is the Reissner–Nordström metric [4]. It should be noted that when the charge is zero, this reduces to the Schwarzschild metric (see appendix). In other words, we describe the motion of the electron lying within the Reissner–Nordström metric of the proton of mass m_p. Linearizing this metric gives:

$$ds = -i\alpha_1 R^{-1}dr - ir(\alpha_2 d\theta - \alpha_3 \sin\theta d\phi) + \alpha_0 Rcdt \tag{8.8.1}$$

where $R(r) = \left(1 - \frac{2Gm_p}{c^2 r} + \frac{Ge^2}{c^4 r^2}\right)^{\frac{1}{2}}$. The corresponding wave equation becomes:

$$\frac{\partial}{\partial s}\psi'(r,t) = \left\{-i\alpha_1 R\frac{\partial}{\partial r} - \frac{i}{r}\left(\alpha_2\frac{\partial}{\partial\theta} + \alpha_3\frac{1}{\sin\theta}\frac{\partial}{\partial\phi}\right) + \alpha_0 R^{-1}\frac{1}{c}\frac{\partial}{\partial t}\right\}\psi'(r,t) \tag{8.8.2}$$

We will find that along a geodesic $\psi = \exp(\int^x p_a dx^a)$ is a solution, provided $p'_a = p_a + \frac{e}{c}A_a$ and A_a represents the vector potential associated with the electromagnetic field of the proton. Also, recall that by definition $p^a = m(dx^a/d\tau)$ (cf. Equation (5.2.27)) along a (local) geodesic, and therefore,

$$p_o = mRc\dot{t} \quad \text{and} \quad p_1 = mR^{-1}\dot{r} \tag{8.8.3}$$

These equations enable us to estimate A_a for this metric. Note also that in the absence of charge that $p_0 = p'_0 = mR(r)\dot{t}$ where $R = (1 - \frac{2Gm_p}{c^2 r})^{\frac{1}{2}}$ is obtained from the Schwarzschild metric. In particular, if we seek solutions of the form $\psi = \exp(\frac{ie}{c\hbar}\int A_a dx^a)\psi'(\frac{i}{\hbar}\int p_a dx^a)$ in tetrad coordinates, then this can be reduced to the following equation:

$$\left[c\sum_{a=1}^{3}\hat{\alpha}_a P'_a + \alpha_0(mc^2 + eV)\right]\psi = E\psi \tag{8.8.4}$$

where $\hat{\alpha}_a = -i\alpha_0\alpha_a$, $P'_a = -i\hbar\partial_a + \frac{e}{c}A_a$, m is the mass of the electron and V is the electrostatic potential in the rest frame of the proton. Note that this can be written in the usual form of the equation for the hydrogen atom $H\psi = E\psi$, provided $H \equiv \left[c\sum_{a=1}^{3}\hat{\alpha}_a P'_a + \alpha_0(mc^2 + eV)\right]$. However, there is also an important difference. In this formulation, there is the presence of a $\alpha_0 eV$, instead of the usual eV, but as we will see in what follows this has advantages.

If we let $\mathcal{G}_a = \frac{\partial V}{\partial x^a}$, assume $A_a = 0$ for $a = 1$, 2, and 3, then the "square" of this equation reduces to

$$\left(-c^2\hbar^2 \sum_1^3 \partial_a^2 + (mc^2 + eV)^2 + i\hbar ec\hat{\alpha}_a\alpha_0\mathcal{G}_a\right)\psi = E^2\psi \qquad (8.8.5)$$

Noting that $-i\hbar\hat{\alpha}_a\alpha_0 = \alpha_a$, which can also be seen as generators of the Clifford algebra, this can be rewritten as follows:

$$\left(-c^2\hbar^2 \sum_1^3 \partial_a^2 + (mc^2 + eV)^2 - \hbar ec\alpha_a\mathcal{G}_a\right)\psi = E^2\psi \qquad (8.8.6)$$

Contrast this with conventional relativistic quantum mechanics where we would have found

$$\left(-c^2\hbar^2 \sum_1^3 \partial_a^2 + m^2c^4 - i\hbar ec\alpha_a\mathcal{G}_a\right)\psi = (E + eV)^2\psi \qquad (8.8.7)$$

Note that the complex number coefficient associated with the electric moment $\mathcal{G}$ in this latter equation has been replaced with a real coefficient in Equation (8.8.6). As discussed in Chapter 6, the complex valued $\frac{\hbar e}{2imc}$ term has been traditionally interpreted as electron spin momentum for s orbitals, *but* as orbital angular momentum for other orbital types. This imaginary term can be now replaced with a real angular momentum term $\frac{\hbar}{2mc}$, thus removing the ambiguity normally associated with its interpretation in the hydrogen atom. Happily, we can say that besides resolving the difficulty associated with the imaginary term, Equation (8.8.6) implies that the $\hbar ec\alpha_a\mathcal{G}_a$ term is always describing orbital angular momentum, even for s orbitals. We have been calculating the quantum mechanical wavefunction, and not the internal magnetic moment of the electron. These results are in line with the conclusions of Chapter 6, and conform to the principle of angular momentum conservation, for instance during $2p$ to $2s$ transitions.

As explained in Chapter 6, the electron is in fact orbiting around the center-of-mass in the electron–nucleus system. The nucleus is also orbiting around the same center of mass, following a geodesic in the same space–time curvature which was calculated above. This nuclear motion can be calculated using the above methodology. Therefore, an analogous $\hbar ec\alpha_a\mathcal{G}_a$ term describes the orbital angular momentum of the nucleus. The angular momentum of the electron and nucleus are entangled; the geometry of this entanglement was discussed in Sections 0.10 and 6.9.

In summary, we successfully derived the spin-independent quantum mechanical wavefunction ψ. How to account for the additional electron spin effect? This was already explained in Chapter 7, where we showed how the quantum mechanical wavefunction factorizes into the herein discussed spinless ψ field and the spin-carrying Proca field $\boldsymbol{a}_{\square}$.

8.9. Light Emission and Absorption

In this section, we consider the interaction between an electromagnetic wave and a quantum mechanical wavefunction. We look at light–matter interaction from three different perspectives: in each case we ask whether experimental phenomena require any additional electromagnetic structure beyond the vector potential field. In each case, we will find that there is no need to introduce any new structure: the electromagnetic vector potential field A perfectly accounts for light–matter interactions. As will be shown in the following paragraphs, the "photon particle" concept is not only superfluous but leads to paradoxes.

8.9.1. *Quantum mechanical state transition*

For an elementary particle to emit electromagnetic wave energy at the angular frequency ω, its charge must be oscillating at this frequency. This statement is a direct consequence of Maxwell's equation, and therefore it remains true under any circumstance. Since the kinetic oscillation of charges is described by the quantum mechanical wavefunction Ψ, light–matter interaction is a process between an electromagnetic angular frequency ω and the same kinetic oscillation frequency $\Omega_{\text{oscillation}}$, i.e., $\Omega_{\text{oscillation}} = \omega$.

We consider a transition from a higher $\hbar\Omega_1$ energy level to a lower $\hbar\Omega_2$ energy level. Since $\hbar\Omega_1$ and $\hbar\Omega_2$ are energy eigenstates, Ω_1 and Ω_2 are non-radiating angular frequencies. Is there any additional $\Omega_{\text{oscillation}}$ angular frequency, which may be associated with the state transition process? Let's assume a single-frequency oscillation. The value of $\Omega_{\text{oscillation}}$ directly follows from the conservation of energy:

$$\Omega_1 = \Omega_2 + \Omega_{\text{oscillation}} \tag{8.9.1}$$

The above relationship has been experimentally validated since more than hundred years. The frequency of light, emitted upon state transitions of the quantum mechanical wavefunction, is precisely given by the above formula. This indicates the presence of a transient wavefunction state during electromagnetic light emission, which simultaneously comprises Ω_2 and $\Omega_{\text{oscillation}}$ angular frequencies.

On the one hand, it is a direct consequence of Maxwell's equation that a quantum mechanical state transition process must last for a certain duration and must involve the presence of $\Omega_{\text{oscillation}} = \omega$ angular frequency wavefunction state. On the other hand, the current axioms of quantum mechanics completely bypass Maxwell's equation. These axioms state that a quantum mechanical state transition is a timewise discontinuous "jump" process, which releases a so-called "photon particle." This hypothetical photon particle is characterized by the frequency ω, which is presumably never present in the photon-emitting particle wavefunction. Such a polar opposite perspective makes one wonder about the origin of the "photon particle" concept, which dates back to more than 100 years ago. In his first quantum mechanics-related publication, Einstein was trying to explain the experimental observation that the probability of a quantum mechanical state transition depends on the frequency of incident light, and not on its amplitude. Regrettably, Einstein chose to go against the historic trend of exploring electromagnetic waves, and introduced the notion of "light particles" which are presently referred to as photons. At that time, most physicists believed in the reality of "aether particles" which supposedly permeate all space. In that pioneering era, the notion of a second type of light carrying particle appeared to be the simplest explanation of experiments. While it was mathematically clear from the start that a single-frequency electromagnetic emission implies an infinite wavefunction size of the hypothetical "photon particle," Einstein's contemporaries chose to ignore this mathematical paradox. For an average scientist of the early 20th century, the study of light–matter interactions meant the study of interactions between electron particles, aether particles, and the "recently discovered" photon particles. Bohr complemented Einstein's postulate by an additional axiom stating that the release of a photon particle coincides with an abrupt jump between quantum mechanical states. But some years later Einstein was taken aback by the paradoxical implications of these axioms, and then fought against them till the end of his career. He felt that there is some hidden reality behind quantum mechanics, which is occulted by all these axioms. Unfortunately for Einstein, his program of finding quantum mechanics' "hidden variable" was a dead-end, as J. S. Bell derived in 1964 the non-existence of such hidden variables. Actually, Einstein's re-introduction

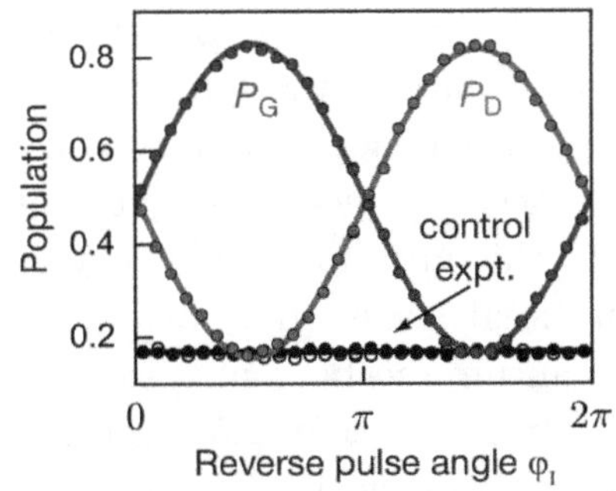

Figure 8.1. External control over the outcome of a quantum mechanical transition, which was applied at mid-point in the transition process. By definition, at this mid-point half of the particles are still in the ground state while the other half are already in the destination state. The applied external pulse angle determines whether the ongoing quantum mechanical transition completes or reverses itself. P_G denotes the probability of finding the particle in the ground state, while P_D denotes the probability of finding the particle in the excited destination state. In the control experiment the same pulse is applied at a random time, i.e., not at the mid-point of an ongoing quantum mechanical transition process. Reproduced from [17].

of the outdated "light particle" concept was itself the problem, and this dogmatized photon axiom continues to confuse generations of physicists up to this day.

Until recently, there was no experimental evidence to determine whether a quantum mechanical state transition is a smooth process of a certain duration or a sudden jump process. In a recent seminal experiment, the authors of [17] experimentally proved that quantum mechanical state transitions do have a characteristic duration. The authors of [17] distinguish the trigger event of quantum mechanical state transitions, which is stochastic, and the subsequent state transition process. They found that a triggered light emission or absorption process is deterministic and has a well-defined duration. Furthermore, Figure 8.1 demonstrates a deterministic control over the completion or reversal of an already ongoing quantum mechanical state transition process. An electromagnetic pulse is employed for gaining such deterministic control over the outcome of an ongoing quantum mechanical state transition process. It is clear from these results that understanding light–matter interactions requires finding analytic solutions of Maxwell's equation in the microscopic limit. *Ad hoc* axioms, which bypass all equations of electromagnetism, can only lead to mis-understandings.

Upon rejecting the axiomatic notion that the emitted electromagnetic frequency is an inherent property of a "light particle," the over 100 years old paradigm of single-frequency light emission during a quantum mechanical state transition brings up a major question. Why is electromagnetic wave energy emitted at a single frequency, i.e., why can the transition between electron orbitals not be accompanied by light emission at two or more frequencies? We look into this question in the following paragraphs.

The key idea of our analysis is that instead of using the laboratory frame, we can study a quantum mechanical state transition in the electron's own frame. Let's recall from Section 2.5 that in the laboratory frame perspective an electron's orbital motion takes place in curved space–time, whose curvature is created by the nuclear charge. The electron is in an electromagnetic free-fall state in this curved space. It is a well-known theorem of general relativity that the reference frame attached to a free-falling particle is a linear space–time, without any curvature. The straight trajectory which the particle follows in its own reference frame becomes a curved and closed geodesic trajectory in the laboratory frame. In the electron's free-falling reference frame, its Hamiltonian is the same as a free particle's Hamiltonian, and depends only on its momentum. As the electron transits from a higher energy eigenstate to a lower energy eigenstate, its entire transition process can be described as a free particle process in a co-moving reference frame.

Based on the above logic, we are interested in the Hamiltonian matrix operator of a free particle. We use Heisenberg's matrix operator approach to quantum mechanics, and in particular we use the mathematical approach of [18]. Although the authors of [18] studied a different problem, their method is perfectly applicable for describing state transitions. During a $\Psi_1 \to \Psi_2$ eigenstate transition, the Hamiltonian has

$$H = \begin{pmatrix} H_{11}(\mathbf{p}) & H_{12}(\mathbf{p}) \\ H_{21}(\mathbf{p}) & H_{22}(\mathbf{p}) \end{pmatrix} \quad \text{form}$$

We explicitly indicated that the Hamiltonian matrix elements depend on momentum only. The traditional approach is to assume $H_{11}\Psi_1 = \hbar\Omega_1\Psi_1$ and $H_{22}\Psi_2 = \hbar\Omega_2\Psi_2$ for the diagonal elements, and to assume $H_{12} = 0$ and $H_{21} = 0$ for the off-diagonal terms which express quantum mechanical correlations. However, such assumption leads to the model of instantaneous $\Psi_1 \to \Psi_2$ transition, and erases any information about the transition process itself. Therefore, we refrain from making any restrictive assumption about the Hamiltonian matrix elements.

In the Heisenberg picture, the $U = e^{\frac{i}{\hbar}Ht}$ time evolution operator is then also a function of the momentum only. Note that we use the same $U = e^{\frac{i}{\hbar}Ht}$ time evolution operator to describe a quantum mechanical state transition that is normally used to describe eigenstates: i.e., we do not invent any new rules or postulates to describe state transitions. The crucial step is to realize that this Hamiltonian can be always written as a sum of projection operators: $H = \sum_{a=1,2} E_a Q_a$ where E_a is an eigenvalue of the Hamiltonian at a given momentum $\mathbf{p}$. The Q_a projection operators are orthogonal, i.e., they satisfy the following relations:

$$Q_a Q_b = \delta_{ab} Q_a, \sum_a Q_a = \begin{pmatrix} 1 & 0 \\ 0 & 1 \end{pmatrix} \tag{8.9.2}$$

In terms of the projection operators, the time evolution operator can be written as $e^{\pm\frac{i}{\hbar}Ht} = \sum_{a=1,2} e^{\pm\frac{i}{\hbar}E_a t} Q_a$.

We are interested in finding the position operator $\mathbf{x}(t)$. At $t = 0$, i.e., at the start of the state transition, our position operator is the same as the position operator in the Schrödinger picture: $\mathbf{x}(0) = i\hbar\frac{\partial}{\partial \mathbf{p}}$ in the mometum representation.

Working in momentum coordinates, we evaluate the $[\mathbf{x}(0), F(\mathbf{p})]$ commutation of operators, where $F(\mathbf{p})$ is any operator that depends on momentum only:

$$[\mathbf{x}(0), F(\mathbf{p})]\,\Psi = i\hbar\frac{\partial}{\partial \mathbf{p}}(F(\mathbf{p})\,\Psi) - F(\mathbf{p})\,i\hbar\frac{\partial \psi}{\partial \mathbf{p}} = i\hbar\frac{\partial F(\mathbf{p})}{\partial \mathbf{p}}\Psi \tag{8.9.3}$$

Finally, we express the position operator $\mathbf{x}(t)$ with the help of the time evolution operator:

$$\mathbf{x}(t) = U^{-1}\mathbf{x}(0)\,U = U^{-1}[\mathbf{x}(0), U] + U^{-1}U\mathbf{x}(0) = \mathbf{x}(0) + i\hbar U^{-1}\frac{\partial U}{\partial \mathbf{p}} \tag{8.9.4}$$

We substitute the orthogonal Q_a projection operators into the above expression, and obtain the following result:

$$\mathbf{x}(t) = \mathbf{x}(0) + \sum_a Z_{aa} + t\sum_a V_a Q_a + \sum_a \sum_{b \neq a} e^{i\omega_{ab}t} Z_{ab} \tag{8.9.5}$$

where $V_a = \frac{\partial E_a(\mathbf{p})}{\partial \mathbf{p}}$ are the velocities corresponding to a given eigenstate, $Z_{ab} = i\hbar Q_a \frac{\partial Q_b}{\partial \mathbf{p}}$ is the off-diagonal correlation amplitude, and $\omega_{ab} = \frac{E_a - E_b}{\hbar}$ is the beating frequency. As illustrated in Figure 8.1, the electron is never measured at any other energy level than E_1 or E_2; these parameters are constants during the state transition. The state transition process therefore involves a continuous evolution of only the Q_a projection operators.

The interesting term of Equation (8.9.5) is the last oscillatory term: it defines a position oscillation at one specific frequency during the entire state transition. In the free particle case, the $\hbar\Omega_{1\text{free}} \to \hbar\Omega_{2\text{free}}$ energy state transition directly involves the $\omega_{12} = \Omega_{1\text{free}} - \Omega_{2\text{free}}$ angular frequency of the charge oscillation which we are looking for. In summary, the energy eigenstate transition of a free particle is always accompanied by a single-frequency position oscillation of the electric charge.

As we switch back from the electron co-moving frame to the laboratory frame, the apparent charge oscillation frequency may change. An $\hbar\Omega_{1\text{free}} \to \hbar\Omega_{2\text{free}}$ energy state transition in the electron co-moving frame may be perceived as a different $\hbar\Omega_1 \to \hbar\Omega_2$ energy state transition in the laboratory frame. Nevertheless, a single-frequency position oscillation in one frame remains a single-frequency oscillation after the frame transformation. Thereby we prove the presence of a single-frequency charge oscillation during a quantum mechanical state transition. This charge oscillation will emit a single-frequency electromagnetic radiation.

Since the Maxwell and Dirac equations are time-symmetric, light absorption can be described as a light emission process played in reverse. Thus, the electromagnetically induced $\hbar\Omega_2 \to \hbar\Omega_1$ energy state transition necessitates the presence of an electromagnetic wave with $\omega = \Omega_1 - \Omega_2$ angular frequency. It follows from the above derivation that neither any different electromagnetic frequency nor a combination of several frequencies can induce the $\hbar\Omega_2 \to \hbar\Omega_1$ energy state transition.

The above results also give us a more precise understanding of Heisenberg uncertainty. An unbound electron freely interacts with the background radiation field, and its momentum uncertainty takes then the $\sigma_p\sigma_x \geq \frac{\hbar}{2}$ form. However, the amount of momentum in a bound electron has no uncertainty: each orbital corresponds to an energy and momentum eigenstate, where the electron takes on very specific energy and momentum values. The remaining uncertainty is only in the direction of the electron's momentum, not its magnitude.

While electron orbital transitions are induced only by a matching electromagnetic wave frequency, an elementary particle also responds to other electromagnetic wave frequencies. The general effect of other electromagnetic frequencies' presence is the "blurring" of the quantum mechanical particle position. This effect is precisely calculated in Chapter 6, where the Lamb shift effect is studied, and it is found that the magnitude of position blurring is proportional to the electromagnetic field amplitude. We emphasize that this Lamb shift is also a direct consequence of Maxwell's equation: there is absolutely no need for Bethe's *ad hoc* vacuum model postulate which he introduced for "explaining" this effect.

8.9.2. *Light detection*

To further demonstrate how unnecessary the photon axiom is, let's think through the electromagnetic double-slit experiment in the high-intensity and low-intensity conditions. These different regimes are illustrated in Figure 8.2. In the high intensity limit, one may calculate the electromagnetic signal intensity by solving Maxwell's equations, and this exactly corresponds to the interference pattern detected by the receiver. In this high intensity limit, the signal is everywhere much higher than the absorbed $\hbar\omega$ energy quanta. In the very low intensity limit, the received electromagnetic signal is

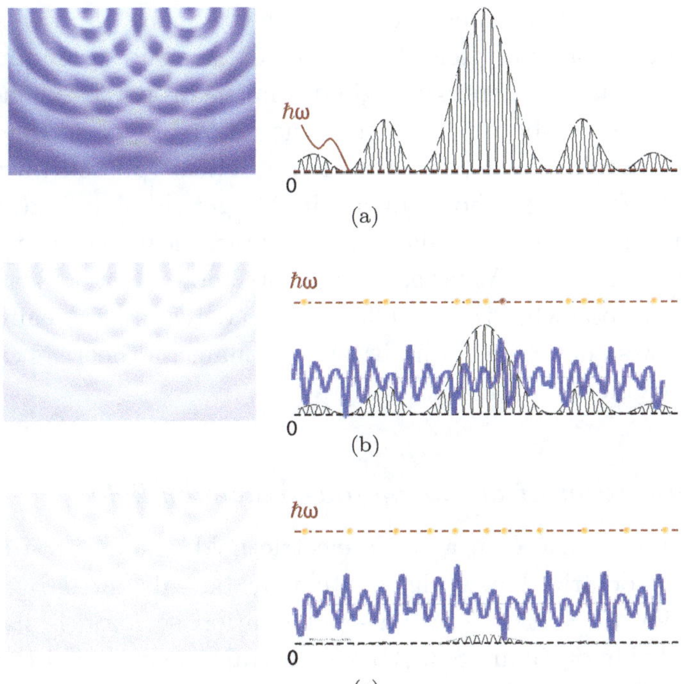

Figure 8.2. An illustration of electromagnetic wave detection behind a double slit. (a) high intensity scenario, where the signal is several orders of magnitude higher than the absorbed $\hbar\omega$ energy quanta. (b) low intensity scenario, where the electromagnetic signal has similar intensity as the thermal noise, and both intensities are on the average lower than the detection limit. (c) very low intensity scenario, where the electromagnetic signal is negligible with respect to the thermal noise. The yellow dots show the aggregate locations of signal detection, and the red dot illustrates the latest appearing signal blip.

negligible with respect to the thermal electromagnetic noise. The magnitude of the thermal noise signal is temperature dependent; the Boltzmann formula describes its stochastic distribution on the basis of entropy considerations. At high temperatures, the detector will continuously measure the illustrated noise signal, but at sufficiently low temperatures there will be spot-like detection blips when the noise signal exceeds the detection threshold. Since the electromagnetic interference pattern is much smaller than the noise, the blips will appear at random locations. Finally, let's consider the scenario shown in Figure 8.2(b), where the electromagnetic signal has similar intensity as the thermal noise. Again, at sufficiently low temperatures there will be spot-like detection blips when the sum of interference signal plus noise signal exceeds the detection threshold. But this time the blip distribution is no longer random because the signal interference pattern significantly influences where the blips appear. One by one the consecutively appearing blips trace out the interference pattern which one may calculate by solving Maxwell's equations.

The above explanation feels almost trivial, once one understands that the quantization of absorbed energy originates in the properties of the absorbing particle. Why do textbooks insist that the above-described detection blips are causality- breaking collapses of the hypothetical "photon wavefunction"? Such state of affairs shows how difficult it is to get rid of an ingrained axiom, despite being unable to formulate a non-paradoxical photon wavefunction after 100 years of trying. As far as the authors know, Carver Mead is the first one who understood what the so-called photon detectors are really measuring in the very low signal intensity limit: "What quantum theorists call statistics is what we engineers call noise."

8.9.3. *Ionization of atoms by low-frequency light*

When an atom is placed in a static electric field, the electron and nucleus polarize. The polarized new eigenstate may be calculated by solving the Dirac equation of Section 7.4. This polarization scenario is illustrated on the left-hand side of Figure 8.3. Above a certain electric field threshold, the atom becomes ionized. Crossing this threshold simply means that the Dirac equation no longer yields a bound eigenstate of the outermost electron. As the applied electric field is static, there cannot be any hypothetical photon particles involved in triggering this quantum mechanical state transition.

Now let's consider a circularly polarized microwave field, whose frequency is much lower than the quantum mechanical frequency of electron orbital. This scenario is illustrated in the middle of Figure 8.3 the polarized electron–nucleus system rotates along with the phase of the circularly polarized

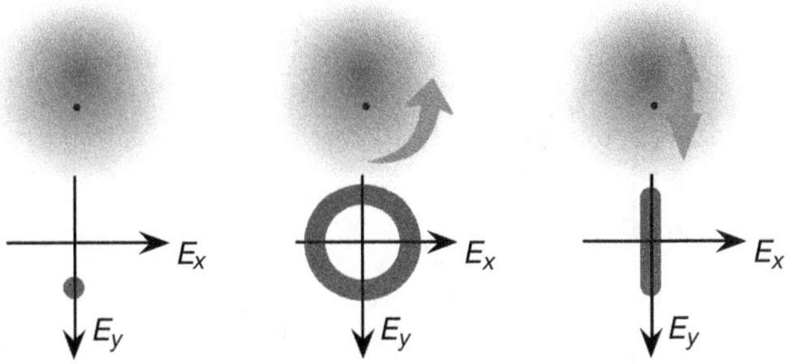

Figure 8.3. Various electrically induced atomic polarizations. Left: permanent polarization under a constant electric field. Middle: rotating polarization under a circularly polarized microwave field. Right: vertically oscillating polarization under a linearly polarized microwave field. The electric field orientations are illustrated by the phase diagrams at the bottom.

microwave field, following the orientation of the electric field. From a rotating perspective of the electron–nucleus system, this scenario is not much different from the static electric field scenario. Therefore, we expect the ionization threshold field to have the same amplitude as in the static field case: this expectation is indeed confirmed by [19]. The measured relationship between the main quantum number of the electron orbital and the ionization threshold field strength is depicted by dots in Figure 8.4

Lastly, we consider a linearly polarized microwave field, whose frequency is the same as in the circularly polarized case. This scenario is illustrated on the right-hand side of Figure 8.3 the electron–nucleus system now oscillates between polarized and non-polarized states. In this case, the author of [19] measured a different ionization threshold field strength than in the circularly polarized case, and it is depicted by diamonds in Figure 8.4

Does it make sense to talk about photons in the context of this microwave ionization experiment? Referring to Figure 8.4, we note that the quantum mechanical state transition from bound to ionized state depends on the amplitude of the employed electromagnetic wave, and does not depend on the $\hbar\omega$ energy quantum of hypothetical photons. Secondly, microwaves' $\hbar\omega$ value is significantly smaller than the binding energy of electron states shown in Figure 8.4. Despite these *a priori* problems, mainstream theorists describe microwave-induce d ionization as a "multi-photon process." In their view, 10–100 photon particles are simultaneously absorbed by an ionized electron. But there is no rational explanation as to why such a process would become independent of the hypothetical photons' $\hbar\omega$ energy level, and why it would become dependent on the overall wave amplitude. The final nail in the coffin of such "multi-photon process" hypothesis is to consider the circularly

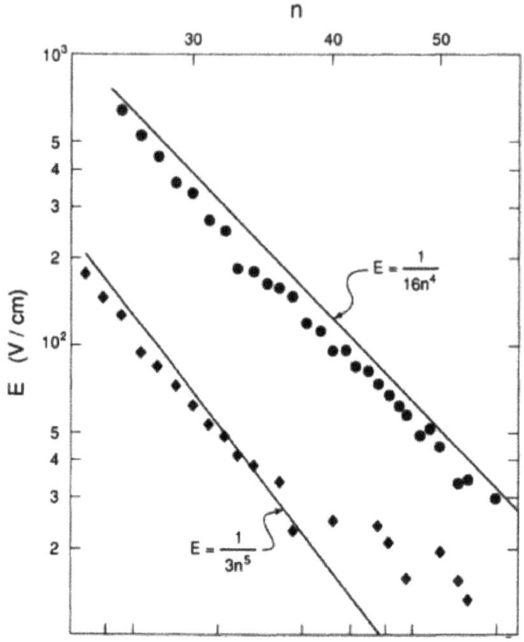

Figure 8.4. Ionization threshold of highly excited sodium atoms in a microwave field. The horizontal scale shows the main quantum number of the outermost electron state, and the vertical scale shows the threshold electric field strength for ionization. Dots: measurements in circularly polarized microwave field. Diamonds: measurements in linearly polarized microwave field. Reproduced from [19].

polarized microwave scenario from the perspective of the rotating electron–nucleus system. As mentioned above, the circularly polarized microwave scenario is conceptually equivalent to the static scenario, and this equivalence was experimentally demonstrated by the author of [19]. The hypothesized "multi-photon process" is thereby proven to be a no-photon process.

In summary, the photon particle conjecture leads to paradoxes. Fortunately, the above ionization scenarios can be explained simply by using the Dirac equation to calculate the electron eigenstate, without any further reference to hypothetical light particles.

8.10. Conclusion

In this chapter, we have attempted to unify general relativity and quantum mechanics by viewing any metric as a dual of a wave equation. We have noted that the resulting wave equation contains the usual Dirac equation of quantum mechanics as a special case. We have also noted that the difference between quantum and classical mechanics seems to lie in boundary conditions, with quantization (as distinct from quantum theory) emerging when the wave function is confined to a finite domain with continuous boundary conditions, and classical mechanics being the result of delta-function type solutions for the wave equation. Indeed, this particular approach also highlights the harmonic oscillator as a natural starting model for quantum field theory, by confining the quantum particle to a box. Of course, our use of the word "confined" differs from the usual non-Abelian gauge theory usage. Here the word confined means considering solutions of the generalized Dirac equation constrained by periodic boundary conditions.

Overall, our approach was able to duplicate the standard results of quantum mechanics, but without any paradoxical postulates or interpretation difficulties. Most importantly, our approach shows that general relativity and quantum mechanics are in fact closely entwined. The obtained hydrogen solution gives a realistic interpretation of electron orbital states.

Finally, we discussed quantum mechanical state transitions. It turned out that the 100 year old notion of considering a state transition to be an abrupt jump leads to paradoxes. We derived that a quantum mechanical state transition is accompanied by a single-frequency emission of electromagnetic radiation. The state transition has an experimentally measurable duration, and we thus removed the paradoxical wavefunction collapse postulate. We did not need to make any new hypothesis about transversal electromagnetic waves, beyond the already known fact that they represent solutions of Maxwell's equations. In agreement with Carver Mead's insights, we refrained from introducing hypothetical "photon particles," thus avoiding all associated paradoxes. An important conclusion is that the quantization of absorbed or emitted electromagnetic energy derives from the properties of elementary particles.

Appendix: The Schwarzschild Metric

The Schwarzschild metric is the simplest space–time solution after the flat-space metric. We directly apply the theory to a test particle of mass m moving along a timelike curve parametrized by τ in a Schwarzschild space, whose metric is defined by [16]:

$$ds^2 = B(r)dt^2 - A(r)dr^2 - r^2 d\theta^2 - r^2 \sin^2 \theta d\phi^2$$

This allows us to define tetrad coordinates $dx^0 = B^{1/2}dt$, $dx^1 = -A^{1/2}dr$, $dx^2 = -rd\theta$, $dx^3 = -r\sin(\theta)d\phi$, which on substituting into Equation (5.2.28) give

$$ds\xi = [\gamma_0 B^{\frac{1}{2}}dt - \gamma_1 A^{\frac{1}{2}}dr - \gamma_2 r d\theta - \gamma_3 r\sin(\theta)d\phi]\xi \tag{8.10.1}$$

This is also equivalent to

$$c\xi = [\gamma_0 B^{\frac{1}{2}}\dot{t} - \gamma_1 A^{\frac{1}{2}}\dot{r} - \gamma_2 r\dot{\theta} - \gamma_3 r\sin(\theta)\dot{\phi}]\xi \tag{8.10.2}$$

which can be seen as the equation of motion along a time-like geodesic. The corresponding wave equation associated with the above metric equation is then given by

$$\frac{\partial\psi}{\partial s} = \gamma^0 B^{-\frac{1}{2}}(r)\frac{\partial\psi}{\partial t} - \gamma^1 A^{-\frac{1}{2}}(r)\frac{\partial\psi}{\partial r} - \gamma^2\frac{1}{r}\frac{\partial\psi}{\partial\theta} - \gamma^3\frac{1}{r\sin\theta}\frac{\partial\psi}{\partial\phi} \tag{8.10.3}$$

To simplify the problem, we assume $\dot{\theta} = \dot{\phi} = 0$. Hence, only the first two terms of the above equation are non-zero, and the equation of motion reduces to

$$\gamma^0 B^{\frac{1}{2}}\dot{t}\xi - \gamma^1 A^{\frac{1}{2}}\dot{r}\xi = c\xi \tag{8.10.4}$$

where

$$\gamma^0 = \begin{pmatrix} 1 & 0 & 0 & 0 \\ 0 & 1 & 0 & 0 \\ 0 & 0 & -1 & 0 \\ 0 & 0 & 0 & -1 \end{pmatrix} \quad \text{and} \quad \gamma^1 = \begin{pmatrix} 0 & 0 & 0 & i \\ 0 & 0 & i & 0 \\ 0 & i & 0 & 0 \\ i & 0 & 0 & 0 \end{pmatrix}$$

Multiplying out gives

$$\begin{aligned} (B^{\frac{1}{2}}\dot{t} - c)\xi_0 - iA^{\frac{1}{2}}\dot{r}\xi_3 &= 0 \\ (B^{\frac{1}{2}}\dot{t} - c)\xi_1 - iA^{\frac{1}{2}}\dot{r}\xi_2 &= 0 \\ -(B^{\frac{1}{2}}\dot{t} + c)\xi_2 - iA^{\frac{1}{2}}\dot{r}\xi_1 &= 0 \\ -(B^{\frac{1}{2}}\dot{t} + c)\xi_3 - iA^{\frac{1}{2}}\dot{r}\xi_0 &= 0 \end{aligned}$$

which can be solved to give

$$(B^{\frac{1}{2}}\dot{t} - c)\xi_0 = iA^{\frac{1}{2}}\dot{r}\xi_3 \tag{8.10.5}$$

and

$$(B^{\frac{1}{2}}\dot{t}-c)\xi_1 = iA^{\frac{1}{2}}\dot{r}\xi_2 \tag{8.10.6}$$

Clearly, many solutions are possible. One solution is given by $\xi_0 = \xi_1 = iA^{\frac{1}{2}}\dot{r}$ and $\xi_3 = \xi_2 = (B^{\frac{1}{2}}\dot{t}-c)$. Another solution is given by $\xi_0 = \xi_1 = (B^{\frac{1}{2}}\dot{t}+c)$ and $\xi_3 = \xi_2 = -iA^{\frac{1}{2}}\dot{r}$. Note that in this latter case, if we substitute into Equations (52) and (53), we obtain the equation of motion $B\dot{t}^2 - A\dot{r}^2 = c^2$. Also, recall from Theorem 17 that $\xi = \frac{\partial\psi^i}{\partial s}e_i$. Therefore,

$$\psi^0 = \psi^1 = \int (B^{\frac{1}{2}}\dot{t}+c)ds \tag{8.10.7}$$

and

$$\psi^3 = \psi^2 = -i\int (A^{\frac{1}{2}}\dot{r})ds \tag{8.10.8}$$

In the case of constant motion along a geodesic, the above reduces to $\psi = \psi(\int p_a dx^a) = \psi(p_0x_0 - p_1x_1)$ where $p_0 = mB^{\frac{1}{2}}\dot{t}$ and $p_1 = mA^{\frac{1}{2}}\dot{r}$ are constants of the motion. In particular, in the case of motion inside a potential well of radius $x_1(r) \le l$, we find that $C\exp[i(p_0x_0 - p_1x_1)] + D\exp[i(p_0x_0 + p_1x_1)]$ is a solution of the squared out Dirac equation (expressed in tetrad coordinates)

$$\hbar^2\frac{\partial^2\psi}{\partial s^2} = \hbar^2\frac{\partial^2\psi}{\partial x_0^2} - \hbar^2\frac{\partial^2\psi}{\partial x_1^2} \tag{8.10.9}$$

Denoting the eigenvalues of $i\hbar\frac{\partial}{\partial s}$ by mc and the eigenvalue of $i\hbar\frac{\partial}{\partial x_0}$ by $\frac{E}{c}$, this equation reduces to

$$(mc)^2\psi = \left(\frac{E}{c}\right)^2\psi + \hbar^2\frac{\partial^2\psi}{\partial x_1^2} \tag{8.10.10}$$

Imposing the boundary conditions $\psi(x_0, x_1(r) = l) = 0$ gives

$$\psi = \exp(i(E/c)x_0)\sin(p_1x_1) = \exp(i(E/c)B^{\frac{1}{2}}t)\sin(p_1x_1) \tag{8.10.11}$$

with $p_1 \equiv mk_n = \frac{n\pi}{l}$. Therefore, the energy levels for the motion of the particle along a geodesic given by $\dot{\theta} = \dot{\phi} = 0$ inside the potential well can be calculated from the equation

$$(mc)^2\psi = \left(\frac{E}{c}\right)^2\psi - (mk_n)^2\psi \tag{8.10.12}$$

In other words,

$$E^2 = m^2c^2(c^2 + k_n^2) = m^2c^2B\dot{t}^2, \quad n \text{ is an integer}$$

Also, if we permit $l \to 2mG/c^2$ (the Schwarzschild radius), in such a way that $p_0 = m(1 - \frac{2Gm}{c^2r})^{\frac{1}{2}}\dot{t}$ and $p_1 = m(1 - \frac{2Gm}{c^2r})^{-\frac{1}{2}}\dot{r}$ remain constant, then we can interpret the above energy levels as the energy levels within a black hole provided we work with space like geodesics and not time like ones.

Finally, to conclude, note that in the case of a photon the wave equation along a null geodesic is given by

$$0 = \gamma^0 B^{-1/2}(r)\frac{\partial\psi}{\partial t} - \gamma^1 A^{-1/2}(r)\frac{\partial\psi}{\partial r} - \gamma^2\frac{1}{r}\frac{\partial\psi}{\partial\theta} - \gamma^3\frac{1}{r\sin\theta}\frac{\partial\psi}{\partial\phi} \qquad (8.10.13)$$

References

[1] É. Cartan, *The Theory of Spinors.* Dover, 1981, p. 134.

[2] G. Duff and D. Naylor, *Differential Equations of Applied Mathematics.* Wiley, New York, 1966, p. 71.

[3] G. Folland, *Real Analysis.* Wiley & Sons, New York, 1984, pp. 170–210.

[4] S. Hawking and G. Ellis, *The Large Scale Structure of Space–Time.* Cambridge University Press, New York, 1995, p. 17.

[5] R. Lindsay and H. Margenau, *Foundations of Physics.* Dover, New York, 1957, pp. 46–55.

[6] R. Lindsay and H. Margenau, *Foundations of Physics.* Dover, New York, 1957, p. 511.

[7] P. W. Milonni, *The Quantum Vacuum.* Academic Press, San Diego, 1994, pp. 305–308.

[8] H. Nikolic, How (not) to teach Lorentz covariance of the Dirac equation (2014), arXiv:1309.7070v2[quant-ph].

[9] P. O'Hara, Rotational invariance and the spin-statistics theorem. *Foundations of Physics*, **33**(9) (2003) 1349–1368.

[10] P. O'Hara, Wave-particle duality in general relativity, *Nuovo Cimento B*, **111**(7) (1996) 799–810.

[11] P. O'Hara, Quantum mechanics and the metrics of general relativity, *Foundations of Physics*, **35**(9) (2005) 1563–1584.

[12] P. O'Hara, Equations of motion in general relativity and quantum mechanics. *Journal of Physics: Conference Series*, **330** (2011) 012013.

[13] L. O'Raifeartaigh, *The Dawning of Gauge Theory.* Princeton University Press, Princeton, 1997, p. 112.

[14] E. Schroedinger, An undulatory theory of the mechanics of atoms and molecules. *The Physical Review*, **28**(6) (1926) 1049–1070.

[15] E. Schroedinger, Quantisierung als eigenwertproblem. *Annalen der Physik*, **81** (1926) 162.

[16] S. Weinberger, *Gravitation and Cosmology: Principles and Applications of the General Theory of Relativity.* Wiley, Hoboken, 1972, p. 179.

[17] Z.K. Minev *et al.*, To catch and reverse a quantum jump mid-flight. *Nature*, **570** (2019) 200–204. https://doi.org/10.1038/s41586-019-1287-z.

[18] Gy. Dávid and J. Cserti, General theory of Zitterbewegung. *Physical Review B*, **81**(12) (2010) 121417.

[19] T.F. Gallagher, Ionization in Linearly and Circularly Polarized Microwave Fields, *The Electron: New Theory and Experiment*, vol. 45. Springer, Dordrecht, 1991, pp. 311–320. DOI: https://doi.org/10.1007/978-94-011-3570-2_15.

https://doi.org/10.1142/9781800611306_0009

Chapter 9

The Pauli Exclusion Principle

Paul O'Hara[*,‡] and András Kovács[†]

[*] *Sophia University Institute, Loppiano, Italy*
[†] *BroadBit Energy Technologies, Metallimiehenkuja 8 C 02150 Espoo, Finland*
[‡] *paul.ohara@sophiauniversity.org*

9.1. Introduction

Since the early days of quantum mechanics, it was recognized that in an atom or molecule two electrons cannot share all the same quantum numbers,[a] or equivalently no more than two electrons may occupy any given atomic orbital. The same pairwise occupancy limit of any orbital state was observed for covalent molecular orbitals and even for delocalized electrons in metals. Without this exclusion principle, all electrons would fall into the same ground state orbital. Understanding the origin of the exclusion principle appears therefore central to understanding the physics of tangible matter. Once we know its origin, we may also understand the boundary conditions under which this principle remains valid. In that regard, although this pairwise occupancy limit has been associated with the Pauli exclusion principle, and it is at the heart of understanding chemistry, nevertheless the historically formulated principle only emphasized the anti-symmetry of wavefunctions, without giving any consideration to particle pairing and entanglement. In this chapter, the novelty of our approach is in recognizing the key role of entanglement in the formulation of the principle.

[a] The solution of the Dirac equation for the hydrogen atom results in four quantum numbers usually denoted by n, l, m_l, s where n describes the principal energy levels, l the azimuthal quantum number associated with total angular momentum, m_l the magnetic quantum number, and s the spin.

Pauli was keen to derive the exclusion principle named after him, so that it would not be a new axiom. His derivation involved a new principle, which has become known as "micro-causality." In the words of Pauli, this principle states that all physical quantities at finite distances exterior to the light cone are commutable [1]. What Pauli failed to note was that this principle as formulated does not apply to entangled particle states. Entangled particles violate micro-causality by definition and for this reason, within the context of the Einstein–Podolski–Rosen (EPR) paradox, Einstein referred to spooky action at a distance [5]. Regarding the current status of the micro-causality principle, it is still debated whether micro-causality is an axiom or a consequence of relativity, and a recent systematic review of pro and con arguments suggests that micro-causality is still in axiom status [2]. Inspired by Pauli's approach, in most works of QFT a distinction is made between a "spinless" Klein–Gordon field and "half-spin" Dirac field, which essentially means distinguishing between a set of commutator relations versus a set of anti-commutator relations. QFT theorists apply commutator relations to a Klein–Gordon field, and apply anti-commutator relations to a Dirac field.

By this point in the book, we can identify a large number of problems with Pauli's approach: (i) invoking "micro-causality" appears not to be a deductive but an axiomatic approach, (ii) Pauli's "micro-causality" principle is experimentally contradicted by entangled particle states, (iii) Pauli bases his reasoning on assigning a special role to $\hbar/2$ electron-internal angular momentum while the preceding chapters prove the electron-internal angular momentum to be $\hbar$, and (iv) we demonstrated in several preceding chapters that making a distinction between Klein–Gordon and Dirac equation solutions is mathematically untenable.

This situation implies that the mystery of tangible matter is still unsolved: what is the physics behind the Pauli exclusion principle? What process is responsible for generating the chemistry of tangible matter? In this chapter, we investigate this mystery. We will use the word "orbital" to refer to any electron state that is a solution of the Dirac equation; it may be atomic or molecular orbital, or even delocalized electron state in metals.

9.2. Isotropic Spin Correlation

In the following, we refer to electrons' Zitterbewegung plane orientation as their "spin orientation." The concept of isotropically spin-correlated (ISC) states is now introduced. Intuitively, n particles are said to be isotropically spin-correlated, if a measurement made in an *arbitrary* direction θ on *one* of the particles allows us to predict with certainty the spin value of each of the other $n-1$ particles for the same direction θ. Again, intuitively we could say that ISC particles are spin polarized in all directions at once.

Definition 34. Let $H_1, \ldots, H_n$ represent n two-dimensional inner product spaces. n particles are said to be in an ISC state $|\psi\rangle \in H_1 \otimes \cdots \otimes H_n$ if we have rotational invariance for any θ value:

$$R(\theta)\,|\psi\rangle = |\psi(\theta)\rangle \tag{9.2.1}$$

where R is the rotation matrix.

As we shall be calculating spin correlation, we are primarily concerned with R acting on singlet states. For this reason, it is sufficient for what follows to take

$$R = \begin{pmatrix} \cos k\theta & \sin k\theta \\ -\sin k\theta & \cos k\theta \end{pmatrix}$$

While the $k = \frac{1}{2}$ value was historically thought to be needed for the Pauli exclusion principle, it is irrelevant for rotational invariance in that any value of k will work. Moreover, in the case of light intensity $k = 1$ is associated with the angle of polarization and it is also associated with the angle of rotation of a Stern–Gerlach apparatus.

It is important to note that although the Pauli spin operators $\{S_x, S_y, S_z\}$ are related to rotations, we are not identifying R with these spin operators. The $\{S_i\}$ are related to an intrinsic property of an electron while the R is related to the orientation of the Stern–Garlach apparatus in the laboratory frame of reference. For example, $\{RS_xR^\dagger, RS_yR^\dagger, RS_zR^\dagger\}$ is a different representation of the spin matrices obtained by rotating the laboratory frame (note that $R^\dagger = R^{-1}$). If we consider the spin eigenstate $\binom{1}{0}$ of S_z during a measurement along the z direction, then a rotation of $\frac{\pi}{2}$ in the $x-z$ plane of the laboratory will correspond to a rotation given by $\frac{\theta}{2} = \frac{\pi}{4}$ for the spinor by taking $k = \frac{1}{2}$ in our choice of R above. It follows that for the spinor

$$S_z \begin{pmatrix} 1 \\ 0 \end{pmatrix} = \frac{1}{2} \begin{pmatrix} 1 \\ 0 \end{pmatrix} \tag{9.2.2}$$

$$\Longrightarrow \quad R^\dagger S_z R R^\dagger \begin{pmatrix} 1 \\ 0 \end{pmatrix} = \frac{1}{2} R^\dagger \begin{pmatrix} 1 \\ 0 \end{pmatrix} \tag{9.2.3}$$

$$\Longrightarrow \quad S_x \begin{pmatrix} \frac{1}{\sqrt{2}} \\ \frac{1}{\sqrt{2}} \end{pmatrix} = \frac{1}{2} \begin{pmatrix} \frac{1}{\sqrt{2}} \\ \frac{1}{\sqrt{2}} \end{pmatrix} \tag{9.2.4}$$

Similarly, if we include the azimuthal angle as we do in the matrix T below, then we can also check that by taking $\phi = \frac{\pi}{4}$ that $TS_xT^\dagger = S_y$. In general, regarding spin rotations in space, the combined product RT will allow us to perform such rotations. For what follows, we take $k = 1$ unless stated otherwise but would also encourage the interested reader to repeat the subsequent arguments using an arbitrary k and verify that the results still hold.

9.2.1. *Rotational invariance in two dimensions*

Firstly, we investigate rotational invariance in two dimensions, where the spin state is rotated via a single rotation matrix R.

The simplest example of a rotationally invariant spin state is a singlet state. Let $\mathbf{a}^T = (a_1, a_2)$ and $\mathbf{b}^T = (b_1, b_2)$ denote any two independent vectors defined with respect to an arbitrary basis such that the area spanned by them is 1. Recall that by the definition of Section 0.5, a singlet state means that

$$\begin{aligned}\mathbf{a} \wedge \mathbf{b} &= \frac{1}{\sqrt{2}}\left[\begin{pmatrix}1\\0\end{pmatrix}\begin{pmatrix}0\\1\end{pmatrix} - \begin{pmatrix}0\\1\end{pmatrix}\begin{pmatrix}1\\0\end{pmatrix}\right]\\ &\equiv \frac{1}{\sqrt{2}}(|+\rangle\,|-\rangle - |-\rangle\,|+\rangle)\end{aligned} \tag{9.2.5}$$

As was shown in Section 0.8.5, the above expression is isotropic and therefore independent of the initial **a** and **b**. For this reason, we will denote this singlet by $|\psi_1\rangle$ and write $|\psi_1\rangle = \mathbf{a} \wedge \mathbf{b}$ to describe two spin-correlation particles that are anti-parallel in all directions at once. Experimental spin measurements reveal that such a state is realized when two electrons share the same atomic or molecular orbital.

Similarly, $|\psi_2\rangle = \frac{1}{\sqrt{2}}(|+\rangle\,|+\rangle + |-\rangle\,|-\rangle)$ is rotationally invariant in two dimensions and corresponds to two spin-correlated particles within a plane that are parallel in all directions at once:

$$R\,|\psi_2\rangle = \frac{1}{\sqrt{2}}\left(R\begin{pmatrix}1\\0\end{pmatrix}R\begin{pmatrix}1\\0\end{pmatrix} + R\begin{pmatrix}0\\1\end{pmatrix}R\begin{pmatrix}0\\1\end{pmatrix}\right) = |\psi_2\rangle$$

We can consider the states $(|+\rangle\,|+\rangle + |-\rangle\,|-\rangle)$ and $(|+\rangle\,|-\rangle - |-\rangle\,|+\rangle)$ as parallel and anti-parallel conjugate states, respectively,[b] and note that the

[b] We could also define $(|+\rangle\,|+\rangle - |-\rangle\,|-\rangle)$ and $(|+\rangle\,|-\rangle + |-\rangle\,|+\rangle)$ as another pair of conjugate states. In that case, the invariance is not under the operation $R \times R$ as above, but under the operation $R \times R^T$.

sum of these two irreducible states is an element of a Clifford algebra over a tensor product space defined by

$$|\psi\rangle = |\psi_2\rangle + |\psi_1\rangle = \frac{1}{\sqrt{2}}\left(|+\rangle\,|+\rangle + |-\rangle\,|-\rangle\right) + \frac{1}{\sqrt{2}}\left(|+\rangle\,|-\rangle - |-\rangle\,|+\rangle\right)$$

Furthermore, it should be noted that the anti-parallel and parallel singlet can be related as follows:

$$\left(|+\rangle\,|-\rangle - |-\rangle\,|+\rangle\right) = \left(|+\rangle\, C\,|+\rangle + |-\rangle\, C\,|-\rangle\right) \tag{9.2.6}$$

where

$$C = \begin{pmatrix} 0 & 1 \\ -1 & 0 \end{pmatrix} \quad \text{and} \quad C\,|\pm\rangle \text{ are conjugate spinors of } |\mp\rangle$$

In the case of parallel ISC spin states, the total spin is non-zero and its existence would need to be experimentally revealed by Zeeman-split measurements.

9.2.2. *Rotational invariance in three dimensions*

We now consider rotational invariance in the actual three-dimensional space, where an arbitrary rotation is a composition of two orthogonal rotations. We may write a general rotation matrix as

$$U = RT = \begin{pmatrix} \exp(-i\phi)\cos\theta & \exp(i\phi)\sin\theta \\ -\exp(-i\phi)\sin\theta & \exp(i\phi)\cos\theta \end{pmatrix}$$

$$T = \begin{pmatrix} \exp(-i\phi) & 0 \\ 0 & \exp(i\phi) \end{pmatrix}$$

where ϕ is the azimuthal angle.

As we saw in Section 0.5, $(|+\rangle\,|-\rangle - |-\rangle\,|+\rangle)$ is invariant not only for the rotation R but for any element of $SL(2, \mathbb{C})$. Therefore, $|\psi_1\rangle = U(|\psi_1\rangle)$. In contrast, $|\psi_2\rangle \neq T(|\psi_2\rangle)$.

While $|\psi_1\rangle$ is rotationally invariant in all three-dimensional spaces, $|\psi_2\rangle$ is rotationally invariant only in a two-dimensional plane. To understand intuitively why this is so, consider what happens when a pair of rotating arrows moves out of an arbitrary chosen plane on which we are looking down. For a single arrow, the invariance will unravel as the tip of the spinning arrow moves toward us or away from us depending on whether it is spin up or spin down, resulting in an extra $\exp(i\phi)$ and $\exp(-i\phi)$ associated with the $|+\rangle$ and $|-\rangle$ terms, respectively. For two arrows, we apply the operation of (U, U) to both. In the case of the $|\psi_1\rangle$, these additional terms cancel and the invariance remains, while in the case of $|\psi_2\rangle$ the additional terms reinforce each other. **This leads us to conclude that the $|\psi_1\rangle$ singlet state is the only ISC state in spaces of dim $n \geq 3$.** A more rigorous treatment of this can be found in [4].

9.2.3. *The physical origin of Pauli exclusion*

Returning to the topic of the Pauli exclusion, we shall argue that the origin of the principle comes from the existence of anti-parallel spin correlations among particles. In the case of atoms or molecules, spin-correlated electrons occupy atomic or molecular orbitals. It follows from the above discussion that the Pauli principle as usually formulated appears to be only a special case of a more general principle associated with ISC states. For example, the positron and an electron in a positronium state (cf Chapter 7) share the same atomic-like orbital and their spins may be correlated either as parallel or anti-parallel, corresponding to para-positronium and ortho-positronium variants. This positronium example demonstrates that two particles' spin correlation is generally not restricted to only the anti-parallel case, even when they share the same orbital. It is another indication that the Pauli principle can be generalized.

With this in mind, let us consider whether the Pauli exclusion may originate from some kind of symmetry principle. On this topic, most quantum theorists engage in a tortured discussion about the supposed "anti-symmetry" of exchanging two electron wavefunctions. If we consider a system comprising two hydrogen atoms, it appears impossible to argue that there would be anything antisymmetric about the exchange of these two electrons. Indeed, given that each electron is in the ground state of its autonomous hydrogen atom there is no reason why each electron would not share the same four quantum numbers, given that they are in the same 1s orbital but of two different atoms. It is only when the two electrons belong to the same atom like helium or share a common ion like hydrogen that we would expect an antisymmetric wavefunction for the two electrons. In effect, this means that the underlying cause of the antisymmetry is related to the anti-parallel electron spin correlation, as all other quantum mechanical parameters in the orbital would be the same. Moreover, as the above-mentioned positronium example indicates, in some circumstances spin orientations can be parallel. Therefore, the requirement as a universal principle that a two-fermion wave function has to be anti-symmetric has to be seriously questioned. As we shall show in what follows, it is valid only under some very strict conditions.

Finally, this leads us to consider whether the Pauli exclusion may originate from the rotational symmetry of spin correlations. When two electrons share the same orbital, it is a well-known experimental observation that we may perform the total spin measurement from any angle and will always observe the same result. Also, for the positronium case, whether we observe para-positronium or ortho-positronium variant does not depend on the observation angle: these two variants are two distinct states and have very different lifetimes. In other words, ISC particles are pure quantum mechanical states where measurements from any observation angle give the same results. The discussion of this section demonstrates that ISC states are composed of parallel or anti-parallel spin correlations. We investigate, in the next section, what this isotropy implies for the allowed number of correlated particles and also derive the Pauli exclusion principle as a consequence of ISC coupling.

9.3. Isotropic Coupling Principle

Essentially, to show that ISC states exist only for $n = 2$, it is sufficient to prove that it is impossible to have three such particles. The impossibility of three ISC particles excludes the possibility of $n \geq 3$ ISC particles. It follows that $n = 2$ is the only allowed value.

Suppose that an ISC state exists for $n = 3$. We demonstrate in the following paragraphs that this assumption leads to a contradiction.

In the interest of clarity, assume without loss of generality that the three ISC particles are such that they are detected to be in $(+, +, +)$ correlation for an arbitrary measurement direction. Later we will generalize the proof to any other correlation type. Define the x axis along this arbitrary direction, and define the z axis in any orthogonal direction to x. We will perform further spin measurements in the $x - z$ plane. In Section 0.8.3, we mathematically defined spin measurements along various axes, and showed that measurements in orthogonal directions are statistically independent. Although we know a given particle spin to be $|+\rangle$ along the x axis, a subsequent spin measurement along the z axis of the apparatus gives $\frac{1}{2}$ probability of measuring $|-\rangle$ state. In general, a spin state in direction θ given that it is in the state $|+\rangle$ with respect to the x axis can be constructed from the rotation R (taking $k = 1$) and is given by $R\,|+\rangle = \cos\theta\,|+\rangle - \sin\theta\,|-\rangle$. Therefore, in direction θ the probability of measuring $|+\rangle$ state is $\cos^2\theta$ and of measuring $|-\rangle$ is $\sin^2\theta$. Taking the (x, θ) direction with respect to two spin-correlated particles, the joint probabilities are $P(+, +) = \frac{1}{2}\cos^2\theta$ and $P(+, -) = \frac{1}{2}\sin^2\theta$. Similarly, for the ket $|-\rangle$, $R|-\rangle = \sin\theta\,|+\rangle + \cos\theta\,|-\rangle$ and the joint probabilities are $P(-,-) = \frac{1}{2}\cos^2\theta$ and $P(-,+) = \frac{1}{2}\sin^2\theta$. In principle, if three ISC particles exist, a sequence of spin-correlated measurements in the directions θ_1, θ_2, θ_3 can be performed on the three entangled particles. Let $(s_1(\theta_1), s_2(\theta_2), s_3(\theta_3))$ represent each particle's observed spin values in the three different directions. Recall that the above stated spin correlation implies that if any particle is measured to be in the $s_i(\theta_i) = |+\rangle$ spin state, the probability of measuring another particle in the $s_j(\theta_j) = |-\rangle$ spin state becomes $\frac{1}{2}\sin^2(\theta_j - \theta_i)$.

Given that $s_n(\theta_n) = |\pm\rangle$ for each n, there exists only two possible values for each measurement, which we associate with "spin-up" and "spin-down," respectively. Hence, for three measurements there are a total of eight possibilities. In particular, following an argument of Wigner [11],

$$\{(+,+,-),(+,-,-)\} \subset \{(+,+,-),(+,-,-),(-,+,-),(+,-,+)\} \tag{9.3.1}$$

implies the following probability relationship:

$$P\{(+,+,-),(+,-,-)\} \leq P\{(+,+,-),(+,-,-),(-,+,-),(+,-,+)\} \tag{9.3.2}$$

Consider the meaning of various subsets in the above inequality:

- The $\{(+,+,-),(+,-,-)\}$ subset is interpreted as follows: we measured the spin of particle 1 to be in $|+\rangle$ state and particle 3 to be in $|-\rangle$ state. The corresponding probability is $\frac{1}{2}\sin^2(\theta_3 - \theta_1)$.
- The $\{(+,+,-),(-,+,-)\}$ subset is interpreted as follows: we measured the spin of particle 2 to be in $|+\rangle$ state and particle 3 to be in $|-\rangle$ state. The corresponding probability is $\frac{1}{2}\sin^2(\theta_3 - \theta_2)$.
- The $\{(+,-,-),(+,-,+)\}$ subset is interpreted as follows: we measured the spin of particle 1 to be in $|+\rangle$ state and particle 2 to be in $|-\rangle$ state. The corresponding probability is $\frac{1}{2}\sin^2(\theta_2 - \theta_1)$.

Substituting the above terms into (9.3.2), we arrive at

$$\frac{1}{2}\sin^2(\theta_3 - \theta_1) \leq \frac{1}{2}\sin^2(\theta_3 - \theta_2) + \frac{1}{2}\sin^2(\theta_2 - \theta_1) \tag{9.3.3}$$

which is Wigner's interpretation of Bell's inequality. Taking $\theta_3 - \theta_2 = \theta_2 - \theta_1 = \frac{\pi}{6}$ and $\theta_3 - \theta_1 = \frac{\pi}{3}$ gives $\frac{1}{2} \geq \frac{3}{4}$, which is a contradiction. Therefore, three particles cannot all be in the same spin state with probability 1, or in other words, isotropically spin-correlated particles must occur in pairs.

The above results can be cast into the form of a theorem (already proven above) which will be referred to as the "coupling principle":

Theorem 35 (The isotropic coupling principle). *Isotropically spin-correlated particles must occur in PAIRS.*

Remark. The proof of the above theorem was worked out for $(+,+,+)$ or $(-,-,-)$ type spin correlation. To generalize the proof, suppose that the ISC particles are measured to be $(+,-,+)$ along an arbitrary measurement direction. Then the spin outcomes in the three different directions θ_1, θ_2, θ_3 can be written as follows:

$$\{(+,-,-),(+,+,-)\} \subset \{(+,-,-),(+,+,-),(-,-,-),(+,+,+)\}$$

Essentially, this means that we flipped the $|+\rangle$ to $|-\rangle$ to represent the state of particle 2. Applying the same probability argument as before, but noting that $P\{(+,-,-),(-,-,-)\} = \frac{1}{2}\cos^2(\theta_3-\theta_2)$, the inequality becomes

$$\frac{1}{2}\sin^2(\theta_3-\theta_1) \leq \frac{1}{2}\cos^2(\theta_3-\theta_2) + \frac{1}{2}\cos^2(\theta_2-\theta_1) \tag{9.3.4}$$

Then taking $\theta_3-\theta_2 = \theta_2-\theta_1 = \frac{\pi}{2}-\frac{\pi}{6}$ and $\theta_3-\theta_1 = \pi-\frac{\pi}{3}$ gives as before $\frac{1}{2} \geq \frac{3}{4}$, which is a contradiction.

As previously noted, from the perspective of rotational invariance the above proof can be worked out for any value of k, which means mathematically there is nothing special about $\frac{\hbar}{2}$ versus $\hbar$ angular momentum values.

9.4. From Isotropic Coupling to Pauli Exclusion

In the previous section, we proved that isotropically spin-correlated particles must occur in pairs. But we don't know yet why spin correlation occurs in the first place. To explore the origin of spin correlation, we recall the Dirac equation's factorization from Chapter 7, which allowed us to calculate the interaction between two particles:

$$\hbar\boldsymbol{\partial}\psi = \boldsymbol{m}_{\|}\psi + \kappa e A_{\text{applied}} \tag{9.4.1}$$

$$\hbar\boldsymbol{\partial}\boldsymbol{a}_{\square} = -\boldsymbol{m}_{\perp}\left(\boldsymbol{a}_{\square} + (1-\kappa)\,A_{\text{applied}}\right) \tag{9.4.2}$$

where A_{applied} is the electromagnetic vector potential generated by the interacting mass M particle, and the κ parameter value depends on the interaction type:

(1) $\kappa = \frac{M}{m+M}$ for an orbital state,
(2) $\kappa = 0$ for a composite particle state.

The results of Section 9.3 apply only to the $\kappa = \frac{M}{m+M}$ case, when we can make distinct measurements on the spin state of individual particles. In the $\kappa = 0$ composite particle case, we can only measure the spin of the composite, not of its components. An example of such a composite particle is the ultra-dense hydrogen structure, which was discussed in Section 4.5. Later in the book we will also show that a deuteron is a spin-correlated composite of three particles: two protons and one nuclear electron.

Taking the $\kappa = \frac{M}{m+M}$ case, let's consider first a hydrogen atom in its ground state. The electron energy level in the 1s hydrogen ground state is known to split into two closely spaced substates, whose energy depends on whether the electron–proton spin correlation is parallel or anti-parallel. Quantum mechanics textbooks refer to this ground state split as the "hyperfine structure" of hydrogen. If the nucleus has a spin, then a spin correlation between the electron and nucleus is always observed in hydrogen-like atoms. Given that the 1s wavefunction is spherically symmetric and that the spin orientation is along the electron's momentum vector, it may seem counterintuitive that the electron spin could interact with the nuclear spin at the center of the sphere. However, we recall from Section 6.9 that the nuclear spin is not at the center of the sphere because both the electron and nucleus are spinning around their center-of-mass. Although the magnetic moments of an electron and a proton are very different, such difference is irrelevant to the equations of Section 9.2, which yield rotational invariance.

Now suppose that the hydrogen captures also a second electron, and turns into a hydride ion. Both electrons are in the same 1s orbital now, and due to their symmetry, their states are described by the same solution of Equations (9.4.1)–(9.4.2). The spin correlation must originate from Equation (9.4.2) since the ψ term in Equation (9.4.1) is spinless. Since the two spin-interacting electrons have the same mass, Equation (9.4.2) can be now written as follows:

$$\begin{aligned} \hbar\boldsymbol{\partial}\boldsymbol{a}_{1_\square} &= -\boldsymbol{m}_\perp\left(\boldsymbol{a}_{1_\square} + \frac{1}{2}\boldsymbol{a}_{2_\square}\right) \\ \hbar\boldsymbol{\partial}\boldsymbol{a}_{2_\square} &= -\boldsymbol{m}_\perp\left(\boldsymbol{a}_{2_\square} + \frac{1}{2}\boldsymbol{a}_{1_\square}\right) \end{aligned} \tag{9.4.3}$$

Table 9.1. An overview of spin correlations between electrons and the nucleus.

	$p^+ + e^-$ (Hydrogen)	$p^+ + 2e^-$ (Hydride anion)	$^3He^{2+} + e^-$ (Helium cation)	$^3He^{2+} + 2e^-$ (Helium)
Hyperfine split of $1s$ state	Yes	No	Yes	No
Spin-correlated electrons	—	Yes	—	yes

Since electrons' magnetic moment is much larger than the nuclear magnetic moment, magnetic interaction is now mainly between the two electrons.

Once the electron–electron spin correlation is established, the theory of Section 9.3 implies that the nucleus can no longer be spin correlated with the electrons. As summarized in Table 9.1, this exactly matches the experimentally observed pattern. Furthermore, the theory of Section 9.3 implies that the spin correlation between two electrons is always anti-parallel, which also matches the experimentally observed pattern.

In summary, the Dirac equation requires that electrons in the same orbital become spin-correlated, which is another word for spin entanglement. According to the isotropic coupling principle, a third electron cannot become isotropically spin entangled with the other two electrons, and this produces the Pauli exclusion principle. Moreover, as the orbital-sharing electron pair forms a coupled solution of Equations (9.4.1)–(9.4.2), perturbations of this orbit jointly effect both electrons, and one may expect to observe two-electron transitions.

9.5. From Pauli Exclusion to Fermi–Dirac Statistics

To derive Fermi–Dirac statistics, we start by distinguishing the spin variable of an electron state from its other quantum mechanical variables:

Definition 36. Two particles whose states are given by $|\psi(q_1, s_1)\rangle$ and $|\psi(q_2, s_2)\rangle$, respectively, are said to be in the same q-orbital when $q_1 = q_2$.

Lemma 37. *Let*

$$|\psi(\lambda_1, \lambda_2)\rangle = c_1 |\psi_1(\lambda_1)\rangle \otimes |\psi_2(\lambda_2)\rangle + c_2 |\psi_1(\lambda_2)\rangle \otimes |\psi_2(\lambda_1)\rangle \quad (9.5.1)$$

represent an indistinguishable two-particle system. If, for this system, ISC states for two indistinguishable particles occur in the same q-orbital, then the system of particles can be represented by the Fermi–Dirac statistics.

Proof. The general form of the non-interacting and indistinguishable two-particle state is given by

$$|\psi(\lambda_1,\lambda_2)\rangle = c_1\,|\psi_1(\lambda_1)\rangle\otimes|\psi_2(\lambda_2)\rangle + c_2\,|\psi_1(\lambda_2)\rangle\otimes|\psi_2(\lambda_1)\rangle \tag{9.5.2}$$

$$= c_1\,|\psi_1(q_1)\rangle\, s_1\otimes|\psi_2(q_2)\rangle\, s_2 + c_2\,|\psi_1(q_2)\rangle\, s_2\otimes|\psi_2(q_1)\rangle\, s_1 \tag{9.5.3}$$

where c_1, c_2 are constants for all λ_1 and λ_2. Let $q_1 = q_2$, then the particles are in the same q-orbital. By invoking the ISC condition, it follows from the results of Sections 9.2–9.3 that $c_1 = -c_2$. Therefore,

$$|\psi(\lambda_1,\lambda_2)\rangle = \frac{1}{\sqrt{2}}[|\psi_1(\lambda_1)\rangle\otimes|\psi_2(\lambda_2)\rangle - |\psi_1(\lambda_2)\rangle\otimes|\psi_2(\lambda_1)\rangle] \tag{9.5.4}$$

by normalizing the wavefunction. The result follows. □

As the next step, we generalize from two-particle systems to n-particle systems.

Theorem 38. *A sufficient condition for a state, representing n-cyclically permutable particles, to exhibit Fermi–Dirac statistics is that it contain spin-coupled q-orbitals.*

Remark. A system of n-cyclically permutable particles will be referred to as n indistinguishable particles.

Proof. It is sufficient to work with three particles because the following argument can be extended by induction to an n-particle system. Consider a system of three indistinguishable particles, containing spin-coupled particles. Using the above notation and applying Equation (9.5.4) to the coupled particles in the second line below gives:

$$\begin{aligned}
|\psi(\lambda_1,\lambda_2,\lambda_3)\rangle &= \frac{1}{\sqrt{3}}\{|\psi(\lambda_1)\rangle\otimes|\psi(\lambda_2,\lambda_3)\rangle + |\psi(\lambda_2)\rangle\otimes|\psi(\lambda_3,\lambda_1)\rangle \\
&\quad + |\psi(\lambda_3)\rangle\otimes|\psi(\lambda_1,\lambda_2)\rangle\} \\
&= \frac{1}{\sqrt{3!}}\{|\psi(\lambda_1)\rangle\otimes[|\psi(\lambda_2)\rangle\otimes|\psi(\lambda_3)\rangle - |\psi(\lambda_3)\rangle\otimes|\psi(\lambda_2)\rangle] \\
&\quad + |\psi(\lambda_2)\rangle\otimes[|\psi(\lambda_3)\rangle\otimes|\psi(\lambda_1)\rangle - |\psi(\lambda_1)\rangle\otimes|\psi(\lambda_3)\rangle] \\
&\quad + |\psi(\lambda_3)\rangle\otimes[|\psi(\lambda_1)\rangle\otimes|\psi(\lambda_2)\rangle - |\psi(\lambda_2)\rangle\otimes|\psi(\lambda_1)\rangle]\} \\
&= \sqrt{3!}\,|\psi_1(\lambda_1)\rangle\wedge|\psi_2(\lambda_2)\rangle\wedge|\psi_3(\lambda_3)\rangle
\end{aligned}$$

where $\wedge$ represents the wedge product. Thus, the wavefunction for the three indistinguishable particles obeys Fermi–Dirac statistics. □

The n-particle case follows by induction. This theorem shows that the Pauli exclusion principle is associated with the Slater determinant for an n-particle system. More details and a more general form of the theorem can be found in [3, 4].

9.6. Experimental Proofs of Isotropic Electron Entanglement

Recent experiments support the above-stated hypothesis that electrons sharing the same quantum mechanical orbital are in isotropically spin-entangled composite state. In contrast, there is nothing in Pauli's theory which implies that electrons in the same orbital would be entangled and pairwise transitioning. We list the most relevant experimental results for various electron orbital types:

- **Atomic orbitals:** Experimental evidence is produced by the study of atomic orbitals which are occupied by two electrons. Although the existence of doubly excited resonant states has been known since the 1940s, the primary role of electron correlations in forming collective states of the electron pair became suddenly evident with the key experiment of Madden and Codling [10] in 1963. This revealed dramatic deviations from the expectations of Hartree–Fock-like models. Using synchrotron radiation and recording the absorption spectrum of helium, they could observe a series of ${}^1p^0$ states converging to the $N = 2$ ionization threshold. ${}^1p^0$ states are strongly favored due to dipole selection rules when probing the ${}^1s^0$ helium ground state by photo-excitation. Within the independent-particle picture, three series were expected. Instead, Madden and Codling observed one dominant, intense series. A second very faint series could only be guessed and the third series was not detected. The difference in the intensities was a clear indication that the electron–electron interaction leads to previously unknown selection rules for radiative excitation. The position of the single resonance indicates that the two electrons are initially in an s-orbital spin-singlet state, and upon excitation they both transition into a p-orbital state.
- **Molecular orbitals:** The authors of [6] studied the molecular photodissociation of H_2 and D_2 under linearly polarized incident light, employing a wavelength corresponding to 33.66 eV photon energy. They measure the angular correlation function of electromagnetic Lyman-α radiation produced by the resulting atom pair in order to find out whether the atom pair is entangled or not. The authors conclude that an entangled pair of hydrogen atoms is spontaneously produced through the photodissociation of a hydrogen molecule, and this entanglement originates from the

molecular state. Another example concerns the molecular orbitals of O_2: its outer electrons are known to be in either triplet state (in different orbitals) or singlet state (in the same orbital). In the triplet state, the electrons react individually, while in the singlet state they react pairwise. When chemists want to simultaneously bind two oxygens to another molecule, they start by inducing a singlet state of O_2 outer electrons [7].

- **Superconducting Cooper-pair orbital:** The authors of [9] placed two indium–arsenide nanowires in contact with a superconducting material. They found that the two nanowires were capable of extracting superconducting Cooper-pairs, and allowed them to prove the entanglement of the two Cooper-paired electrons. We also note that all superconductor experiments to date found pairwise coupled electrons, comprising the so-called superconducting Cooper-pair state. The presently mainstream theory of superconductors is the BCS theory, named after its authors. There is nothing in BCS theory which would explain why electrons couple exclusively pairwise into a Cooper-pair, and never as a group of three or four electrons.
- **Delocalized conduction-band orbitals in metals:** The authors of [8] present a listing of low-intensity electron emission experiments, where electrons are emitted from various metallic electrodes. If N conduction-band electrons are simultaneously emitted from a metal, a detection equipment will detect the simultaneous arrival x electrons, where x is $1 \leq x \leq N$. The average detection of N electrons corresponds to a perfect capture of all simultaneously emitted electrons. As shown in Table 9.2, various studies measured the simultaneous arrival of $1 \leq x \leq 2$ electrons. These results prove that in most cases more than 1 electrons are emitted simultaneously. If there were, e.g., 3 or 4 electrons simultaneously emitted, it would be very unlikely that no experiment would detect the simultaneous arrival of more than 2 electrons. Therefore, we are lead to conclude that electrons are emitted in pairs from double-occupied delocalized states. A single electron is emitted from a single-occupied delocalized state, which may be found at the top of the occupied band of electron states. The authors of [8] compared thermally induced versus electrostatically induced electron emission, and found the same results. Such pairwise emission of metallic electrons is a consequence of their entangled state. We note that there is nothing in Pauli's theory which implies that metallic band electrons would be pairwise emitted.

Altogether, the above-listed experiments conclusively prove the electron entanglement hypothesis presented in this chapter. This isotropic spin

Table 9.2. An overview of studies on voltage-driven electron emission from metals, reproduced from [8].

Experimenters	T_c (kV)	Z	n
Herrmann[c]	15	W	1
Gazier[d]	40	W(Th)	1.3
Gazier	40	W	1.5
Gazier	40	Pt	1.5
Gazier	40	Cu	1.24
Fursei[c]	12	Si	1.41
Fursei[f]	12	W	1.03
Fursei[g]	17	W	1
Fursei[h]	10	W	1
Fursei[i]	20	Ceramic $YBa_2Cu_3O_{7-\delta}$	1.1
Ebert[j]	30	W	1.7
Ebert	40	W	1.4
Ebert	40	W	1.5
James[k]	9–30	W	1.4
This work	50	W	1.6

Note: The rightmost column shows the number of electrons simultaneously detected in these low-intensity emission experiments.

entanglement principle is very universal, and applies to any type of electron orbital. A bright theorist could have figured it out already during the 1960s. In light of this accumulating experimental counter-evidence to Pauli's theory, and also considering the associated conceptual problems which we listed in the introduction, it is time to conclude that Pauli's exclusion theory is not universal but a special case of a more general principle associated with ISC states. Consequently, all related QFT speculations about "spinless" Klein–Gordon fields versus "half-spin" Dirac fields become invalid as well.

9.7. Antisymmetric Versus Symmetric Spin Entanglement

We established that three-dimensional ISC states are anti-symmetric, with oppositely oriented spins. Since the Pauli exclusion principle applies also to any two electrons sharing the same orbital, their spin correlation must be isotropic in this case.

Considering now the bonding between two particles of equal mass, experiments reveal two types of stable orbitals: the anti-symmetric spin entanglement (also referred to as "para state"), and the symmetric entanglement into parallel spin orientations (also referred to as "ortho state"). The positronium particle and the hydrogen molecule are two simple examples. Both particle types have para and ortho versions. The para- and ortho-positronium differ in the spin orientation between its electron and positron constituents. The para- and ortho-hydrogen differ in the spin orientation between its proton constituents, as illustrated in Figure 9.1.

As discussed in Section 9.2, the $\frac{1}{\sqrt{2}}(|+\rangle|+\rangle + |-\rangle|-\rangle)$-type symmetric spin entanglement is rotationally invariant only within a two-dimensional plane. It appears that the wavefunction of two equally massive bound particles is rotationally invariant only within a plane.

This suggests several open questions about spin entanglement, which require further studies:

- Why is three-dimensionally isotropic spin correlation required only for stable electron orbitals, but not for the stable orbital of the two protons in a hydrogen molecule, or the two constituents of positronium? One may consider the results of Section 8.8: the spacetime curvature around a heavy nucleus is described by the Reissner–Nordström metric, which is spherically symmetric. A light electron's stable wavefunction adheres to this spherical symmetry. In contrast, two equally massive particles create spacetime curvature with two equal dips around the two particles. Intuitively, the Kepler orbit model remains relevant even in the quantum mechanical regime. Beyond this intuitive picture, a precise calculation is needed to determine the condition under which the isotropy requirement is relaxed to rotational invariance only within a plane.

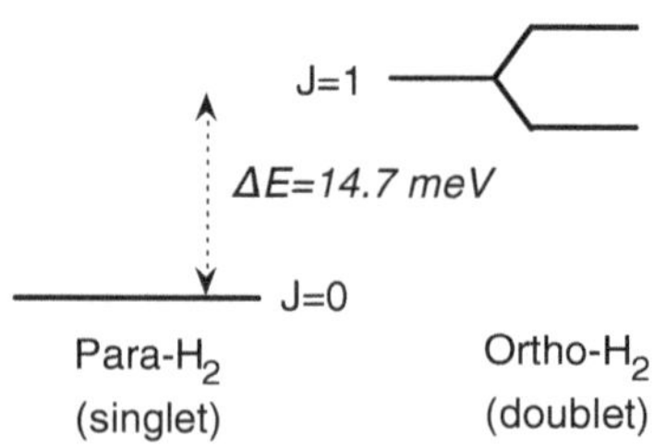

Figure 9.1. A spectroscopic diagram of proton spin correlation in a hydrogen molecule. Para-H_2 is a singlet spin state (no Zeeman effect), while ortho-H_2 is a doublet spin state (anomalous Zeeman effect).

- When a hydrogen molecule forms, its N_{ortho}:N_{para} ratio is 3:1. We note that, with reference to Figure 9.1, there is no triplet state involved in ortho-hydrogen. When a positronium particle forms, its N_{ortho}:N_{para} ratio is again 3:1. There appears to be a universal rule which determines this ratio. How could one calculate this ratio?
- In the lowest energy state, the hydrogen molecule's protons are in the singlet (para) spin state. Its ortho spin state is at 15 meV higher energy [12]. When cooled to near absolute zero, the ortho-to-para transition time is on the order of days, i.e., it is a slow process. When para-hydrogen is heated above 15 meV average energy, the para-to-ortho transition is again a slow process. In contrast, a thermally induced break-up and establishment of its protons' orbital angular momentum entanglement is a fast process [12]. What makes spin entanglement so resilient that it only reacts slowly to thermal excitations?

Resolving the above questions will lead to a deeper understanding of spin entanglement.

9.8. Conclusions

Our main conclusion is that spin correlations between particles sharing the same orbital arise naturally and are related to the Dirac spinors. The rotational invariance of these spin correlations restricts the occupancy of any orbital to $n \leq 2$ electrons, from which the Pauli exclusion principle follows as a corollary. Isotropic entanglement is at the heart of the exclusion principle and not spin values.

References

[1] W. Pauli, The connection between spin and statistics. *Physics Review*, **58** (1940) 716.
[2] J. Wright, Quantum field theory: Motivating the axiom of microcausality, Waterloo, Ontario, Canada, 2012.
[3] P. O'Hara, Rotational invariance and the spin-statistics theorem. *Foundation of Physics*, **33**(9) (2003) 1349.
[4] P. O'Hara, A generalized spin statistics theorem. *Journal of Physics Conference Series*, **845** (2017) 012030.
[5] P. O'Hara, The Einstein-Podolsky-Rosen paradox and SU(2) relativity. *Journal of Physics: Conference Series*, **1239** 012021.
[6] Y. Torizuka *et al.*, Entangled pairs of 2p atoms produced in photodissociation of H_2 and D_2. *Physical Review A*, **99**(6) (2019) 063426.
[7] M. Laing, The three forms of molecular oxygen. *Journal of Chemical Education*, **66**(6) (1989) 453.
[8] M.A. Piestrup *et al.*, Enhanced correlated-charge field emission (1998).

[9] K. Ueda *et al.*, Dominant nonlocal superconducting proximity effect due to electron–electron interaction in a ballistic double nanowire. *Science Advances*, **5**(10) (2019) eaaw2194.
[10] R.P. Madden and K. Codling, *Physical Review Letters*, **10** (1963) 516.
[11] E.P. Wigner, On hidden variables and quantum mechanical probabilities. *American Journal of Physics*, **38**(8) (1970) 1005.
[12] E. Ilisca, Hydrogen conversion in nanocages. *Hydrogen*, **2**(2) (2021) 160–206.

https://doi.org/10.1142/9781800611306_0010

Chapter 10

Electron Dynamics in Metals

András Kovács
BroadBit Energy Technologies, Metallimiehenkuja 8 C
02150 Espoo, Finland
andras.kovacs@broadbit.com

10.1. Introduction

We studied the internal structure of the electron in preceding chapters, and then two-body dynamics between an electron and a nucleus. We identified a new interaction mode of two-body systems, which we refer to as Compton-scale composites. One may naturally expect that many-body systems would exhibit richer complexity, but would be difficult to solve analytically. Nevertheless, our world comprises many-body solid-state systems. One aspect of technological progress is to apply our understanding of fundamental interactions for the gradual exploration of solid-state structures.

Metals are a particularly important class of solid state material. Most chemical elements are metals in their pure state. The milestones of gaining the ability to utilize the electronic and mechanical properties of metals are in fact the defining milestones of technological progress. Metals are defined by the delocalization of their outermost electrons, which then form a standing wave across the entire solid-state body.

This standing wave is quite analogous to an electron state in a square potential well, which is feasible to solve analytically. Besides the potential step at the metal boundary, the electron also experiences a periodic pattern of potential dips at the nuclear sites. The strict periodicity of this pattern again makes the problem tractable. Thus, metals are a class of solid-state materials which stand out both in terms of their technological importance and in terms of the feasibility of analytical methods.

In this chapter, we introduce analytic methods for studying electron dynamics in metals. We start from the basics of solid-state theory, and gradually move toward asking questions which were not asked before.

10.2. The Drude–Sommerfeld Model of Delocalized Electrons

Delocalized electrons' current understanding was principally developed by Sommerfeld in 1927, and became known as the Drude–Sommerfeld model. A delocalized electron wavefunction is modeled as a standing wave, analogous to the electron wavefunction in a square potential well. This wavefunction vanishes at the metal boundary, and it is in a free-particle state within the interior of the metal. Since this model fits reasonably well the main properties of metals, we also start from this simple model.

Consider a cube-shaped metal with dimensions L_x, L_y, L_z. A standing wave's wavenumbers may take on the values illustrated in Figure 10.1, specifically:

$$k_x = l\frac{2\pi}{L_x}, \quad k_y = m\frac{2\pi}{L_y}, \quad k_z = n\frac{2\pi}{L_z}$$

where $l, m, n = 0, \pm 1, \pm 2, \ldots$.

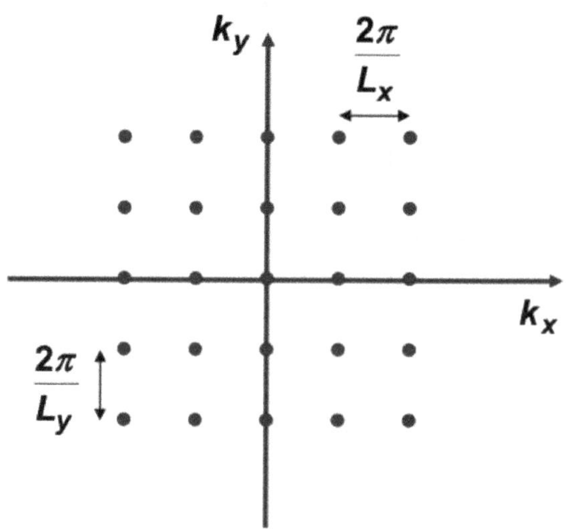

Figure 10.1. A k-space representation of delocalized electron states.

When the electron movement is restricted to a two-dimensional plane, such as in graphite, the allowed wavenumbers are as follows:

$$k_x = l\frac{2\pi}{L_x}, \quad k_y = m\frac{2\pi}{L_y}$$

with $l, m = 0, \pm 1, \pm 2, \ldots$.

Each gridpoint of Figure 10.1 represents an allowed state, and the Pauli exclusion principle implies that at most two electrons may occupy any state. In the k-space of Figure 10.1, there is only one state per every $\frac{(2\pi)^3}{L_x L_y L_z}$ volume element. Thus, the k-space density of allowed states is

$$\varrho_k = \frac{L_x L_y L_z}{(2\pi)^3} \tag{10.2.1}$$

Similarly, the k-space density of two-dimensional states is

$$\sigma_k = \frac{L_x L_y}{(2\pi)^2} \tag{10.2.2}$$

The kinetic energy of any given state is

$$E = \frac{1}{2}mv^2 = \frac{p^2}{2m} = \frac{(\hbar k)^2}{2m} \tag{10.2.3}$$

where $k^2 = k_x^2 + k_y^2 + k_z^2$ for the three-dimensional case and $k^2 = k_x^2 + k_y^2$ for the two-dimensional case.

Suppose there are N delocalized electrons in the metal. According to the energy minimization principle, they will occupy the lowest k states. In the k-space representation, they will occupy a Fermi-sphere, with radius k_F. We account for N delocalized electrons within k_F radius:

$$N_{3d} = 2\varrho_k \frac{4\pi}{3} k_F^3 \tag{10.2.4}$$

The number two appearing in the above equation accounts for two electrons occupying any k state. We denote the total density of electrons by $n_{3d} = \frac{N_{3d}}{V}$. The relationship between the electron density and k_F is

$$n_{3d} = \frac{k_F^3}{3\pi^2} \tag{10.2.5}$$

In the case of a two-dimensional metallic plane, delocalized electrons occupy a Fermi-disc, with radius k_F. The relationship between $n_{2d} = \frac{N_{2d}}{A}$

and k_F becomes

$$n_{2d} = \frac{k_F^2}{2\pi} \tag{10.2.6}$$

The energy of electrons at k_F is called the Fermi energy:

$$E_F = \frac{(\hbar k_F)^2}{2m} \tag{10.2.7}$$

Now we calculate the density of electrons as a function of electron energy. Using Equation (10.2.3), we get $k = \frac{\sqrt{2mE}}{\hbar}$ and the following expression for $g_{3d}(E)$:

$$\begin{aligned} g_{3d}(E) &\equiv \frac{dn_{3d}}{dE} = \frac{dn_{3d}}{dk}\frac{dk}{dE} = \frac{k^2}{\pi^2}\frac{\sqrt{2m}}{\hbar}\frac{1}{2\sqrt{E}} \\ &= \left(\frac{\sqrt{m}}{\hbar}\right)^3 \frac{\sqrt{2}}{\pi^2}\sqrt{E} \end{aligned} \tag{10.2.8}$$

In an ordinary metal, the density of electron states grows with the square root of electron energy. The analogous calculation for $g_{2d}(E)$ yields

$$\begin{aligned} g_{2d}(E) &\equiv \frac{dn_{2d}}{dE} = \frac{dn_{2d}}{dk}\frac{dk}{dE} = \frac{k}{\pi}\frac{\sqrt{2m}}{\hbar}\frac{1}{2\sqrt{E}} \\ &= \frac{m}{\hbar^2}\frac{1}{\pi} \end{aligned} \tag{10.2.9}$$

The above equation means that the density of electron states is constant in materials which comprise two-dimensional metallic planes.

10.3. Thomas–Fermi Screening

Now suppose that one of the lattice ions is replaced by a more positively charged ion. How do delocalized electrons respond to this more positive site? As they are attracted to it, the electron density increases around this more positive site.

In the strict sense of the Drude–Sommerfeld model, single-frequency electron waves uniformly fill the interior of a metal, and thus cannot respond to any local positive charge. We need to employ a careful approximation, which allows to estimate the local electron response without giving up the Drude–Sommerfeld model.

The main idea is illustrated in Figure 10.2. The zero level of Figure 10.2 corresponds to the lowest k value in the interior of the metal, and the horizontal lines above it are the higher k value states. The potential

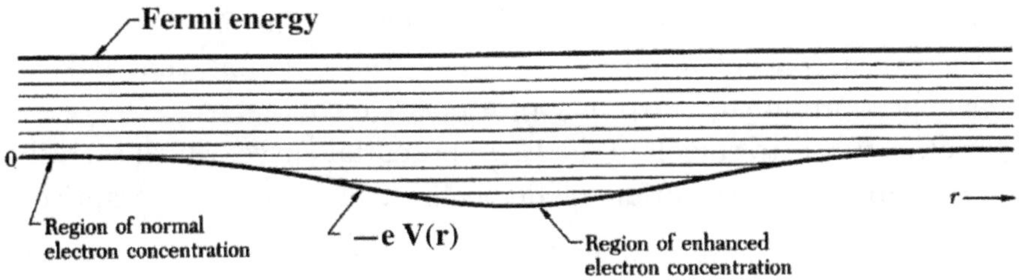

Figure 10.2. The electron screening around a positive charge is modeled as additional standing waves at a lower energy level around that charge.

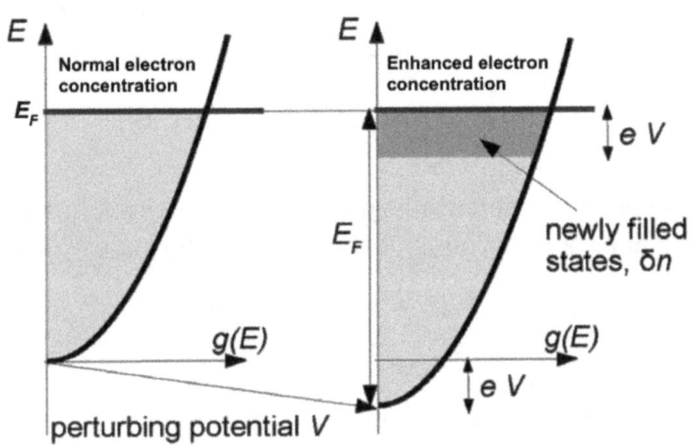

Figure 10.3. The appearance of newly filled electron states near the positive charge.

dip around the positive charge shifts down this lowest energy state; the delocalized electrons are attracted to the more positively charged region. Since the electrons minimize their total energy, the top of the electron sea remains flat at the Fermi energy level.

Although delocalized electrons are single-frequency standing waves only in the region of normal electron concentration, the simplifying assumption of the Thomas–Fermi screening theory is to model them as single-frequency standing waves also in the sloping region of Figure 10.2. Then the net effect is the appearance of additional electron states, as shown in the pink-shaded region of Figure 10.3.

Another way of thinking about the Thomas–Fermi screening theory is to imagine that the potential slope of Figure 10.2 is approximated by a staircase-like sequence of flat segments, which are in thermodynamic balance with each other. Note that the Fermi energy level has higher k_F in the segments which are closer to the positive charge than in the segments of normal electron concentration. This reflects the dynamic that delocalized electrons speed up in the proximity of a positive charge.

Our objective is to calculate the $V(r)$ potential curve, which corresponds to the equilibrium distribution of delocalized electrons. It comprises two parts:

$$V(r) = V_{\text{ext}}(r) + \delta V(r)$$

where $V_{\text{ext}}(r) = \frac{1}{4\pi\varepsilon_0}\frac{Ze}{r}$ is the unscreened Coulomb potential of the positive charge, and the $\delta V(r)$ screening potential is calculated from the increased electron density by using Poisson's equation:

$$\nabla^2 \delta V(r) = \frac{e}{\varepsilon_0}\delta n = \frac{e}{\varepsilon_0} g(E_F)[eV(r)] \tag{10.3.1}$$

The $eV(r)$ energy difference is graphically illustrated in Figure 10.3. We guess that the solution is of the following form:

$$V(r) = \frac{1}{4\pi\varepsilon_0}\frac{Ze}{r}e^{-rk_{TF}} \tag{10.3.2}$$

where k_{TF} is the Thomas–Fermi wavenumber, whose $r_{TF} = k_{TF}^{-1}$ inverse is also referred to as the Thomas–Fermi screening length. The resulting expression for $\delta V(r)$ is then the following:

$$\delta V(r) = \frac{1}{4\pi\varepsilon_0}\frac{Ze}{r}\left(e^{-rk_{TF}} - 1\right) \tag{10.3.3}$$

In spherical coordinates, the left-hand side of Equation (10.3.1) evaluates to:

$$\begin{aligned}
\left[\nabla^2 \delta V(r)\right]_{3d} &= \frac{1}{r^2}\frac{\partial}{\partial r}\left(r^2 \frac{\partial \delta V(r)}{\partial r}\right) \\
&= \frac{Ze}{4\pi\varepsilon_0}\frac{1}{r^2}\frac{\partial}{\partial r}\left(-\left(e^{-rk_{TF}} - 1\right) - rk_{TF}\left(e^{-rk_{TF}}\right)\right) \\
&= \frac{Ze}{4\pi\varepsilon_0}\left[\frac{k_{TF}}{r^2}e^{-rk_{TF}} - \frac{k_{TF}}{r^2}e^{-rk_{TF}} + \frac{k_{TF}^2}{r}e^{-rk_{TF}}\right] \\
&= \frac{k_{TF}^2}{4\pi\varepsilon_0}\frac{Ze}{r}e^{-rk_{TF}}
\end{aligned} \tag{10.3.4}$$

In the three-dimensional case, the right-hand side of Equation (10.3.1) evaluates to:

$$\frac{e^2}{\varepsilon_0} g_{3d}(E_F) \frac{1}{4\pi\varepsilon_0} \frac{Ze}{r} e^{-rk_{TF}} = \left(\frac{e^2}{4\pi\varepsilon_0^2} \left(\frac{\sqrt{m}}{\hbar} \right)^3 \frac{\sqrt{2}}{\pi^2} \sqrt{E_F} \right) \frac{Ze}{r} e^{-rk_{TF}} \tag{10.3.5}$$

The correspondence between the above two equations shows that we guessed the correct form of the solution, and the fulfillment of Equation (10.3.1) yields the following result for k_{TF}:

$$\left[k_{TF} = e\sqrt{\left(\frac{\sqrt{m}}{\hbar} \right)^3 \frac{\sqrt{2}}{\varepsilon_0 \pi^2} \sqrt[4]{E_F}} \right]_{3d} \tag{10.3.6}$$

In common metals, $E_F \approx 10\,\mathrm{eV}$. Since k_{TF} only weakly depends on the precise E_F value, we can evaluate k_{TF} by using $E_F = 10\,\mathrm{eV}$. The resulting Thomas–Fermi screening length is as follows:

$$\left[k_{TF} \approx 3.1 \times 10^9\, m^{-1} \right]_{3d} \tag{10.3.7}$$

$$\left[r_{TF} \approx 0.32\, nm \right]_{3d}$$

We note that this electron screening length is approximately the same as the inter-atomic distance in the lattice.

In the two-dimensional case, the left-hand side of Equation (10.3.1) is difficult to evaluate for a single plane, because the charge distribution is no longer spherically symmetric. In practice, two-dimensional metals comprise stacked planes, with graphite-like structure. For stacked planes, we can estimate that the screening charge distribution is still spherically symmetric, with the strongest screening in the plane where the positively charged ion is, and gradually weaker screening in the neighboring planes. The left-hand side of Equation (10.3.1) then evaluates in the same way as for the three-dimensional case:

$$\left[\nabla^2 \delta V(r) \right]_{2d,\mathrm{stack}} \approx \left[\nabla^2 \delta V(r) \right]_{3d} \tag{10.3.8}$$

We proceed to evaluate the right-hand side of Equation (10.3.1):

$$\frac{e^2}{\varepsilon_0} \left[g_{2d}(E_F) \frac{1}{d_{il}} \right] \frac{1}{4\pi\varepsilon_0} \frac{Ze}{r} e^{-rk_{TF}} = \left(\frac{e^2}{4\pi\varepsilon_0^2} \frac{m}{\hbar^2} \frac{1}{\pi d_{il}} \right) \frac{Ze}{r} e^{-rk_{TF}} \tag{10.3.9}$$

where d_{il} is the inter-layer spacing of the two-dimensional stack. As $g_{2d}(E_F)$ counts the density of states within one plane, we need to use the $g_{2d}(E_F)\frac{1}{d_{il}}$

expression to get the volumetric density of states. In graphite, d_{il} =0.33 nm, and we use this value for estimating k_{TF}.

The above equations lead to the following result for k_{TF}:

$$\left[k_{TF} = \sqrt{\frac{me^2}{\varepsilon_0 \hbar^2} \frac{1}{\pi d_{il}}}\right]_{2d,\text{stack}} \tag{10.3.10}$$

In these three-dimensional stacked materials, the Thomas–Fermi screening length mainly depends on the inter-layer spacing parameter. In graphite, $d_{il} = 0.33\,\text{nm}$, and we get $k_{TF} = 2.4 \times 10^9\,\text{m}^{-1}$ and $r_{TF} = 0.4\,\text{nm}$. That is a very similar screening value as in ordinary metals. The most studied high-temperature superconductor material is $YBa_2Cu_3O_{6.92}$, which also comprises a stack of two-dimensional metallic planes. According to measurements, $YBa_2Cu_3O_{6.92}$ has 1.17 nm lattice parameter along its c-axis, which contains two conducting planes. Therefore, $d_{il} = 0.585\,\text{nm}$ in $YBa_2Cu_3O_{6.92}$, and therefore its Thomas–Fermi screening length is $r_{TF} = 0.53\,\text{nm}$, which is slightly larger than in ordinary metals.

10.4. Orbitals Under Electron Screening Effect

Looking at Figure 10.2, one may have the impression that the region of enhanced electron concentration contains standing waves which are localized around the nucleus. Should we consider such localized standing waves as a bound electron state? The methodology of the previous section allows us to calculate the charge contained in any specific localized k value, and the result shows a fractional charge. Therefore, one must not think of an apparently localized standing wave as a real electron orbital. Its fractional charge is just taking into account the higher probability of finding an electron near a highly charged site. In other words, a delocalized electron wavefunction remains spread through the interior of the metal.

A natural question is to ask what prevents some delocalized electron from forming a localized orbit around a positively charged nucleus. The main hindrance is that any unbound electron sees only a fraction of the nuclear charge, screened by other delocalized electrons. The key point is to distinguish between hindrance versus impossibility.

Suppose an electron does form a localized Bohr orbit around a Z-charged site, at a mean orbit radius r_{scr}, where the scr index denotes that it is a screened orbit. We calculate the positive charge which the localized electron experiences at the r_{scr} distance. In the unscreened case, an electron localized at mean distance r experiences $\frac{1}{4\pi\varepsilon_0}\frac{Ze}{r}$ potential and $\frac{-1}{4\pi\varepsilon_0}\frac{Ze}{r^2}$ electric field. In the actual screened case, the localized electron experiences the following potential:

$$U = \frac{1}{4\pi\varepsilon_0}\frac{Ze}{r_{\text{scr}}}e^{-r_{\text{scr}}/r_{TF}}$$

The experienced electric field is the radial derivative of the screened potential:

$$E = \frac{-1}{4\pi\varepsilon_0}\frac{Ze}{r_{\text{scr}}}e^{-r_{\text{scr}}/r_{TF}}\left(\frac{1}{r_{\text{scr}}} + \frac{1}{r_{TF}}\right)$$

The ratio of screened and unscreened charges corresponds to the ratio of screened and unscreened electric fields:

$$Z_{\text{scr}} = Ze^{-r_{\text{scr}}/r_{TF}}\left(1 + \frac{r_{\text{scr}}}{r_{TF}}\right) \tag{10.4.1}$$

At the same time, the eigenstate solution of the Dirac equation tells us that the mean orbital radius is inversely proportional to the experienced charge:

$$r_{\text{scr}} = \frac{a_0}{Z_{\text{scr}}} \tag{10.4.2}$$

where $a_0 = 52.9\,\text{pm}$ is the Bohr radius. Combining the above two equations, we may calculate r_{scr}:

$$\frac{a_0}{r_{\text{scr}}} = Ze^{-r_{\text{scr}}/r_{TF}}\left(1 + \frac{r_{\text{scr}}}{r_{TF}}\right) \tag{10.4.3}$$

The above equation gives two bounded solutions for r_{scr}, which we observe by numerical computation. The first one is very close to the ordinary Bohr orbit value, and we interpret it as ordinary bound electron orbitals. In this case, the electron screening has only very minor effect. The second solution is a surprisingly large orbital radius: the electron feels only a fraction of the positive ion charge from this screened orbital.

Equation (10.4.3) is a first approximation, which calculates the charge that an electron sees from its mean orbit radius. This calculation assumes that the screened s-orbital has the same radial distribution as an ordinary s-orbital, while in reality the Dirac equation's eigenstate will be somewhat different because the experienced charge has a radial dependence. Nevertheless, the simple Equation (10.4.3) can be used as an order of magnitude estimate, and it illustrates the principles involved.

Let's take the example of $YBa_2Cu_3O_{6.92}$ material, where we estimated $r_{TF} = 0.53\,\text{nm}$. Studies have shown the yttrium sites to be in $Z = 3$ state; these sites are more positively charged than others, and will be screened by the delocalized electrons. The second solution of Equation (10.4.3) yields then $r_{\text{scr}} = 4\,\text{nm}$. Such a mean radius parameter corresponds to a binding energy of just 2.3 meV. Another well-known example is $TlBa_2CaCu_2O_8$, whose structure is shown in Figure 10.5: it has $d_{il} = 0.73\,\text{nm}$ spacing between its metallic planes, and therefore its Thomas–Fermi screening length is $r_{TF} = 0.59\,\text{nm}$. The Tl sites are also in $Z = 3$ state. The second solution of Equation (10.4.3) yields $r_{\text{scr}} = 4.56\,\text{nm}$ and a binding energy of 1.8 meV. Orbits with such low binding energy could not be realized at room temperature.

At the beginning of the 20th century, ordinary electron orbitals were explored via their spectroscopic signatures. It is therefore rational to look for experimental evidence of single-frequency light emission and absorption, which must correspond to a quantum mechanical state transition of weakly bound electron orbitals. Such single-frequency light emission is indeed observed in superconductor research, and it is referred to as the Josephson–Plasma resonance (JPR) effect [1, 2]. We consider that subset of experiments where a sharp single-frequency light emission peak was observed, which must be the signature of a quantum mechanical state transition. Figure 10.4 shows an example of a sharp emission peak, which we are looking for. The name of the JPR effect does not have any significance because up to now there was no satisfactory explanation for the appearance of a sharp THz

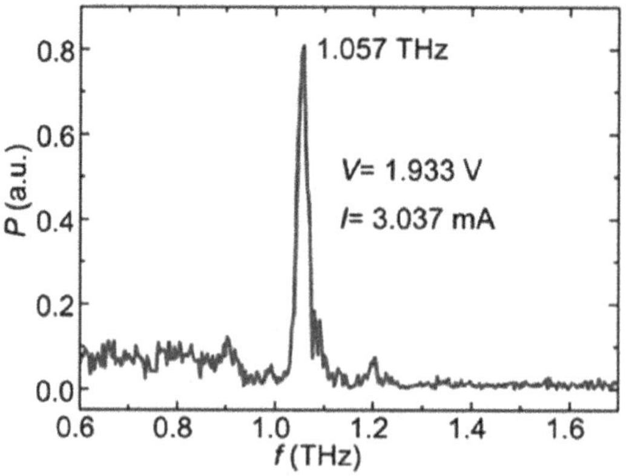

Figure 10.4. The spectrum of light emission from a superconductor, reproduced from [1]. The emission is induced by applying DC voltage bias on a thin superconductor layer. The 1.057 THz single-frequency peak is clearly observable.

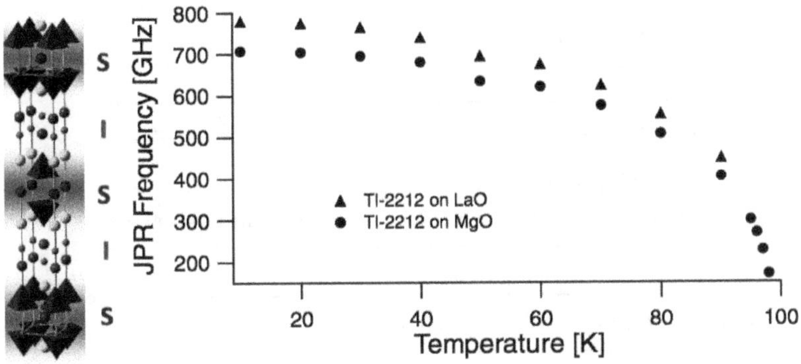

(a) (b)

Figure 10.5. Single-frequency light emission from $TlBa_2CaCu_2O_8$. (a) An illustration of the crystal structure, where the S letters represent superconducting planes and the I letters represent insulating planes. (b) The temperature dependence of the JPR emission frequency, reproduced from [2]. The JPR frequency goes to zero at T_c, which is 103 K for this superconductor.

emission peak in response to applying a step-function voltage. Figure 10.5 shows the frequency of light emission from a thin film of $TlBa_2CaCu_2O_8$ superconductor, observed after its optically induced excitation. Similar light emission is described in [1], but induced by applying voltage bias on a thin superconductor film.

Assuming that the frequency seen in Figure 10.5 is the light emission frequency from the re-establishment of weakly bound electron orbitals, this data give direct information on the involved binding energy. In the absence of thermal noise, the average emission frequency from $TlBa_2CaCu_2O_8$ is 740 GHz, which corresponds to 3 meV energy transition. $La_{1.9}Ba_{0.1}CuO_4$ is another material where the JPR frequency is experimentally studied: Ref. [6] estimates that its low-temperature JPR frequency is 500 GHz, which corresponds to 2 meV energy transition. Based on crystal structure data, Table 10.1 shows the calculated binding energy values for these superconductors. The experimental values are reasonably well estimated by our binding energy calculation. In particular, the calculation is more precise for $La_{1.9}Ba_{0.1}CuO_4$ than for $TlBa_2CaCu_2O_8$. The reason for this difference will be addressed in the next section.

With this interpretation of the JPR emission frequency, Figure 10.5 reveals the effect of thermal noise on screened electron orbitals. As temperature grows, the decreasing emission frequency implies a decreasing binding energy. Correspondingly, the radius of screened orbital grows larger. These data show that the binding energy approaches zero exactly at the

Table 10.1. Calculated and experimental binding energies of electron-screened orbitals. r_{scr} is calculated from Equation (10.4.3).

Superconductor	Z	d_{il}	r_{TF}	r_{scr}
$\mathbf{La_{1.9}}Ba_{0.1}CuO_4$	+3	0.656 nm	0.56 nm	4.3 nm
$\mathbf{Tl}Ba_2CaCu_2O_8$	+3	0.73 nm	0.59 nm	4.56 nm

Superconductor	E_b calculated	ν_{JPR}	E_b experimental
$\mathbf{La_{1.9}}Ba_{0.1}CuO_4$	2.1 meV	500 GHz	2 meV
$\mathbf{Tl}Ba_2CaCu_2O_8$	1.8 meV	740 GHz	3 meV

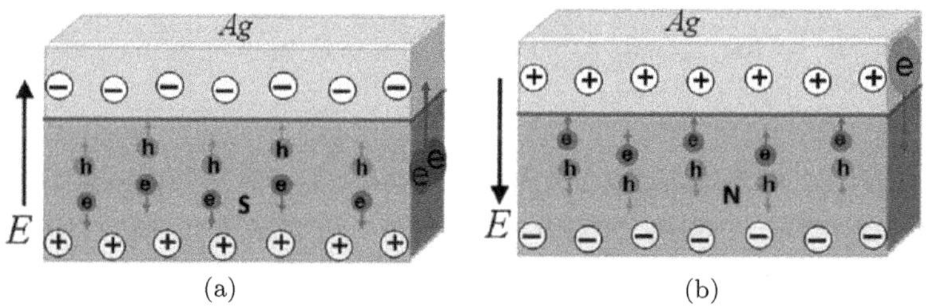

Figure 10.6. Fermi-level depletion upon transition from normal to superconducting state, manifesting as interface polarity change. (a) Ag-MgB_2 interface below T_c, MgB_2 is in superconducting (S) state. (b) Ag-MgB_2 interface above T_c, MgB_2 is in normal (N) state. The polarity is measured by laser-induced production of electron–hole pairs, which create photovoltaic effect.

superconducting critical temperature, and therefore the radius of screened orbital grows to infinity at that transition temperature. The disappearance of screened electron orbitals exactly at the superconducting transition temperature proves that the presence of screened electron orbitals is required for superconductivity. This discovery is a radical departure from the traditional understanding of the superconductivity phenomena.

One may ask what happens to weakly bound electrons' wavefunction above T_c. Weakly bound electrons must transition then into the Dirac–Fermi sea of delocalized electron waves. In other words, a transition from superconducting to normal state increases the Fermi energy level, while a transition from normal to superconducting state decreases the Fermi energy level. A direct signature of this Fermi level change is reported in Refs. [3–5]: these works measured MgB_2, Bi – 2223, and $YBa_2Cu_3O_{6.96}$ materials' Fermi level relative to a silver contact. As shown in Figure 10.6, the Fermi level depletion upon entering superconducting state manifests as a polarity change in the superconductor–silver interface. The interface polarity was detected by laser illumination technique, and the same effect was found for all superconducting materials. These experimental results confirm our hypothesis on the transition from delocalized electron state into weakly bound localized state, taking place near T_c. In contrast, conventional superconductivity theories consider superconducting electrons to remain delocalized waves, i.e., they imply no Fermi level change upon superconducting state transition, which is contradicted by experiments.

10.5. Screening in Weakly Metallic Materials

Let's review qualitatively the range of materials between insulating and metallic types. In insulating materials, the localized orbit of outermost electrons is strongly favored: the energy of delocalized electron states is several eVs higher. In semiconductors, the energy of delocalized electron states is only around 1 eV higher than the energy of bound states: this is the bandgap energy value. At room temperature, thermal fluctuations energize some electrons across this small bandgap, thus creating a small conductivity. The lattice site from which an electron was energized into conducting state is referred to as a "hole." At a given temperature, the average number of holes is constant, but their corresponding lattice sites are variable; this is referred to as "hole mobility." When the bandgap value shrinks to zero, the material is referred to as a semimetal. Going further, metallic materials have negative bandgap, i.e., delocalized orbitals are energetically favored. When there is less than an eV energy difference between a localized valence orbital and the Fermi level of delocalized orbitals, the material is said to be weakly metallic. Chemists also refer to such materials as "metavalent." In this case, thermal fluctuations energize some electrons into a localized valence orbital: the charge of the corresponding lattice site goes from z to $z - 1$ upon a full electron transfer. In practice, the valence state does not need to be an integer, but can be a probabilistic fractional value. The screening effect of such "inverse hole" sites will be the focus of this section. When the energy gap from the Fermi level to a localized valence orbital becomes large, the material is said to be strongly metallic and "inverse hole" sites practically disappear.

At room temperature, the average concentration of hole or inverse hole sites is described by the Boltzmann distribution: $\bar{n}_h \sim e^{-\frac{-\triangle E_{CV}}{k_b T}}$ and $\bar{n}_{ih} \sim e^{-\frac{-\triangle E_{VF}}{k_b T}}$, respectively. Here, $\triangle E_{CV}$ is the energy gap between the conduction and valence bands, and $\triangle E_{VF}$ is the energy gap between the valence band and Fermi level. When temperature approaches absolute zero, these concentrations "freeze out": $\bar{n}_h \to 0$ and $\bar{n}_{ih} \to 0$. For simplicity, we focus here on this low-temperature region.

For weakly metallic materials, the bound valence state may become energetically more favored in the vicinity of a screened positive charge. The physical basis of this effect is illustrated in Figure 10.3: the Fermi energy is locally raised by $eV(r)$, and inverse hole formation is favored when $eV(r) > \triangle E_{VF}$. Suppose that the distance between the Z charged site and nearest inverse hole candidate sites is approximately d_{ih}. Within this d_{ih} distance, there is only Thomas–Fermi screening and the electrostatic potential is given by Equation (10.3.2). At the hole candidate sites inverse holes are formed till the following balance is reached:

$$eV(r) = \triangle E_{VF} \tag{10.5.1}$$

where $\triangle E_{VF}$ is the gap energy at normal electron concentration. In the absence of thermal fluctuations, the above balance is exact. Beyond the d_{ih} distance, the screened charge value is reduced by the inverse hole charges. We thus work with a reduced Z' charge value, and the resulting screened potential is again given by Thomas–Fermi screening:

$$V(r) = \frac{1}{4\pi\varepsilon_0}\frac{Z'e}{r}e^{-rk_{TF}} \tag{10.5.2}$$

At $r = d_{ih}$, the above two equations both hold, and we may calculate Z' as follows:

$$Z' = 4\pi\varepsilon_0\frac{\triangle E_{VF}}{e^2}d_{ih}e^{d_{ih}k_{TF}} \tag{10.5.3}$$

Finally, one may calculate the radius of the screened orbital from Equation (10.4.3), but using Z' instead of the unscreened Z value.

The obtained results demonstrate that there is additional screening effect in weakly metallic materials, and we need to know the $\triangle E_{VF}$ and d_{ih} parameters to quantify its effect.

Let's take the prototypical $YBa_2Cu_3O_{7-z}$ superconductor as an example to illustrate weakly metallic screening. The yttrium sites are the highly charged ones, and inverse holes are on the copper sites, which are 0.33 nm away. One must also add the 0.09 nm ionic radius of Cu^{2+} because the inverse hole screening effect is realized only after passing beyond the outermost

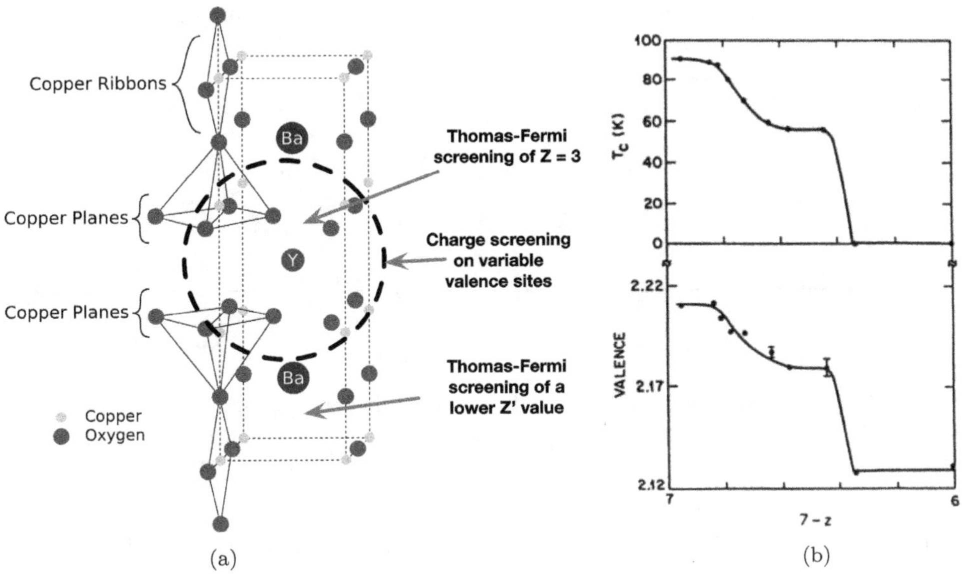

Figure 10.7. Illustration of weakly metallic screening in hole-doped $YBa_2Cu_3O_{7-z}$ superconductor. (a) The valence band energy of Cu sites at the dashed line is only slightly above the Fermi energy level. (b) A close correlation between the T_c of $YBa_2Cu_3O_{7-z}$ and the valence state of Cu sites, reproduced from [7].

valence electron shell. Thus we estimate $d_{ih} = 0.424\,\text{nm}$. Figure 10.7 illustrates the geometry of weakly metallic screening in this material.

As calculated previously, $k_{TF} = 1.89\,\text{nm}^{-1}$. Plugging in the d_{ih} and k_{TF} values into Equation (10.5.3), we get the following Z' value:

$$Z'_{\text{YBCO}} = 0.645\,\text{eV}^{-1} \triangle E_{VF}$$

The right-hand side of Figure 10.7 shows a direct proof of our screening theory: the valence state of Cu sites grows larger as $7 - z \to 7$. As Cu sites get more positively charged, their electron affinity increases, i.e., $\triangle E_{VF}$ decreases toward zero. The close correlation between the superconducting temperature of $YBa_2Cu_3O_{7-z}$ and the valence of Cu sites is a direct signature of the weakly metallic screening effect. In the following, we calculate this effect in terms of the $\triangle E_{VF}$ parameter.

We use the above Z'_{YBCO} value in Equation (10.4.3): Figure 10.8 shows the resulting screened orbital radius and binding energy values, as a function of $\triangle E_{VF}$. Binding energy indeed increases as $\triangle E_{VF}$ goes to zero. As noted above, the $\triangle E_{VF}$ parameter depends on the valence state of Cu sites. What is the $\triangle E_{VF}$ range where this calculation is valid? The $Z' < Z$ condition

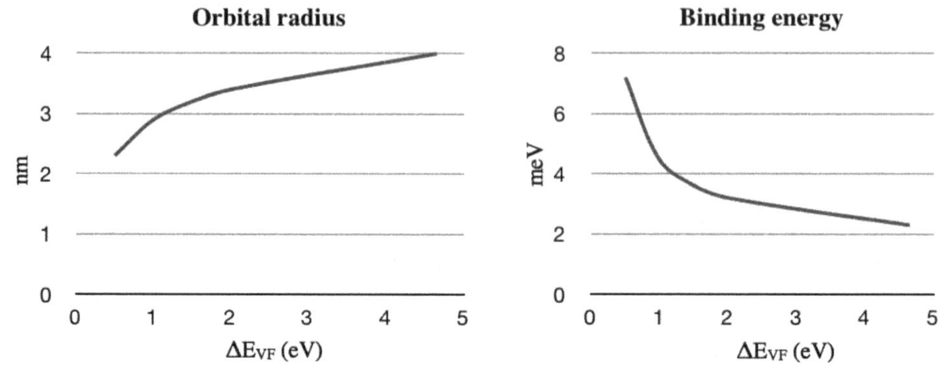

Figure 10.8. Electron screened orbital radius and binding energy values in $YBa_2Cu_3O_{7-x}$, as a function of $\triangle E_{VF}$. The weakly metallic screening effect is realized when $\triangle E_{VF} <$ 4.65 eV. In the $\triangle E_{VF} >$ 4.65 eV range, there is only Thomas–Fermi screening.

yields $\triangle E_{VF} < 4.65\,\text{eV}$. Inverse hole screening will not occur at larger gap values.

We now discuss experimental validation of the above calculation. An interesting system to look at is the interface between the $LaAlO_3$ and $KTaO_3$ oxides. Although both oxides are insulators, a two-dimensional electron gas is formed at their interface. This interface is weakly metallic, and becomes superconducting at low temperature. The authors of Ref. [10] control the Fermi level of this metallic interface by controlling a gate electrode voltage, which is situated at some distance above it. That is, their gate voltage V_g is proportional to the $\triangle E_{VF}$ parameter. Insofar as the superconducting current is generated by weakly bound electrons, a measurement of the superconducting layer thickness along the two sides of this interface is a measurement of weakly bound electrons' orbital radius. As shown in Figure 10.9, the measurements of [10] demonstrate the same dynamics as in Figure 10.8 and imply a similar orbital radius, in the 2-6 nm range. Our calculation of screened orbitals under weakly metallic screening is thus supported by experimental data. We note that the spectroscopic method of JPR emission frequency measurement is the precise means of orbital radius determination. In the $TlBa_2CaCu_2O_8$ case the JPR measurement gives 3 meV binding energy: according to Figure 10.8 this corresponds to 3.3 nm orbital radius. The weakly screened radius value is therefore smaller than the 4.56 nm value shown in Table 10.1, which was calculated for the normal metallic screening condition.

To explore the lower limit on $\triangle E_{VF}$, we consider Ref. [8], whose authors mapped the relationship between the Cu-O bond length and T_c

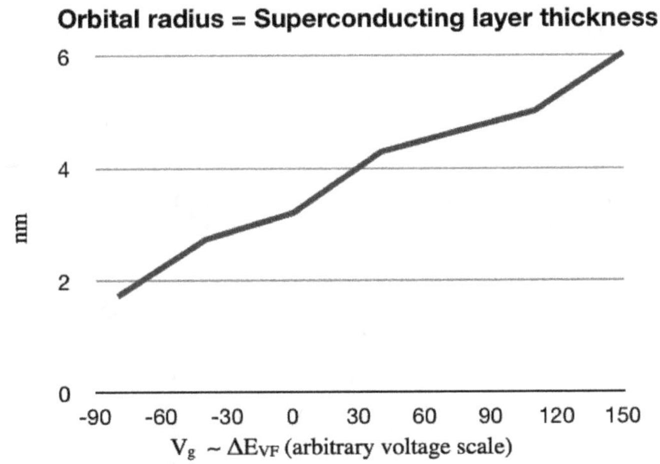

Figure 10.9. Experimental electron screened orbital radius values, deduced from the superconducting layer thickness at the weakly metallic $LaAlO_3$-$KTaO_3$ interface, as a function of gate voltage V_g. Note, that $\triangle E_{VF}$ is proportional to V_g. Measurement data taken from [10].

for many families of optimally doped superconductors. Its main finding is shown in the left-hand side of Figure 10.10, and we attempt to interpret this data.

As the valence of Cu sites increases, there is more electrostatic attraction between the oppositely charged Cu and O sites, which shortens the Cu–O bond length. This dynamic defines the curve in the $R\,(Cu - O) > 0.1922\,\text{nm}$ region. Why does T_c drop when the Cu–O bond length shrinks below 0.1922 nm? Some mechanism must begin hindering the electron uptake of Cu sites in this region. We interpret this hindrance as covalent Cu–O bond formation in the <0.1922 nm region, which becomes a barrier to delocalized electron uptake onto Cu sites. The pressure dependence of cuprate superconductivity corroborates this interpretation. As shown in the right-hand side of Figure 10.10, T_c initially increases with pressure. This is anticipated from Equation (10.5.3) because the shrinking crystal structure decreases d_{ih} and the resulting decrease of Z' enhances screening. However, T_c starts decreasing beyond a certain pressure, which we again interpret as covalent Cu–O bond formation. We note that the maximum T_c occurs at the same pressure, irrespective of the doping value. Altogether, the data of Figure 10.10 implies that the 0.1922 nm Cu–O bond length is a universal threshold in all cuprate materials: covalent Cu-O bond forming is favored at shorter bond lengths. This threshold sets a lower bound on the $\triangle E_{VF}$ parameter.

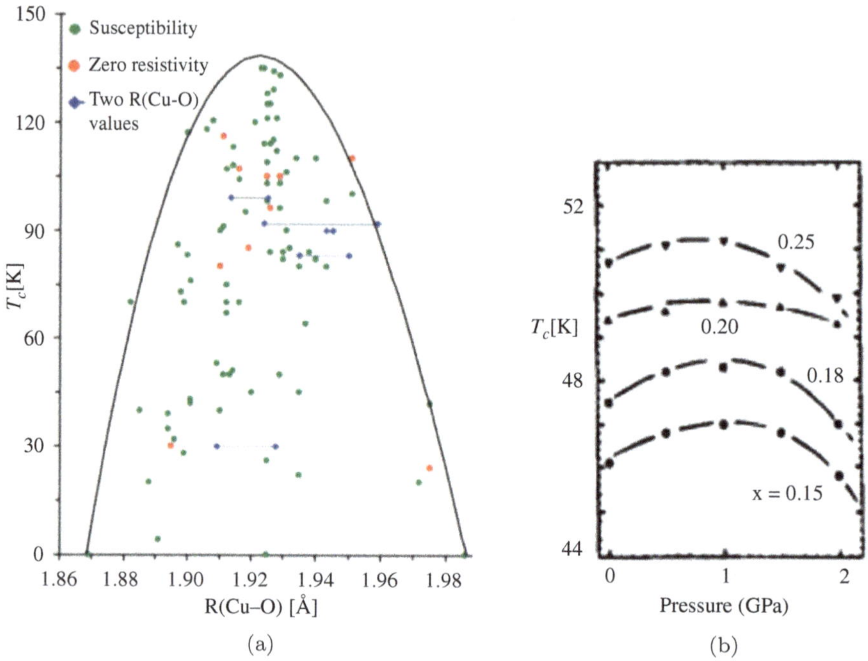

Figure 10.10. In superconducting cuprates, the Cu–O bond length sets a limitation on T_c. (a) A mapping of T_c in various optimally doped cuprate types, as a function of the Cu–O bond length, reproduced from [8]. Dots' colors represent different measurement techniques. The optimal T_c value requires 0.1922 nm Cu–O bond length. (b) A mapping of T_c in $La_{2-x}Ca_{1+x}Cu_2O_6$, as a function of the applied pressure, reproduced from [9].

At last, we can make sense of the two classes of cuprate materials which are referred to as "electron-doped" and "hole-doped" types. Their distinction now gains a physical meaning:

- **Electron-doped cuprates:** These feature strong metallicity in their Cu–O planes, and hence only Thomas–Fermi screening. An exemplary compound is $La_{1.9}Ba_{0.1}CuO_4$: as shown in Table 10.1, the calculated and experimental binding energies are well matching. The highest T_c in this family of cuprates is approximately 40 K. This temperature value coincides with the 39 K temperature limit of MgB_2, which is the highest value among simple materials with strongly metallic planes.
- **Hole-doped cuprates:** Under optimal hole doping, these materials feature weak metallicity in their Cu–O planes. Hence, the effective screened charge is calculated from Equation (10.5.3). $TlBa_2CaCu_2O_8$ is an exemplary compound: as shown in Table 10.1, the experimental binding energy of its screened orbital is 1.7 times larger than the value calculated

from a strong metallicity model. With reference to Figure 10.8, such enhancement of binding energy implies $\triangle E_{VF} \approx 1\,\text{eV}$. The highest T_c in this family of cuprates is 135 K.

In summary, experimental data clearly reveals that the Cu valence state and Cu–O bond length are the main control parameters of cuprate superconductors. Mainstream superconductivity theory cannot make sense of these parameters, and simply ignores them. As far as the author knows, the present chapter is the first attempt to understand the role of these parameters. Since weakly metallic screening is responsible for high-temperature superconductivity, its further study may allow rational design of high-temperature superconductors.

10.6. Conclusions

The calculations of this chapter established **the existence of a new type of bound orbital in metals, which has significantly larger radius than ordinary electron orbitals**. Nature tends to realize what is mathematically feasible, and thus one may expect that at a sufficiently low temperature some delocalized electrons will lower their energy by transitioning into electron-screened orbitals. The reviewed experimental data indeed prove the existence of electron-screened orbitals, whose binding energy is in the few meV range. The appearance of electron-screened orbitals coincides with the appearance of superconductivity, proving their key role in superconductivity.

In this chapter, we did not address how weakly bound electron orbitals relate to the phenomenology of superconductivity. That topic will be discussed in the following chapter.

References

[1] M. Ji *et al.*, $Bi_2Sr_2CaCu_2O_8$, intrinsic Josephson junction stacks with improved cooling: Coherent emission above 1 THz. *Applied Physics Letters*, **105**(12) (2014) 122602.

[2] V.K. Thorsmolle *et al.*, C-axis Josephson plasma resonance observed in $Tl_2Ba_2CaCu_2O_8$ superconducting thin films by use of terahertz time-domain spectroscopy. *Optics letters*, **26**(16) (2001) 1292–1294.

[3] Z. Chu *et al.*, Photovoltaic effect in Ag/MgB_2 heterostructure. *Journal of Alloys and Compounds*, **793** (2019) 662–671.

[4] Z. Chu *et al.*, Origin of photovoltaic effect in $(Bi, Pb)_2Sr_2Ca_2Cu_3O_{10+d}$/Ag heterostructure. *Journal of Alloys and Compounds*, **724** (2017) DOI: 10.1016/j.jallcom.2017.07.007.

[5] F. Yang *et al.*, Polarity switching of the photo-induced voltage in YBCO/Ag heterojunction. *Journal of Physics and Chemistry of Solids*, **138** (2020). Article id: 109232.

[6] S. Rajasekaran *et al.*, Parametric amplification of a superconducting plasma wave. *Nature Physics Letters*, **12** (2016) 1012–1016.

[7] H. Asano, Crystal Structure of High-Tc Superconductors. *Nihon Kessho Gakkaishi*, **36**(2) (1994) 94–99.

[8] K. Fijalkowski *et al.*, The "magic" electronic state of high-T(c) cuprate superconductors, *Dalton Transactions*, **40** (2008) 5447–5453.

[9] J.K. Burdett, High-temperature superconductors: A structural-electronic model. *Accounts of Chemical Research*, **28**(5) (1995) 227–231.

[10] Z. Chen *et al.*, Electric field control of superconductivity at the $LaAlO_3/KTaO_3(111)$ interface, *Science*, **372**(721) (2021) 721–724.

https://doi.org/10.1142/9781800611306_others01

Part 2

Experimental Validation and Practical Applications

The second part of the book goes through practical applications which become feasible with the theory discussed in the first part. Our theoretic results are applied to make sense of experimental data on superconductors, nuclear structures, and Compton-scale composite particles.

We validate our theory against experiments by deriving a consistent interpretation to thousands of measurements, across a broad range of experiments. On several topics we find the same pattern: what is being claimed as "knowledge" in present-day textbooks is in fact just *ad hoc* models and hand-waving. Upon replacing the shifting sand of curve-fitting models by the rock-solid physics foundation described in the first part of the book, rational engineering becomes feasible.

The selected applications represent a major step forward for human civilization: Chapter 11 describes the physics of high-temperature superconductivity which enables the efficient transportation and storage of electric energy, Chapter 15 describes a transmutation technology which may allow the production of currently rare industrial materials from more abundant resources, and finally Chapter 17 describes energy production experiments involving abundant and highly concentrated energy sources.

Our theory is of course not restricted to these selected applications — it may eventually spark new engineering solutions in many different technologies.

https://doi.org/10.1142/9781800611306_0011

Chapter 11

Superconductivity

András Kovács*,‡ and Giorgio Vassallo†,§
*BroadBit Energy Technologies, Metallimiehenkuja 8 C, 02150 Espoo, Finland
†Engineering Department, University of Palermo
viale delle Scienze, 90128 Palermo, Italy
‡andras.kovacs@broadbit.com
§giorgio.vassallo@unipa.it

11.1. The Bose–Einstein Condensation of Weakly Bound Electrons

Superconductivity is not rare: the majority of metals and semi-metals become superconducting when sufficient cooling and pressure is applied. Besides elementary metals, there are many other types of simple structured metallic materials that are superconductors at ambient pressure: for example, boron-doped diamond, inter-metallic compounds (e.g., magnesium di-boride), nitrides (e.g., niobium nitride), selenides (e.g., iron selenide), intercalated graphite (e.g., C_6Ca), or hydrated metals (e.g., palladium hydride). Therefore, the causation of superconductivity cannot be related to some special constellation or complex microscopic model: the causation factor must be something universally available in metals. Accordingly, this chapter builds onto the foundations established in Chapter 10, which are applicable to any metallic material.

Superconducting electron pairs are referred to as "Cooper pairs" in the literature. Their pairwise constellation is a strict limit, i.e., Cooper triplets are never observed. Mainstream theory considers superconducting electrons to be in a completely unbound state, but struggles to explain key experimental findings: (i) how do repulsive unbound electrons form bound pairs?,

(ii) where does the pairwise coupling limit originate from?, and (iii) how do Cooper-paired electrons become entangled, as discussed in Chapter 9? We note that all these experimental findings are universal properties of bound electron states. Therefore, these properties naturally emerge from the theory of weakly bound electron states, which was introduced in Chapter 10.

While finding an explanation for the above-listed Cooper-pair properties is inspiring, one would not expect a weakly bound electron to conduct electricity without resistance. What could be the physical mechanism behind zero-resistance conductivity? Historically, theorists have focused on constructing a phenomenological Hamiltonian, which has become known as the BCS theory of superconductivity. This BCS Hamiltonian was initially formulated over 60 years ago. The author of [2] emphasizes that BCS theory is phenomenological, and as such it does not yet provide any physical explanation: "We have all been taught that it [the BCS theory] is a marvellous success of quantum theory, accounting for persistent currents, Meissner effect, isotope effect, Josephson effect, etc. Yet, on examination one realizes that the model Hamiltonian is phenomenological, chosen not from first principles but from trial-and-error so as to agree with just those experiments. Then in what sense can one claim that the BCS theory gives a *physical explanation* of superconductivity? Surely, if the Meissner effect did not exist, a different phenomenological model would have been invented that does not predict it; one could have claimed just as great a success for quantum theory whatever the phenomenology to be explained."

It is anticipated that the predictions of a phenomenological model may diverge from later experimental discoveries. This is exactly what happened to the BCS model upon the discovery of new classes of high-temperature superconductors in the 1980s and 1990s. Reference [8] outlines some of these divergences, and proceeds to formulate an even more elaborate phenomenological Hamiltonian. The authors of [8] refer to their study as an "exact theory," but that is a complete misnomer for the activity of building increasingly complex phenomenological models. Many additional inconsistencies of the BCS theory are explained in Ref. [9]. In summary, the BCS theory is limited in its ability for modeling superconductivity.

Regarding an actual physical explanation, two models emerged about 70 years ago:

- The phonon-mediated model claims that superconducting electrons are bound to positively charged phonon oscillations, and not bound to any lattice site. Although the phonon is a dispersed wave, i.e., it does not have any specific location, it is modeled as a wavefunction of a virtual particle. In this model, a superconducting electron is bound to such a virtual particles, and the delocalized electron–phonon composite wavefunction fills the metallic interior.

- The Bose–Einstein condensate (BEC) model claims that superconducting electrons are in a BEC state, where all of them occupy the same quantum mechanical wavefunction.

The phonon-mediated model was embraced by quantum field theorists for the following main reasons: (i) their mathematical tools are well suited for modeling phonons and free electrons, (ii) the isotope effect, discovered for mercury in 1950, was interpreted as a phonon interaction signature, and (iii) positively charged phonons were postulated to provide a binding mechanism between two mobile electrons. The isotope effect refers to the empirically observed $T_c \sim M^{-\frac{1}{2}}$ relationship, where T_c is the superconducting critical temperature and M is the nuclear mass. This $M^{-\frac{1}{2}}$ factor was thought to demonstrate that the binding energy of a Cooper-pair is derived from phonon oscillations. Although this favored theory did not succeed to predict any higher-temperature superconductor, quantum field theorists remain convinced in their theory, and recent quantum field theory textbooks [3] still present the hypothesis of phonon-mediated superconducting electrons as "knowledge."

If the phonon-mediated superconductivity model is deemed to be valid, it must be particularly accurate for modeling simple metals. The simplest metal is lithium, and the $T_c \sim M^{-\frac{1}{2}}$ relationship suggests that its superconducting temperature must be particularly high. Based on this anticipation, a high T_c was predicted for lithium in the 1960s. However, eventual measurements revealed that lithium is barely superconducting: its T_c is only 0.0004 K. Experimenters also found that the T_c of lithium strongly depends on the applied pressure [10]; T_c rises as high as 15 K above 30 GPa, but the phonon-mediated model did not predict any significant role of pressure. Furthermore, experimenters compared ^6Li against ^7Li to see whether the $T_c \sim M^{-\frac{1}{2}}$ relationship holds. To their surprise, experimenters found [10] overly strong $T_c \sim M^{-4}$ isotope relationship in the <21 GPa region, and $T_c \sim M^2$ type inverse isotope effect in the >21 GPa region. Thus, the phonon-mediated superconductivity model spectacularly fails with the simplest metal. Regarding high-temperature superconductors, recent electron–phonon coupling studies [5] and inverse isotope effect measurements [24] also conclusively proved that the positively charged entities, which hold a

Cooper-pair together, are not phonons. Altogether, the evidence for rejecting the phonon-mediated superconductivity model is rather conclusive.

This leads us to consider the BEC model of superconductivity. The BEC idea is that electron wavefunctions become overlapping. As electron wavefunctions become indistinguishable, all BEC-involved electrons start to simultaneously occupy the same quantum mechanical ground state. Historically, scientists thought to apply the BEC model to unbound electrons, but struggled to explain why unbound electrons would pair up. In contrast, we apply Bose–Einstein condensation to the weakly bound electrons described in Chapter 10. As noted above, the properties of Cooper-pairs follow naturally. Although various weakly bound electrons are centered around different lattice sites, the exponential tails of their wavefunctions can and do overlap. It is intuitive that weakly bound electrons, which have large radius, will have much stronger overlap than electrons in ordinary orbitals. A direct consequence of this model is that T_c is limited either by the BEC temperature or by the electron-screened bond-forming temperature:

$$T_c = \min\left(CE_b, T_{\text{BEC}}\right) \tag{11.1.1}$$

where T_{BEC} is the Bose–Einstein condensation temperature, E_b is the binding energy of electron-screened orbitals, and C is some constant for a given family of materials.

If the BEC comprises non-interacting particles, T_{BEC} is known to be given by the following formula:

$$\left[T_{\text{BEC}} = \left(\frac{n}{\zeta\left(\frac{3}{2}\right)}\right)^{\frac{2}{3}} \frac{2\pi\hbar^2}{mk_b} \sim n^{\frac{2}{3}}\right]_{\text{3D}} \tag{11.1.2}$$

where n is the particle density, m is the particle mass, ζ is the Riemann zeta function, and k_b is the Boltzmann constant. The question arises whether T_{BEC} of weakly bound electrons might be defined by Equation (11.1.2). Assuming that Equation (11.1.2) is applicable, it implies that T_{BEC} becomes the limiting factor as $n \to 0$. A reasonable test of this question is to look for a superconductor where n is a known and small value: we must have $T_c \approx T_{\text{BEC}}$ in such material. Boron-doped diamond is a simple structured material meeting this criteria; the boron-doping concentration is a small and controlled value. The boron sites are more highly charged than the carbon sites, and thus electron-screened orbitals will form around them. Therefore, n is defined by the boron concentration. As shown in Figure 11.1, there is indeed a $T_c \sim n^{\frac{2}{3}}$ relationship in boron-doped diamond.

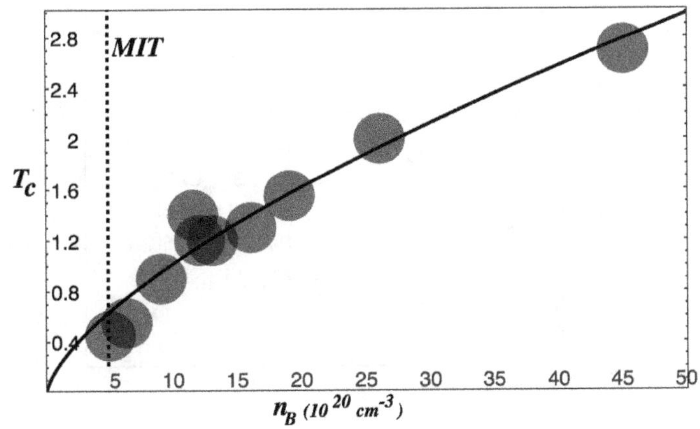

Figure 11.1. The T_c (K) of boron-doped diamond, as a function of boron concentration. The data are taken from [18], disc sizes correspond to the error bars. The solid line represents $\sim n^{\frac{2}{3}}$ scaling. The dashed line indicates metal-insulator transition.

The derivation of Equation (11.1.2) considers the pressure of a freely moving ideal gas. Superconducting electrons are weakly bound around their respective nucleus sites, and thereby transfer momentum to the involved nuclei. Their momentum transfer strength depends on the screened charge value appearing in Equations (10.4.3) and (10.5.3). In ordinary gas, when atoms collide via the intermediation of their outermost electron shell, we have $Z \approx 1$. The calculations of Chapter 10 yield a much lower screened charge; in the range of $Z_{\text{scr}} \approx 0.01$. Considering correspondingly scaled momentum transfer, the "effective Cooper pair mass" is $m = 0.01 M_B \approx 200 m_e$. Using Equation (11.1.2) with $m = 200 m_e$ yields $T_{\text{BEC}} = 1.5\,\text{K}$ for $n = 10^{21}\ \text{cm}^{-3}$, which agrees well with data of Figure 11.1.

In case of a two-dimensional BEC, the relationship between n and T_{BEC} becomes

$$\left[T_{\text{BEC}} \sim n^{\frac{2}{2}} = n\right]_{\text{2D, stack}} \tag{11.1.3}$$

With all else being equal, the T_{BEC} has larger scaling for a two-dimensional material than for a three-dimensional one. Not surprisingly, all high-temperature superconductors have stacked two-dimensional structure.

In some materials, the weakly bound electrons of Chapter 10 may persist still above T_c. We are interested to know if there are any physical signatures related to their presence. Figure 11.2 reveals a universal relationship of residual diamagnetism in superconducting materials, extending to twice the superconducting temperature in some cases. Therefore, diamagnetism

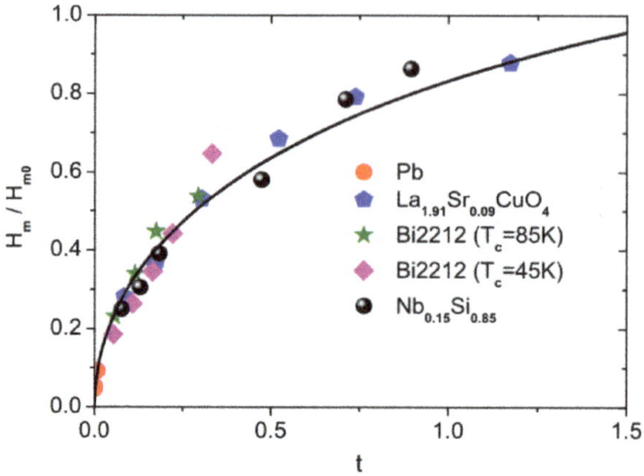

Figure 11.2. Above-T_c residual diamagnetism in various superconductors. The horizontal axis measures $t = \frac{T}{T_c} - 1$, and thus $T = T_c$ at the origin. The vertical axis shows magnetization H_m at a given temperature, as a fraction of the room temperature magnetization H_{m0}. The zero of the vertical axis represents perfect diamagnetism. Reproduced from [4].

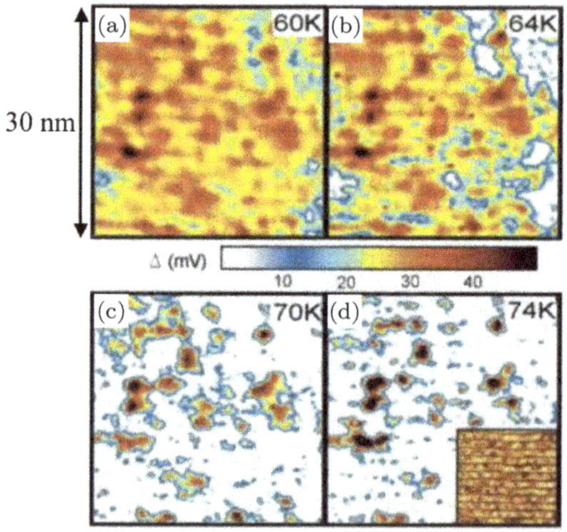

Figure 11.3. An atomic-resolution spectroscopic map of superconductivity signatures in Bi2212 with $T_c = 65\,\text{K}$ value, reproduced from [41]. The four charts show measurements taken at the indicated temperatures. Colored areas show spectroscopic signatures associated with superconductivity.

does not just suddenly appear at T_c, but has a precursor above T_c which might be related to weakly bound electrons. Figure 11.3 shows an atomic-resolution spectroscopic map of superconductivity signatures in the Bi2212 material of Figure 11.2. Figure 11.3 demonstrates that superconductivity

Table 11.1. The relationship between $E_b(0)$ and binding formation onset temperature.

Superconductor	Z	$E_b(0)$ experimental	T_{onset}	$\frac{k_b T_{\text{onset}}}{E_b(0)}$
$La_{1.9}Ba_{0.1}CuO_4$	+3	2 meV	70 K	3
$TlBa_2CaCu_2O_8$	+3	3 meV	103 K	3

signatures indeed do not disappear at T_c; the loss of superconductivity is characterized by the break-up of a continuous superconducting domain into separated islands, which are a few nm in size.

We recall from Chapter 10 the experimental E_b values for $La_{1.9}Ba_{0.1}CuO_4$ and $TlBa_2CaCu_2O_8$ superconductors, obtained from low-temperature JPR frequency measurements. $La_{1.9}Ba_{0.1}CuO_4$ has the same structure as $La_{1.9}Sr_{0.1}CuO_4$, and Figure 11.2 indicates that the onset of diamagnetism is at 2.2 times T_c. Considering the $T_c = 32\,\text{K}$ value of $La_{1.9}Ba_{0.1}CuO_4$, the onset of diamagnetism is at $T_{\text{onset}} = 70\,\text{K}$. As shown in Figure 10.5, the $T_c = 103\,\text{K}$ value of $TlBa_2CaCu_2O_8$ coincides with the onset of electron-screened orbital formation, and therefore $T_{\text{onset}} = 103\,\text{K}$. Table 11.1 summarizes these data, and suggests the following phenomenological relationship:

$$k_b T_{\text{onset}} = 3E_b(0) \tag{11.1.4}$$

How to make sense of Equation (11.1.4)? Figures 11.2 and 11.3 indicate that the number of electrons participating in diamagnetism decreases as temperature goes higher. It is known from the study of superfluid helium that BEC formation requires at least $N = 60$ particles [40]. If diamagnetism is a signature of BEC electrons, then the onset of diamagnetism requires a BEC cluster of at least 60 electrons, i.e., $T_{\text{onset}} = T_{\text{BEC}}(N = 60)$. Reference [42] derives T_{BEC} for N particles which are bound in an isotropic harmonic potential. Equation (62) of [42] gives the T_{BEC} formula in terms of the $\hbar\omega$ energy of a harmonic oscillator. For a bound electron orbital, the $\hbar\omega$ oscillation energy equals the binding energy. Upon replacing $\hbar\omega$ by $E_b(0)$, Equation (62) of [42] yields

$$\frac{k_b T_{\text{BEC}}}{E_b(0)} = \left(N^{\frac{1}{3}} - 0.7275\right)(\zeta(3))^{-\frac{1}{3}} \tag{11.1.5}$$

At $N = 60$, the above equation evaluates to exactly $\frac{k_b T_{\text{BEC}}}{E_b(0)} = 3$, which explains Equation (11.1.4). The diamagnetic onset of superconductors thus reveals a direct relationship between the binding energy of weakly bound electrons, which we studied in Chapter 10, and their BEC transition temperature. A remaining issue is to justify replacing $\hbar\omega$ by $E_b(0)$ at $T > 0$. As we saw in Figure 10.5, the binding energy goes to zero as temperature increases. This justifies that one should keep a constant $E_b(0)$ in the denominator because otherwise the condensation temperature would go to infinity. Perhaps the physical meaning of a constant $E_b(0)$ in the above equation is that thermal fluctuations reduce only the binding energy to zero, while the $\hbar\omega$ oscillation energy remains unchanged.

Inspired by the above data, the methodology of this chapter is based on the following assumption: **superconducting electrons are weakly bound electrons in the BEC state.** This means that superconducting electrons all share the same quantum mechanical wavefunction, which becomes macroscopic. They are in a weakly bound orbital, which was calculated in Chapter 10, but the binding lattice site becomes indistinguishable. In other words, a superconducting electron is simultaneously bound to all weakly binding sites of the superconducting material.

11.2. The London Equation

The London equation was proposed in the 1930s as a phenomenological equation for describing the interaction between magnetic fields and superconductors, which are known for their perfect diamagnetism. One may try to understand such an interaction by using Maxwell's equation to calculate the currents induced by a time-varying magnetic field, which is gradually ramped up from zero to its final value. Such a calculation is presented, for example, in Chapter 6 of [30]. The calculation of induced currents reveals that a perfect conductor indeed diamagnetically responds to a time-varying magnetic field; the induced surface currents expel the magnetic field from its interior. Unfortunately, such calculation does not yet explain how diamagnetism works in superconductors. As illustrated in Figure 11.5, a superconductor expels magnetic fields even when $\frac{dB}{dt} = 0$, e.g., when its temperature is cooled below T_c. A more general approach is to solve for the global minimum of free energy. A recent textbook recognizes that the London equation gives an energy minimizing solution [31], but omits a proper derivation. This same recognition appears in several articles [32, 33], but always without mathematical derivation, as far as the authors know. We endeavor to derive the London equation in the following paragraphs.

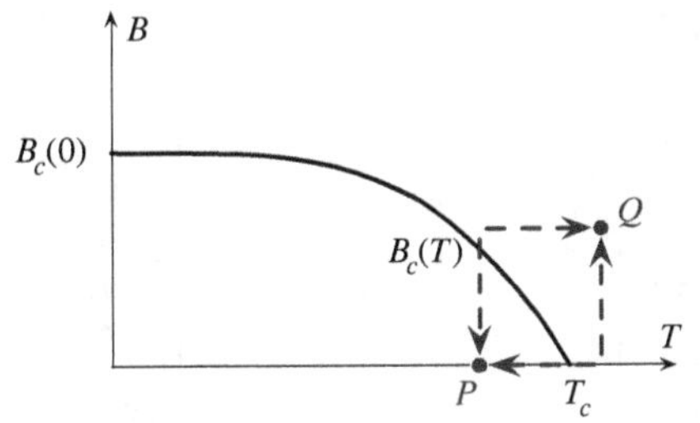

Figure 11.4. The phase diagram for a type I superconductor. B is the externally applied magnetic field, and T is the temperature. In the figure, state "P" is in the superconducting phase and state "Q" is in the normal phase. There is a reversible state change between "P" and "Q," irrespective of the path taken between them.

We firstly consider type-I superconductors, which have isotropic structure. We note that a series of works [34–37] demonstrated already in the 1930s that the superconducting-normal state transition in the presence of a magnetic field occurs without sudden energy dissipation; calorimetry indicates that the superconducting current stops without any heat spike. Figure 11.4 illustrates the $B - T$ phase diagram of a type-I superconductor. Starting from state "P" in Figure 11.4, a reversible superconducting-normal state transition may be induced either by raising temperature or by raising magnetic field strength. According to the traditional theory of superconductivity, superconducting electrons flow without energy dissipation, and they are abruptly converted to normal conduction band electrons once the external magnetic field reaches a critical value. Thus, traditional theory predicts that the superconducting "Meissner flow" inevitably produces Joule heating upon superconducting-to-normal phase transition due to the transition of superconducting electrons into normal state, which flow with dissipation. The predicted Joule heating, which should be generated during the phase transition, will make it an irreversible transition. But this prediction contradicts the experimentally observed reversible phase transition in a magnetic field. This 90-year-old contradiction indicates that serious revisions are needed in the standard theory.

The superconducting "Meissner flow," which causes the perfect diamagnetism of superconductors, comprises induced electron currents on the superconductor's surface. This flow is illustrated in Figure 11.5, and it carries an associated kinetic energy. What happens to superconducting electrons' kinetic energy beyond the superconducting-to-normal phase transition in a type-I superconductor? The example of Pb in Figure 11.2 already revealed

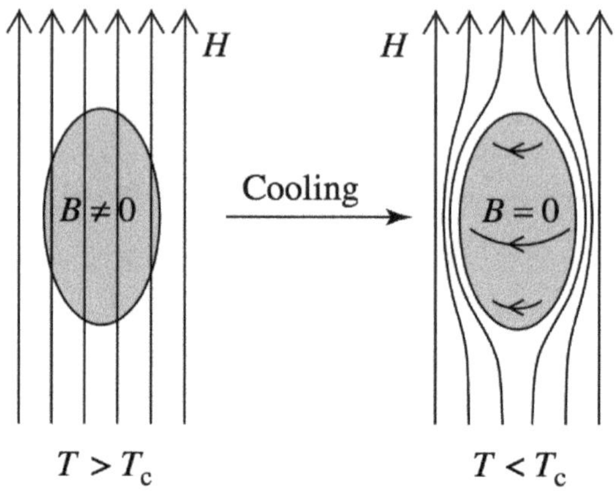

Figure 11.5. An illustration of the Meissner flow, which arises below T_c and expels magnetic field from the superconductor's interior. Horizontal arrows indicate Meissner flow currents in the superconductor's surface region.

that their kinetic energy remains nearly the same also on the normal side of the phase transition. According to Figure 11.2, the only change is that the perfect diamagnetism of a BEC is replaced by a residual diamagnetism. In the reverse direction, this residual diamagnetism converges to perfect diamagnetism as temperature is lowered to T_c. Perfect diamagnetism is the vertical axis' zero value in Figure 11.2. It is seen from Figure 11.2 that there is a smooth convergence to perfect diamagnetism, which means that the system is in a thermodynamic equilibrium at any point and thus state changes are reversible. One may also say that the application of external magnetic field lowers the effective T_c. As the magnetic field strength is raised from state P to B_c, the temperature at B_c indeed becomes the new T_c: any further increase of temperature triggers superconducting-to-normal phase transition.

Let's consider what thermodynamic equilibrium looks like in our superconducting material, when it is in superconducting phase. Perfect diamagnetism means that magnetic energy density is zero in the interior. In contrast, there is a magnetic energy density ϱ_m in the exterior, while superconducting electrons' translational kinetic energy density ϱ_k is zero in that region. The challenge is to understand what happens in the transitionary surface region between the exterior and interior zones, where ϱ_m and ϱ_k are both non-zero.

Anywhere in the surface region, magnetic energy density is given by

$$\varrho_m = \frac{1}{2\mu_0}\boldsymbol{B}^2 \tag{11.2.1}$$

The kinetic energy density of superconducting electrons is

$$\varrho_k = \frac{m_e}{2}\boldsymbol{v}_s^2 n_s \tag{11.2.2}$$

where n_s is the density of superconducting electrons and $\boldsymbol{v}_s$ is their velocity. The corresponding current can be written as

$$\boldsymbol{J} = e\boldsymbol{v}_s n_s$$

Maxwell's equation defines the relationship between $\boldsymbol{J}$ and $\boldsymbol{B}$, as follows:

$$\mu_0 \boldsymbol{J} = \nabla \times \boldsymbol{B}$$

We note that the above equation merely defines a relationship between $\boldsymbol{J}$ and $\boldsymbol{B}$, but does not mean that $\boldsymbol{v}_s$ would be induced by the applied magnetic field! This is an important distinction between a superconductor and a hypothetical perfectly conducting metal. Although the above equation implies that macroscopically $\boldsymbol{v}_s = 0$ when $\boldsymbol{B} = 0$, microscopically $\boldsymbol{v}_s$ is never zero. In a perfect conductor, the microscopic value of $\boldsymbol{v}_s$ is defined by the kinetic energy of electrons at the Fermi surface, n_s does not depend on the magnitude of $\boldsymbol{B}$, and the macroscopic value of $\boldsymbol{v}_s$ is generated by the gradually ramped up magnetic fields when $\frac{dB}{dt} \neq 0$. In a superconductor, the microscopic value of $\boldsymbol{v}_s$ is defined by the kinetic energy of weakly bound electrons, while the macroscopic value of $\boldsymbol{v}_s$ is pre-determined by the BEC condition. Consequent to the $\mu_0 \boldsymbol{J} = \nabla \times \boldsymbol{B}$ condition, n_s becomes dependent on the magnitude of $\boldsymbol{B}$. In the following paragraphs, we will discuss the proper calculation of $\boldsymbol{v}_s$ and n_s. The main point here is to emphasize that it is misleading to think of a superconductor as a perfectly conducting metal, because the BEC state has different dynamics.

Substituting the above two equations into (11.2.2), we express kinetic energy density in terms of magnetic field:

$$\varrho_k = \frac{m_e}{2e^2\mu_0^2 n_s}(\nabla \times \boldsymbol{B})^2 \tag{11.2.3}$$

Using Equations (11.2.1) and (11.2.3), we can express the total energy density as a function of magnetic field:

$$\varrho(\boldsymbol{B}) = \varrho_m + \varrho_k = \frac{1}{2\mu_0}\boldsymbol{B}^2 + \frac{m_e}{2e^2\mu_0^2 n_s}(\nabla \times \boldsymbol{B})^2 \tag{11.2.4}$$

Suppose there is a disturbance or variation of the magnetic field; we denote this magnetic field variation as vector field $\boldsymbol{G}$. The derivative of energy density with respect to infinitesimal magnetic field variations is

$$\varrho'(\boldsymbol{B},\boldsymbol{G}) = \frac{\partial}{\partial\varepsilon}\varrho(\boldsymbol{B}+\varepsilon\boldsymbol{G})\,|_{\varepsilon\to 0} \tag{11.2.5}$$

We now evaluate the above differential by substituting Equation (11.2.4):

$$\varrho'(\boldsymbol{B},\boldsymbol{G}) = \frac{1}{\mu_0}(\boldsymbol{B}\cdot\boldsymbol{G}) + \frac{m_e}{e^2\mu_0^2 n_s}(\nabla\times\boldsymbol{B})\cdot(\nabla\times\boldsymbol{G}) \tag{11.2.6}$$

where we obtained the last term by switching the order of the two differential operators:

$$\frac{\partial}{\partial\varepsilon}(\nabla\times(\boldsymbol{B}+\varepsilon\boldsymbol{G})) = \nabla\times\frac{\partial}{\partial\varepsilon}(\boldsymbol{B}+\varepsilon\boldsymbol{G}) = \nabla\times\boldsymbol{G}$$

Thermodynamic equilibrium requires that the total system energy remains invariant with respect to small changes in the magnetic field, i.e., the system occupies the lowest energy state:

$$\int_\Omega \varrho'(\boldsymbol{B},\boldsymbol{G})\,dV = 0$$

for any choice of $\boldsymbol{G}$, keeping in mind that the variational field $\boldsymbol{G}$ vanishes at the boundary. Substituting Equation (11.2.6) into the above equilibrium requirement, we obtain

$$\int_\Omega \left[\frac{1}{\mu_0}(\boldsymbol{B}\cdot\boldsymbol{G}) + \frac{m_e}{e^2\mu_0^2 n_s}(\nabla\times\boldsymbol{B})\cdot(\nabla\times\boldsymbol{G})\right] dV = 0$$

We use the $\int_\Omega(\nabla\times\boldsymbol{B})\cdot(\nabla\times\boldsymbol{G})\,dV = \int_\Omega(\nabla\times(\nabla\times\boldsymbol{B}))\cdot\boldsymbol{G}\,dV$ identity, which is proven in the appendix. Note that we substituted $\nabla\times\boldsymbol{B}$ for the vector field $\boldsymbol{F}$ that is used in the appendix. The above equation can therefore be written as follows:

$$\int_\Omega \left[\frac{1}{\mu_0}\boldsymbol{B} + \frac{m_e}{e^2\mu_0^2 n_s}\nabla\times(\nabla\times\boldsymbol{B})\right]\cdot\boldsymbol{G}\,dV = 0$$

Since the above equation must remain zero for any choice of $\boldsymbol{G}$, the equilibrium condition is given by

$$\boldsymbol{B} + \frac{m_e}{\mu_0 n_s e^2}\nabla\times(\nabla\times\boldsymbol{B}) = 0 \tag{11.2.7}$$

which is known as the London equation. Our simple derivation shows that the London equation yields the solution of lowest energy state.

Since magnetic fields have zero divergence, we get $\nabla \times (\nabla \times \boldsymbol{B}) = -\nabla^2 \boldsymbol{B}$, and the London equation may be written as

$$\nabla^2 \boldsymbol{B} = \frac{\mu_0 n_s e^2}{m_e} \boldsymbol{B} \tag{11.2.8}$$

The solution of the above equation is a $\boldsymbol{B} = \boldsymbol{B}_0 e^{-x/\lambda_L}$ type exponential decay of $\boldsymbol{B}$ from the surface toward the interior, over the characteristic distance λ_L:

$$\frac{1}{\lambda_L} = \sqrt{n_s \frac{\mu_0 e^2}{m_e}} \tag{11.2.9}$$

Thus, we obtain a relationship between λ_L and n_s:

$$\frac{1}{\lambda_L} \sim \sqrt{n_s} \tag{11.2.10}$$

Regarding the reversibility of superconducting-normal state transition, a question still remains: what happens to superconducting electrons' kinetic energy beyond B_c? The smooth convergence to zero in Figure 11.2 reveals that there is no sudden disappearance of n_s during superconducting-normal state transition. Rather, the transition process involves a gradual depletion of n_s, which gradually enlarges λ_L. When λ_L reaches the physical dimensions of the superconductor, diamagnetism becomes less than perfect. Thermodynamic equilibrium is maintained throughout the process, i.e., there is no abrupt kinetic energy dissipation.

Equation (11.2.9) does not have an explicit temperature dependence. However, we saw in Chapter 10 that the radius of a weakly bound electron increases when temperature increases. Considering that our superconductor material is densely filled by weakly bound electron orbits, n_s goes down as temperature increases. This process explains the shape of $B_c(T)$ curve in Figure 11.4.

Our analysis demonstrates that the London equation may be derived via a purely classical approach, by minimizing the energy of magnetic fields and superconducting electrons. One may ask about the value of $\boldsymbol{v}_s$. It is important to keep in mind that $\boldsymbol{v}_s$ is the speed of electrons in BEC state, and NOT the speed of free-flowing individual electron pairs. Chapter 6 of [30] calculates how free-flowing electrons would respond to a time varying magnetic field. Its author finds that $\boldsymbol{v}_s = -\frac{e\lambda_L}{m_e c} B$. That is a paradoxical result; Figure 11.6 illustrates that $\lambda_L \to \infty$ as $T \to T_c$, which would mean that $\boldsymbol{v}_s \to \infty$. To determine $\boldsymbol{v}_s$, one needs quantum mechanical

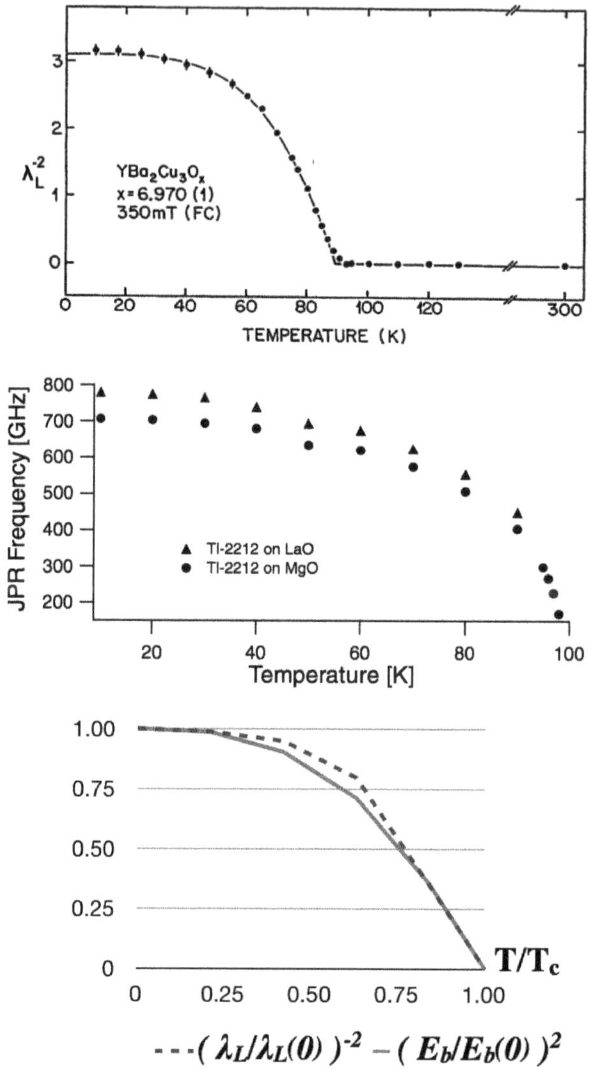

Figure 11.6. A comparison between λ_L^{-2} ($T_c = 95\,\mathrm{K}$, from [6]) and E_b ($T_c = 103\,\mathrm{K}$, from [23]). λ_L^{-2} is evaluated via muon spin relaxation measurement, and E_b is evaluated via JPR frequency measurement. The chart at the bottom compares the two data sets, revealing that $E_b \sim \lambda_L^{-1}$.

considerations. Knowing $\boldsymbol{v}_s$, we can determine the frequency of the electron wavefunction and *vice versa*. In the following, we approach superconductors' diamagnetism quantum mechanically.

So far we studied the diamagnetism of type-I superconductors, which have isotropic structure. Beyond the B_c threshold, the whole interior of a type-I superconductor is filled by the externally applied magnetic field. High-temperature superconductors are however two-dimensional layered materials, and their diamagnetism qualitatively differs from three-dimensional superconductors. Firstly, λ_L is not isotropic in two-dimensional superconductors: its value is several times larger orthogonally to the two-dimensional plane than in parallel directions. Secondly, when the external magnetic field exceeds a critical threshold value, it still does not completely fill the interior of a two-dimensional superconductor, but only penetrates in spatially confined flux lines, which are illustrated in Figure 11.11. How to derive then relation (11.2.10) for two-dimensional layered materials?

We start by noting that Equation (11.2.9) has analogous form to the Langmuir frequency formula of plasma oscillations:

$$\omega_{\text{plasma}} = \sqrt{n\frac{e^2}{m_e\varepsilon_0}} = \sqrt{c^2}\sqrt{n\frac{\mu_0 e^2}{m_e}} = \frac{c}{\lambda_L} \tag{11.2.11}$$

A BEC state is considered to be a condensation in momentum space. For superconducting electrons in the BEC state, the frequency of their "plasma oscillation" is their quantum mechanical wavefunction frequency. Therefore, λ_L is proportional to the wavelength of the quantum mechanical wavefunction:

$$\lambda_L \sim \lambda_{\text{BEC}} \tag{11.2.12}$$

Regarding the binding energy of a weakly bound orbital, we can write the $E_b \sim \frac{Z_{\text{scr}}}{r_{\text{rsc}}}$ relationship, where r_{rsc} is the mean orbital radius and Z_{rsc} is the screened charge value. Since superconducting electrons are in bound state, their quantum mechanical wavelength is 2π times their mean orbit radius:

$$\lambda_{\text{BEC}} = 2\pi r_{\text{scr}} \tag{11.2.13}$$

Based on the above relationships, we expect to see the $\lambda_L^{-1} \sim E_b$ relationship for superconducting electrons. Indeed, the data of Figure 11.6 demonstrate that λ_L^{-1} has the same temperature dependence as E_b, which is measured through the JPR frequency. Thus, the data of Figure 11.6 validates that, within measurement error, $\lambda_L^{-1} \sim E_b$.

When a two-dimensional plane is densely filled by electron orbits, the superconducting electron density n_s is proportional to r_{scr}^{-2}. Equation (11.2.13) then gives $n_s \sim \lambda_L^{-2}$, which leads to Equations (11.2.9) and (11.2.10).

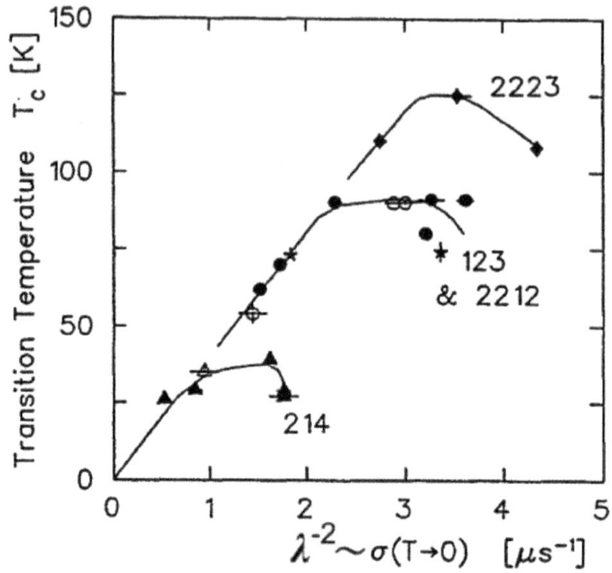

Figure 11.7. The $T_c \sim \lambda_L^{-2}(0)$ relationship, which is universally valid in the underdoped region of cuprates. The labels indicate various cuprate materials. Reproduced from [6].

When the superconducting temperature is limited by T_{BEC}, Equation (11.1.3) tells us that $T_c \sim n$ for two-dimensional superconductors. Using Equation (11.2.10), we expect to see the $\lambda_L^{-2} \sim T_c$ relationship in two-dimensional superconductors, as long as $T_c \approx T_{\text{BEC}}$.

Figure 11.7 shows that the $\lambda_L^{-2} \sim T_c$ relationship indeed universally holds up for the underdoped region of cuprate superconductors. The data of Figure 11.7 were obtained via muon spin relaxation measurement, where the muon spin relaxation rate σ is proportional to λ_L^{-2}. Experimentalists noticed this universal linear relationship in the underdoped region already during the 1990s, but up to now mainstream theorists considered the straight line in Figure 11.7 to be coincidental. In our interpretation, the under-doped region is characterized by $T_c \approx T_{\text{BEC}}$, and the eventual departure from linearity implies $T_c < T_{\text{BEC}}$ near the optimal temperature point.

Altogether, the above data give conclusive evidence concerning the physical origin of the London penetration depth λ_L. We may finally determine the speed of superconducting electrons in the surface current. The quantum mechanical relationship between particle momentum and wavefunction frequency is

$$m_e v_s = \hbar \frac{\omega}{c}$$

Using Equation (11.2.11) we substitute $\omega = \frac{c}{\lambda_L}$, and obtain

$$v_s = \frac{\hbar}{m_e \lambda_L} \tag{11.2.14}$$

As seen in (11.6), $\lambda_L \to \infty$ as $T \to T_c$, and thus $\boldsymbol{v}_s \to 0$ at that limit. This result resolves the above-mentioned paradox of diverging $\boldsymbol{v}_s$, which appears in some works.

Considering v_s as a function of the applied magnetic field, the behavior of $v_s(B)$ is opposite in a hypothetical perfectly conducting metal than in a superconductor. Suppose the applied magnetic field strength is gradually ramped up from zero to B_c, which is the critical magnetic field value. While $\frac{dB}{dt} \neq 0$, there is an induced electric field near the surface, i.e., $\nabla \times \boldsymbol{E} \neq 0$ in that region. This induced electric field exerts on electrons and positively charged ions, and accelerates delocalized electrons with respect to the lattice sites. In a hypothetical perfectly conducting metal, the acceleration of delocalized electrons is calculated from Newton's law according to $m_e \frac{\partial v_s}{\partial t} = eE$. This leads to $v_s \sim B$ in a perfectly conducting metal, and the highest current speed is reached at B_c. In a superconductor, the induced electric field polarizes the weakly bound electron orbitals in the surface region. Referring back to Figure 11.4, we observe that a given temperature becomes the critical temperature value as B_c is reached. Therefore $\lambda_L \to \infty$ as $B \to B_c$, and the superconducting current speed decreases toward zero at B_c. In this aspect, our model of superconductivity predicts opposite dynamics with respect to traditional models, which consider superconducting electrons to be in a pairwise free-flowing state.

Among high-temperature superconductors, MgB_2 has one of the lowest λ_L value with $\lambda_L(0) = 90$ nm. The corresponding $\boldsymbol{v}_s$ value is 1300 m/s. This superconducting current speed is much higher than electrons' propagation speed in a normal metal, which is on the order of mm/s. Among elementary metals, aluminum and niobium have even lower $\lambda_L(0) = 50$ nm value. Their corresponding $\boldsymbol{v}_s$ value is 2300 m/s.

11.3. T_c optimization

As illustrated in Figure 11.7, the $T_c < T_{\text{BEC}}$ condition arises near the T_c optimum of any given superconductor. Therefore, understanding this regime is essential for the discovery of higher T_c materials. In this section, we consider all experimental data from the past 100 years, aiming to identify the fundamental limitations on T_c.

Table 11.2. T_c prediction for two-dimensional superconductors near their T_c optimum. T_c is predicted from the $\lambda_L^{-1} \sim T_c$ relationship and MgB_2 values are used as a reference. The bold highlighting indicates the highest charged element, whose charge is screened by metallic electrons.

Superconductor	$\lambda_L(0)$	T_c predicted	T_c measured	Reference
MgB_2	90 nm	reference	39 K	[15]
FeSe	357 nm	9.8 K	9.1 K	[16]
LiFeAs	210 nm	17 K	17 K	[17]
$BaFe_2\,(As_{0.7}P_{0.3})_2$	109 nm	32 K	30 K	[25]

The $T_c < T_{\text{BEC}}$ condition implies that T_c is in fact determined by the binding energy of weakly bound orbitals; this statement follows from Equation (11.1.1) and is experimentally supported by Table 11.1 of Section 11.1. Based on the $\lambda_L^{-1} \sim E_b$ relationship, which was identified in the previous section, we expect to see a $\lambda_L^{-1} \sim T_c$ relationship in this region. To validate this expectation, we compare the low-temperature λ_L values for some widely studied two-dimensional superconductors, which are listed in Table 11.2. Table 11.2 values indeed show the anticipated $\lambda_L^{-1} \sim T_c$ relationship, implying that the T_cs of materials listed in the table are limited by E_b. Many attempts at enhancing MgB_2's superconducting temperature via doping are reported in the literature: its T_c decreases in all cases, which is characteristic of being at or beyond the optimum point.

We note some parallels between FeSe-based and cuprate-based superconductors. As shown in Table 11.2, the ambient pressure T_c of FeSe is 9.1 K. With applied pressure, its T_c rises to 37 K. We interpret this optimum point as the full realization of +2 valence state at Fe sites. This interpretation is supported by the similar T_c value in pressurized FeSe as in the Mg-based and Ba-based superconductors of Table 11.2, which easily go to +2 valence state already at ambient pressure. FeSe has a two-dimensional metallic structure. It is possible to intercalate weakly-metallic planes between the Fe and Se planes; Refs. [20–22] accomplish this by intercalating it with solvated lithium. Metallic lithium can be dissolved, e.g., in ammonia, ethylenediamine, or pyridine; all cases give solvated electrons in the liquid phase and weak metallicity upon freezing. The dissolved lithium is intercalated into FeSe in the liquid phase, and the resulting frozen material thus comprises weakly metallic planes. The weak metallicity of lithium-intercalated FeSe is therefore analogous to hole-doped cuprates' weak metallicity. As shown in Figure 11.8, the $T_c \sim \lambda_L^{-2}(0)$ relationship of FeSe/solvated lithium fits nicely onto

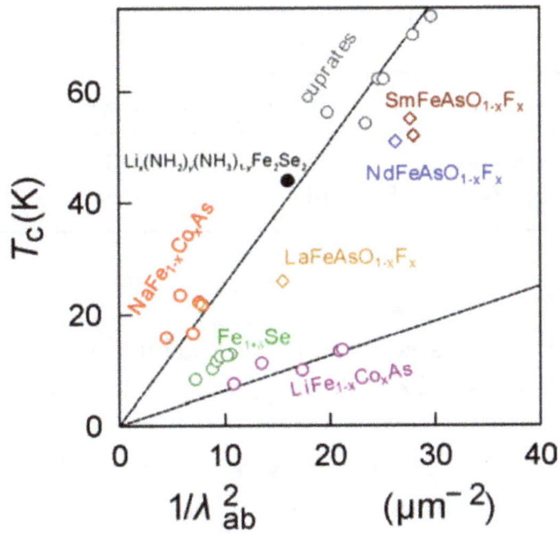

Figure 11.8. The $T_c \sim \lambda_L^{-2}(0)$ relationship for FeSe with intercalated weakly metallic interlayer. Its $T_c \sim \lambda_L^{-2}(0)$ scaling fits on the same line with cuprate superconductors. Reproduced from [22].

the trendline of cuprate superconductors. These features make lithium-intercalated FeSe and cuprates well comparable. The T_c of FeSe/solvated lithium is in the 43–46 K range, independently of the solvent choice.

Regarding the $T_c \sim \lambda_L^{-2}(0)$ scaling in the $T_c \approx T_{\text{BEC}}$ regime, the cuprate line seen in Figures 11.7 and 11.8 has the largest known steepness. Therefore, the phenomenological limit for the $T_c \approx T_{\text{BEC}}$ regime is

$$T_c \leq a\lambda_L^{-2}(0) \tag{11.3.1}$$

where $a_{\text{max}} = 2.53\,\text{K}\,\mu\text{m}^2$.

Our most important goal is to find the limiting value for the T_c optimum point, when superconducting temperature becomes limited by the binding energy of electron-screened orbitals. We consider all available literature data, and group them according to the Z value of the highest charged site. As shown in Table 11.3, for $Z = 2$ the highest reported T_c is 48 K, while for $Z = 3$ the highest reported T_c is 114 K.

Based on the above data, we make the $T_c \sim Z^2$ hypothesis for the optimum transition temperature. Specifically, the data of Table 11.3 fit into

$$T_c \leq bZ^2 \tag{11.3.2}$$

where $b = 12\,K$.

Table 11.3. The highest known T_c values for the $Z = 2$ and $Z = 3$ cases.

Z	Optimum material	T_c ambient	Pressurized T_c	Reference
+2	FeSe/solvated Li	46 K	48 K	[20, 21]
+3	$Tl_2Ba_2CaCu_2O_8$	109 K	114 K	[19]

If the above hypothesis is correct, then achieving a high Z value is the key factor for high T_c. In the following, we consider relevant experimental data.

Among cuprates, record T_c is achieved in mercury-based cuprates, which stands out from the cuprates of all other known cations. Specifically, $HgBa_2Ca_2Cu_3O_{8+x}$ achieves a T_c of 138 K at ambient pressure [27], and a T_c of 166 K at higher pressure [29]. What is the valence state of Hg in this material? For a long time, +2 was thought to be the highest valence state of Hg. Recently, metastable HgF_4 was reported, where Hg is in +4 valence state. Cuprates are highly oxidizing structures: Figure 10.7 shows Cu sites to be in the unusually high +2.2 valence state. Possibly, some Hg sites of $HgBa_2Ca_2Cu_3O_{8+x}$ are in +4 valence state. This idea is strongly supported by Ref. [27], which reports achieving a T_c of 138 K upon fluorine treatment of $HgBa_2Ca_2Cu_3O_{8+x}$, while achieving only a T_c of 134 K with oxygen treatment, which is less strongly oxidizing.

With $Z = 4$, Equation (11.3.2) implies $T_c \leq 192\,K$. The 166 K achieved with $HgBa_2Ca_2Cu_3O_{8+x}$ under pressure is already at 86% of this limit. Another element which can take on many valence states is sulfur: it has +4 state in SO_2, +5 state in SO_2Cl, and even +6 state in SO_2F_2. Pressurized H_2S was discovered to be superconducting, reaching a T_c of 190 K at 155 GPa [15]. This value is very close to the above-mentioned 192 K limit; it means that S sites may be in +4 valence state, thus minimizing volume at this high pressure. Recently, Ref. [28] claims to accomplish room-temperature superconductivity, and its authors studied a sulfur-based pressurized hydrate. As shown in Figure 11.9, initially the T_c versus pressure curve converges to the 194 K temperature limit, which practically coincides with the 192 K limit for the $Z = 4$ state. In case of $Z = 5$, Equation (11.3.2) yields $T_c \leq 300\,K$, and that is indeed convergence temperature for the extrapolation of Figure 11.9 curve. Therefore, the sulfur sites of [28] material are in +4 valence state at the 200 GPa pressure range, and are in +5 valence state at the 500 GPa pressure range.

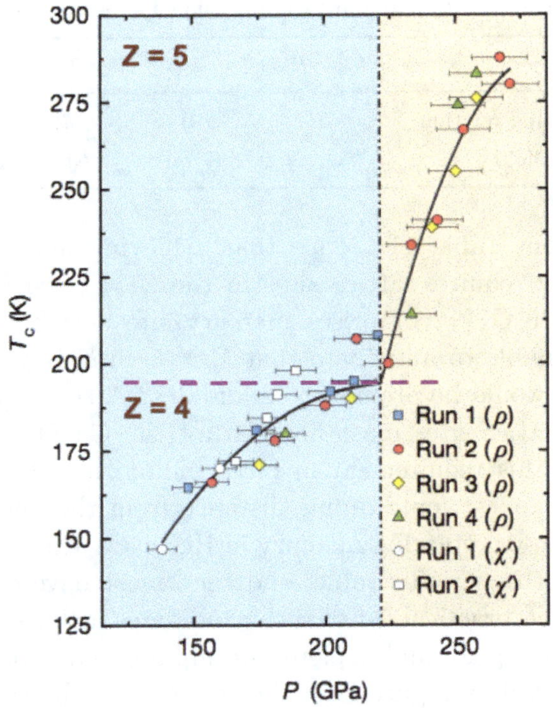

Figure 11.9. T_c as a function of pressure in carbonaceous sulfur hydride (CSH_x), reproduced from [28]. In the blue background region, the curve converges to the overlaid dashed line at 194 K. In the yellow background region, the extrapolated curve converges to about 300 K.

Based on the above experimental data, we conclude that Equation (11.3.2) cannot be a coincidence. Considering Equation (11.3.2) together with the $T_c < T_{\mathrm{BEC}}$ condition in Equation (11.1.1), we arrive at the following relationship between the Z value and binding energy:

$$T_{c,\mathrm{optimum}} = bZ^2 \sim E_{b,\max} \tag{11.3.3}$$

The above relationship would be well understood for ordinary electron orbitals, whose binding energy is indeed proportional to Z^2. But we don't understand yet the reason behind the same relationship for screened electron orbitals in the weakly metallic materials of Section 10.5, under the $T_{c,\mathrm{optimum}}$ condition. Nevertheless, it is not too surprising that the limiting case of very weak metallicity would asymptotically approach the properties of ordinary electron orbitals.

Table 11.4 summarizes the experimentally identified T_c limitations for two-dimensional superconductors.

Table 11.4. T_c limits in two-dimensional superconductors.

Regime	$T_c \approx T_{\mathrm{BEC}}$	$T_c < T_{\mathrm{BEC}}$
Limiting relation	$T_c \leq a\lambda_L^{-2}(0)$	$T_c \leq bZ^2$
Threshold	$a_{\max} = 2.53\,\mathrm{K}\,\mu\mathrm{m}^2$	$b_{\max} = 12\,\mathrm{K}$

It follows from Table 11.4 data that achieving ambient temperature superconductivity requires lattice sites in the $Z = 5$ valence state. Given that $\mathrm{HgBa_2Ca_2Cu_3O_{8+x}}$ comprises mercury sites in the $Z = 4$ valence state, it is reasonable to anticipate that $Z = 5$ sites may also be feasible. Which materials would be promising candidates? A key requirement is that highly charged lattice sites must not disrupt the Cu–O planes' structure. Accommodating this requirement involves minimizing the electric field at the cation edge, i.e., at ionic radius distance from the nucleus. Therefore, suitable materials for replacing mercury in $\mathrm{HgBa_2Ca_2Cu_3O_{8+x}}$ are from the last two rows of the periodic table, and they must have chemically stable +5 valence state. A practical list of such candidates includes praseodymium, tantalum, tungsten, gold, and depleted uranium-based cuprates.

As demonstrated by Figure 11.9, the (CSH_x) type hydride materials are in the $T_c < T_{BEC}$ regime of Table 11.4. Certain hydride materials, such as YH_9 and LaH_{10}, superconduct at nearly 270 K under approximately 200 GPa pressure. Since the highest oxidation state of Y and La is +3, YH_9 and LaH_{10} must be in the $T_c \approx T_{BEC}$ regime of Table 11.4. The structure of these materials is rather similar to metallic hydrogen. With reference to Equation 11.1.2, the rather high T_c in these metallic hydrogen type materials is related to the light mass of the proton lattice sites, implying that weakly bound electron orbits form around proton sites in such materials.

Metallic hydrogen compounds have been researched since the 1980s, anticipating their high-temperature superconductivity. The authors of [39] report an ambient pressure metallic hydrogen compound, described by the C_4KH_x formula, where potassium hydride is intercalated between graphite planes. In such C_4KH_x material, metallic hydrogen comprises negatively charged hydride planes. Unfortunately, C_4KH_x is not a high-temperature superconductor. Based on our theory, it is straightforward to understand the difference between LaH_{10} and C_4KH_x: weakly bound electron orbits do not form around negatively charged hydride sites in C_4KH_x, but they do form around positively charged proton sites in pressurized LaH_{10}. As far as the authors know, preceding superconductivity theories cannot explain the enormous T_c differences between various forms of metallic hydrogen.

11.4. Rotating Superconductors

When a superconducting cylinder is being rotated, a magnetic field appears in its interior. Traditionally, this magnetic field was thought to originate from the inertia of free-flowing electrons. The phenomenology of rotating superconductors is elaborately reviewed in Ref. [1].

Let's consider what happens to hypothetical free-flowing electrons at the start of rotation. Since the already superconducting electrons are supposedly completely free to flow in the body, their inertia keeps them at the same location at the moment when the body starts to rotate. During this start of rotation, a time-dependent magnetic field is generated by the diverging motion of positive ions and superconducting electrons, which induces a Faraday electric field pushing the superconducting electrons in the same direction as the ions. Consequent to this induced Faraday electric field, the superconducting electrons attain the same velocity as the ions in the interior of the cylinder but lag slightly behind in the outer surface layer, thus giving rise to the static magnetic field configuration shown at the top of Figure 11.10.

Measurements show that the uniform magnetic field which fills all the interior has $B = -\frac{2m_e}{e}\omega$ strength, where ω is the angular frequency of the cylinder's rotation. Considering an electron current at the outer surface

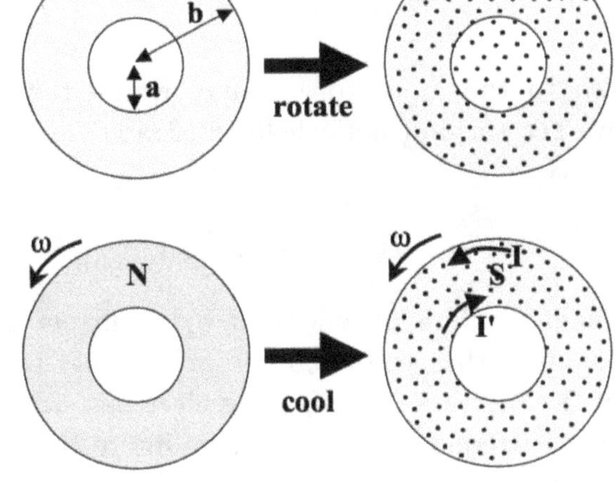

Figure 11.10. The rotating and cooling of a superconducting hollow cylinder. (a) Cylinder cooled into the superconducting state and then set into rotation. (b) Cylinder set into rotation while in the normal state and then cooled into the superconducting state. Dots represent magnetic field lines which are perpendicular to the figure, "N" represents normal state, and "S" represents superconducting state. Reproduced from [1].

of the superconducting cylinder, we give here a simple derivation of this magnetic flux density formula. From the perspective of the rotating cylinder, the above-mentioned superconducting current flows around its surface. A surface electron rotating with angular frequency ω_s at distance R from the rotation axis has a tangential velocity $v = \omega_s R$ and a momentum

$$p_e = m_e v = m_e \omega_s R = eA_{||} \tag{11.4.1}$$

where $A_{||}$ is the component of the vector potential seen by the electron charge, parallel to velocity vector v (see Section 3.4). Now, the line integral of $A_{||}$ over a closed circular path of radius R gives the magnetic flux Φ_s:

$$\begin{gathered}\Phi_s = \oint \boldsymbol{A} \cdot \boldsymbol{dr} = \int_0^{2\pi R} A_{||} dr \\ \Phi_s = 2\pi R A_{||}\end{gathered} \tag{11.4.2}$$

Substituting $A_{||} = \frac{m_e c \omega_s R}{e}$ in Equation (11.4.2), we have

$$\Phi_s = 2\pi R^2 \frac{m_e \omega_s}{e}$$

but

$$\Phi_s = B\pi R^2$$

and consequently

$$B = \frac{2m_e \omega_s}{e}$$

Switching back to the laboratory frame, where the cylinder rotates with $\omega = -\omega_s$, we get the magnetic field illustrated in the upper part of Figure 11.10:

$$B = -\frac{2m_e}{e}\omega \tag{11.4.3}$$

Normally, a magnetic flux is generated by a current $I = dQ/dt$, and at first it seems obvious that the magnetic field would be generated by a current caused by the inertia of free-flowing electrons, which do not follow the rotation of the superconductive body. It is interesting to note that the above formula does not depend on the number of charge carriers involved in the generation of the magnetic field B, but depends on the electron mass m_e. In current textbooks, this is the end of the story about rotating superconductors.

As discussed in Section 11.2, the dynamics of electron currents are dictated by the principle of magnetic energy minimization. Therefore, it should be possible to derive the $\boldsymbol{B} = -\frac{2m_e}{e}\boldsymbol{\omega}$ formula from the London equation. The London equation's scaling factor, which is given by Equation (11.2.9), implies $\lambda_L \sim \sqrt{\frac{m_e}{n_s}\frac{1}{e}}$ relationship. There is no obvious way to eliminate the dependence on superconducting electron density, as needed to arrive at Equation (11.4.3). Nevertheless, Ref. [38] claims to derive Equation (11.4.3) from the London equation, and Ref. [1] lists several studies which contain similar derivations. All these works struggle with the challenge of reconciling Equation (11.4.3) with the London equation. The problem is very deep-rooted. Consider a long superconductive cylinder, which is rotating. The magnetic field in its interior is $\boldsymbol{B} = -\frac{2m_e}{e}\boldsymbol{\omega}$. We cut the cylinder into two shorter cylinders. Each rotating cylinder still has $\boldsymbol{B} = -\frac{2m_e}{e}\boldsymbol{\omega}$ magnetic field in its interior, and it retains the same value regardless of whether the cylinders are on top of each other or far apart. But this situation is totally contrary to the basic principles of current-generated magnetic fields, which dictate that magnetic fields must add up when the two pieces are placed on top of each other — just like magnetic fields of solenoid segments add up. We come to a strange conclusion: the magnetic field of a rotating superconductor cannot be generated by free-flowing currents! This conclusion is further corroborated by the situation depicted in the bottom half of Figure 11.10. In that scenario, the free-flowing current model would dictate outer surface superconducting electrons lagging behind the main body rotation, but inner surface superconducting electrons rotating faster than the main body. The traditional explanations of the free-flowing current model, which are electron inertia based, break down once again.

Given that the magnetic field of Equation (11.4.3) cannot be related to free-flowing currents, and given that this magnetic field is not related to the number of superconducting electrons, we investigate whether it is generated from electron spin effect. As discussed in Chapters 4 and 7, the Dirac equation describes electron dynamics, and it can be factorized into two parts. One of these parts, which describes electron spin dynamics, turns out to be the Proca equation. When weakly bound electrons condense into BEC state, i.e., a single wavefunction fills the superconductor's interior, this Proca equation becomes macroscopically relevant. Let's think of the implications.

Considering a single electron, its momentum p_e can be written as a function of De Broglie wavelength λ:

$$p_e = m_e v = \hbar k = \frac{h}{\lambda}$$
$$\left[m_e v = k = \frac{2\pi}{\lambda}\right]_{\text{NU}} \tag{11.4.4}$$

This implies a phase shift φ after a full body rotation at angular speed ω_s:

$$\varphi = k2\pi R = \frac{2\pi m_e \omega_s R^2}{\hbar} \tag{11.4.5}$$

This phase shift can be also derived from the magnetic Aharonov–Bohm equation, which was studied in Chapter 4:

$$\varphi = \frac{e}{\hbar}\oint \boldsymbol{A}\cdot \boldsymbol{dr} = \frac{eA_{\parallel}}{\hbar}\int_0^{2\pi R} ds = \frac{2\pi R m_e v}{\hbar} = e\frac{\Phi_s}{\hbar} \tag{11.4.6}$$
$$[\varphi = e\Phi_s]_{\mathrm{NU}}$$

It is clear that Equations (11.4.5) and (11.4.6) give the same phase shift value for a single electron.

Considering that $\Phi_s = B\pi R^2$, the above Aharonov-Bohm equation leads to the $B = \frac{2m_e\omega_s}{e}$ result when $v = \omega_s R$. From the perspective of the Proca equation, Figure 11.10 is interpreted as follows:

- The top of Figure 11.10 corresponds to a $-\varphi$ phase shift of the Proca field at the outer surface only. The phase shift value is derived from Equation (11.4.6).
- The bottom of Figure 11.10 corresponds to a $-\varphi$ phase shift of the Proca field at the outer surface, and a $+\varphi$ phase shift of the Proca field at the inner surface. The phase shift values are derived from Equation (11.4.6).

What could dictate the difference between the two configurations of Figure 11.10? The bottom of Figure 11.10 represents a scenario where the superconductor is mechanically isolated from its environment, and therefore it must conserve its total angular momentum. The phase shift obtained by the Proca field implies pushing against the superconducting body. If the Proca field gained only a $-\varphi$ phase shift at the outer surface, it means that it would push the lattice body to rotate faster. However, the actual situation depicted in Figure 11.10 means that the lattice body does not gain any angular momentum during normal-to-superconductor transition, i.e., its angular speed does not change. This corresponds to the symmetry principle during normal-to-superconductor state transition: such a transition must not lead to an angular momentum asymmetry between superconducting electrons and the lattice body.

In contrast, the top of Figure 11.10 depicts a situation where external rotating force is applied, which sets the superconductor into rotating motion. This external force is responsible for generating the $-\varphi$ phase shift at the outer surface only, as the superconductor is set into rotation.

11.5. Magnetic Flux Quantization

Considering electron-screened orbitals, let v_s denote the superconducting electron's kinetic speed, $A_{\|}$ is the v_s-parallel component of the vector potential seen by the electron charge, r_e is the electron's Zitterbewegung radius, and s is the ratio between the electron-screened orbital radius R_s and an ordinary Bohr orbit ground state radius, i.e., $s = R_s/R_B$. The superconducting electron's orbital momentum is then given by the following expression:

$$eA_{\|} = m_e v_s = m_e \omega_s s R_B$$

$$\left[R_B = \frac{r_e}{\alpha} = \frac{1}{\alpha m_e}\right]_{\text{NU}}$$

$$\left[\omega_B = \frac{\alpha}{R_B} = \frac{\alpha^2}{r_e} = \alpha^2 m_e\right]_{\text{NU}}$$

An electron-screened orbital ground state has the same amount of angular momentum as an ordinary Bohr orbital ground state. To see this, we recall from Chapter 10 that the process of transitioning from R_s to R_B is equivalent to a gradual increase of the amount of charge in the center; all forces remain centrally directed in such a process. Therefore, vR is constant, and consequently:

$$m_e \omega_s R_s^2 = m_e \omega_B R_B^2$$

$$\left[\omega_s = \frac{\omega_B R_B^2}{R_s^2} = \frac{\alpha^2 m_e}{s^2}\right]_{\text{NU}}$$

Now we address magnetic flux calculation. In a classical Kepler orbit, magnetic flux is calculated from a line integral over a closed circular path:

$$\left[\Phi_k = A_{\|} 2\pi R_s = \tfrac{m_e \omega_s 2\pi R_s^2}{e} = \tfrac{2\pi}{e}\right]_{\text{NU}} \qquad (11.5.1)$$

$$\Phi_k = \tfrac{h}{e}$$

A classical Kepler orbit model of the ground state, i.e., the Bohr model, corresponds to $\hbar$ angular momentum. However, Sections 0.10 and 6.9 demonstrated that a ground state electron has $\hbar/2$ orbital angular momentum, which is isotropically coupled to the $\hbar/2$ angular momentum of the nucleus. Therefore, it is reasonable to assume that the electron's corresponding

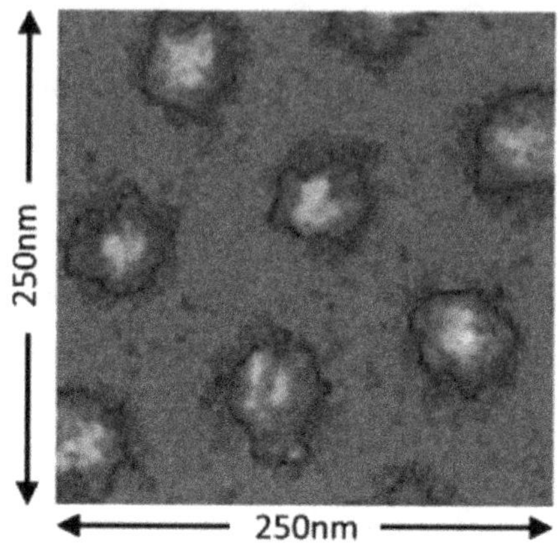

Figure 11.11. A lattice of quantized magnetic flux lines in a superconducting MgB_2 film, obtained through scanning tunneling spectroscopy imaging. Reproduced from [13].

magnetic flux would be given by

$$\Phi_s = \frac{\Phi_k}{2} = \frac{h}{2e} \tag{11.5.2}$$

We note that the magnetic flux value does not depend on the scaling factor s: it is a constant given by Equation (11.5.2).

The flux value given by Equation (11.5.2) has the same value as the magnetic flux value in each penetrating flux line, which appears when two-dimensional superconductors are placed under a strong magnetic field. Figure 11.11 shows such quantized flux lines.

The results of this section are a second method of showing that external magnetic fields interact with weakly bound electron orbitals. The other derivation was given in Section 11.2. It is no coincidence that both methods uncover the same phenomenon.

11.6. Conclusions

We developed a realistic understanding of superconductivity, which matches experimental data. We demonstrated that superconductivity phenomena originates from weakly bound electrons' BEC state. The apparent mysteries of superconductivity relate to the presently poor understanding of Bose–Einstein condensates and electron-screened orbitals. Nevertheless, some puzzling aspects of superconductors have now gained rational explanations:

- **Strictly pairwise Cooper electron coupling:** Conventional superconductivity theories consider superconducting electrons to be in fully delocalized state, but lack a proper explanation for the observed strictly pairwise occupancy rule of Cooper-paired electrons. In other words, there is no good reason for the pairwise electron coupling limit for the hypothetical interaction between negatively charged electrons and positively charged phonons. We describe superconducting electrons to be in BEC state of weakly bound orbitals. Therefore, the same pairwise orbital occupancy limit applies which we proved in Chapter 9. Moreover, Cooper-pairing appears spontaneously, in analogy to the appearance of electron pairing in any ordinary electron orbital. There is no further need to "explain" the production of Cooper-pairing.

- **The entanglement of separating Cooper-paired electrons:** As mentioned in Chapter 9, recent experiments demonstrated the spin entanglement of Cooper-paired electrons after their separation. This experimental property follows from the theory of Chapter 9, which demonstrates that any two electrons sharing the same orbital are in isotropically entangled state.
- **The origin of the critical temperature value:** A first-principles calculation of T_c is a main goal of all superconductivity theories. A proper first-principles calculation is impossible under traditional superconductivity theories because their basic premise is phonon-mediated electron coupling, which is incorrect. Mainstream efforts are thus doomed to become a phenomenological "epicycles building" exercise, involving empirically adjusted parameters. The calculations of [11] represent the latest mainstream accomplishment, and it operates with at least three empirically adjusted parameters. Reference [12] reviews past attempts at finding simple formulas for the calculation of T_c: empirically adjusted parameters have been present in all the T_c estimations. In contrast, our analysis reveals that T_c may be bounded by either the BEC condensation temperature or by the electron-screened orbital forming temperature. Thus, the overall T_c is derived as the minimum value between two different formulas, which can never be expressed as one single formula. The limiting relationships appearing in Table 11.4 have a well-understood physical meaning.
- **Magnetic flux quantization:** Low-intensity magnetic fields are expelled from the interior of superconductors. In type-II superconductors, quantized magnetic flux lines, which punctuate the interior, arise above a certain external field strength. With further increasing external magnetic field strength, the growing number of flux lines starts to organize into a lattice structure. The measurement of such a lattice of flux lines is shown

in Figure 11.11. In all type-II superconductors, the penetrating magnetic flux has $\Phi_o = \frac{h}{2e}$ value in each flux line. Mainstream theorists explain this quantized magnetic flux value by imposing a periodicity condition on the wavefunction of superconducting electrons: it must be continuous along any loop, which encircles a flux line. To derive magnetic flux quantization, this wavefunction continuity requirement is then used together with the so-called minimal coupling Ginzburg–Landau equation, which is thought to describe the interaction between the superconducting wavefunction and the external electromagnetic field. The Ginzburg–Landau equation is derived from the BCS theory through lengthy calculations, which are valid only in the $T \rightarrow T_c$ limit. However, the phenomenon of magnetic flux quantization is measured also at temperatures much lower than T_c. Thus, recent efforts [14] aim to extend the temperature range where the derivation of the Ginzburg–Landau equation is mathematically valid, but the complexity of this theory grows enormously. Moreover, the above-discussed problems associated with the BCS theory are then also inherited by the Ginzburg–Landau equation. Another major problem is that the Ginzburg–Landau equation implies $\lambda_L^{-2} \sim 1-(\frac{T}{T_c})$ relationship [26], which is in contradiction with the experimentally measured $\lambda_L^{-2} \sim 1-(\frac{T}{T_c})^4$ relationship [6, 7]. How to find then a more reasonable explanation, which would be favored by Occam's razor? We calculated the magnetic flux in electron-screened orbits in Section 11.5. Our calculation exactly matches the experimentally measured $\Phi_o = \frac{h}{2e}$ value. We suggest therefore that this $\Phi_o = \frac{h}{2e}$ value originates from the interaction between the external magnetic field and just one electron-screened orbital.

We anticipate that the results of this chapter will eventually lead to a rational design of higher temperature superconductors.

Appendix: A Useful Vector Field Identity

Let $\boldsymbol{F}, \boldsymbol{G}$ be vector fields, and let Ω be a volume region in space. Let σ denote the surface of Ω. The field $\boldsymbol{G}$ will serve as "variation field," and therefore we require it to vanish on the surface of Ω, i.e., $\boldsymbol{G}\mid_\sigma = 0$. We start from the following well-known vector field identity:

$$\nabla \cdot (\boldsymbol{F} \times \boldsymbol{G}) = (\nabla \times \boldsymbol{F}) \cdot \boldsymbol{G} - \boldsymbol{F}\,(\nabla \times \boldsymbol{G})$$

We integrate the above equation over Ω:

$$\int_\Omega \nabla \cdot (\boldsymbol{F} \times \boldsymbol{G})\, dV = \int_\Omega (\nabla \times \boldsymbol{F}) \cdot \boldsymbol{G}\, dV - \int_\Omega \boldsymbol{F}\,(\nabla \times \boldsymbol{G})\, dV$$

Upon rearranging terms, we get

$$\int_{\Omega} \boldsymbol{F}\,(\nabla \times \boldsymbol{G})\; dV = \int_{\Omega} (\nabla \times \boldsymbol{F}) \cdot \boldsymbol{G}\, dV - \int_{\Omega} \nabla \cdot (\boldsymbol{F} \times \boldsymbol{G})\; dV \tag{A.1}$$

We evaluate the last term of the above equation by changing from volume integration to bounding surface integration, as follows:

$$\int_{\Omega} \nabla \cdot (\boldsymbol{F} \times \boldsymbol{G})\; dV = \int_{\sigma} \boldsymbol{F} \times \boldsymbol{G}\, dA = 0$$

The above equation evaluates to zero because $\boldsymbol{G}$ vanishes on the surface of Ω. Therefore, Equation (A.1) simplifies to

$$\int_{\Omega} \boldsymbol{F}\,(\nabla \times \boldsymbol{G})\; dV = \int_{\Omega} (\nabla \times \boldsymbol{F}) \cdot \boldsymbol{G}\, dV \tag{A.2}$$

Acknowledgments

The authors thank Jan von Pfaler for clarifications about the London equation's mathematical derivation.

References

[1] J.E. Hirsch, Defying inertia: How rotating superconductors generate magnetic fields. (2019) arXiv:1812.06780v2.

[2] E.T. Jaynes, Scattering of light by free electrons as a test of quantum theory. *The Electron — New Theory and Experiment*, vol. 45. Chapter 1, Springer, Dordrecht, 1991.

[3] T. Lancaster and S.J. Blundell, *Quantum Field Theory for the Gifted Amateur.* Oxford University Press, New York, NY, USA, 2014.

[4] T. Schneider *et al.*, Diamagnetism, Nernst signal, and finite-size effects in superconductors above the transition temperature. *Physical Review B*, **83**(14) (2011) 144527.

[5] T. Valla *et al.*, Disappearance of superconductivity due to vanishing coupling in the overdoped $Bi_2Sr_2CaCu_2O_{8+d}$. *Nature Communications*, **11**(1) (2020). DOI: 10.1038/s41467-020-14282-4.

[6] H. Keller, Positive muons as probes in high-T_c superconductors. *Exotic Atoms in Condensed Matter.* Springer, Springer Berlin Heidelberg, 1992.

[7] L. Krusin-Elbaum *et al.*, Direct measurement of the temperature-dependent magnetic penetration depth in Y-Ba-Cu-O crystals. *Physical Review Letters*, **62**(2) (1989) 217–220.

[8] P.W. Phillips *et al.*, Exact theory for superconductivity in a doped Mott insulator. *Nature Physics Letters*, **16** (2020) 1175–1180.

[9] J.E. Hirsch, BCS theory of superconductivity: It is time to question its validity. *Physica Scripta*, **80**(3) (2009) 035702.

[10] A.M. Schaeffer *et al.*, High-pressure superconducting phase diagram of ^{6}Li: Isotope effects in dense lithium. *Proceedings of the National Academy of Sciences*, **112**(1) (2015) 60–64.

[11] A. Sanna *et al.*, Combining eliashberg theory with density functional theory for the accurate prediction of superconducting transition temperatures and gap functions. *Physical Review Letters*, **125**(5) (2020) 057001.
[12] M. Surma, A new semiempirical formula for superconducting transition temperatures of metals and alloys. *Physica Status Solidi B-basic Solid State Physics B*, **116**(2) (1983) 465–474.
[13] M.R. Eskildesn *et al.*, Vortex imaging in the π band of magnesium diboride. *Physical Review Letters*, **89** (2002) 187003.
[14] L. Xu *et al.*, From BCS theory for isotropic homogeneous systems to the complete Ginzburg-Landau equations for anisotropic inhomogeneous systems. *Physical Review B*, **57**(18) (1998) 11654.
[15] E.F. Talantsev *et al.*, Universal scaling of the self-field critical current in superconductors: From sub-nanometre to millimetre size. *Nature Scientific Reports*, **7** (2017) 10010.
[16] A.E. Koshelev *et al.*, Melting of vortex lattice in the magnetic superconductor $RbEuFe_4As_4$. *Physical Review B*, **100**(9) (2019) 094518.
[17] D.S. Inosov *et al.*, Weak superconducting pairing and a single isotropic energy gap in stoichiometric LiFeAs. *Physical Review Letters*, **104**(18) (2010) 187001.
[18] T. Klein *et al.*, Metal-insulator transition and superconductivity in boron-doped diamond. *Physical Review B*, **75**(16) (2007) 165313.
[19] J.B. Zhang *et al.*, Pressure tuning of superconductivity independent of disorder in $Tl_2Ba_2CaCu_2O_{8+d}$. (2015) arXiv:1502.02227.
[20] T. Hatakeda *et al.*, New Li-Ethylenediamine-Intercalated Superconductor Lix(C2H8N2)yFe2-zSe2 with Tc = 45 K. *Journal of the Physical Society of Japan*, **82**(12) (2013) 123705.
[21] L. Zhao *et al.*, Structural evolution and phase diagram of the superconducting iron selenides. *Physical Review B*, **99**(9) (2019) 094503.
[22] M. Burrard-Lucas *et al.*, Enhancement of the superconducting transition temperature of FeSe by intercalation of a molecular spacer layer. *Nature Materials*, **12**(1) (2013) 15–19.
[23] V.K. Thorsmolle *et al.*, C-axis Josephson plasma resonance observed in $Tl_2Ba_2CaCu_2O_8$ superconducting thin films by use of terahertz time-domain spectroscopy. *Optics letters*, **26**(16) (2001) 1292–1294.
[24] P.M. Shirage *et al.*, Inverse iron isotope effect on the transition temperature of the (Ba, K)Fe_2As_2 superconductor. *Physical Review Letters*, **103**(25) (2009) 257003.
[25] K. Hashimoto *et al.*, A sharp peak of the zero-temperature penetration depth at optimal composition in $BaFe_2(As_{1-x}P_x)_2$. *Science*, **336**(6088) (2012) 1554–1557.
[26] C.E. Gough, High-T_c superconductors: A background for μSR measurements. *Exotic Atoms in Condensed Matter*. Springer, 1992.
[27] K.A. Lokshin *et al.*, Enhancement of T_c in $HgBa_2Ca_2Cu_3O_{8+d}$ by fluorination. *Physical Review B*, **63**(6) (2001) 064511.
[28] E. Snider *et al.*, Room-temperature superconductivity in a carbonaceous sulfur hydride. *Nature*, **586**(7829) (2020) 373–377.
[29] M. Monteverde *et al.*, Fluorinated Hg-1223 under pressure. *Physica C*, **408–410** (2004). DOI: 10.1016/j.physc.2004.02.020.
[30] J.E. Hirsch, *Superconductivity begins with H.* World Scientific, 2020 arXiv:2001.09496.
[31] K. Fossheim *et al.*, *Superconductivity — Physics and Applications*, Chapter 6.12. John Wiley & Sons, 2004.

[32] A. Badia-Majos, Understanding stable levitation of superconductors from intermediate electromagnetics. *American Journal of Physics*, **74**(12) (2006) 1136.

[33] H. Koizumi *et al.*, Theory of supercurrent in superconductors. *International Journal of Modern Physics B*, **34**(31) (2020) 2030001.

[34] W. Keesom *et al.*, Measurements of the latent heat of thallium connected with the transition, in a constant external magnetic field, from the supraconductive to the non-supraconductive state. *Physica*, **1**(1) (1934) 503–512.

[35] W. Keesom *et al.*, Measurements of the latent heat of tin in passing from the supraconductive to the non-supraconductive state. *Physica*, **3** (1936) 371–384.

[36] W. Keesom *et al.*, Measurements of the latent heat of tin while passing from the superconductive to the non-superconductive state at constant temperature. *Physica*, **4** (1937) 487–493.

[37] W. Keesom *et al.*, On the reversibility of the transition processs between the superconductive and the normal state. *Physica*, **5** (1938) 993–998.

[38] H. Essén *et al.*, The Darwin-Breit magnetic interaction and superconductivity (2013) arXiv:1312.1607.

[39] S. Miyajima *et al.*, Two-dimensional metallic hydrogen in the potassium-hydrogen-graphite ternary intercalation compound. *Physical Review Letters*, **64**(3) (1990) 319.

[40] S. Grebenev *et al.*, Superfluidity within a small helium-4 cluster: The microscopic Andronikashvili experiment. *Science*, **279**(5359) (1998) 2083–2086.

[41] K.K. Gomes *et al.*, Visualizing pair formation on the atomic scale in the high-Tc superconductor $Bi_2Sr_2CaCu_2O_{8+d}$. *Nature*, **447**(7144) (2007) 569+.

[42] M. Xie, Bose–Einstein condensation temperature of finite systems. *Journal of Statistical Mechanics*, **5** (2018) 053109.

https://doi.org/10.1142/9781800611306_0012

Chapter 12

Compton-Scale Electron–Proton Composite

András Kovács
BroadBit Energy Technologies, Metallimiehenkuja 8 C, 02150 Espoo, Finland
andras.kovacs@broadbit.com

12.1. Introduction

In this chapter, we discuss the physics of a highly localized electron–proton composite, whose radius is the electron Zitterbewegung radius. If such an electron state exists, it shall have an impact on the probability of nuclear reactions. It is indeed known that the electron environment can impact, e.g., the rate of nuclear electron capture; this effect has been most extensively studied for ^{7}Be. For instance, Table 1 of Ref. [2] shows the half-life ^{7}Be in different environments, where the electron capture rate of ^{7}Be can indeed be changed by its environment. An electron configuration involving close electron–nucleus proximity would enhance the probability of electron capture. Most importantly, a close proximity electron–proton or electron-deuteron configuration would allow small inter-nuclear distance between such quasi-neutron and some other nucleus, thereby enabling catalyzed fusion reactions.

A Compton-scale electron–nucleus proximity also implies high probability of nucleus-to-electron energy transfer, which is a pre-condition for the production of energetic electrons. In muon-catalyzed $p-D$ fusion experiments, it has been observed that the muon carries away the fusion energy in most reactions, suggesting that a similar effect might arise under close electron–nucleus proximity. Historically, no similar effect has been observed with electrons. However, it was recently observed that certain environments cause a shift in the deexcitation pathway of a fused ^{3}He nucleus from gamma photon emission toward electron acceleration [1]. Specifically, all of the nuclear reaction energy of some $p-D$ fusion events has been carried away by energetic electrons. The results presented in Ref. [1] therefore point to the existence of an electron orbital with close electron–nucleus proximity, resulting in strong enough electron–nucleus interaction for electrons to carry away the nuclear excitation energy. In fact, we can estimate the electron–nucleus proximity of this sought for electron state. In muon-catalyzed $p-D$ fusion events, the muon carries away the reaction energy in 15% of cases, and has 250 fm average distance from the nucleus. According to the measurements of [1], the electron carries away the reaction energy of $p-D$ fusion events in $7 \pm 2\%$ of cases. Therefore, as an order of magnitude estimation, this special electron state has ~500 fm average distance from the nucleus. This estimation well matches the 386 fm electron Zitterbewegung radius.

Based on the above motivation, we investigate such a proton–electron state, where an electron is in Compton-scale proximity with respect to the $Z = 1$ nucleus.

12.2. The Theory of Close Proximity Electron–Nucleus Composite

12.2.1. *Characterization of the electron state*

In Chapters 3 and 4, we referred to ultra-dense hydrogen experiments [8], which established the existence of a two-electron plus two-proton structure with 2.3 pm inter-nuclei distance. As was shown in Figure 4.3, this inter-nuclei distance corresponds to $d_c = cT = \lambda_c \approx 2.42 \cdot 10^{-12}$ m distance between electrons' charges, where λ_c is the Compton wavelength. In other words, each electron's Zitterbewegung motion is permanently centered around the nucleus.

Let's consider a single electron–proton system. Shall we understand this state as a "deep orbit" of the electron, or as a Compton-scale pseudo-neutron type composite particle?

In a hypothetical "deep orbit" case, the virial theorem applies to the electron wavefunction. At $R_0 = 386.16$ fm Zitterbewegung distance from the nucleus, the electron's potential is $U_p = -3.728$ keV. According to the virial theorem, its kinetic energy then must be $T = -U_p \frac{\gamma_L}{\gamma_L+1}$. If such a state were feasible, its binding energy must not exceed the 13.6 eV binding energy of the ordinary hydrogen ground state because otherwise it would become the actual ground state of hydrogen. This condition implies $-\frac{T}{U_p} > 0.99635$, which in turn implies $\gamma_L > 274$. The available 3.728 keV potential energy cannot produce such enormous γ_L value, and therefore such a "deep orbit" state is impossible.

In contrast, the results of Section 4.5 imply that the electron does not have any kinetic energy with respect to protons in ultra-dense hydrogen composites. This leads us to consider the Compton-scale composite as a neutron-like particle, which only has a single quantum mechanical wavefunction for the entire particle. The electron has then no quantum mechanical motion relative to the nucleus, i.e., its Zitterbewegung structure becomes localized with respect to the nucleus. The existence of such a composite particle is a building block for the formation of a two-electron plus two-proton structure discussed in Section 4.5, which are observed to have 2.3 pm inter-nuclei distance. The energy balance of this composite state may be calculated by using the electric Aharonov–Bohm formula, given by Equation (4.3.7). At a constant potential of the electric charge surface, this formula reduces to the following form:

$$\hbar\omega = eV \tag{12.2.1}$$

where ω is the de Broglie frequency, e is the elementary charge, and V is the potential at the charge surface. In its rest frame, the electron's de Broglie frequency corresponds to $\hbar\omega = m_0c^2 = 511\,\text{keV}$. Within the Compton-scale composite, the electron is also at rest with respect to the nucleus.

At a given Zitterbewegung radius R of the composite state, the electrostatic potential at the spherical electron charge surface is $m_0c^2\frac{R_0}{eR}$, where R_0 is the Zitterbewegung radius of a free electron. This electron-internal electrostatic potential is lowered by the electron-external electric field of the nucleus: the electron surface potential is thus lowered by $\alpha m_0c^2\frac{R_0}{eR}$. The condition of composite particle formation, i.e., the $\kappa = 0$ condition of Section 7.5, is that the electron surface remains at the same potential as in the free electron state:

$$\frac{R_0}{R} - \alpha\frac{R_0}{R} = 1 \tag{12.2.2}$$

The above condition leads to $R = R_0\,(1-\alpha) = 383.34\,\text{fm}$ Zitterbewegung radius. In other words, the lowered potential energy which the electron experiences due to the proton's field is counterbalanced by its own electromagnetic field energy increase. The total electron energy remains

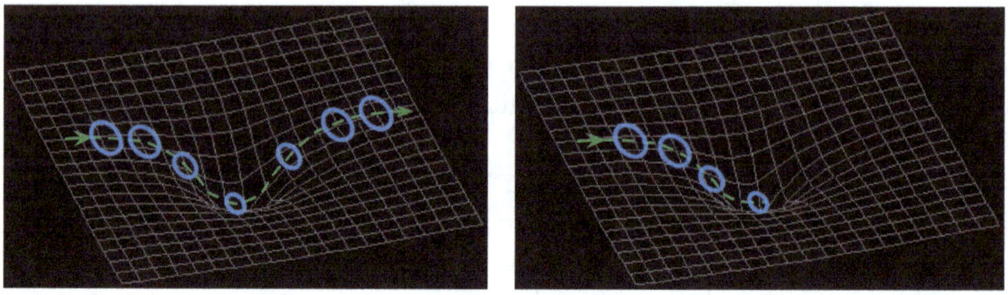

(a) (b)

Figure 12.1. Illustration of the electron–proton composite establishment process. The proton is located at the bottom of the spacetime curvature dip. The grid represents spacetime curvature, the electron path is represented by the dashed curve, the rings represent electron Zitterbewegung. (a) The electron bypasses the proton without being captured. (b) Formation of an electron–proton composite when the nucleus "measures" a free electron's position within its wavefunction.

unchanged with respect to a stationary free electron state. The above equation may be of course applied to any slow-moving electron, the electron does not need to be stationary.

Figure 12.1 illustrates the process of electron–proton composite establishment. Mathematically, this process may be understood from the results of Section 7.5, in particular Equations (7.5.5) and (7.5.6). In an orbital-forming interaction, the electron dynamics is characterized by the $\kappa = \frac{M}{m_e+M}$ parameter of Equations (7.5.5) and (7.5.6): the electron wavefunction moves right through the proton as illustrated in the Figure 12.1(a). In a composite-forming interaction the electron dynamics is characterized by the $\kappa = 0$ parameter of Equations (7.5.5) and (7.5.6): the proton and electron interact via their quantized magnetic fluxes, which drag the nucleus along with the electron as shown in Figure 12.1(b). This magnetic interaction energy is a few electronvolts, and will be discussed in the next section.

The above paragraphs demonstrate that the process of electron–proton composite establishment cannot be studied from the perspective of a point-particle model. It would appear from the point-particle perspective as though the electron gained 3.728 keV kinetic energy just before the composite formation, which then disappeared into the composite, while the actual magnetic binding energy is just few eV. In order to understand the Compton-scale electron–proton composite formation process, it is essential to work with the correct electron structure, which we developed in Chapters 3–7.

Table 12.1 summarizes the main properties of various fundamental electron states.

Table 12.1. Electron properties in basic electron states.

	Free electron state	1s orbital state around a proton	Compton-scale electron-proton composite
Electromagnetic angular momentum	$\hbar$	$\hbar$	$\hbar$
Mean distance from proton		$\alpha^{-1}R_0 = 52.9\,\text{pm}$	$R_0\,(1-\alpha) = 383\,\text{fm}$
Kinetic electron speed	Frame dependent	αc with respect to the proton	0 with respect to the proton
Uncertainty of the electron speed	$\sigma_p\sigma_x \geq \frac{\hbar}{2}$	0 with respect to the proton	0 with respect to the proton

12.2.2. *Magnetic electron–proton and electron–electron interactions*

The above-discussed $\kappa = 0$ case of zero electrostatic binding energy means that so far there is nothing to stabilize the neutron-like Compton-scale particle. We now take magnetic electron–nucleus interactions into account.

To minimize the magnetic potential, the electron's and proton's magnetic moments align their directions. The strong magnetic field at the center of the electron Zitterbewegung interacts with the proton's Zitterbewegung motion, causing it to precess around the magnetic field lines. This type of nuclear precession is equivalent to the spin precession described in Chapters 3 and 4, and is routinely exploited in nuclear magnetic resonance imaging devices. The induced precession of the proton's Zitterbewegung orientation is depicted by the green cone in Figure 12.2: it causes a Zeeman split in the proton's energy levels, and the proton assumes the lower energy level. This lowered proton energy level creates a restoring force for maintaining the equilibrium state, i.e., the neutron-like composite particle is now in a magnetically stabilized state. Figure 12.2 illustrates this Compton-scale particle system. Since the magnetic binding energy of this state is less than the 13.6 eV binding energy of the hydrogen ground state orbital, it is metastable with respect to the ordinary hydrogen atom state.

The proton's precession axis is always orthogonal to the Zitterbewegung plane. We derived in Chapter 3 that in the electron's rest frame the magnetic field strength at the center of Zitterbewegung is 32.21055 MT. The "gyromagnetic ratio" of the proton is measured to be 42.58 MHz/T. It

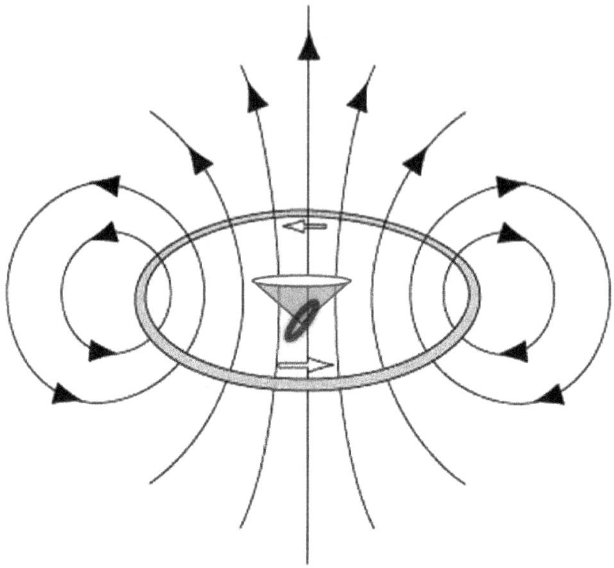

Figure 12.2. Illustration of the stabilized electron orbit. The gray ring represents the electron's Zitterbewegung current, the oriented lines represent magnetic field lines, the central tilted ring represents the proton's Zitterbewegung current, and the cone represents the Larmor precession cone of the proton's spin.

means that it has, e.g., 100 MHz NMR resonance under 2.3488 T magnetic field. Under 32.21055 MT field, the proton's NMR resonance frequency is 1371.525 THz, and it would therefore realize $\frac{1371.525}{2} = 685.763$ THz Zeeman split under a magnetic field of fixed orientation.

This 685.763 THz Zeeman split corresponds to emitted energy and wavelength of $\varepsilon_{\text{split}} = 2.836$ eV and $\lambda_{\text{split}} = 437.16$ nm, respectively. However, this estimation is valid with respect to a known electron orientation. Since the proton measures the electron wavefunction, it also measures the electron's spin value. Before the transition, the electron is in the $|\psi\rangle = \frac{1}{\sqrt{2}}(|+\rangle + |-\rangle)$ state, where $|+\rangle$ means that the electron spin is parallelly aligned with respect to the nuclear spin and $|-\rangle$ means that the electron spin is aligned anti-parallel to the nuclear spin. As we will see in what follows, the $\frac{1}{\sqrt{2}}$ probability factor appears in the emitted energy, which thus becomes $E = \frac{1}{\sqrt{2}}\varepsilon_{\text{split}}$. Therefore, the wavelength of emitted light will be $\lambda_{\text{emitted}} = \sqrt{2}\lambda_{\text{split}} = 618.24$ nm.

Now we consider that multiple Compton-scale particles may form a stable composite along their axis, as was illustrated in Figure 5.3. Each constituent is an electrically neutral Compton-scale particle, and the stabilizing force is magnetic interaction between electrons. Within a Compton-scale composite chain, the 1st and Nth particle are separated by $d = (N-1)\,\lambda_c\sqrt{1-\frac{1}{\pi^2}}$ distance, where λ_c is the Compton wavelength. See Section 4.5 for the derivation of this formula. The magnetic binding energy between them is as follows:

$$U_m = \frac{\mu_0}{4\pi}2\frac{(2\mu_b)\,(2\mu_b)}{d^3} = (N-1)^{-3} \times 35.24\ eV \tag{12.2.3}$$

where μ_b is the Bohr magneton, and the $2\mu_b$ factor expresses that the electron's internal magnetic dipole moment is twice the Bohr magneton. Upon the establishment of such magnetic binding, each of the involved electrons emits half of the binding energy, and also the above-mentioned $\frac{1}{\sqrt{2}}$ factor applies. Therefore, the energy quantum and wavelength of emitted light will be $E = \frac{1}{\sqrt{2}}\frac{U_m}{2}$ and $\lambda_{\text{emitted}} = (N-1)^3 \times 99.51\,\text{nm}$.

There is spectroscopic evidence for the above-calculated light emission. Slobodan Stankovic recently measured the emission spectrum of plasmas which were generated from oxyhydrogen fuel combustion or from high-voltage arc discharge. These measurements are documented in Ref. [15], and shown in Figure 12.3. These proton-containing plasmas were observed to cause nuclear transmutations [15], and therefore fit well the above-described model of Compton-scale composite particle formation from a low-temperature electron–proton plasma state. Figure 12.3 shows the presence of three well-defined emission peaks, which have exactly the same wavelength in both oxyhydrogen fuel combustion and from high-voltage arc discharge experiments. The peaks have 618.22, 720.9, and 800.4 nm wavelength. These emission lines are very sharp and do not match the emission of any known chemical transitions. The lowest wavelength of this triplet fits very well the expected 618.24 nm value for electron–proton interaction. Regarding electron-electron interaction, light emission is in the visible range for $N = 3$, and the above dipole approximation formula gives $\lambda_{\text{emitted}} = 796.1\,\text{nm}$. That is 99.5% match against the measured 800.4 nm peak. These results are summarized in Table 12.2.

Finally, we calculate the magnetic interaction for the deuteron case. The ratio of deuteron:proton NMR frequencies is $6.535\,\text{MHz/T}/42.58\,\text{MHz/T} = 0.1535$. This difference is related to the different spins of the proton and deuteron nuclei; the nuclear spin dependence of the NMR frequency was given in Equation (4.4.4). Thus, in the deuteron case the Zeeman split energy is 0.1535 times lower than in the proton case. Applying 1/0.1535 scaling,

(a)

(b)

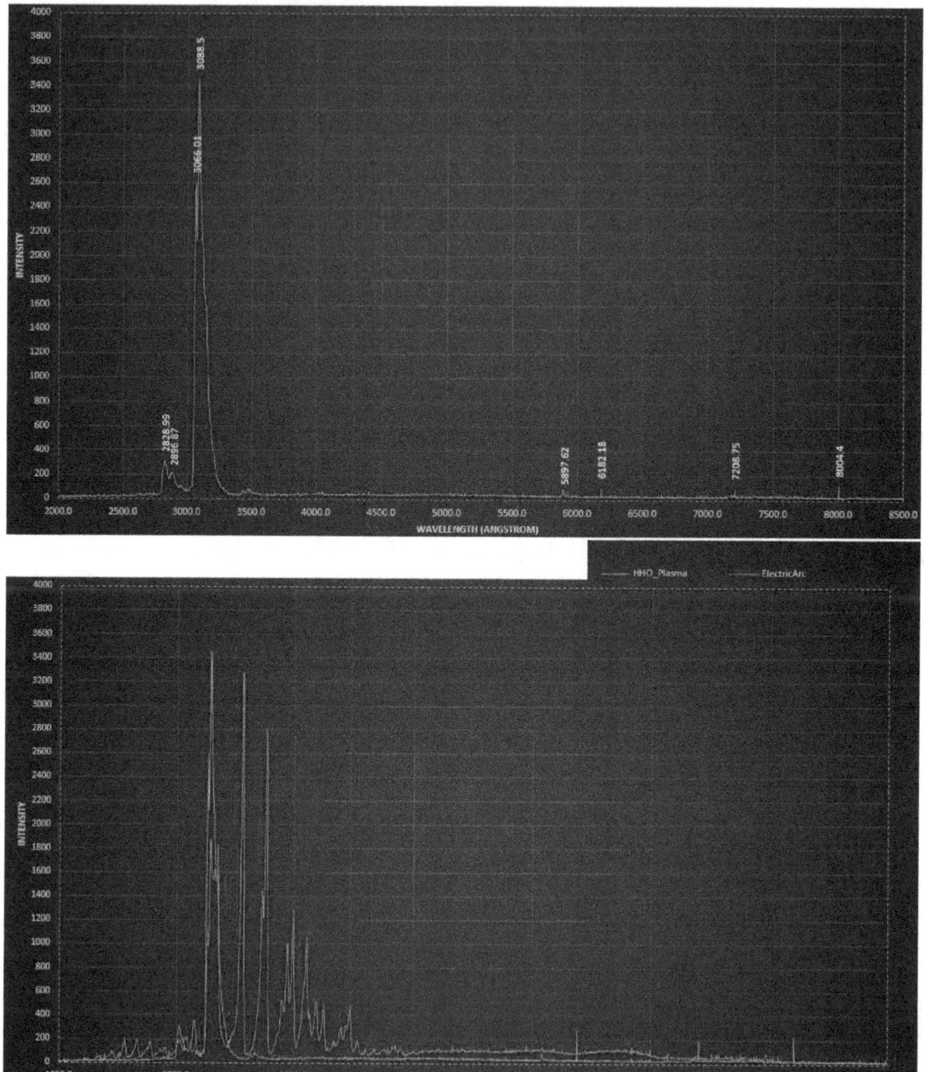

Figure 12.3. Oxyhydrogen flame and arc-discharge plasma spectrum measurements, provided by S. Stankovic. (a) Flame spectrum of oxyhydrogen fuel combustion. (b) The plasma spectrum of a high-voltage electric arc (orange), overlaid with the above-shown oxyhydrogen flame spectrum (green).

wavelength of the above discussed 618.24 nm emission line becomes 4.027 microns, which is in the infrared range.

References [13, 14] indeed report non-thermal infrared emission during a deuterium electrolysis-based experiment. The intensity of the emission

Table 12.2. Visible spectrum peaks of Compton-scale composites' magnetic binding.

	Calculated wavelength	Measured wavelength
Magnetic Zeeman split of a proton	618.24 nm	618.22 nm
Magnetic binding of stacked composites	796.1 nm	800.4 nm

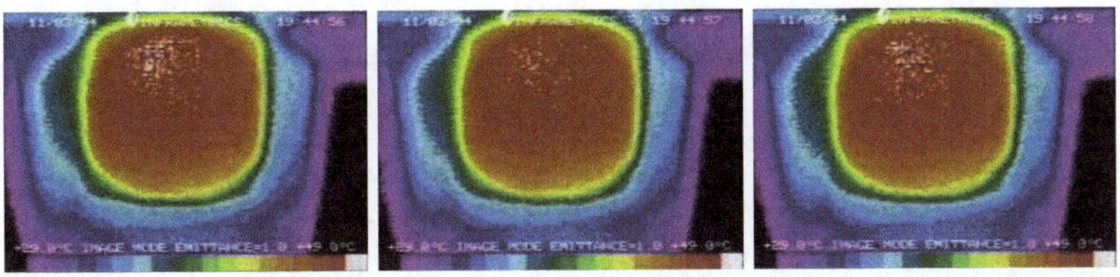

Figure 12.4. Thermal camera imaging of the near-IR emission during electrochemical Pd-D co-deposition. The corresponding experimental report is published in [14]. The white spots are the non-thermal infrared emission sources, and the shown images are taken one second apart.

was found to be correlated with the measured non-chemical power output, which demonstrates that the source of this infrared emission is related to the catalysis of nuclear reactions. Figure 12.4 shows the visual recording of this phenomenon.

Altogether, there is strong spectroscopic support for our Compton-scale composite model.

12.3. Transition to Compton-Scale Composite State

As established in the preceding section, a free electron and a nucleus may react according to the pathways shown in Figure 12.5. The Compton-scale composite is seen as a metastable state, which has a certain probability of formation from the free particle state. In the following sections, we discuss various processes which validate such transition from free particle state to Compton-scale composite state.

In the absence of an externally applied magnetic field, the well-known atomic hydrogen formation is the main reaction pathway. Under the $B = 0$ condition, the author of [1] measured only a few signals of Compton-scale composite formation. Under a strong B field, e.g., inside a tokamak reactor,

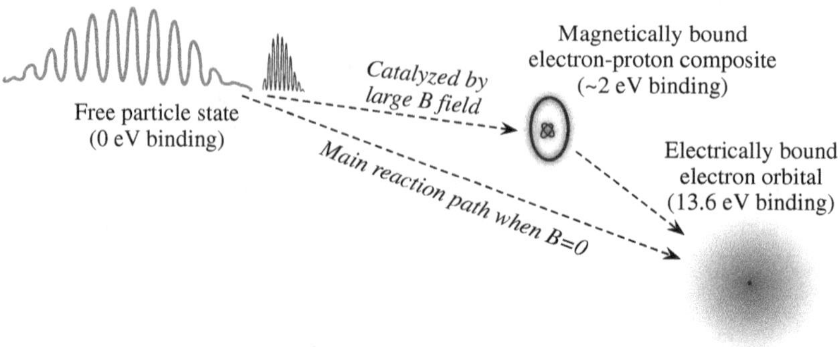

Figure 12.5. Illustration of free electron–proton reaction pathways.

the electron–deuteron recombination process yields a very strong signature of Compton-scale composite formation. Figure 12.6 shows such a signal, where the neutron production rate is 100 times higher for the recombining deuteron plasma ($t = 1.04\,\mathrm{s}$) than in the preceding case of hot deuteron–electron plasma ($t = 1.0\,\mathrm{s}$). Therefore, a strong magnetic field catalyzes the formation of Compton-scale composites.

12.3.1. *Cooling deuterium plasma*

Hot fusion experimenters have observed for over 40 years that deuterium plasmas enter into a state characterized by the emission of energetic electrons upon cooling. To illustrate this effect, Figure 12.6 shows the appearance of a "runaway current plateau," which suddenly appears after a thermal cooling phase, where the plateau duration coincides with a neutron production process. The measured neutrons are produced by the impact of very high energy electrons [11], i.e., they are energetic electron signatures. The first indication suggesting that we might be looking at Compton-scale composite formation is that this unexpected production of energetic electrons starts up when tokamak plasma is cooled down. Such cooling is usually done either by cold argon injection or by electromagnetically induced disruption of the plasma circulation.

When an electron–deuteron pair forms a Compton-scale composite, the resulting neutral particle catalyzes D-D fusion since it can approach another deuteron to 0.38 pm proximity before any Coulomb barrier arises. In analogy with the observations of Ref. [1], when catalyzed D-D fusions take place, the involved electron must carry away all the nuclear reaction energy in some percentage of cases. Thus, we expect the emission of 24 MeV electrons. As tokamak plasmas are in magnetic enclosure, the produced energetic

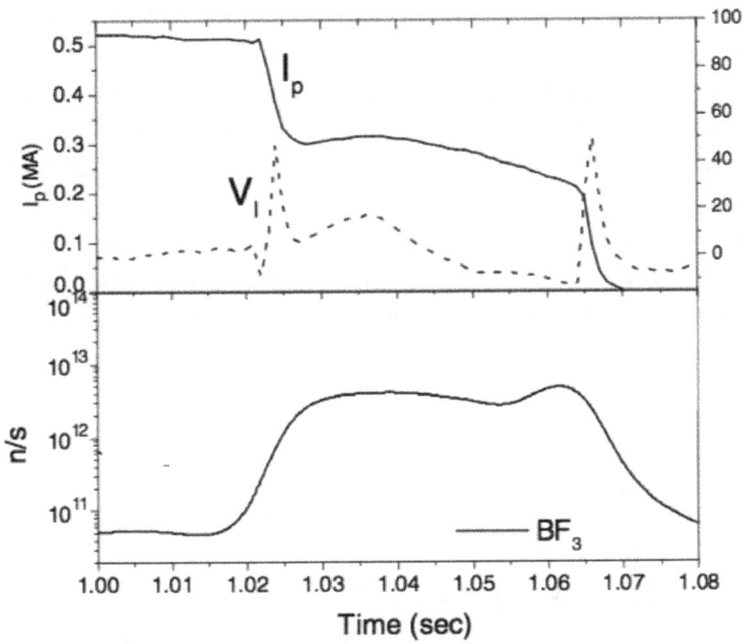

Figure 12.6. Energetic electron production in a tokamak reactor during a post-disruption current plateau, reproduced from [11]. I_P indicates the plasma current, V_I indicates the tokamak loop voltage, and the bottom part shows the amount of neutrons measured by a BF_3 detector. The plasma disruption takes place at about 1.02 s.

electrons escape only after a certain time, and lose energy via braking radiation during this magnetic entrapment. Therefore, we expect to detect a continuous energy spectrum of electrons, all the way up to 24 MeV. As shown in Figure 12.7, the measured electron energy spectrum exactly matches our expectation. The production of ~20 MeV electrons in few eV plasmas cannot be rationally explained by other means.

12.3.2. *Decelerating particles*

The rate of p or D involving nuclear fusion reactions has been extensively studied in various materials. The fusion enhancement rate is generally characterized by the screening energy parameter U_e, characterizing the fusion rate enhancement over a range of incoming ion beam energies. Up to now, the observed fusion rate enhancement was generally thought to be a consequence of delocalized electron screening, which is described by the Thomas–Fermi screening model and characterized by the screening energy parameter.

Table 12.3 shows the screening energy parameter in various background materials of interest, from data reported in [3, 5, 6]. A very strong fusion enhancement is seen in hydrogenated graphite; in this material, the 5.6 MeV electron-assisted nuclear de-excitation has been also observed [1]. Strong fusion rate enhancements are observed for hydrogenated Ti, Pd, and graphite materials.

A significantly higher fusion rate enhancement is seen in molten lithium than in solid lithium or lithium fluoride. Interestingly, this liquid phase rate enhancement appears to be specific to lithium, and not a general

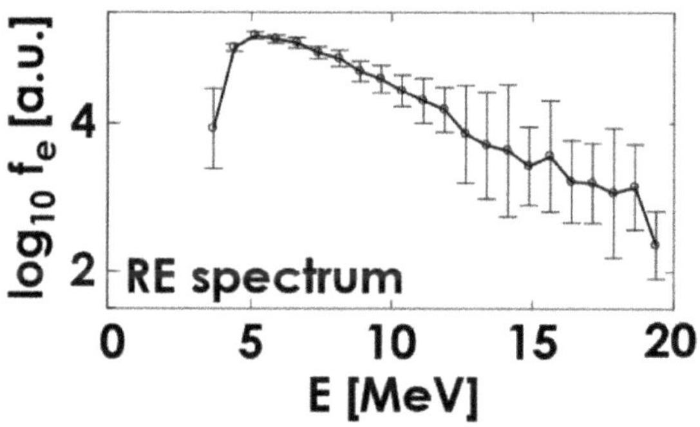

Figure 12.7. Energy spectrum of runaway electrons in a typical "runaway electrons" event. The fuel is deuterium plasma.

Table 12.3. Fusion rate enhancement characterized by the screening energy parameter for the p+^{7}Li (top left), D+^{6}Li (top right), and p+^{19}F (bottom) reactions.

Material	H stoichiometry	U_e
TiH	1.03	3.9 keV
$PdH_{0.2}$	0.21	3.6 keV
Graphite	0.06	10.3 keV

Material	U_e
LiF	350 eV
Solid Li	400 eV
Liquid Li	700 eV

Material	U_e
Kapton	0 keV
Amorphous carbon (hydrogenated)	36 keV
TiH	73 keV
Graphite (hydrogenated)	115 keV

solid–liquid difference in metals. In related experiments [3, 4], the authors performed deuterium bombardment of solid/molten indium and ^{6}Li metals, respectively. These studies show that at 10 keV bombardment, the reaction rate is about two orders of magnitude higher in solid indium than in liquid indium. In contrast, the reaction rate has been about 40% higher in liquid lithium than in solid lithium.

Altogether, the screening energy data show several major inconsistencies with the delocalized electron screening model: (i) The fusion rate enhancement is three to four orders of magnitude higher in some cases, for example, in hydrogenated graphite, than the theoretically predicted Thomas–Fermi screening model-based screening energy parameter; (ii) The solid–liquid phase transition has strong influence on the screening energy, which is unexpected; and (iii) Ref. [9] reports 600 eV screening energy for D-D fusion in non-metallic PdO, in contrast to the 300 eV screening energy for metallic Pd, measured using the same methods.

In our view, measuring the fusion enhancement rate is a suitable proxy for measuring the production rate of Compton-scale composites. The correspondence between the observation of surprisingly strong screening energy parameter in graphite and the observation of electron-assisted nuclear de-excitation indicates that the screening energy parameter is in fact measuring the rate of highly localized electron production. At a given proton beam energy, the theoretical rate of p+^{7}Li fusion is much higher than the rate of p+^{19}F fusion due to the Coulomb barrier difference. Since the rate of highly localized electron production in a given material is the same in either case, the apparent screening energy parameter becomes an order of magnitude higher for the p+^{19}F fusion case, which is indeed seen in the data of Table 12.3. This correspondence is yet another evidence in favor of our interpretation of Table 12.3. The most direct evidence for our interpretation is provided by the previously mentioned Ref. [1], which discovered electron-assisted nuclear de-excitation.

In summary, measurements of the screening energy parameter show results which are incompatible with the delocalized electron screening concept and appear to be quantifying the production rate of Compton-scale composites by energetic particles.

The possibility of a positive feedback loop between the energetic electrons produced by a nuclear reaction and the production of more Compton-scale composites during the braking of these energetic electrons may explain the burst-like reaction dynamics, which was observed in some experiments [7].

12.4. Summary of Experimental Signatures

We analyzed data from a broad range of experiments. Here, we summarize those experimental signatures which give a rather direct evidence for the formation of Compton-scale electron–proton and electron–deuteron composites:

(1) **618 nm visible light signature of proton–electron composite formation and 800 nm visible light signature of such composites' stacking process.** Regarding an electron–proton composite, the nuclear Zeeman split calculation of Section 12.2.2 appears to be validated by spectroscopic measurements of Ref. [15]. The corresponding nuclear transmutations reported in [15] provide evidence for such small electron–proton distance which catalyzes nuclear transmutations. Regarding an electron-deuteron composite, the nuclear Zeeman split calculations of Section 12.2.2 predict infrared light emission. Reference [13] indeed reports unexplained non-thermal infrared emission in a deuterium electrolysis-based experiment, and mentions several other experiments documenting the same phenomenon. The intensity of the emission was found to be correlated with the measured non-chemical power output, which demonstrates that the Compton-scale composite state catalyzes nuclear reactions. Figure 12.4 shows the visual recording of this infrared light emission phenomenon. In addition, the calculation of Section 12.2.2 predicts approximately 800 nm light emission during the stacking of Compton-scale composites with 2.3 pm internuclear distance. This calculation also appears to be validated by spectroscopic measurements of Ref. [15].

(2) **Ultra-dense hydrogen composite material with 2.3 pm internuclear distance.** The inter-nuclei distance calculations of Section 4.5 appear to be validated by the 2.3 pm inter-nuclei distance measurement reported in [8], which corresponds to the theoretically expected value.

(3) **Energetic electron emission, with energies up to the nuclear reaction energy.** The 0.38 pm electron–nucleus distance implies that the catalyzed nuclear reactions may de-excite via energetic electron emission. The energy of such electrons corresponds to the nuclear reaction energy level, i.e., 5.6 MeV for p-D fusion and 24 MeV for D-D fusion. For p-D fusion, this prediction is compatible with the observation of 5.6 MeV electron energy [1], and for D-D fusion this prediction is compatible with the observation of up to 20 MeV electron energies [12],

where the upper energy threshold was limited by the sensitivity of the detecting equipment.

(4) **Excessive screening energy measurement in p+^{7}Li and p+^{19}F fusion.** The accelerator experiments summarized in Table 12.3 indicate up to 10 keV and 100 keV apparent screening energy for p+^{7}Li and p+^{19}F fusion, respectively. Ordinary electron orbitals cannot account for such extraordinary high screening energy values, while our Compton-scale electron–proton composite model predicts this effect.

12.5. Conclusions

We demonstrated the existence of a highly localized electron state, which is at 0.383 pm average distance around a proton or deuteron. This metastable Compton-scale composite state has a certain probability of formation during any recombination of free electron–proton or electron–deuteron pairs. Its probability of formation is catalyzed by strong magnetic fields.

The existence of Compton-scale composite state has profound consequence for the correct interpretation of Heisenberg uncertainty: understanding such an electron state is very relevant for understanding nuclear bonds, which will be discussed in Chapter 14.

Acknowledgments

The author thanks Slobodan Stankovic for providing oxyhydrogen flame spectrum measurements.

References

[1] M. Lipoglavšek *et al.*, Observations of electron emission in the nuclear reaction between protons and deuterons. *Physics Letters B*, **773**(10) (2017) 553–556, DOI: 10.1016/j.physletb.2017.09.004.

[2] T. Ohtsuki *et al.*, Enhanced electron-capture decay rate of ^{7}Be encapsulated in C_{60} cages. *Physical Review Letters*, **93** (2004) 11.

[3] J. Kasagi, Screening potential for nuclear reactions in condensed matter. In *Proceedings of the ICCF-14 International Conference on Condensed Matter Nuclear Science*, Washington, DC, 2008.

[4] J. Kasagi *et al.*, Screening energy of the d+d reaction in an electron plasma deduced from cooperative colliding reaction. *Journal of Condensed Matter Nuclear Science*, **19** (2016) 127–134.

[5] A. Cvetinovic *et al.*, Molecular screening in nuclear reactions. *Physical Review C*, **92**(6) (2015). DOI: 10.1103/PhysRevC.92.065801.

[6] M. Lipoglavšek, Catalysis of nuclear reactions by electrons. *EPJ Web of Conferences*, **165** (2017). DOI: 10.1051/epjconf/201716501035.

[7] Y. Iwamura *et al.*, Replication experiments at Tohoku university on anomalous heat generation using nickel-based binary nanocomposites and hydrogen isotope gas. *Journal of Condensed Matter Nuclear Science*, **24** (2017) 191–201.

[8] L. Holmlid and S. Zeiner-Gundersen, Ultradense protium p(0) and deuterium D(0) and their relation to ordinary Rydberg matter: A review. *Physica Scripta*, **94**(7) (2019) 075005.

[9] J. Kasagi, Low-energy nuclear reactions in metals, *Progress of Theoretical Physics Supplement*, **154**(1) (2004) 365–372.

[10] K. Czerski *et al.* Screening and resonance enhancements of the 2H(d, p)3H reaction yield in metallic environments. *Europhysics Letters*, **113**(2) (2006) 22001.

[11] J.R. Martin-Solis *et al.*, Enhanced production of runaway electrons during a disruptive termination of discharges heated with lower hybrid power in the Frascati Tokamak Upgrade. *Physical Review Letters*, **97**(16) (2006) 165002.

[12] C. Paz-Soldan *et al.*, Spatiotemporal Evolution of Runaway Electron Momentum Distributions in Tokamaks. *Physical Review Letters*, **118**(25) (2017) 255002.

[13] M. Swartz *et al.*, Non-Thermal Near-IR Emission from High Impedance and Codeposition LANR Devices. In *Proceedings of the ICCF-14 International Conference on Condensed Matter Nuclear Science*, Washington, USA, 2008.

[14] S. Szpak *et al.*, Polarized $D^+/Pd–D_2O$ system: Hot spots and mini–explosions. In *Proceedings of the ICCF-10 International Conference on Condensed Matter Nuclear Science*, Cambridge, USA, 2003.

[15] S. Stankovic, Nuclear transmutation with carbon and oxyhydrogen plasma. In *Proceedings of the ICCF-22 International Conference on Condensed Matter Nuclear Science*, Assisi, Italy, 2019.

https://doi.org/10.1142/9781800611306_0013

Chapter 13

Electron-Mediated Nuclear Fusion

András Kovács*,¶ and Giorgio Vassallo†
*BroadBit Energy Technologies, Metallimiehenkuja 8 C, 02150 Espoo, Finland
†Engineering Department, University of Palermo
viale delle Scienze, 90128 Palermo, Italy
¶andras.kovacs@broadbit.com

13.1. Electron-Mediated Fusion Signatures

This chapter is dedicated to the exploration of electron-mediated deuterium fusion. Based on the previous chapter's analysis of Compton-scale composites (CSCs), we look at experimental methods for achieving CSC catalyzed fusion, which is essentially electron-mediated nuclear fusion. CSC states can be thought of as pseudo-neutrons: their 0.38 pm-sized electron structures around a $Z = 1$ nucleus catalyze nuclear fusion reactions by reducing the Coulomb barrier. In this context, nuclear energy production requires forming electron–proton or electron–deuteron type CSC at such efficiency that the needed input energy would be less than the produced nuclear fusion energy.

CSC catalyzed D–D fusion has three reaction pathways:

$$(D+e)_{\mathrm{CSC}} + D \rightarrow \begin{cases} {}^{4}\mathrm{He} + e\ (24\,\mathrm{MeV}) \\ {}^{3}\mathrm{He}\ (0.8\,\mathrm{MeV}) + n\ (2.5\,\mathrm{MeV}) + e \\ T\ (1\,\mathrm{MeV}) + {}^{1}H\ (3\,\mathrm{MeV}) + e \end{cases}$$

The simultaneous occurrence of the above three reaction pathways is a signature of CSC catalyzed D–D fusions. We recall from Chapter 12 that an electron carries away all the fusion reaction energy in the ^{4}He branch, while in other cases the reaction energy is dissipated along the usual nuclear fragmentation pathway. Thus, CSC catalyzed fusion is NOT a radiationless process: its energetic reaction products generate braking radiations. In fact, the ^{4}He branch produces the most energetic reaction products: electrons with over 20 MeV energy. In Chapter 12, we identified the ^{4}He branch with the "runaway electron" phenomenon of tokamak reactors.

According to the observations discussed in Chapter 12, CSC formation happens during electron–deuteron recombination events. In the following sections, we look at experimental methods which realize the electron–deuteron recombination, and show signatures of CSC catalyzed D–D fusion.

13.2. Degassing of Metal Deuterides

In metal deuterides, the amount of absorbed deuterium is temperature-dependent. At room temperature, palladium absorbs deuterium to nearly 1:1 ratio of Pd:D. Within the PdD material, the deuteron sites are surrounded by delocalized electrons. Upon heating, PdD releases deuterons, which recombine with delocalized electrons at the metal surface. This $D^+ + e^- \to D$ recombination process is referred to as degassing. According to the above stated reaction scheme, we expect to detect neutron, ^{1}H, T, ^{3}He, and ^{4}He production.

Reference [1] reports neutron measurements during PdD degassing: its results are shown in Figure 13.1. The neutron count is persistently above-background during degassing, and the measured proton recoil spectrum is consistent with the expected 2.5 MeV neutron energy.

13.3. Electrochemically Driven Deuteron–Electron Recombination

In electrochemically driven experiments, CSC formation may occur at the negative electrode, where $D^+ + e^- \to D$ recombinations take place.

Firstly, we discuss metal-deuteride co-deposition, where a reductive deposition of PdD material is the main reaction at the negative electrode. As a side reaction, $D^+ + e^- \to D$ recombination always takes place as well.

Figure 12.4 of Chapter 12 shows infrared emissions during such co-deposition, which is a direct signature of CSC formation. During this PdD co-deposition experiment, the ^{4}He producing fusion branch is revealed by particle track detectors: Figure 13.2 shows a "triple track" pattern in a particle detector. Such a pattern corresponds to the break-up of a ^{12}C nucleus into three ^{4}He particles, and requires >7 MeV energy of the detected

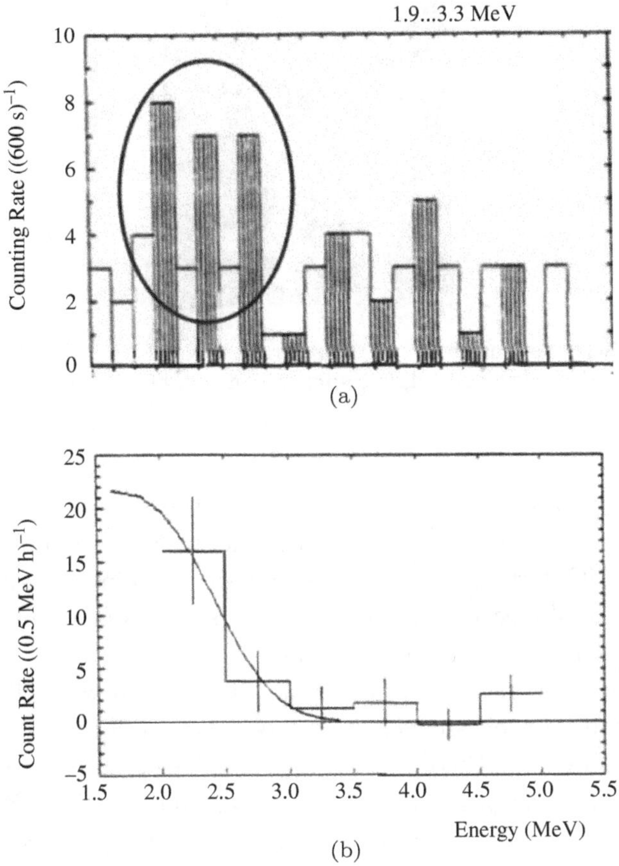

Figure 13.1. Neutron measurement during PdD degassing. (a) Consecutive measurements of neutrons. Blank bars indicate measurements without the sample, shaded bars indicate measurements with the sample present. The circled measurements are made during degassing. (b) Neutron energy measurement during degassing, using proton recoil technique. The crosses are measurement data, while the curve indicates theoretic expectation for 2.5 MeV neutron energy. Reproduced from [1].

energetic particle. The source of over 7 MeV energy must be a particle which carries away the energy of D–D to ^{4}He fusion reaction, i.e., it is the signature of the 24 MeV electrons produced in the ^{4}He reaction branch.

The production of tritium during the co-deposition of PdD is also well documented in Ref. [3, 4]. After three days of co-deposition, the authors of [4] measured about 10 times the initial tritium concentration.

Secondly, we discuss the electrolysis method, where the main reaction is $D^+ + e^- \to D$ recombination on the negative electrode. Such electrolysis

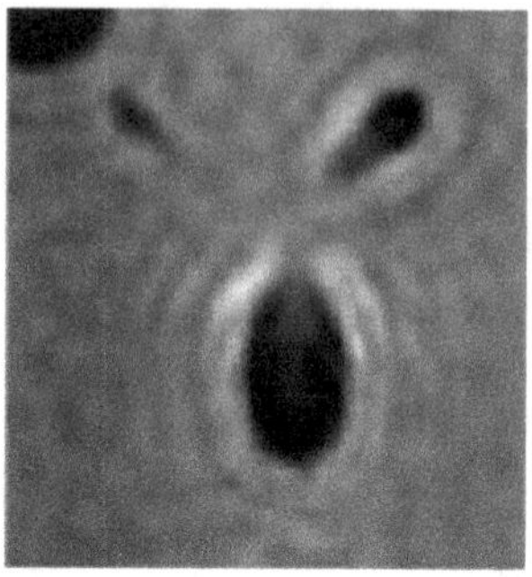

Figure 13.2. Microphotograph of a "triple-track" event in a CR-39 detector which occurred during electrochemical Pd/D co-deposition, reproduced from [2].

experiments typically use PdD material for the negative electrode, as in the pioneering Fleischmann–Pons experiment [5]. The infrared emission signature of CSC formation was detected during electrolysis on PdD [8]. After three days of electrolysis on PdD, the authors of [4] measured nearly 17 times the initial tritium concentration.

The author of [6, 7] discovered that PdD electrodes facilitate not only D_2 gas evolution on their surface, but also accommodate atomic D in lattice vacancies. Recalling the theory of electron–nucleus coupling from Chapter 9, the hyperfine splitting of atomic $1s$ orbital around the D nucleus yields the signature of atomic D states: the corresponding RF emission frequency is 327.38 MHz. As shown in Figure 13.3, this signature of atomic D clearly appears upon the application of electric power onto the Pd electrode. Such atomic D state is metastable because it disappears within few minutes after the termination of applied voltage. Thus, the presence of the 327.38 MHz signal is witness to a continuous atomic $D^+ + e^- \rightarrow D$ recombination process.

The energetic fusion reaction products lose their kinetic energy via collisions, and emit braking radiation in this process. Since the initial energy of reaction products is in the MeV range, the emitted braking radiation power has a slope in the gamma spectrum, and then a flat power spectrum in the RF to X-ray frequency range. The measurements of [6, 7] indeed show RF signatures of fusion energy dissipation. As shown in the bottom of Figure 13.3, RF emission was measured in the 327.3–327.45 MHz range. This measurement excluded the 327.38 MHz peak and all of its sidebands. It is seen from Figure 13.3 that the detected RF emission timewise coincides with the presence of $D^+ + e^- \rightarrow D$ recombination. Therefore, this RF

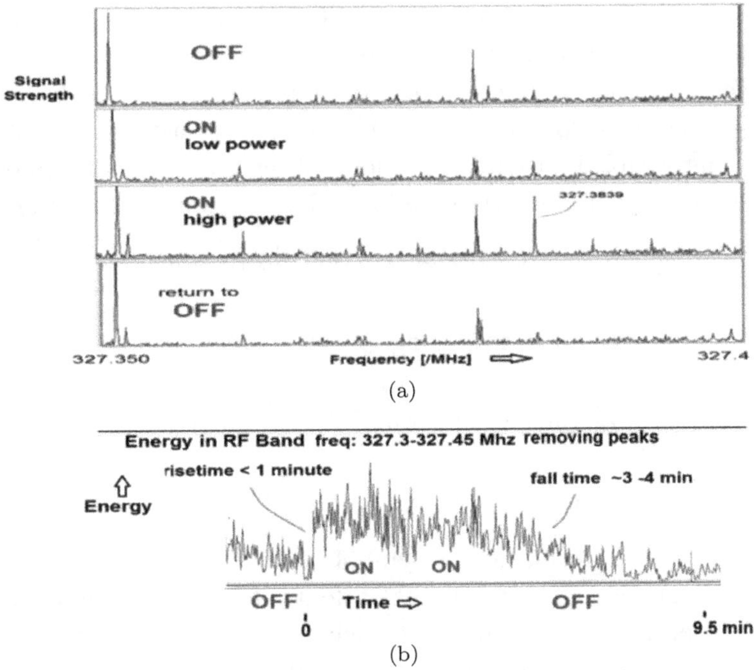

Figure 13.3. Measurement of the electrochemically driven 327.38 MHz atomic D-line, reproduced from [6, 7]. (a) Four snapshots of the frequency spectrum around the D-line, showing the dependency of the 327.38 MHz D-line peak on the applied electric power. (b) Time evolution of RF emission in the 327.3–327.45 MHz range, excluding the D-line signal, and overlayed with the on/off state of electrochemical driving voltage.

emission is perhaps the braking radiation signature of energetic fusion products.

13.4. Deuterium Diffusion Across Heterogeneous Nanolayers

The low-energy nuclear transmutation of certain elements was observed in replicated experiments [9–11]. The transmuted elements were cesium, strontium, or barium, and they were implanted into palladium. Such a transmutation occurs when the system comprises alternating thin layers of Pd and calcium oxide (CaO) and a flow of deuterium diffuses across these layers. It is, therefore, necessary to find a mechanism that explains the role of the CaO layer and the overcoming of the Coulomb barrier by deuterium nuclei. An interesting hypothesis is to consider the formation of at the interface between calcium oxide and palladium: a location where a high work function difference between Pd and CaO causes high electron density. During diffusion, a CSC composite of several deuterium nuclei might form at this interface. Therefore, deuterium aggregates of CSC could be, according to this hypothesis, responsible for the observed transmutation of Cs into Pr and Sr into Mo. The transmutations observed in [9–11] appear to be fusion-like, and may be described as the following *many-body reactions*:

$$^{133}_{55}\mathrm{Cs} + 4\,(D+e)_{\mathrm{CSC}} \rightarrow {}^{141}_{59}\mathrm{Pr} + 4e$$

$$^{88}_{38}\mathrm{Sr} + 4\,(D+e)_{\mathrm{CSC}} \rightarrow {}^{96}_{42}\mathrm{Mo} + 4e$$

$$^{137}_{56}\mathrm{Ba} + 6\,(D+e)_{\mathrm{CSC}} \rightarrow {}^{149}_{62}\mathrm{Sm} + 6e$$

$$^{138}_{56}\mathrm{Ba} + 6\,(D+e)_{\mathrm{CSC}} \rightarrow {}^{150}_{62}\mathrm{Sm} + 6e$$

In the above equations, the symbols $4\,(D+e)_{\mathrm{CSC}}$ and $6\,(D+e)_{\mathrm{CSC}}$ represent picometric, coherent chains of deuterium particles in a CSC state, which was discussed in Section 3.6. The short distance between deuterons in such hypothetical structures may possibly allow their simultaneous fusion, as required by the above-listed nuclear transmutations.

A depth profiling of Cs to Pr transmutation is given in [11], where Pd metal is doped by Cs, and it raises several unanswered questions in the context of CSC-catalyzed deuterium fusion: (i) why is the Cs transmutation rate higher than the Pd transmutation rate?, (ii) why does the Cs transmutation involve 4 deuterium nuclei, but not 1, 2, or 3?, (iii) why does Cs transmutation occur only near the outer metal surface but not near the alternating Pd–CaO layers?, and (iv) according to the data of [11], Cs to Pr transmutation accounts for less than 50% of Cs transmutations — what happens to the transmuted Cs in the remaining cases? In light of these pending questions, in Chapter 15 we introduce an alternative hypothesis regarding such transmutations.

13.5. Conclusions

A small percentage of $D^+ + e^- \rightarrow D$ recombinations lead to CSC formation, which in turn catalyzes D–D fusion. This conclusion is based on the combined signatures of neutrons, tritium, high-energy electrons, and IR emissions. These are direct signatures of electron-catalyzed fusion, and are obtained under conditions where the Coulomb barrier prevents any direct fusion of bare nuclei.

Naturally, this reaction mechanism is not restricted to D–D fusion: it may be exploited also for the catalysis of various aneutronic fusion schemes, such as proton–lithium or proton–boron fusion.

In order to determine whether CSC formation via $D^+ + e^- \to D$ recombinations may lead to a net energy producing fusion device, the $D^+ + e^- \to D$ recombinations should be performed under a strong magnetic field. This recommendation is based on the results of Chapter 12.

References

[1] M. Bittner *et al.*, Observation of D-D fusion neutrons during degassing of deuterium loaded palladium. *Fusion Technology*, **23**(3) (1993) 346–352.

[2] P.A. Mosier-Boss, It is not low energy — But it is nuclear. *Journal of Condensed Matter Nuclear Science*, **13** (2014) 432–442.

[3] S. Szpak *et al.*, On the behavior of the Pd/D system: Evidence for tritium production. *Fusion Technology*, **33**(1) (1998) 38–51.

[4] K.-H. Le *et al.*, A change of tritium content in D_2O solutions during Pd/D co-deposition. *Journal of Condensed Matter Nuclear Science*, **13** (2014) 294–298.

[5] M. Fleischmann, S. Pons, and M. Hawkins, Electrochemically induced nuclear fusion of deuterium. *Journal of Electroanalytical Chemistry*, **261** (1989) 301 and errata in **263** (1989).

[6] M.R. Swartz, Atomic deuterium in active LANR systems produces 327.37 MHz superhyperfine RF maser emission. In *Proceedings of the ICCF-22 International Conference on Condensed Matter Nuclear Science*, Assisi, Italy, 2019.

[7] M.R. Swartz, Superhyperfine structure of the deuteron line emission from active ZrO_2-PdD heralds an FCC vacancy. In *Proceedings of the ICCF-22 International Conference on Condensed Matter Nuclear Science*, Assisi, Italy 2019.

[8] M. Swartz *et al.*, Non-thermal near-IR emission from high impedance and codeposition LANR devices. In *Proceedings of the ICCF-14 International Conference on Condensed Matter Nuclear Science*, Washington, USA 2008.

[9] Y. Iwamura *et al.*, Elemental analysis of Pd complexes: Effects of D_2 gas permeation. *Japanese Journal of Applied Physics*, **41**(7R) (2002).

[10] T. Hioki *et al.*, Inductively coupled plasma mass spectrometry study on the increase in the amount of Pr atoms for Cs-Ion-implanted Pd/CaO multilayer complex with deuterium permeation. *Japanese Journal of Applied Physics*, **52**(10R) (2013).

[11] S. Tsuruga *et al.*, Transmutation reaction induced by deuterium permeation through nanostructured multi-layer thin film. *Mitsubishi Heavy Industries Technical Review*, **52**(4) (2015) 106–110.

https://doi.org/10.1142/9781800611306_0014

Chapter 14

Nuclear Forces and Occam's Razor

András Kovács
BroadBit Energy Technologies, Metallimiehenkuja 8 C, 02150 Espoo, Finland
andras.kovacs@broadbit.com

14.1. Introduction

In the first part of the book, we described the unified physics of electromagnetism, quantum mechanics, and general relativity. This unification was achieved by relaxing the Lorenz gauge of electromagnetism, by establishing a relationship between electromagnetism and mass, and by noting that the quantum wave equation is the dual of the metric of general relativity. We showed in Chapter 1 that the electromagnetic Lagrangian density is $\mathcal{L} = \frac{1}{2}\sqrt{g}G\widetilde{G}$, where g is the determinant of the metric tensor, G is the electromagnetic field, and $\tilde{}$ is the Clifford reversion operator. In flat space–time, the electromagnetic Lagrangian is simply $L = \frac{1}{2}G\widetilde{G}$. Mathematically, electromagnetism is the simplest possible field theory.

The first hint of this is given in Chapter 6, where we could apply the same formula for calculating the "anomalous" magnetic moments of the electron and proton. Since our calculation was based on recognizing particle mass as electromagnetic field energy, it implies that the proton mass at least partly originates from electromagnetic field energy.

We encountered the next nucleus-related result in Section 7.9, which demonstrated that the nuclear isospin appears to be an electromagnetic phenomenon. That result contradicts present-day nuclear theories, which consider the isospin to be a conserved current of non-electromagnetic interactions.

The author of Ref. [21] noticed that various particle lifetimes are also scaled by powers of the electromagnetic fine structure constant α.

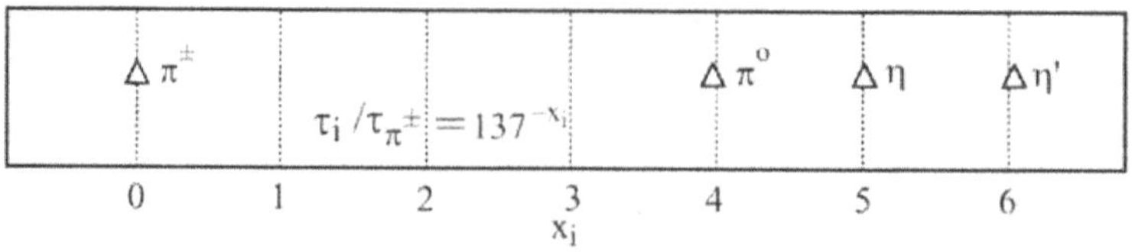

Figure 14.1. The geometric scaling of nuclear meson lifetimes, where the scaling factor is α^x.

This scaling is shown in Figure 14.1. Moreover, experimenters were surprised to find that the mass difference between the charged $\pi^{\pm}$ and the neutral π^0 particle is exactly nine electron masses. These electromagnetic patterns give further inspiration to consider the role of electromagnetism in nuclear structures.

If nuclear forces and structures also turn out to be described by the simple $L = \frac{1}{2}G\widetilde{G}$ Lagrangian, it will bring a major simplification to our understanding of fundamental forces and will open up major technological prospects. Therefore, it is worth to make a serious analysis of nuclear data. As in the preceding chapters, we use Occam's razor principle to look for the simplest explanation.

14.2. What is the "Strong Nuclear Force"?

Nuclear force is what holds the nucleons together. Several researchers noticed that important properties of nuclei can be better explained by a model where protons and neutrons are arranged in a solid-state structure [3, 4], i.e., a well-defined spatial configuration of nucleons. As discussed in [3], most scientific reviewers plainly refused to engage in a rational discussion or criticism of such a nuclear model, and rejected it simply because it is different from the currently prevailing liquid-drop nuclear model. The few rational critics of a solid-state nuclear model objected to its apparent contradiction to the Heisenberg uncertainty principle, meaning that a well-defined spatial configuration of nucleons would imply too large a momentum uncertainty, which would destabilize the nucleus.

In this chapter, we will frequently refer to the physics of composite particle formation, which was discussed in Chapters 7 and 12. One must not misapply the Heisenberg uncertainty principle, and we must consider the possibility of composite particle formation.

Let us begin by taking a closer look at the textbook nuclear model, which is briefly reviewed in [1]. The basic premise of currently prevailing nuclear models is that protons and neutrons are swimming in a liquid droplet type nuclear soup, and are held together by the "spill-over" effect of the attractive nuclear "strong force," which spills over the boundary of individual nucleons. From the liquid drop model, the nuclear binding energy is derived via the Bethe–Weizsäcker formula, which is a curve-fitting equation fitting of five variables that are empirically adjusted to fit the experimental data. Statistically, it cannot be used to predict anything. A further problem is that there is zero correlation between the binding energy variations in light elements ($Z = 1 - 10$) and the Bethe–Weizsäcker formula. To explain binding energy variations in light elements, it is postulated that the spill-over of nuclear strong force is most effective for the ^{4}He nucleus structure, and therefore the liquid drop tends to be composed of alpha particle fragments. This postulate is referred to as the "alpha correlations" within the nucleus. The still significant discrepancies between the resulting model and experimental data prompted nuclear scientists to propose a further add-on postulate, the theory of nuclear shells which are based on so-called "magic" numbers. This model assumes that there is a passive nuclear core of nucleons, beyond which only the valance nucleons contribute to the nuclear behavior. Theoretically, to get these magic numbers, the shell model starts from what is called the "Woods-Saxon" potential well. This potential energy well lies somewhere between the abrupt square well and the smooth harmonic-oscillator well. To this Woods-Saxon potential, a spin–orbit term is added, in an attempt to duplicate the observed magic numbers. However, the result does not coincide with these magic numbers unless an empirical spin–orbit coupling, named the Nilsson Term, is also added. The Nilsson Term has at least two or three different values for its coupling constant, depending on the nucleus being studied. The overall situation with the currently prevailing nuclear model is eerily reminiscent of the pre-Newtonian attempt of explaining the observed celestial motions of planets and their moons by increasingly elaborated epicycloid models. Despite their increasing sophistication to match the observed planetary motions, the epicycloid models did not represent any actual progress in the understanding of planetary physics. Building increasingly sophisticated approximations is a valid exercise as long as no simpler explanation is available, but we believe that there should be a clear separation between the concept of phenomenological approximations and the concept of theoretical understanding.

Even before considering any electromagnetic aspects, the phenomenology of real nuclear data rules out the liquid drop model. Bethe's original liquid drop model postulates a soup of protons and neutrons. However,

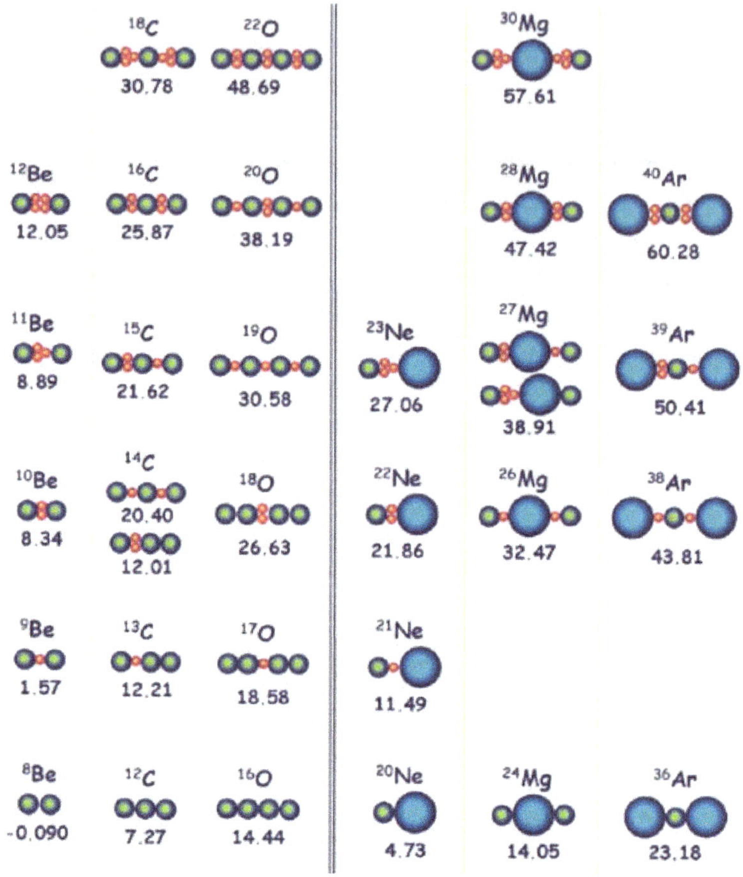

Figure 14.2. The clustered model of nuclei. Green circles represent alpha fragments, blue circles represent oxygen fragments, and the numbers indicate the overall nuclear binding energy of the fragments in MeV. Reproduced from [40].

nuclear scattering experiments and binding energy measurements give strong evidence that a nucleus comprises clusters of alpha-particle fragments, and in case of heavier elements, four alpha particle fragments combine to form clusters of "oxygen fragments." The clustered structure of nuclei is illustrated in Figure 14.4. The evidence of clusters is so overwhelming that even mainstream theorists conceded the existence of clustered structure and started referring to it as "deformed nuclei," i.e., deformed from spherical shape. What they really mean is that each cluster comprises a liquid proton–neutron soup, and that these clusters are somehow stuck together to form "deformed nuclei." To investigate what goes on within these small clusters, let's take a look at the excitation energy level of the simplest light nuclei.

Table 14.1. The excitation levels of $Z = 1$ and $Z = 2$ nuclei.

Nucleus	Excitation	Comment
^2H	None	
^3H	None	
^3He	None	
^4He	20.2 MeV, 21 MeV	Nuclear break-up

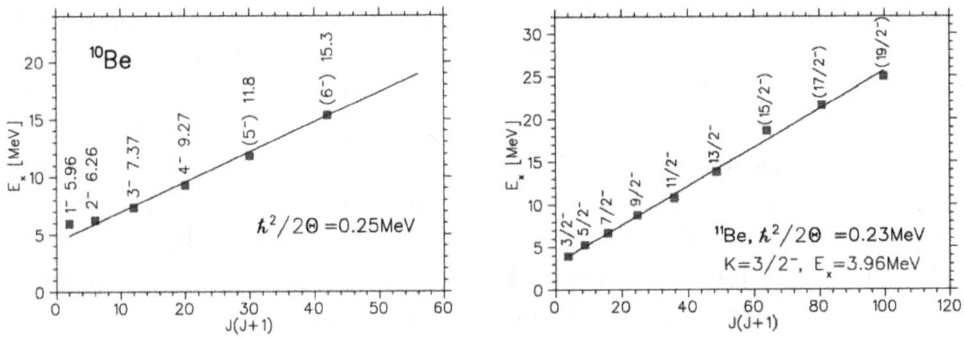

Figure 14.3. The nuclear excitation energies of Be isotopes, as a function of the nuclear angular momentum $J(J+1)$. Reproduced from [40].

As shown in Table 14.1, none of the $Z = 1$ or $Z = 2$ nuclei have any stable excitation level. For example, alpha fragments have nuclear excitations only at such high energy levels that it would cause it to fragment. The lack of any excitation level does not seem congruent with the physics of liquid droplets, which are well known to have various oscillation modes. Going further, let us look at the excitation modes of relatively simple nuclei comprising two-alpha fragments. Figure 14.3 shows the nuclear excitation energy of beryllium isotopes. It is clearly seen that the excitation energy is proportional to the angular momentum of the excited state. Since the alpha fragments do not have excited states with angular momentum, this angular momentum must come from the particles positioned between the alpha fragments. In other words, $E_{\text{excitation}} \sim J(J+1)$ for the particles between the alpha fragments: the authors of [40] identify the same relationship in molecular wavefunctions, where the spatial constellation is fixed, and it appears to have nothing to do with liquid droplet oscillations. Considering that physics is an experimental science and the liquid drop nuclear model directly contradicts experimental data, we must conclude that the epicycloid-style liquid drop model is has been falsified.

14.2.1. *Maxwell's equations and the binding energy*

In contrast, we look for a simple correspondence between the Maxwell equation and nuclear binding energy, without any *ad hoc* assumptions or statistical chicanery. The simplest stable nuclear bond is between the two nucleons of ^{2}H. We estimate nuclear binding energy and bond length of ^{2}H from basic electromagnetic principles. The symmetric bond structure resulting from the neutron's charge polarization is illustrated in Figure 14.4, showing two protons and the negative elementary charge centered in the middle between them. Such polarized charge separation is justified by experimental measurements of the neutron's electric polarizability parameter Λ. In an external electric field E, the neutron's induced electric dipole moment becomes ΛE, giving an interaction potential of $-\frac{1}{2}\Lambda E^2$. Reference [13] lists three independent measurements of Λ, using different methods. The average value from these measurements is $\Lambda = 1.1 \times 10^{-3}$ fm^3. Regarding magnetic interactions, the proton's magnetic moment is $\mu_p = 2.793\mu_N$, while the neutron's magnetic moment is $\mu_n = -1.913\mu_N$. The magnetic moment of ^{2}H is $0.857\mu_N$. The negative sign of the neutron's magnetic moment means that neutron's negative charge is, on the average, further away from the neutron's axis than its positive charge.

The magnetic moment of ^{2}H is $0.857\mu N$, which closely matches the magnetic moment summation of a proton and a neutron. This match between the magnetic moments also justifies modeling the deuteron as a proton — polarized neutron composite.

As discussed in the preceding chapters, when two particles form a phase coherent composite, their relative position becomes constant. Phase coherence is the only way for two protons to be in the close proximity required for nuclear bonding, and the above described magnetic moment match supports our phase coherent model. Thus, we see the physics of ultra-dense hydrogen composites also at play in the nuclear bond. As in Section 4.5, we require the spatial distance between the protons to be N times the Compton-wavelength of protons, i.e., $d = 2\pi R_p$, where $R_p = 0.21$ fm is their Zitterbewegung radius. In Figure 14.4 shows the resulting distances within the ^{2}H nucleus. The constant distance between the two protons means that the deuteron has a rigid structure, without an oscillation of its inter-nucleon distance. Thus the phase coherence requirement explains the above-mentioned absence of any nuclear excitation level in a deuteron.

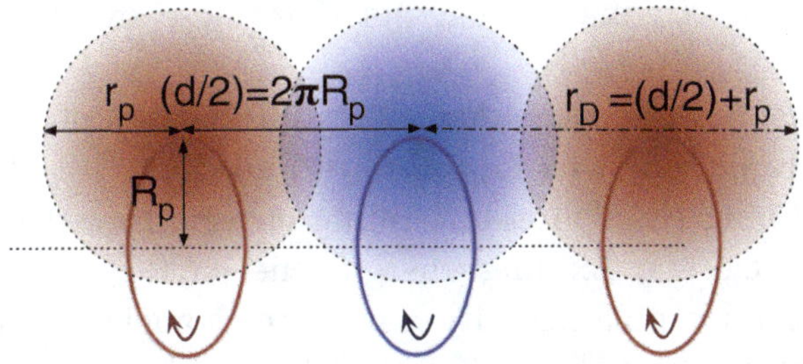

Figure 14.4. A phase coherent composite formation with two protons and a nuclear electron between them. The red spheres illustrate the spherical proton charge, and the red dashed ellipses illustrate its Zitterbewegung. A nuclear electron is located between them, at the energetically most favorable position. The shaded sphere illustrates that the nuclear electron is not in a phase-coherent state, and the blue dashed ellipse illustrates that the protons' magnetic fields define its Zitterbewegung radius.

The CODATA value for the deuteron charge radius is $r_D = 2.14\,\text{fm}$. With reference to Figure 14.4, we observe that for $N=2$ the deuteron radius becomes:

$$r_D = \frac{d}{2} + r_p = 2\pi R_p + r_p = 1.32 + 0.85\,\text{fm} = 2.17\,\text{fm}$$

where r_p is the proton charge radius. In other words, the deuteron structure of Figure 14.4 is 98% accurate with respect to the measured deuteron charge radius.

With reference to Figure 14.3, we estimate the involved electrostatic interactions. The electric repulsion between the two protons is $F_e = \frac{1}{4\pi\varepsilon_0}\frac{e^2}{d^2}$. Considering the parallel Zitterbewegung orientation between the two protons, the magnetic attraction between them is $F_m = -ec \times B = \frac{-\mu_0}{4\pi}\frac{e^2c^2}{d^2}$. Interestingly, the ratio of these expressions yields $F_e/F_m = -1$, i.e., the magnetic attraction balances out the electric repulsion between the two protons. The remaining electrostatic potential is between the negative nuclear charge and the two protons:

$$U_p = -2\frac{e^2}{4\pi\varepsilon_0}\frac{1}{(d/2)} \tag{14.2.1}$$

The above formula yields $U_p = -2.18$ MeV potential energy estimation. This simple estimation is very close to the $U_b = 2.22$ MeV experimental deuteron binding energy value. The still missing deuteron binding energy terms are the negative charge's kinetic energy and its magnetic potential energy. The $U_p \approx -U_b$ match suggests that these two terms cancel each other out.[a]

At this point, our goal is to demonstrate the effect of neutron polarization. Based on Equation (14.2.1), we do not need any hypothetical "strong force" for explaining the deuteron binding energy. In order to dispute the above-estimated electric binding energy, one would have to argue that the neutron's electric polarizability parameter somehow "switches off" in the proximity of a proton, and that the above calculated deuteron radius match is a mere coincidence.

In the following, we refer to the negative elementary charge shown in Figure 14.4 as a "nuclear electron," but emphasize that this electron-like particle is not the same as an ordinary electron. This electron-like particle will be characterized in Section 14.5.

There are further experimental observations which are logical consequences of the above-outlined electromagnetic binding model, but have absolutely no explanation under the "strong force spill-over" model:

- The deuteron nucleus has no nuclear excitation levels. Similarly, the ^{3}H and ^{3}He nuclei have no nuclear excitation levels either. Why do these simplest nuclei have no excitation levels? Under the "strong force spill-over" model, the nucleons form an oscillator, and nuclear excitation levels are interpreted as meta-stable excited states of this oscillator. However, the absence of excitation levels in these simplest nuclei contradicts this model. Under the herein discussed electromagnetic model, the absence of excitation levels is a direct consequence of the absence of transversal oscillations with respect to the nucleons' Zitterbewegung planes, i.e., the deuteron bond structure is closely related to the Compton-scale composite structure discussed in Section 4.5. In these simplest nuclei, there is only one configuration in which the nuclear bond can be constructed.
- The neutron capture cross-section of a proton is 0.3 barns. This cross-section is related to the proton–neutron bond establishment probability. The resulting proton–neutron binding is 1.1 MeV/nucleon. The subsequent capture of one more neutron leads to the ^{3}H nucleus formation, where the binding is 2.8 MeV/nucleon. Under the "strong force spill-over" model,

[a] The cancellation between the kinetic energy and magnetic potential directly follows from the virial theorem, at low energies. At an orbital radius R, the radially acting Lorentz force is $\sim R^{-3}$, and the resulting magnetic potential is $\sim R^{-2}$. The virial theorem states that the kinetic energy and magnetic potential are related by a factor of $\frac{2}{2} = 1$ at non-relativistic energies.

the increased "strong force spill-over" of the liquid-state ^{3}H nucleus should be reflected in higher neutron capture probability of the deuteron in comparison to the proton. However, experimental data show that the neutron capture cross-section of a deuteron drops three orders of magnitude from the proton's value, to just 0.0005 barns. Under the herein discussed electromagnetic model, this cross-section drop indicates that the neutron capture by a deuteron represents not just an additional nucleon bond, but also the rearrangement of the existing proton–neutron bond. This needed bond rearrangement reduces the probability of ^{3}H formation. The rearrangement of the proton–neutron bond upon ^{3}H formation can be inferred by comparing the magnetic moments of ^{2}H and ^{3}H. If a neutron would simply add onto ^{2}H, the nuclear magnetic moment would change from $0.857\mu_N$ to $-1\mu_N$. However, the magnetic moment of ^{3}H is $3\mu_N$, which indicates the rearrangement of the existing proton–neutron bond.

14.2.2. *Further evidence from pions*

The nuclear interaction of a negatively charged pion also validates our deuteron model. When a proton captures a π^- particle, the occurring reaction is a nuclear electron transfer onto the proton, and in most cases the remaining neutral pion can be detected [41]:

$$\pi^- + p^+ \rightarrow \begin{cases} \pi^0 + n\ (61\%) \\ \gamma + n\ (39\%) \end{cases}$$

In either case, the proton turns into a neutron upon the nuclear electron transfer. The 3.31 MeV reaction energy follows from the $m_{\pi^-} - m_{\pi^0}$ versus $m_{p^+} - m_n$ mass differences.

In contrast, when a deuteron captures a π^- particle, the probability of nuclear electron transfer is only 1.45×10^{-4} [41]. We explain in what follows how this experiment also directly contradicts the strong force hypothesis. The mainstream deuteron model states that a deuteron comprises a proton and a neutron, which are held together by a strong force spill-over from these two nucleons. This model disregards the neutron polarizability parameter, and postulates that there is essentially no difference between the nucleons' structure in a free particle state or in a bound state. Thus, one would expect the π^- particle to turn a deuteron into a di-neutron, which is contrary to experimental data. We note that the strong-force spillover theory postulates that a proton–neutron structure has the same binding energy as a di-neutron structure since the hypothetical strong-force does not distinguish between nucleon types. In other words, the existing deuteron binding energy also should not hinder the $\pi^- + p^+ \rightarrow \pi^0 + n$ reaction. However, it is presently under debate whether a di-neutron particle even exists. At the very least, one may state that the difficulty of detecting a di-neutron means that it is either weakly bound or non-existent. To explain the non-existence of a di-neutron particle, nuclear theorists postulate that the deuteron is in an isospin singlet state while the di-neutron is in an isospin triplet state, and then give a mathematical argument for the unbound state of an isospin triplet. This postulate does not have any experimental proof, and presumes that isospin is generated by the hypothetical nuclear strong force. But in Chapter 7 we showed that isospin is an electromagnetic quantity. Even after all these problematic postulates, the 3.31 MeV reaction energy should still allow di-neutron production. In summary, the outcomes of nuclear π^- capture experiments appear to contradict the nuclear strong force theory.

Altogether, our analysis suggests that the "strong force spill-over" hypothesis-based mainstream nuclear binding model may be untenable.

14.2.3. *More evidence from scattering experiments*

The nature of nuclear interactions can be also studied through the analysis of scattering interactions in high-energy collisions. As illustrated in Figure 14.5, the experimental alpha particle scattering cross-section is well matched by the model of electrostatic Coulomb repulsion below 25 MeV alpha particle energy, and by the combination of electrostatic Coulomb repulsion plus magnetic Poisson repulsion at alpha particle energies beyond 25 MeV. This simple electromagnetic model fits well the experimental data even up to 40 MeV impact energy, without any need for the introduction of hypothetical new forces. We note the repulsive magnetic interaction term, which is in line with the above-identified magnetic repulsion in the deuteron bond.

Studying the momentum and angle distribution of particles emerging from high-energy collisions provides a means for experimentally measuring the relative position of nucleons inside the nucleus. In solid-state physics, X-ray diffraction is used to determine its crystal structure. A nuclear equivalent of the crystal structure measurement technique seems to have been discovered and reported in Ref. [19]. The authors of [19] analyzed post-collision momentum distribution data to reconstruct the solid-state structure of ^{2}H and ^{3}He nuclei; their main result is shown in Figure 14.6. The reported data clearly shows the triangular structure of ^{3}He and the well-defined inter-nucleon distance in ^{2}H. In the future, this methodology may

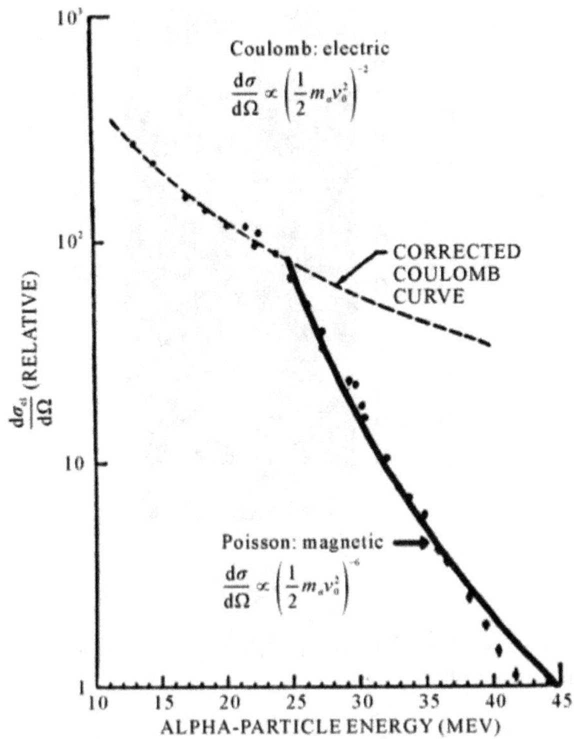

Figure 14.5. Straightforward electromagnetic interpretation of high-energy alpha particle scattering data, reproduced from [2].

be used to experimentally map out the solid-state structure of other light nuclei as well. In our interpretation, the data presented in [19] is a rather direct experimental confirmation of the above-discussed solid-state nuclear structure, and is irreconcilable with the liquid-drop model of the nucleus. Figure 14.6 also demonstrates the similar size of ^{2}H and ^{3}He nuclei, which are measured to have 2.14 fm and 1.97 fm radii, respectively.

In the context of our electromagnetic nuclear model, one issue is to make sense of the hypothetical fractionally charged particles known as quarks. The nature of hypothetical quarks is puzzling because no fractionally charged elementary particles have ever been experimentally observed, neither in cosmic rays nor in high energy collisions. The existence of quarks has been suggested initially in the 1960s, based on theoretical efforts by Gell-Mann to model baryons and mesons [14]. It took 10 years for the quark hypothesis to progress from fringe theory to consensus theory status, and by the mid-1970s quarks were considered to be elementary particles by most theoretical physicists, despite the fact that such fractionally charged

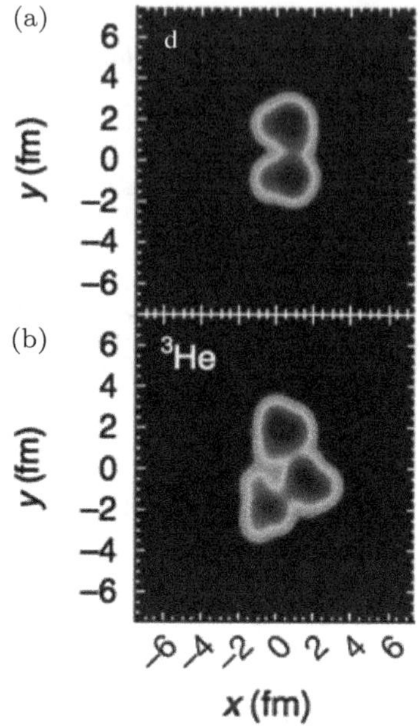

Figure 14.6. Visualization of the ^{2}H (a) and ^{3}He (b) solid-state structures, based on the experimental measurements of [19].

elementary particles have never been observed in nature. History might have been different if the fractional quantum hall effect was known at the time, but it was only discovered in 1982. The essence of the fractional quantum hall effect is the discovery of fractionally charged apparent particles, which also have quark-like $\frac{1}{3}$ elementary charge. Solid-state physicists of that time prudently refrained from announcing the discovery of a fractionally charged elementary particle, and eventually determined that the fractional charge originates from the quasi-particle type excitation of electron–hole pairs [7]. This analysis of the fractional quantum hall effect might have served as an inspiration for the nuclear physics community, but the belief in quarks as real particles was not revised because electromagnetism was thought to be irrelevant for the internal proton structure. History might have been also different if the fundamental reasons for charge quantization, which we discussed in Chapter 4, were known in the 1970s. In that case, physicists of the 1970s would have had a strong theoretic ground for rejecting a fractionally charged elementary particle model.

High-energy scattering data contradicts the hypothesis of quarks as elementary particles. The momentum distribution of particles inside the proton is characterized by the F_2 structure function. As shown in Figure 14.7,

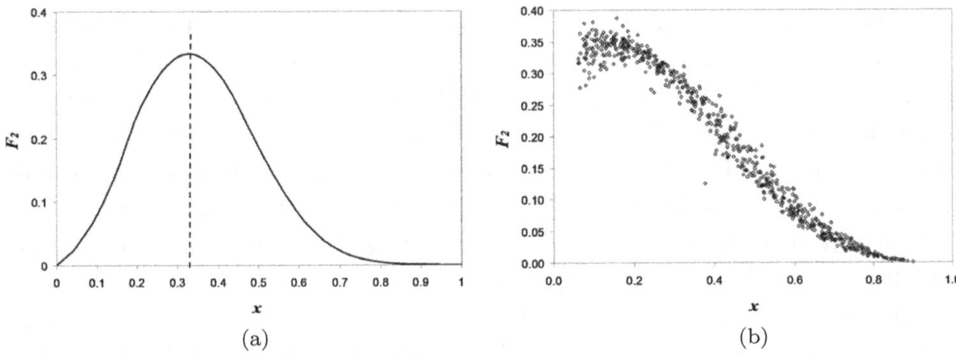

Figure 14.7. The theoretical F_2 momentum distribution function for 3 quark constituents inside the proton (a) versus SLAC measurements of F_2 (b), reproduced from [25].

the F_2 momentum distribution function in proton scattering experiments deviates from the expected F_2 curve for a proton made of just three quarks. The current consensus explanation for this deviation is the hypothesis that the three quarks originally thought to form the proton are the so-called "valence quarks," which are swimming in the background of "sea quarks" [15]. These so-called sea quarks are a collection of quark–antiquark pairs, radiated by the three valence quarks. But even with this speculative hypothesis, calculations showed that the valence quarks together with the sea quarks only accounted for 54% of the proton's momentum [8]. A further hypothesis was added to supplement the momentum shortfall of the quarks; chargeless particles called gluons were introduced into the proton model [16]. Since gluons have no electric charge, the thinking was that they are there, but the electrons probing the proton in deep inelastic scattering cannot see them. These hypothesized gluons were assigned the missing proton momentum, and the resulting proton model became the quark–gluon model that it is today. But even with this new hypothesis about "valence quarks" swimming in the background of "sea quarks and gluons," there seemed to be an angular momentum deficit with respect to the measured angular momentum of the proton, necessitating an additional hypothesis to be introduced in the 1990s, proposing the presence of "virtual strange quarks" [31]. These additional "virtual strange quarks" were thought to carry the proton's missing angular momentum, and the internal momentum distribution among valence quarks, sea quarks, virtual strange quarks, and gluons was re-adjusted to match experimental data.

Let's compare the supposedly three-quark structured proton, which has 938 MeV mass, against the supposedly two-quark structured pion, which has

140 MeV mass. From the quark model, one would expect the proton:pion mass ratio to be 3:2, but the actual mass ratio is over 13:2. Mainstream theorists claim that valence quarks account for only 1% of the proton mass, while the remaining 99% is a foam of "sea quarks and gluons and virtual strange quarks." The proton:pion mass ratio implies that a three-quark structure is accompanied by a much heavier foam of "sea quarks and gluons and virtual strange quarks" than a two-quark structure. But there is no explanation for this large discrepancy.

How much binding energy holds the proton's hypothetical three quarks together? Mainstream nuclear scientists claim that quarks are held together by hypothetical gluons, which are supposedly massless particles, analogous to photons. They attribute the bulk of the proton mass to the "chromodynamic binding energy," i.e., to the energy of the gluon field [27]. However, this is a nonsensical binding energy concept. The stable binding of any particles always releases energy, and thereby creates a mass defect. Such mass defect can be directly measured, e.g., upon nuclear fusion reactions. A stable binding of particles can never create an excess mass. It is hard to take seriously models that suggest otherwise.

In conclusion the three-quark structured proton and neutron particles, and models a proton and neutron to comprise *uud* and *udd* quark types, respectively. Regarding basic nuclear bond properties, the quark model does not facilitate any comprehensive calculation of deuteron binding energy or charge radius. In contrast, modeling the neutron as a proton–nuclear electron composite does allow the above-presented calculation of both deuteron binding energy and charge radius. Based on the experimental science, we find it difficult to retain the quark model in its current form and therefore in the next section we take the liberty of proposing an alternative model.

14.2.4. *A new proton model*

We prefer a direct and speculation-free interpretation of the experimental high energy scattering data, where the momentum distribution data is interpreted according to its actual definitions. Reference [25] presents a detailed and thorough analysis of high energy scattering data from measurements performed at the Stanford Linear Accelerator Center (SLAC), Thomas Jefferson National Accelerator Facility (JLAB), and Hadron Electron Ring Accelerator (HERA). We describe these results in the following paragraphs. The data analysis of Ref. [25] reveals that the proton comprises internal constituents having approximately $\frac{1}{9}$ times the proton mass. The momentum

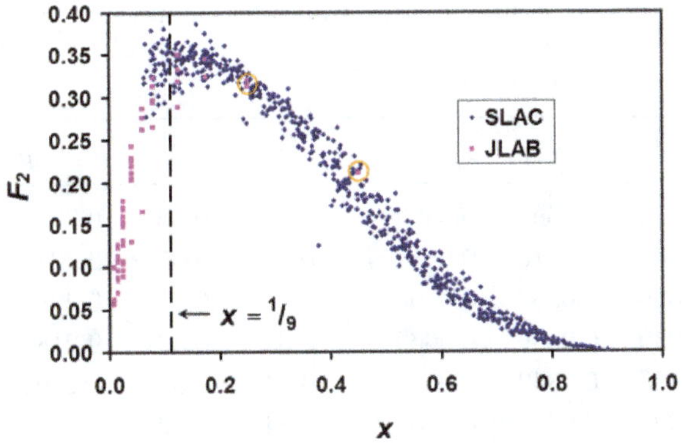

Figure 14.8. Combined SLAC and JLAB data of F_2 momentum distribution measurements from electron–proton scattering, reproduced from [25]. This scattering data shows the same F_2 momentum distribution as in the electron-proton case.

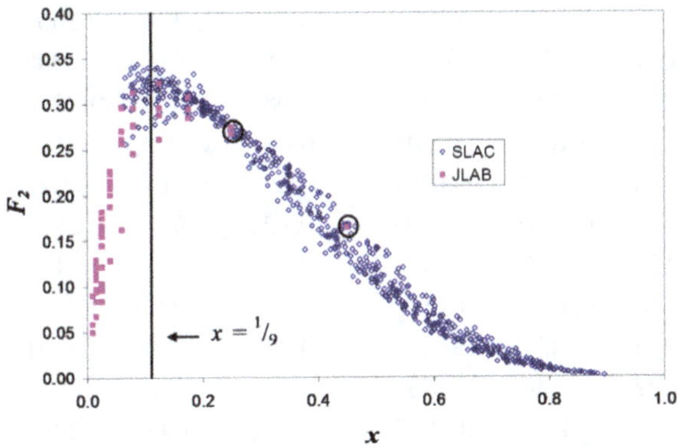

Figure 14.9. Combined SLAC and JLAB data of F_2 momentum distribution measurements from electron–deuteron scattering, reproduced from [25]. This scattering data show the same F_2 momentum distribution as in the electron–proton case.

distribution function supporting this interpretation is shown in Figure 14.8. The absence of $\frac{1}{3}$ proton mass constituents shows that quarks are not real particles.

The proton–electron scattering graphs of Figure 14.8 show clearly, without any curve fittings, that their F_2 values peak in the vicinity of $x = 0.11$, i.e., $x = \frac{1}{9}$. The JLAB F_2 values at $x = 0.45$ and $x = 0.25$ (circled) show that JLAB data integrates well with the original SLAC F_2 data. Is the use of single-variable $F_2(x)$ distribution justified, i.e., are the energy and momentum exchange sufficiently high for convergence? The use of $F_2(x)$ rather than $F_2(x, Q^2)$ is justified because the SLAC data of Figure

Table 14.2. F_2 structure function values for the proton (F_2–p) and the deuteron (F_2–d) as a function of x and Q^2 from the JLAB E99-118 deep inelastic scattering experiments [24].

x	Q^2	$F_2 - p$	$F_2 - d$	x	Q^2	$F_2 - p$	$F_2 - d$
0.009	0.034	0.056	0.0492	0.04	0.287	0.2027	0.2002
0.009	0.051	0.0616	0.058	0.04	0.353	0.2244	0.2077
0.009	0.086	0.0997	0.0896	0.04	0.37	0.2288	0.2139
0.015	0.059	0.0696	0.0669	0.04	0.371	0.2186	0.2155
0.015	0.095	0.0842	0.0831	0.04	0.38	0.2102	0.2231
0.015	0.098	0.0961	0.0935	0.04	0.421	0.2416	0.2268
0.015	0.112	0.0876	0.0966	0.06	0.18	0.1641	0.1616
0.015	0.127	0.1058	0.1073	0.06	0.479	0.2622	0.2563
0.015	0.144	0.1114	0.1129	0.06	0.491	0.2617	0.2702
0.015	0.151	0.1216	0.1186	0.06	0.543	0.2751	0.2609
0.015	0.164	0.1253	0.1227	0.06	0.633	0.2863	0.2958
0.015	0.172	0.1118	0.1286	0.08	0.456	0.265	0.2451
0.025	0.067	0.0883	0.0834	0.08	0.617	0.2935	0.2752
0.025	0.092	0.096	0.0953	0.08	0.619	0.296	0.2767
0.025	0.104	0.104	0.0994	0.08	0.799	0.3128	0.295
0.025	0.113	0.1069	0.1024	0.08	0.818	0.3227	0.3122
0.025	0.14	0.1251	0.1208	0.125	0.588	0.2876	0.2609
0.025	0.186	0.1469	0.1441	0.125	0.797	0.3179	0.2873
0.025	0.195	0.1312	0.1439	0.125	1.032	0.3319	0.2952
0.025	0.212	0.1675	0.1545	0.125	1.056	0.3491	0.3228
0.025	0.222	0.1593	0.1568	0.175	1.029	0.3242	0.2846
0.025	0.24	0.178	0.1656	0.175	1.045	0.3235	0.2939
0.025	0.252	0.1696	0.1745	0.175	1.365	0.3447	0.3072
0.025	0.253	0.153	0.1601	0.25	1.332	0.3126	0.2673
0.025	0.287	0.1669	0.1814	0.25	1.761	0.3183	0.2744
0.04	0.133	0.1295	0.128	0.45	2.275	0.2104	0.1638
0.04	0.273	0.2038	0.1876				

14.7 were shown to satisfy Bjorken scaling, i.e., for $x > 0.2$, the F_2 values are essentially the same for a given x regardless of the Q^2 value. In this $x > 0.2$ region, Q^2 values range from 0.6 to about 30 GeV^2. Regarding the $x < 0.2$ region, the JLAB Q^2 values shown in Table 14.2 are nearly the same as the SLAC Q^2 values listed in Appendix A.1 of [25], which makes the two results directly comparable.

Both the JLAB and SLAC experiments carried out also electron–deuteron scattering experiments. Their combined results shown in Figure 14.9, which again reveals the F_2 peak in the vicinity of $x = \frac{1}{9}$. The analogous results of the electron–proton and electron–deuteron scattering experiments are consistent with the above-described electromagnetic model, where the positively charged nucleons are protons in either case.

One may wonder why this 9-constituent parton model doesn't show up in any other literature? By 1973, mainstream theorists had essentially embraced the quark-gluon model as adequately describing the structure of the proton. Most attempts to explain the SLAC scattering results any other way had ended, and the bulk of the theoretical effort focused on enhancing the quark model. Different versions of the quark-gluon plasma model predict either constant or rising F_2 values as $x \to 0$. In the 1990s, when HERA experiments began producing data in this $x < 0.1$ range, it was assumed that HERA filled the low-x gap left by SLAC, even though its data were generated from scatterings with Q^2 values tens to hundreds of times higher than the SLAC data. Mainstream theorists could not resist the temptation of mixing non-comparable data in this low-x region of the F_2 curve, and proclaimed experimental support for their quark-gluon model. Reference [28] is a typical example of such erroneous data analysis. Around 2000, the JLAB experiment began producing scatterings with comparable Q^2 values to the SLAC experiment. By that time, the erroneous blending of high-Q^2 HERA data with low-Q^2 SLAC data was already a consensus procedure for obtaining the proton's F_2 curve, and mainstream theorists had no interest in pointing out their colleagues' mistakes or discussing the implications of the JLAB experiment.

Based on the data of Figures 14.8 and 14.9 alone, one may not determine with high certainty whether the proton comprises 8, 9 or 10 subparticles. In either case, the mass of the subparticle constituents would be ~100 MeV without binding energy consideration, and >100 MeV when considering that there must be binding energy holding them together within the proton. The two candidate particles in this mass range are the muon and pion. Does the proton comprise muon or pion subparticles? The muon can be excluded because in Section 6.2 we proved that the muon's charge radius is much smaller than the proton's charge radius. This leads us to conclude that the proton is a composite particle, comprising pion-type subparticles.

Can we determine more precisely the number of subparticles in a proton? As discussed in Section 6.4, the ratio of pion and proton charge radii is $r_\pi/r_p = 0.657/0.875 = \frac{3}{4}$. Since the charge radius is inversely proportional to the particle mass, the protonic pion mass is only $\frac{3}{4}$ times the free pion

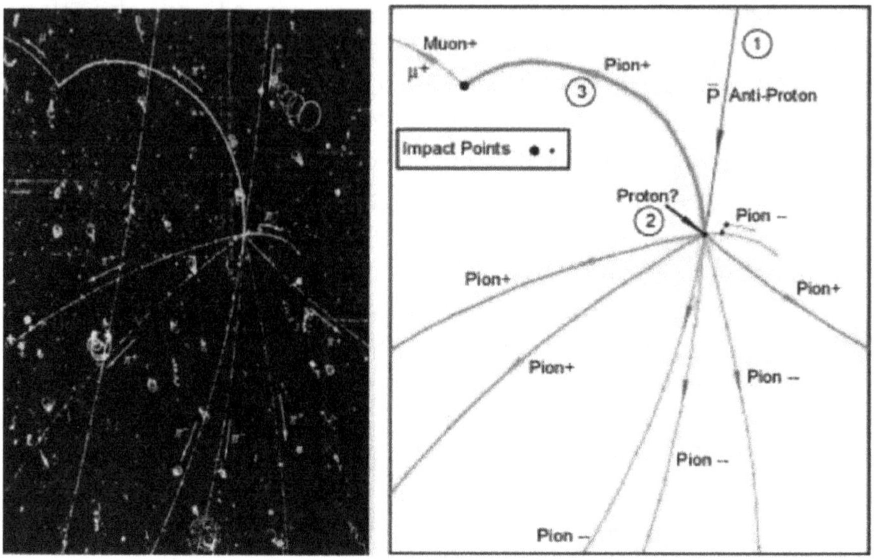

Figure 14.10. Proton–antiproton annihilation event. (a) Bubble chamber photograph. (b) Diagram of the photo, identifying the particles created by the annihilation event. Reproduced from [26].

mass. In other words, $\frac{1}{4}$ of the pion mass becomes the binding energy of proton formation. Using $m_\pi = 139.57\,\text{MeV}$ charged pion mass, we calculate the number of pions within the proton:

$$N_{\text{parton}} = \frac{m_p}{\frac{3}{4}m_\pi} = 8.96 \approx 9 \tag{14.2.2}$$

Employing this charge radius analysis, one may determine with high certainty that a proton comprises nine subparticles. In principle, there are two possibilities for its internal structure: (i) one positively charged pion and eight neutral pions or (ii) five positively charged pions and four negatively charged pions. The above-calculated parton number has 99.6% match when using charged pion mass, but much less precise match when using neutral pion mass. Furthermore, as shown in Figure 14.10, proton–antiproton annihilation events produce mainly charged pions, indicative of their internal subparticle types. This leads us to conclude that a proton comprises five positively charged pion-type constituents and four negatively charged pion-type constituents. This idea of oppositely charged subparticles, which exist without annihilating each other, is unexpected in the context of electron-like particles because an electron–positron pair quickly annihilates.

Table 14.3. Pion-related antiparticle events

Particles	Interaction event
$\pi^- - \pi^+$	Pionium formation, then annihilation
$\pi^- - p^+$	Nuclear electron transfer, yielding π^0, n
$p^- - p^+$	Partial annihilation, release of several π^+, π^-
Subparticles of p^+	Stable composite formation

However, the pion is probably not an elementary particle; its internal structure will be further discussed in Section 14.6. At this point, the only thing we know is that the proton's stability is related to the stabilization of its pion-type subparticles against annihilation or decay. Table 14.3 lists the pion-related antiparticle event types, which further illustrate how special the proton's stability is.

These results give inspiration to consider the meaning of a well-known experimental fact: a proton has never been observed to break up into hypothetical "quarks." When a proton is broken up by a high-energy collision, the observed collision products comprise pions, which then further decay to muons, but never "quarks." Such high-energy collisions between cosmic protons and the atmosphere are the origin of the observable muon background. When a proton annihilates with an antiproton, the decay products are once again pions. Figure 14.10 shows traces of a proton-antiproton annihilation event, highlighting the produced pion tracks.

In the context of our electromagnetic particle model, which was introduced in Chapters 3–4, half of the particle mass comprises magnetic field energy and the Zitterbewegung radius thus becomes inversely proportional to the particle mass. A simple scaling from electron mass to proton mass implies a proton Zitterbewegung radius of 0.21 fm. Using this proton Zitterbewegung radius value, we calculated in Section 6.4 the correct proton magnetic moment. Moreover, this 0.21 fm value is also an input parameter for the above-discussed deuteron radius calculation. These independent calculations ascertain that the proton Zitterbewegung radius is indeed 0.21 fm. Consequently, the nine pion sub-particles must have their magnetic moments aligned along the same axis, which adds up their individual magnetic fields.

Altogether, experimental evidence indicates that the proton is a composite of nine pionic subparticles, while quarks are just figments of imagination. The arising challenge is to understand the binding process among the nine pion constituents, which produces a stable proton composite. The proton stability is perhaps related to the meaning of Table 14.3 values in Section 6.4.

Until recently, the proton was thought to always remain stable in a low-energy environment. Surprisingly, Ref. [5] reports a decay process of those protons, which are inside Compton-scale electron–proton composites. Although the authors of Ref. [5] use different terminology to describe the produced electron–proton structure, their 2.3 pm inter-proton distance measurement leaves no doubt that they produced Compton-scale composites which we discussed in Section 4.5 and Chapter 12. The observed proton decay would not depend on electromagnetic Compton-scale composite formation if the proton's subparticles were held together by a nuclear strong force, because the postulated strong force should be independent of the surrounding electron configuration. Therefore, Ref. [5] confirms that the stabilizing force of the proton's internal structure is electromagnetic. We emphasize that Ref. [5] reports the outcome of multiple experiments, replicated in two different labs.

Having demonstrated that the "strong force" hypothesis is fundamentally flawed, one may wonder whether the theoretical assumption of $SU(3)$ type symmetry for nuclear forces is realistic. Regarding spin 3/2 baryons, the group-theoretically inspired assumption is that they are comprised of three isotropically spin-correlated quarks. However, by using the methodology of Chapter 9, O'Hara has proven that spin 3/2 baryons cannot be composed of three isotropically spin-correlated sub-particles [6]. Regarding spin 1/2 baryons, the exploration of proton versus neutron differences is a central theme in this chapter. Since $SU(3)$ is the symmetry group of a linear isotropic harmonic oscillator in three dimensions, it should become relevant for classifying particles having isotropic radial oscillations.

In summary, the data presented in this section demonstrate that the "strong nuclear force" is most probably just electromagnetic force.

14.3. What is the "Weak Nuclear Force"?

The "weak nuclear force" has been originally proposed to explain the endothermic capture of energetic electrons by protons, as well as the reverse process of electron emission by some unstable nuclei. On the basis of the currently mainstream "weak nuclear force" theory, Ref. [10] derives the Lagrangian expression of "weak nuclear" and electromagnetic interactions. Figure 14.11 shows the contrast between the Lagrangian expressions of electromagnetic theory versus electro-weak theory. Given the incredible

$$
\begin{aligned}
\mathcal{L} = {} & \frac{1}{2}(\partial_\mu\varphi)^2 - \lambda v^2\varphi^2 - \lambda v\varphi^3 - \frac{\lambda}{4}\varphi^4 + W^+_\mu W^{-\mu}\left(M_W + \frac{g}{2}\varphi\right)^2 - \frac{1}{2}|\partial_\mu W^+_\nu - \partial_\nu W^+_\mu|^2 \\
& + \frac{1}{2}(Z_\mu)^2\left(M_Z + \frac{\sqrt{g^2+g'^2}}{2}\varphi\right)^2 - \frac{1}{4}(\partial_\mu Z_\nu - \partial_\nu Z_\mu)^2 - \frac{1}{4}(\partial_\mu A_\nu - \partial_\nu A_\mu)^2 \\
& + \frac{ig}{2}(W^{+\mu}W^{-\nu} - W^{-\mu}W^{+\nu})[\partial_\mu(Z_\nu\cos\theta_W + A_\nu\sin\theta_W) - \partial_\nu(Z_\mu\cos\theta_W + A_\mu\sin\theta_W)] \\
& + \frac{ig}{2}(\partial^\mu W^{+\nu} - \partial^\nu W^{+\mu})[W^-_\mu(Z_\nu\cos\theta_W + A_\nu\sin\theta_W) - W^-_\nu(Z_\mu\cos\theta_W + A_\mu\sin\theta_W)] \\
& - \frac{ig}{2}[(Z_\nu\cos\theta_W + A_\nu\sin\theta_W)W^+_\mu - (Z_\mu\cos\theta_W + A_\mu\sin\theta_W)W^+_\nu](\partial^\mu W^{-\nu} - \partial^\nu W^{-\mu}) \\
& + \frac{g^2}{4}(W^+_\mu W^-_\nu - W^+_\nu W^-_\mu)^2 - \frac{g^2}{2}|W^+_\mu(Z_\nu\cos\theta_W + A_\nu\sin\theta_W) - W^+_\nu(Z_\mu\cos\theta_W + A_\mu\sin\theta_W)|^2 \\
& + \bar{\mathcal{E}}(i\gamma^\alpha\partial_\alpha - m_{\mathcal{E}})\mathcal{E} + \bar{\nu}_{\mathcal{E}}i\gamma^\alpha\partial_\alpha\left(\frac{1-\gamma_5}{2}\right)\nu_{\mathcal{E}} - eA_\alpha\left\{\bar{\mathcal{E}}\gamma^\alpha\mathcal{E} - \frac{2}{3}\bar{\mathcal{U}}\gamma^\alpha\mathcal{U} + \frac{1}{3}\bar{\mathcal{D}}\gamma^\alpha\mathcal{D}\right\} + \\
& \frac{g}{2\sqrt{2}}\left[W^+_\alpha\left\{\bar{\nu}_{\mathcal{E}}\gamma^\alpha(1-\gamma_5)\mathcal{E} + \bar{\mathcal{U}}\gamma^\alpha(1-\gamma_5)V_{\mathrm{KM}}\mathcal{D}\right\} + W^-_\alpha\left\{\bar{\mathcal{E}}\gamma^\alpha(1-\gamma_5)\nu_{\mathcal{E}} + \bar{\mathcal{D}}\gamma^\alpha(1-\gamma_5)V^+_{\mathrm{KM}}\mathcal{U}\right\}\right] \\
& - \frac{\sqrt{g^2+g'^2}}{4}Z_\alpha\left[\bar{\nu}_{\mathcal{E}}\gamma^\alpha(1-\gamma_5)\nu_{\mathcal{E}} - \bar{\mathcal{E}}\gamma^\alpha(1-\gamma_5)\mathcal{E} + 4\sin^2\theta_W\bar{\mathcal{E}}\gamma^\alpha\mathcal{E}\right] \\
& - \frac{\sqrt{g^2+g'^2}}{4}Z_\alpha\left[\bar{\mathcal{U}}\gamma^\alpha(1-\gamma_5)\mathcal{U} - \bar{\mathcal{D}}\gamma^\alpha(1-\gamma_5)\mathcal{D} - 4\sin^2\theta_W\left\{\frac{2}{3}\bar{\mathcal{U}}\gamma^\alpha\mathcal{U} - \frac{1}{3}\bar{\mathcal{D}}\gamma^\alpha\mathcal{D}\right\}\right] \\
& - \frac{m_{\mathcal{E}}}{v}\varphi\bar{\mathcal{E}}\mathcal{E} - \frac{m_{\mathcal{Q}}}{v}\varphi\bar{\mathcal{Q}}\mathcal{Q}\left(+i\frac{m_{\mathcal{E}}}{v}\varphi_3\bar{\mathcal{E}}\gamma_5\mathcal{E} - i\frac{\sqrt{2}m_{\mathcal{E}}}{v}[\bar{\nu}_L\mathcal{E}_R\phi^+ - \bar{\mathcal{E}}_R\nu_L\phi^-]\text{etc.}\right)
\end{aligned}
$$

$$\mathcal{L} = \tfrac{1}{2}\sqrt{g}G\widetilde{G} \Longleftrightarrow$$

Figure 14.11. Contrast between the Lagrangian of the electromagnetic theory (left) and the Lagrangian of the "weak nuclear force" theory (right, reproduced from [10]).

complexity of the "electro-weak" Lagrangian, a rational scientist would carefully consider any possible electromagnetic solutions to the electron capture and emission phenomena before replacing the simple electromagnetic Lagrangian with the incredibly complex "electro-weak" Lagrangian.

What direct evidence is there for the existence of the weak nuclear force? One cited evidence is the measured anapole moment of some nuclei, which is claimed to be a signature of "weak nuclear" interactions [11]. Geometrically, an anapole moment means the simultaneous presence of toroidal and poloidal charge currents over the surface of a torus. How could anyone seriously claim that a toroidal-shaped electric current constellation can only be explained by the introduction of a hypothetical new force, along with the incredibly complex Lagrangian of Figure 14.11?

A second issue is "parity non-conservation," which is based on the assumption that a nuclear spin orientation does not influence the direction into which decay products are emitted. Experimental evidence of parity non-conservation during electron-emitting nuclear decays was first observed already in 1957. At that time, the relationship between nuclear spin, Zitterbewegung, and nuclear electrons was not well understood yet. By this point in the book, it is clear that these are all electromagnetic phenomena, and it is normal to expect that the orientation of nuclear Zitterbewegung would influence the direction into which nuclear electrons are emitted.

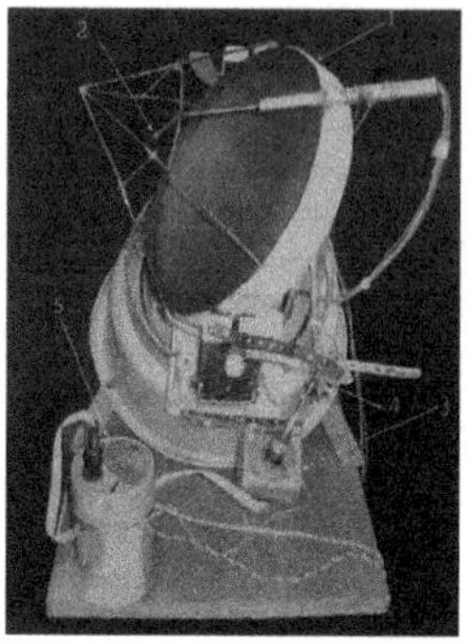

(a)

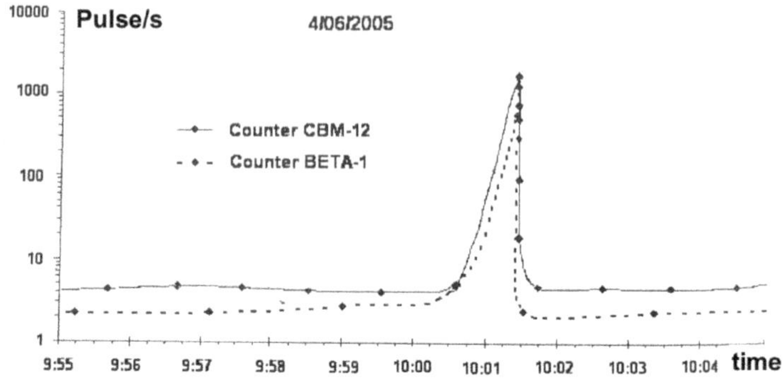

(b)

Figure 14.12. Measurement of neutrinos as electromagnetic waves. (a) Parabolic metal antenna, with beta emitting ^{60}Co material at its focal point. (b) ^{60}Co decay rate during the experiment, as the antenna sweeps a small angle on the sky. Reproduced from [30].

The whole discussion on so-called "parity non-conservation" is thus a misnomer, and was mistakenly assumed to be the signature of a new force.

The most frequently claimed "weak force" evidence is the experimentally observed emission/absorption of neutrinos during nuclear capture/emission of electrons. Neutrinos' low interaction probability with other particle types is claimed to be showing that the neutrino interacts via a different force than electromagnetic force. We discussed in Section 7.9 that such logic is similar to claiming that gamma rays would not be electromagnetic because a radio antenna does not pick them up. Most nuclear physicists also assume that neutrinos are massive particles, which nevertheless always travel at the speed of light, without ever slowing down. This hypothesis is a clear impossibility, as it directly contradicts the principles of relativity. On the basis of both theoretical arguments and experimental observations, we showed in Section 7.9 that a neutrino is nothing else than longitudinally polarized electromagnetic radiation.

In 2005, A. Parkhomov performed a pioneering series of experiments [30], which directly prove that certain electromagnetic waves generate "weak-force" associated beta decay. He constructed a metallic parabola shown in Figure 14.12, which has beta emitting ^{60}Co material at its focal point. This parabola was aimed at the sky in a fixed position, scanning for low-frequency cosmic neutrinos. It sweeps the sky as the Earth rotates. As shown in Figure 14.12, in a certain orientation the ^{60}Co decay rate increases by three orders of magnitude. This strong beta decay enhancement was simultaneously measured by two detectors, and the same effect was observed at several celestial orientations. In contrast, when an alpha emitting material was placed at the focal point, no change in the alpha decay rate was observed in these same orientations. Therefore, two conclusions can be drawn from this experiment: (i) the detected cosmic wave is electromagnetic radiation because it is collected by a metallic parabola, and (ii) this electromagnetic radiation strongly enhances the "weak force" associated nuclear beta decay rate while leaving the "strong force" associated alpha decay rate unchanged. We emphasize that Parkhomov's measurements contradict all weak force-related hypotheses of nuclear theorists.

The measurements of Parkhomov [30] and the results of Section 7.9 lead us to conclude that a neutrino is a longitudinal electromagnetic wave, which enhances the emission of nuclear electrons in beta-decaying materials. Regarding electromagnetic wave detection, it is well understood that the appropriate measuring device depends on the electromagnetic wavelength. For transversal electromagnetic waves, very different equipment is used to detect RF, visible, or gamma radiation. Similarly, metallic parabolas are only suitable for collecting low-frequency longitudinal electromagnetic waves, while a different method is needed for detecting high-frequency longitudinal electromagnetic waves. With reference to Section 7.9, the need for scintillators for detecting high-frequency neutrinos must not be thought as evidence against detecting low-frequency neutrinos with the help of metallic parabolas.

Another supposed "weak force" evidence is the detection of the 80 GeV mass W boson and the 91 GeV mass Z boson particles, which theorists proclaim to be the "mediating particles" of the weak force. The nuclear beta-emission of electrons and antineutrinos is hypothesized to be mediated by the decay of 80 GeV mass W boson particles, i.e., an $X_Z^A \rightarrow X_Z^{A+1} + W^-(80\,\text{GeV}) \rightarrow X_Z^{A+1} + e^- + \bar{\nu}_e$ process which temporarily violates energy conservation. In plain words, the presence of the above-mentioned longitudinal electromagnetic wave is supposed to trigger the birth of an 80 GeV mass particle according to nuclear theorists.

Presently, contemplations about the structure and interactions of 80 GeV particles with 10^{-25} sec half-life are highly speculative. Claiming that the emission and absorption of neutrinos is mediated by 80 GeV virtual particles is a particularly extraordinary claim, and should require extraordinary evidence. Besides the lack of any direct experimental evidence for an electron-neutrino pair being born from such virtual particle decay, it turned out that this theory must also allow a temporary violation of the momentum conservation principle. Figure 14.13 shows the energy spectrum of neutron decay

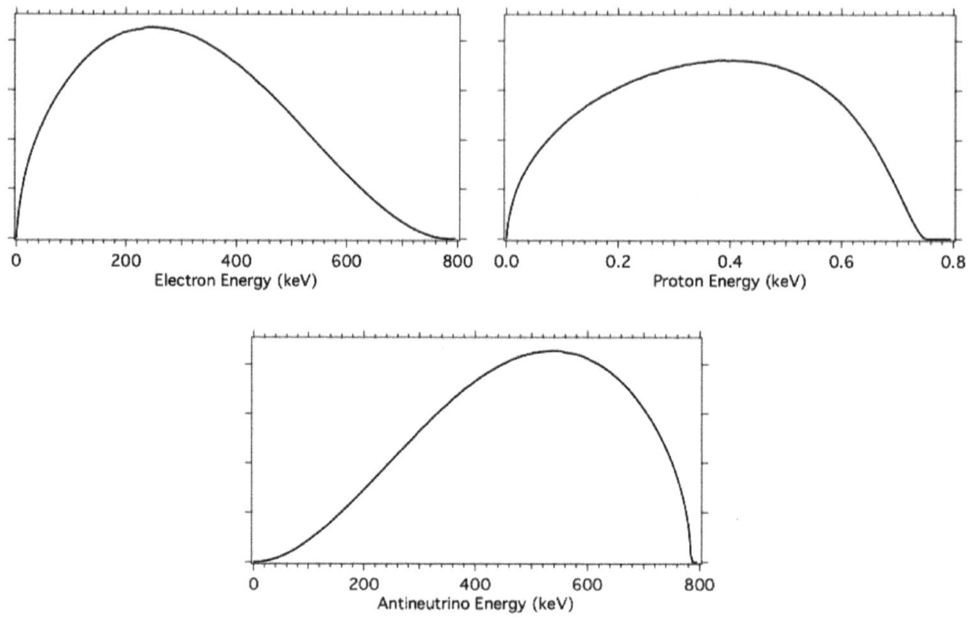

Figure 14.13. Kinetic energy spectra for the electron, proton and antineutrino products of neutron decay (reproduced from [22]).

products. The energy of the electron+antineutrino products always sums up to 782 keV, but individually their energy varies between 0 keV and 782 keV. At one extreme of the distribution, nearly all the decay energy is carried away by an electron. At the other extreme of the distribution, all the decay energy is carried away by an antineutrino. In both cases, the hypothetical intermediating W boson carries 782 keV energy, but its momentum varies between these two cases by a factor of $\sqrt{\frac{m_e}{m_{\bar{\nu}}}}>1000$. How could a particle have always the same energy but wildly varying momentum? This is allowed by the "off mass shell" hypothesis, which means that the "weak nuclear" force theory is also a temporary violation of the momentum conservation principle. Given that the presence of a W or Z boson was never experimentally observed during neutron decay, claims of energy–momentum conservation violating 80 GeV particles' presence in electron–neutrino processes appear to be non-scientific.

Curiously, the W and Z boson particles fit into a geometric $\times 3$ mass scaling pattern of nuclear particles. This pattern is shown in Figure 14.14. Since the supposedly "weak-force bearing" W and Z bosons fit into the mass scaling pattern of supposedly "quark–antiquark" type hadrons, it becomes quite apparent that these are all related particles. In order to understand the

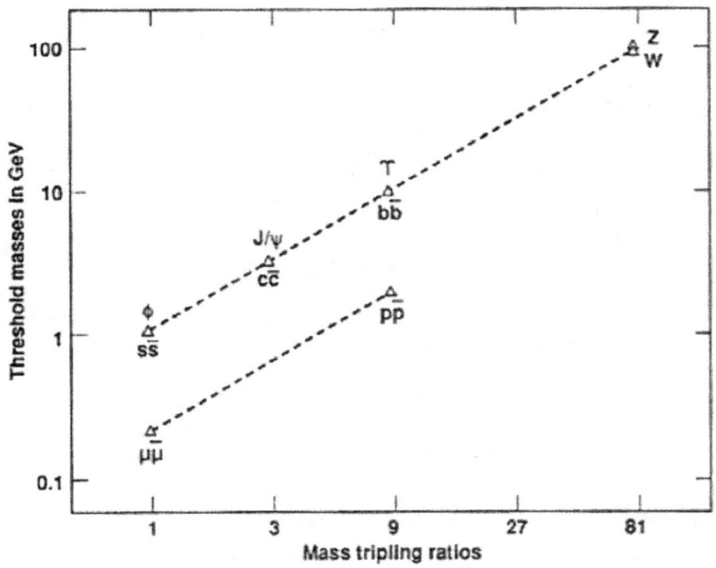

Figure 14.14. Thresholds for the production of various types of elementary particle states, revealing the very accurate $\times 3$ mass scaling pattern (reproduced from [21]).

correct structure of W and Z bosons, one must understand the mathematical background of this mass-tripling phenomenon.

Under the "weak nuclear" force hypothesis, the energy distribution of beta decay products is thought to be meaningless random data. Looking from the perspective of electromagnetic analysis, could the energy distribution of beta decay products reveal useful information about nuclear structures? The antineutrino energy now gains meaning: it is the energy difference between the e_n^- state at the moment of its break-up and the final e^- state, as observed from a co-moving frame. We give two examples to illustrate how this new insight helps to analyze nuclear structures. During a neutron decay, separating proton and e_n^- must maintain zero net momentum, the ratio between their kinetic energies equals the inverse ratio of their respective masses. Figure 14.13 shows that the ratio of proton and electron energy peaks is $\frac{250\,keV}{0.4\,keV} = 625 \approx \frac{1836.15}{3} = \frac{m_{p^+}}{3m_{e^-}}$. This data reveals that the neutron is inherently 3 electron masses heavier than the proton. Based on the neutron mass value, approximately 240 keVp^+–e^-n binding energy would be released if the starting particles are p^+ and e^-n. Conversely, if the starting particles are an ordinary proton and an electron, the required relativistic electron mass is 240 keV less than 3 electron masses; i.e., the electron must have at least 782 keV energy for neutron formation. It is

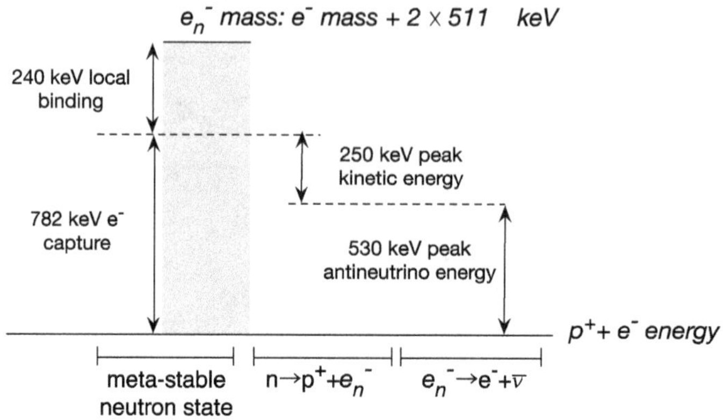

Figure 14.15. The neutron energy diagram.

very interesting to see this connection between the neutron–proton mass difference and the electron mass, and to properly understand the neutron's metastable state. Figure 14.15 illustrates the resulting neutron energy diagram.

A second example is the beta decay of the ^{163}Dy nucleus. It is known at least since 1992 that the ^{163}Dy nucleus beta decays in a fully stripped Dy^{66+} state, while the same ^{163}Dy nucleus is stable in a neutral atom [43]. Such beta decay of Dy^{66+} produces a bound state electron, which has a negative energy state with respect to a free electron. While nuclear theorists insist that any beta decay reaction is completely independent from electromagnetism, it is clear that the electric potential is the controlling parameter of this beta decay process.

In summary, this section demonstrates that the "weak nuclear" force probably has electromagnetic origin, and it is closely related to the energy diagram of Figure 14.14. There is no rationale for replacing the simple electromagnetic Lagrangian by the incredibly complex Lagrangian expression shown in Figure 14.11. Under the correct beta decay process model, the energy spectrum of emitted reaction products becomes useful information for the analysis of nuclear structures.

14.4. What Particles are Released During Nuclear Fission?

The long-standing model of nuclear fission states that a fission event fragments a nucleus into two pieces, possibly accompanied by the release of a few neutrons. Considering that the nucleus comprises protons and nuclear electrons, a more generic view of a nuclear fission process can be represented

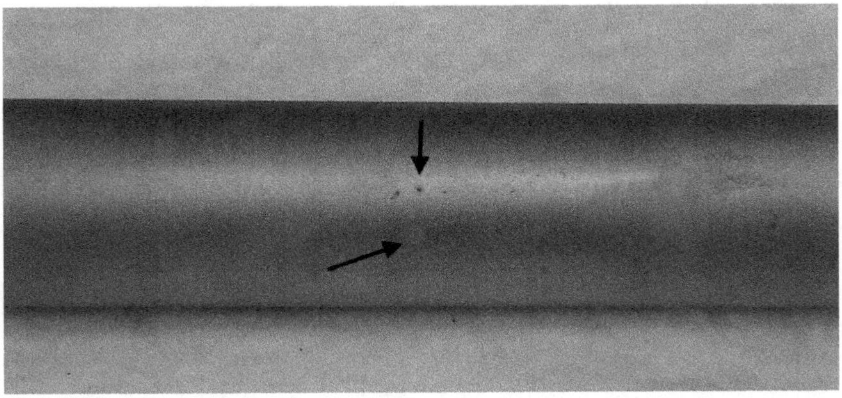

Figure 14.16. Transmutation spots appearing on the surface of a molybdenum cylinder which underwent an experiment described in Section 17.2. The arrows point at the spot locations.

by the following formula:

$$(Z, A) \to (Z_1, A_1) + (Z_2, A_2) + xn^0 + ye_n^- + zp^+$$

where Z is the nuclear charge, and thus $Z = Z_1 + Z_2 - y + z$, and where A is the nuclear mass number, and thus $A = A_1 + A_2 + x + z$. In other words, unlike prior nuclear fission models where it is assumed that $y = 0$ and $z = 0$, we make no such assumption.

In the well-studied case of uranium fission, about 8 MeV of its fission energy is released in the form of prompt neutrino radiation. What could be the source of these immediately released neutrinos? As we saw in the preceding section, the fast decay of nuclear electrons into ordinary electrons produces neutrino radiation. The release of a few nuclear electrons therefore probably accompanies uranium fission reactions, but they have remained unnoticed up to now.

We obtained a more direct evidence of an electron-releasing fission reaction during our own experiments, which are described in Section 17.2. After one run of the experiment, which was performed at the University of Uppsala under the direction of Bo Höistad, approximately 1 mm sized spots appeared on the surface of our metallic container. Figure 14.16 shows a photograph of these spots. The container is made of TZM alloy, whose specified composition is 99.5% Mo, 0.4% Ti, 0.1% Zr, 0.01% Si, 0.01% Fe. We verified by elemental analysis of container samples that the Ti concentration is indeed 0.4% in our container. Two methods were used

Table 14.4. The XRF measurement around a spot shown in Figure 14.16, and the corresponding calculation of the spot composition.

	Mo	Ti	Zr	Si	W
XRF reading	98.41%	1.27%	0.13%	0.12%	0.08%
Difference from TZM		0.9%	≈0%	0.11%	
Spot composition[a]		89%		11%	

Note: [a]other elements than Mo.

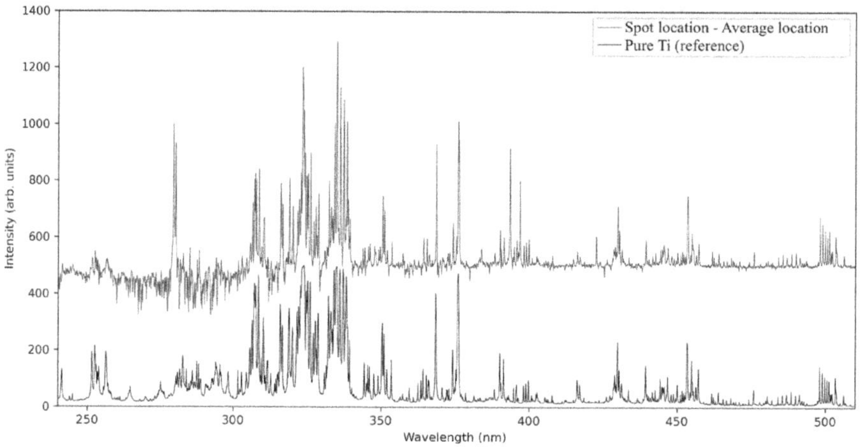

Figure 14.17. The LIBS measurement of a spot composition, showing that the newly created elements are mainly titanium atoms. Upper spectrum: the LIBS spectrum difference between the spot location and an average location on the container. Lower spectrum: the LIBS spectrum of a pure titanium reference.

to analyze the material of the spots: X-ray fluorescence (XRF) and laser-induced breakdown spectroscopy (LIBS). The XRF instrument measures the average composition over a 1 cm^2 area, and its reading around a spot is shown in Table 14.4. Considering that the spot is 1 mm sized, we interpret this XRF data as a near complete transmutation of Mo within the spot, with little or no transmutation on other surface areas. The 0.08% tungsten appears to be a contaminant in our TZM material, and we note that its producer also makes tungsten cylinders.

We used LIBS as a second method to analyze the spot composition. In order to have a clear signal of the newly created elements, we looked at the difference between the spectrum at the spot and the spectrum at an average location. This difference corresponds to the excitation spectrum of newly created elements. As seen in Figure 14.17, the excitation spectrum of new elements is practically the same as titanium's spectrum; all the titanium peaks are well recognizable. Therefore, the LIBS measurement confirms that the newly created elements are mainly titanium atoms. There are a few additional peaks which can be seen only at the spot location but not in titanium; these should correspond to the other elements listed in Table 14.4.

Is there a non-nuclear explanation for the appearance of titanium-rich spots? We did calibration experiments, during which an empty TZM cylinder underwent the same heat treatment program as the above described TZM container.[b] No spots appeared on the empty TZM cylinder, i.e., this phenomenon cannot be explained by a spontaneous concentration of titanium on the TZM surface. It is relevant to keep in mind that TZM's melting point is 2620°C, titanium's melting point is 1670°C, while our reactor's highest temperature was only 1300°C. Moreover, a spontaneous concentration of titanium into such spots would be lowering entropy. Therefore, the remaining explanation that we can think of is an *in situ* fission process.

Based on the above data, we are observing molybdenum to titanium fission, which involves a splitting of molybdenum into two approximately equal halves. Such fission of molybdenum into two titaniums is exothermic by 8 MeV. Molybdenum is element number 42 while titanium is element number 22; thus, the conservation of total charge requires that two electrons are also emitted during the fission process. The overall fission reaction may therefore be written as follows:

$$^{96}\text{Mo} \rightarrow 2 \times^{48} \text{Ti} + 2 \times e^-$$

The fission of other molybdenum isotopes proceeds analogously, and may release neutrons as well. If molybdenum nuclei were to fission without the release of nuclear electrons, the fission product would have been scandium, or a mixture of titanium and calcium. However, the absence of scandium and calcium indicates that molybdenum fission is always accompanied by the release of nuclear electrons in our experiment.

Is it surprising that molybdenum splits into two equal halfs, rather than a mixture of fission products, as in the case of uranium fission? Considering that uranium fission is triggered by neutron absorption, it may not be surprising to find that uranium fragments vary according to the location of neutron impact. In contrast, molybdenum fission is triggered by a different process, which apparently splits the nucleus along its mid-point.

We made radiation measurements during the molybdenum to titanium fission process, but did not detect any above-background gamma emission. However, we measured radio frequency emissions from the reactor. Note that nuclear electrons are the lightest particles among the fission products,

---

[b]See Section 17.2 for the description of materials inside the TZM container.

and their decay into ordinary electrons releases neutrinos. This process is analogous to a neutron decay, and therefore most of the fission energy is expected to go into neutrino radiation.

The accumulation of Si may correspond to a further fission of titanium, such as Ti → Si + O type fission.

Reference [39] describes yet another experiment where the release of nuclear electrons is implied. Its author observes the plasma emission spectrum during electric discharges in water–argon plasma. Besides the spectral lines of hydrogen, water, and argon, the appearance of a strong sodium line is detected. With reference to the clustered nuclear structure shown in Figure 14.2, the fission of argon is expected to release an oxygen fragment. Considering the sodium and oxygen fission fragments, the overall fission reaction may be written as one of the following processes:

$$^{40}\text{Ar} \rightarrow {}^{23}\text{Na} + {}^{16}\text{O} + e^- + n^0$$

$$^{40}\text{Ar} \rightarrow {}^{23}\text{Na} + {}^{16}\text{O} + 2 \times e^- + p^+$$

In either case, the release of nuclear electrons is required by charge conservation.

The authors of [33] conducted a series of experiments with electrically exploded titanium wires. Besides fission fragments, they detected an excess of hydrogen, suggesting that protons are released during the observed fission process. This particular experiment will be further discussed in Chapters 15 and 16.

In summary, **nuclear fission reactions simultaneously release neutrons, nuclear electrons, and sometimes even protons**. The quantitative ratio of these fission products depends on the particular nuclear process. But qualitatively, a detection of neutrons is an indicator of the presence of nuclear electrons.

14.5. The Nuclear Electron Particle

14.5.1. *A precise measurement of the nuclear electron mass*

The preceding chapters focused on finding an analytical solution which describes the electron's internal structure. The capture or emission of a neutrino changes the electron structure, and creates the e_n^- particle. This e_n^- particle was shown in Section 14.2 to be positioned between the two protons comprising a deuteron nucleus, and a neutron was shown to be an e_n^-–p^+ composite particle. We already found in Section 14.3 that nuclear

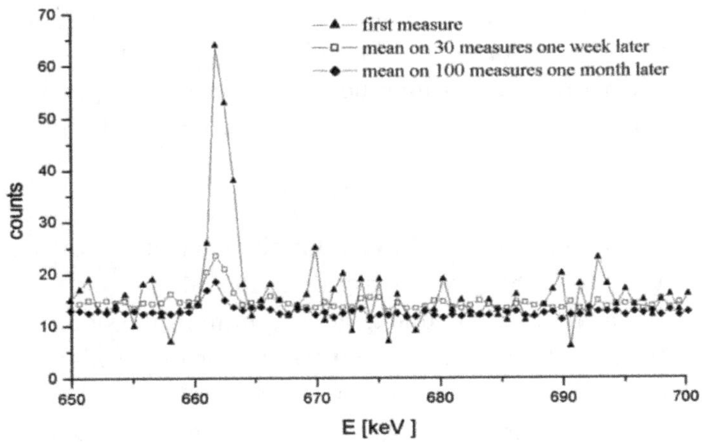

Figure 14.18. The gamma peak corresponding to a nuclear electron capture by ^{58}Ni. Measurement data from [34].

electron mass is approximately three times the ordinary electron mass. Here, we determine its mass more precisely.

It follows from the results of the previous section that nuclear electrons are emitted during fission reactions. What happens if a nuclear electron is captured by another nucleus, prior to its decay into e^-? The nuclear capture of ordinary electrons generates neutrino emission, which is a consequence of the isospin change between the e^- and e_n^- particle states. However, the e_n^- particle's isospin does not change during nuclear capture, and therefore the binding energy of the capture process must be emitted as gamma radiation. The measurement of gamma radiation peaks during the nuclear capture of e_n^- is therefore a direct measurement of their binding energy to a given nucleus. On this basis, we seek experimental data of gamma emission from electron capture capable nuclei.

The authors of [34] observed an unexpected gamma emission from hydrated nickel, and they could not explain the source of the obtained gamma peak. As shown in Figure 14.18, this gamma peak is at 661.5 keV. They also observed a simultaneous emission of neutrons, at low intensity. With reference to Section 14.4, the presence of neutrons signals an ongoing fission process, which must release nuclear electrons as well. It is therefore reasonable to interpret the origin of the gamma peak shown in Figure 14.18 as e_n^- capture by certain nuclei.

The ^{58}Ni isotope of nickel is capable of electron capture via the $^{58}\text{Ni} + e_n^- \to {}^{58}\text{Co}$ reaction. According to Figure 14.18, the binding energy of the e_n^- capture is $E_b = 661.5$ keV. In the case of an ordinary electron, the

^{58}Ni + e$^-$ $\rightarrow$ ^{58}Co reaction is endothermic by $E_{\mathrm{ec}} = -381.6$ keV. Because of the same reaction end product, the difference between these two energies must correspond to the mass difference of the incoming particles. We may therefore determine the e_n^- mass from the following equation:

$$m_{\mathrm{en}}c^2 - m_e c^2 = E_b - E_{\mathrm{ec}} \tag{14.5.1}$$

The above equation yields 1554 keV for the nuclear electron mass, which well matches the $m_{en} \approx 3m_e$ result from Section 14.3.

A second measurement of unexpected gamma emission is reported in Ref. [44]. Its authors experimented on water vapor under high voltage spark discharges. The measurements were done under air atmosphere, and a lead-shielded chamber was employed to minimize the background noise. The experimental details are given in Ref. [44]. This experiment is in fact a laboratory analogue of a natural lightning discharge. The physical processes during lightning discharges have been rather well studied, and several authors noted the production of neutrons by lightnings [45–47]. The author of [46] discusses similar neutron production by artificial lightning discharges in the laboratory. Via a detailed examination of the lightning process signatures, the authors of [47] determined that the produced neutrons originate from the $^{14}N \rightarrow {}^{13}N + n^0$ fission reaction. The resulting ^{13}N isotope then further decays to ^{13}C. The measurement of such neutrons yet again signals the presence of nuclear electrons as well. We therefore interpret the origin of the gamma peak shown in Figure 14.19 as e_n^- capture by some nucleus. This interpretation is further corroborated in Ref. [44] by a time correlation analysis between the spark discharges and gamma photon emissions, which shows that the peak of Figure 14.19 occurs right after each spark discharge, and has microsecond scale duration. Therefore, the gamma peak cannot originate from any radioactive isotope contamination. Such short duration of the gamma peak emission is characteristic of a free nuclear electron's short half-life.

Protons are the main electron capture capable nuclei in this experiment, and therefore the capture of nuclear electrons may be written as $p^+ + e_n^- \rightarrow n$. Figure 14.19 shows the measured gamma peak, which implies E_b=259.6 keV binding energy between a proton and a nuclear electron. For ordinary electrons, the $p^+ + e^- \rightarrow n$ reaction is endothermic by $E_{ec} = -782.4$ keV. We again use Equation 14.5.1 to calculate the nuclear electron mass, and obtain 1553 keV.

Our final estimate for the e_n^- mass is the average of the above two measurements, that is $m_{en}c^2 = 1553$ keV. Knowing this nuclear electron

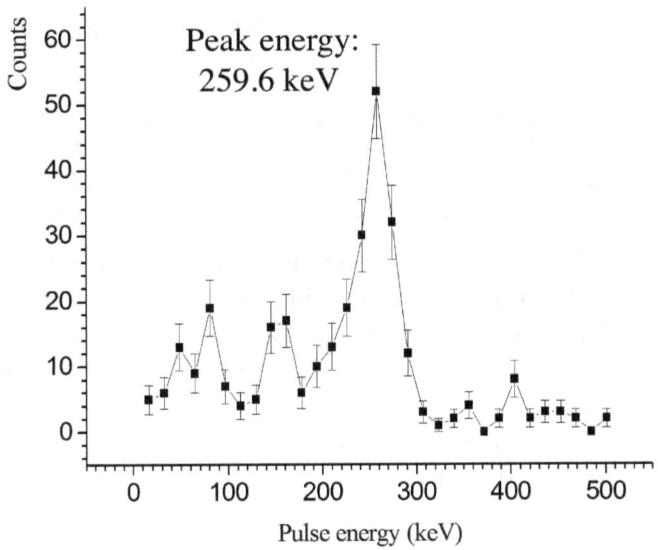

Figure 14.19. The gamma peak corresponding to a nuclear electron capture by a proton. Measurement data provided by V. Zatelepin.

mass value and the 260 keV binding energy between the e_n^- and p^+ particles is a new foundation for understanding the neutron's structure.

14.5.2. *A further characterization of the nuclear electron*

We do not know yet how to explain the very close proximity binding between a proton and a nuclear electron. Nevertheless, the single-frequency emission peaks shown in Figures 14.18–14.19 indicate that we are observing an ordinary transition process between two quantum mechanical eigenstates, which was discussed in Chapter 8. The neutron therefore also corresponds to a particular solution of the Dirac equation, and this solution might be discovered eventually. The binding energy shown in Figure 14.19 is α^{-2} times scaled from the 13.6 eV proton–electron binding energy in a hydrogen atom.

The most recent measurement of the radial distance between the neutron's positive and negative charges yields $\triangle r^2 = -0.1101$ fm [36]. This value is very similar to other recent measurements, for example the authors of Ref. [18] also found $\triangle r^2 = -0.11$ fm. The negative value of $\triangle r^2$ means that the negative charge has a slightly larger radius than the positive charge. The methodology of Ref. [36] actually measures the following value:

$$\triangle r^2 = (r_p + R_{\mathrm{ZBW}})^2 - (r_e + R_{\mathrm{ZBW}})^2$$

Table 14.5. A comparison between the e^- and e_n^- particles.

	e^-	e_n^- in bound state
Electric charge	-1	-1
Intrinsic angular momentum	$\hbar$	$\frac{3}{2}\hbar$
Isospin	0	-1
Mass	511 keV	1553.5 keV $\approx 3\times511$ keV
Charge radius	2.82 fm	0.899 fm $\approx \frac{2.82}{3}$ fm

The $R_{\mathrm{ZBW}} = 0.21$ fm value is common for both particles: this value is set by the proton's strong magnetic field. Thus, r_e is the only unknown in the above equation, and its value comes out to be $r_e = 0.899$ fm. In a free particle state, the nuclear electron's charge radius may slightly differ from this bound-state value.

We now calculate the e_n^- particle's angular momentum. The magnetic moment of e_n^- can be calculated as $\mu_e = \mu_n - \mu_p = -4.706\mu_N$, where μ_N is the nuclear magneton.

From the $g_e = (1 - \frac{r_e}{2\pi R_{\mathrm{ZBW}}})^{-1}$ formula of the anomalous magnetic moment, we get the following value for the intrinsic magnetic moment:

$$\mu_e\left(1 - \frac{r_e}{2\pi R_{\mathrm{ZBW}}}\right) = -4.706\mu_N\left(1 - \frac{0.899}{2\pi \times 0.21}\right) = -1.4996\mu_N$$

In this formula, R_{ZBW} is the Zitterbewegung radius, and r_e parameter is the charge radius of e_n^-. This obtained intrinsic magnetic moment differs by just 0.024% from $-1.5\mu_N$. Since the μ_N intrinsic magnetic moment corresponds to $\hbar$ angular momentum value, we conclude that the angular momentum of the bound e_n^- particle is $\frac{3}{2}\hbar$. The e_n^- particle decays into an ordinary electron by emitting a neutrino. According to Section 7.9 such neutrino emission is required by isospin conservation. Since the nuclear electron carries the isospin difference between a proton and a neutron, its isospin value is -1 in the standard notation. While we already understand neutrino radiation, the physical meaning of isospin remains to be explored in the context of the e_n^- particle.

The e_n^- particle demonstrates that the stability of a particle depends on the electromagnetic environment in which it is embedded; the e_n^- particle is stable when located between the two protons of a deuteron nucleus, but very short-lived in the free particle state. For many decades, scientists assumed that the nuclear environment "stabilizes neutrons," but in reality, **nuclear electrons are stabilized by bonding with one or more protons**.

Table 14.5 compares the two electron-like particles. We don't know yet why the e_n^- particle has three times the mass of a free electron. Nevertheless, it follows the mass tripling pattern seen on Figure 14.14. Understanding how

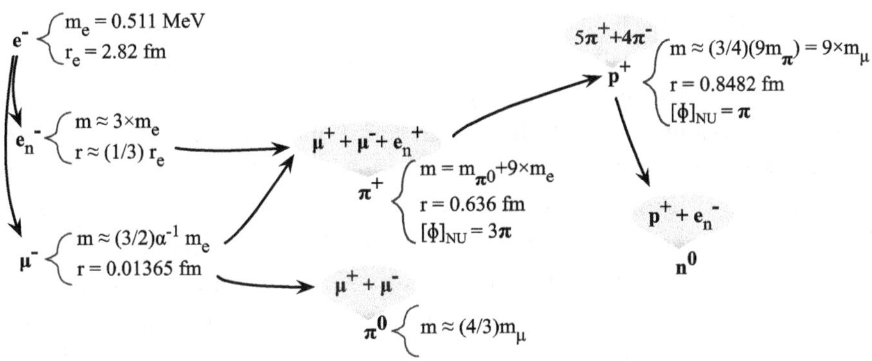

Figure 14.20. The inter-relation among elementary particles.

the presence of isospin alters the electron's internal structure would be an important next step.

14.6. A New Landscape of Elementary Particles

Building upon the above insights, we now attempt to redraw the landscape of elementary particles. In the previous sections we discussed how a free electron turns into a nuclear electron by tripling its mass and gaining isospin.

An electron may also be energized into a muon, upon gaining $\frac{3}{2}\alpha^{-1}$ times its rest mass energy. How do we know that the muon is essentially an energized electron? In Section 6.2, the analysis of the muon's anomalous magnetic moment revealed that it has nearly the same $\frac{R_{\mathrm{ZBW}}}{r_{\mathrm{charge}}}$ ratio as the electron. Thus, the muon can be thought of as a scaled-down electron, having only 0.01365 fm charge radius. The muon's $2.2\,\mu s$ half-life is a long one by nuclear physics standards. Upon decay, the muon transforms back into an electron plus neutrino radiation. Therefore, the muon also has isospin, and its magnetic moment becomes slightly different from the electron. The origin of the $\frac{3}{2}\alpha^{-1}$ mass scaling is yet unknown. The particles discussed so far are illustrated on the left-hand side of Figure 14.15.

Let's consider now the neutral pion particle, whose half-life is only $8.4 \times 10^{-17}\,s$. Upon decaying, the neutral pion emits two "photons," i.e., it decays into purely transversal electromagnetic waves, similarly to an electron–positron annihilation event. This tells us that a neutral pion is a short-lived composite of a particle and its antiparticle. Which particle type might the pion be built from? The charged pion is nine electron masses heavier than the neutral one: this suggests that an electron-derived third subparticle enters into the pion composite. This idea is well confirmed by

Figure 14.14 and [41], which document the charged pion's ability to shed a nuclear electron. This third particle extends the neutral pion's lifetime by nine orders of magnitude: therefore, it somehow prevents the annihilation between the particle–antiparticle constituents of the neutral pion. As shown in Figure 14.1, this half-life extension factor is α^{-4}: a ratio which implies electromagnetic origin. Indeed the charged pion has only 10^{-4} probability of decaying by particle–antiparticle annihilation: in these rare cases, the remaining e_n^- sub-particle de-excites into an electron. For a negatively charged pion, the dominant decay mode must be then the annihilation between a positively charged constituent of the neutral pion and the e_n^- type constituent. This dominant decay mode thus reveals the remaining negatively charged pion constituent. The particle emerging from the charged pion's main decay mode is a muon: an example of such pion-to-muon decay can be seen in the top-left corner of Figure 14.10. This reveals that the neutral pion is in fact a short-lived muon-antimuon composite. The charged pion's $\pi^- \rightarrow \mu^-$ decay reaction also emits a "muon neutrino," i.e., a very high-frequency neutrino radiation, which is generated in the annihilation between the e_n^- type and μ^+ type subparticles.

The mass of two free muons is $2 \times 105.66 = 211.32\,\text{MeV}$, while the neutral pion mass is 134.98 MeV. Upon the $\mu^+ + \mu^- \rightarrow \pi^0$ reaction, each muon constituent must lose 38.17 MeV mass, and the overall mass loss becomes the binding energy of the neutral pion composite. This means that a colliding muon–antimuon pair does not always annihilate: there must also be $\mu^+ + \mu^- \rightarrow \pi^0$ reaction events, producing 38 MeV "di-photons." Interestingly, such 38 MeV di-photon events were found in recent years by two independent laboratories: Figure 14.21 shows the measured signals. There is some controversy around this data, as it was presumed to be the signature of a 38 MeV mass particle, but no trace of such particle was ever seen. Now we may understand the true meaning of this signal, and it directly supports our pion structure hypothesis.

There are several fascinating questions that one may ask about $\pi^\pm$ particle. Why does the presence of its e_n^- type subparticle cause a mass difference of exactly 9 electron masses relative to π^0? How does the e_n^- type subparticle hinder the annihilation between the muon-antimuon type subparticles? How to understand the annihilation between the e_n^- type and antimuon type subparticles, and how does the remaining subparticle gain the free muon mass? Answering these questions will bring us closer to understanding the pion's internal structure.

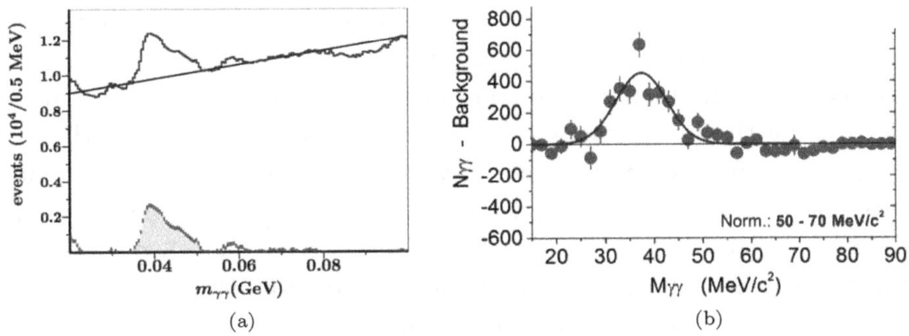

Figure 14.21. 38 MeV di-photon event signatures of the $\mu^+ + \mu^- \to \pi^0$ process. (a) COMPASS collaboration data [37] analyzed by Ref. [38], where the shaded area shows above-background events. (b) JINR data of above-background events, reproduced from [39].

We may however make some conclusions about the pion's magnetic moment. In a π^0 particle, the spin of its muon and antimuon subparticles must be aligned in order to form a composite particle. Consequently, the net magnetic moment of π^0 is zero, i.e., it is a spin 0 particle. On the other hand, the $\pi^\pm$ particles certainly have non-zero spin, and we prove this statement in two different ways. As far as the author knows, no experimental measurement of the $\pi^\pm$ magnetic moment was attempted over the past 60 years, as it was presumed to be "already known" that $\pi^\pm$ particles are spinless. Where does the idea of spin-0 $\pi^\pm$ particle come from? The only related experimental report we can find is [42] from 1951. The authors of [42] give very indirect evidence, and examine the $p^+ + p^+ \to \pi^+ + d^+$ reaction in their article. We have given a strong argument in the preceding sections that the deuteron comprises two protons and a nuclear electron having $\frac{3\hbar}{2}$ angular momentum. By conservation of angular momentum, the produced π^+ particle must have then $\frac{3\hbar}{2}$ angular momentum.

In the sea of unstable heavy particles, the unique proton stability is a fascinating puzzle. In Chapter 7 we showed that, as a consequence of momentum conservation, a quantum mechanical bond establishment decreases mainly the mass of the heavier particle. In the context of nuclear bond establishment, an $\frac{m_p}{m_{en}+m_p}$ ratio of the binding energy is radiated from the proton mass, and only $\frac{m_{en}}{m_{en}+m_p}$ ratio is radiated from the nuclear electron mass. In the case of an iron nucleus, its protons have radiated away on average nearly 8.8 MeV of binding energy, i.e., about 1% of their mass. Even after radiating away 1% of their mass, protons remain stable particles.

The proton stability is somehow related to stabilizing its pionic subparticles against $e_n^- - \mu^+$ or $e_n^+ - \mu^-$ type annihilation of their internal structure. In any case, the first step in understanding the proton stability involves understanding the mechanism by which a charged pion gains vastly longer lifetime than a neutral pion. Then as a second step one may try to derive the conditions for further stabilization. The proton's magnetic moment value reveals that its pion subparticles form a composite structure which constructively adds up their magnetic fields. Why does stability require such alignment of exactly nine subparticles? One may note that the mass of the proton's pionic subparticles almost coincides with the mass of free muons. Alternatively, is proton stability somehow related to its magnetic flux value of π, as calculated in Chapter 6?

At this point, we may formulate a hypothesis on truly elementary particles. Ultimately, all observed particles are derived from electrons and positrons. Such major simplification of the elementary particle landscape is favored by Occam's razor principle. A remaining challenge is to understand the asymmetry between matter and antimatter, and to understand the yet unclassified particles which will be introduced in Chapter 16.

14.7. Conclusions

The conclusions of this chapter are all grounded in state-of-the-art experimental data, and according to Occam's razor principle we applied the most direct interpretation of these data. Regrettably, 100 years ago most theoreticians decided to bridge the gap between new experimental data and their lack of explanation by proposing additional elementary forces and particles. As discussed in this chapter, these newly introduced nuclear forces don't really fit experimental data, make a large set of nuclear data appear as "meaningless random coincidences," and necessitate an incredible pile-up of *ad hoc* postulates about the structure of nuclear matter, which we compared to the pre-Newtonian exercise of Earth-centered epicycloid "explanation" for the motion of planets. The *ad hoc* introduced new elementary particles, e.g., quarks and gluons, have never been observed in experiments. In contrast, the herein reviewed experimental and theoretical results demonstrate an electromagnetic signature to all nuclear interactions and structures. We reach the inevitable conclusion that "strong" and "weak" nuclear forces and their corresponding mediating particles are just figments of imagination, or at best an approximation to the correct solution of electromagnetic equations.

Our results imply a great reorganization on the map of fundamental elementary particles: photons, gluons, quarks, and neutrinos do not exist as elementary particles. On the other hand, a nuclear electron does exist as a particle.

We recognize that all forces and matter originate from a single electromagnetic vector potential field. The dynamics of this field is defined by the simple $\mathcal{L} = \frac{1}{2}\sqrt{g}G\widetilde{G}$ Lagrangian density. The impact of such major simplification on our understanding of elementary forces and interactions cannot be overstated: it gives new meaning to the idea of "beyond standard model physics."

Acknowledgments

The author thanks William Stubbs for insightful discussions on high-energy electron–proton scattering data and review comments, and thanks Nancy L. Bowen for insightul discussions regarding prevalent nuclear models, and thanks Heikki Sipilä and Tuomas Pylkkänen for helping with the transmutation analysis.

References

[1] N.L. Bowen, The electromagnetic considerations of the nuclear force. *Advanced Electromagnetics*, **6**(4) (2017) 70–82.

[2] B. Schaeffer, Anomalous Rutherford scattering solved magnetically. *World Journal of Nuclear Science and Technology*, **6**(2) (2016) 96–102.

[3] N. Cook, *Models of the Atomic Nucleus*, 2nd edition Springer, 2010.

[4] P. Hatt, Atomic nuclei binding energy. *Journal of Condensed Matter Nuclear Science*, **29** (2019) 260–274.

[5] L. Holmlid and S. Zeiner-Gundersen, Ultradense protium p(0) and deuterium D(0) and their relation to ordinary Rydberg matter: A review. *Physica Scripta*, **94**(7) (2019).

[6] P. O'Hara, *Bell's inequality and the structure of spin-3/2 baryons, Spin 96 Proceedings*, World Scientific, 1997, pp. 483–485.

[7] R.B. Laughlin, Anomalous quantum Hall effect: An incompressible quantum fluid with fractionally charged excitations. *Physical Review Letters*, **50**(18) (1983) 1395–1398.

[8] F. Halzen *et al.*, *Quarks & Leptons: An Introductory Course in Modern Particle Physics*, John Wiley and Sons, 1984.

[9] Y. Wang *et al.*, Observing atomic collapse resonances in artificial nuclei on graphene. *Science*, **340**(6133) (2013) 734–737.

[10] M. Nishikawa, Natural beauty of the standard model — A derivation of the electro-weak unified and quantum-gravity theory without assuming a Higgs particle, arXiv:hep-th/0407057v5.

[11] C.S. Wood *et al.*, Measurement of parity nonconservation and an anapole moment in Cesium, *Science*, **275** (1997) 1759–1763.

[12] M.H. MacGregor, The α-quantized systematics of quark-dominated elementary particle lifetimes.

[13] H. Abele, The neutron. Its properties and basic interactions. *Progress in Particle and Nuclear Physics*, **60** (2008) 1–81.
[14] M. Gell-Mann, A schematic model of baryons and mesons. *Physics Letters*, **8**(3) (1964) 214–215.
[15] J.D. Bjorken *et al.*, Inelastic electron-proton and γ-proton scattering and the structure of the nucleon. *Physical Review*, **185**(5) (1969) 1975–1982.
[16] J. Kuti *et al.*, Inelastic lepton-nucleon scattering and lepton pair production in the relativistic quark-parton model. *Physical Review D*, **4**(11) (1971).
[17] D. Bedoya Fierro *et al.*, Lorentz contracted proton. *Journal of High Energy Physics*, **2015**(9) (2015).
[18] H. Atac *et al.*, Measurement of the neutron charge radius and the role of its constituents, *Nature communications*, **12**(1) (2021).
[19] PHENIX Collaboration, Creation of quark–gluon plasma droplets with three distinct geometries. *Nature Physics Letters*, **15** (2019) 214–220.
[20] G. Lochak, A new electromagnetism based on 4 photons: Electric, magnetic, with spin 1 and spin 0. *Annales de la Fondation Louis de Broglie*, **35** (2010).
[21] M.H. MacGregor, *The Enigmatic Electron*, 2nd edition, El Mac Books, 2013.
[22] J.S. Nico, Neutron beta decay. *Journal of Physics G: Nuclear and Particle Physics*, **36**(10) (2009).
[23] G. Rousseaux, The gauge non-invariance of classical electromagnetism. *Annales Fond. Broglie*, **30** (2005) 387–397 arXiv:physics/0506203v1.
[24] V. Tvaskis *et al.*, Proton and deuteron F_2 structure function at low Q^2. *Physical Review C*, **81**(5) (2010).
[25] W.L. Stubbs, *The Nucleus of Atoms: One Interpretation.* CreateSpace Independent Publishing Platform, 2018.
[26] W.L. Stubbs, The particles inside the proton, (2019).
[27] Wikipedia, Quantum chromodynamics binding energy.
[28] E. Perez *et al.*, The quark and gluon structure of the proton. *Reports on Progress in Physics*, **76**(4) (2013) 046201.
[29] I. Stepanov *et al.*, Coherent electron Zitterbewegung, arXiv:1612.06190 (2016).
[30] A.G. Parkhomov, Bursts of count rate of beta-radioactive sources during long-term measurements. *International Journal of Pure and Applied Physics*, **1**(2) (2005) 119–128.
[31] R.L. Jaffe, Where does the proton really get its spin? *Physics Today*, **48**(9) (1995). DOI: https://doi.org/10.1063/1.881473.
[32] A. Klimov, *Decay-Instability of Transmutated Chemical Elements Obtained in LENR Experiment*, presentation at the ICCF-23 International Conference on Condensed Matter Nuclear Science, Xiamen, China, 2021.
[33] L.I. Urutskoev *et al.*, Detection of Abnormal Quantity of Hydrogen upon Electrical Explosion of Titanium Foil in a Liquid, *Journal of Condensed Matter Nuclear Science*, **4** (2011) 106–118.
[34] S. Focardi *et al.*, Evidence of electromagnetic radiation from Ni-H Systems. *Proceedings of the ICCF-11 International Conference on Condensed Matter Nuclear Science*, Marseille, France, 2004.
[35] D.S. Baranov *et al.*, Transfer of "dark hydrogen" and methods of its diagnostics at high-voltage spark discharge in water-air environment, *Proceedings of the 26th Russian conference on Cold Nuclear Transmutation and Ball Lightning*, 2020.
[36] B. Heacock *et al.*, Pendellösung interferometry probes the neutron charge radius, lattice dynamics, and fifth forces, *Science*, **373**(6520) (2021).

[37] COMPASS Collaboration data, eConf C 110613, **83** (2011) [arXiv:1108.6191].
[38] E. van Beveren *et al.*, Material evidence of a 38 MeV boson, arXiv:1202.1739 (2012).
[39] K. Abraamyan *et al.*, Check of the structure in photon pairs spectra at the invariant mass of about 38 MeV/c^2. *EPJ Web of Conferences*, **204** (2019) 08004.
[40] W. von Oertzen *et al.*, Covalently bound molecular states in beryllium and carbon isotopes, *Comptes Rendus Physique*, **4** (2003). DOI: 10.1016/S1631-0705(03)00052-5.
[41] D.S. Armstrong, Capture and transfer of pions in hydrogeneous materials. *Exotic Atoms in Condensed Matter*, Springer 1992.
[42] R. Durbin *et al.*, The spin of the pion via the reaction $p^+ + p^+ \leftrightarrow \pi^+ + d^+$. *Physical Review*, **83**(3) (1951) 646–648.
[43] M. Jung *et al.*, First observation of bound-state β^- decay. *Physical Review Letters*, **69**(15) (1992) 2164–2167.
[44] A. Kovacs, V. Zatelepin, D.S. Baranov, The Discovery of the Nuclear Electron and its Mass Measurement, in publication process.
[45] A.V. Gurevich *et al.*, Strong flux of low-energy neutrons produced by Thunderstorms. *Physical Review Letters*, **108** (2012) 125001.
[46] L.P. Babich, Thunderstorm neutrons. *Physics-Uspekhi*, **62**(10) (2019) 976–999.
[47] T. Enoto *et al.*, Photonuclear reactions triggered by lightning discharge. *Nature*, **551** (2017) 481–484.

https://doi.org/10.1142/9781800611306_0015

Chapter 15

Transmutations by Evanescent Neutrinos

András Kovács
BroadBit Energy Technologies, Metallimiehenkuja 8 C, 02150 Espoo, Finland
andras.kovacs@broadbit.com

15.1. Introduction

As we saw in Chapter 2, light waves and phonon waves realize the simplest mathematical solutions of Maxwell's equation. In this sense, light waves and phonon waves may be regarded as the best understood natural phenomena.

Unexpectedly, reports of light-induced and phonon-induced nuclear transmutations emerged in recent years. Although such a phenomenon is very unexpected, the involved experiments are rather simple and well-replicable. Since we already understand the physics of electromagnetic waves, these unexpected transmutations mean that nuclear interactions are still not completely understood.

An exploration of this phenomenon has not only theoretic implications but also practical aspects: it may allow a large-scale production of rare industrial materials from more abundant precursors.

We thus review reports of light-induced and phonon-induced transmutations, aiming to find common patterns. The common patterns, which will be identified across numerous experiments, can be understood with the help of Chapter 2 and Chapter 14 results.

Once the involved reaction process becomes clear, we find experimental results that reveal the precise mass of a nuclear electron particle.

15.2. Fission-like Transmutations

In the following paragraphs, we review fission-like nuclear transmutations. Although the involved experimental setups are very diverse, they demonstrate similar transmutation patterns, thus suggesting a common underlying physical process.

References [4–6] describe nuclear reactions in titanium wires, which are subjected to intense electric discharge in water. In this case, the illumination effect arises only for a very short time, while the effect of particle collisions and oscillations is more significant during the ensuing electric explosion of wires. The depletion of ^{48}Ti isotope ratio — relative to all other titanium isotopes — was observed. The extent of ^{48}Ti depletion was 10% on average, but highly varying: only 1–2% in some experiment runs, and even 33% in one experiment run. Remnants from the electrically exploded titanium wires comprise spherical objects, which are shown in Figure 15.1. The analysis of these micro-spheres reveals the accumulation of a high amount of oxygen: in terms of mass percentage, the spheres comprise Ti (59.4 wt.%) and O (39.7 wt.%), and thus the Ti:O atomic ratio is 1:2.45. We note that this Ti:O atomic ratio is higher than the normal chemical limit of 1:2 in TiO_2, and reaching even the ordinary 1:2 limit would be unexpected with this underwater experimental setup. Besides titanium and oxygen, the amount of newly synthesized other elements varied between 1% and 8% across 21 experiment runs. The amount of transmutations and the rate of ^{48}Ti isotope depletion has a low correlation, which means that other titanium isotopes also randomly contribute to the transmutations. These newly synthesized other elements comprise Al (13%), Si (8 %), Ca (6%), Cr (10%), Fe (30%), Cu (9%), and Zn (8%). Identical experiments were

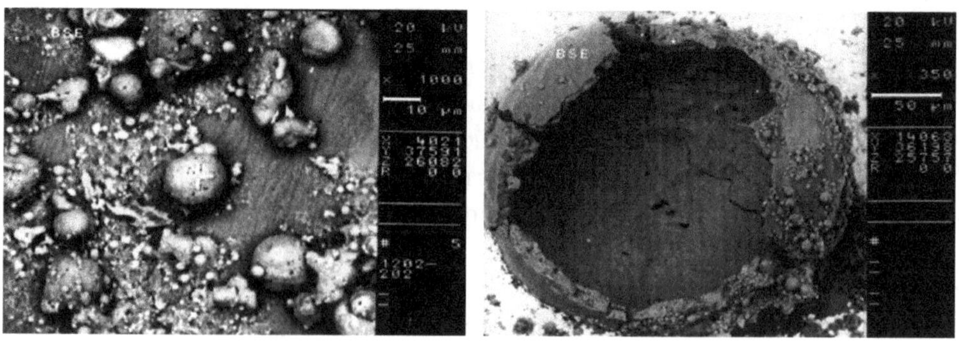

Figure 15.1. Photos of the spherical remnants from an exploded titanium wire. Some remnants are hollow spheres, whose cross-section is shown on the right.

repeated with zirconium wires: the newly appearing elements, other than zirconium and oxygen, comprise Ca (9 %), Ti (17%), Cr (23%), Fe (26%), Ni (4%), Cu (11%), and Zn (5%). This is a surprising outcome because the newly synthesized elements are centered around iron in both cases, even though the starting material is twice as heavy in the zirconium case. Apparently, transmutations are driving primarily toward oxygen production and secondarily toward the production of iron-like elements, which have optimal binding per nucleon.

In [7], the experimenter deposited a 250 nm thick palladium film onto a silicon-dioxide substrate, and placed it under hydrogen atmosphere. Palladium absorbs hydrogen to nearly 1:1 ratio of Pd:H, turning into palladium-hydride. This palladium-hydride film was illuminated for several weeks by a 633 nm laser pointer beam. Subsequently, circular spots appeared on its surface where the laser light was illuminating it. An elemental analysis of these spots showed significant transmutations from the original palladium material. There was a continuous monitoring for gamma and neutron emission during the laser illumination, but nothing above background was detected. This experiment was replicated by Jean-Paul Biberian, using the same 250 nm palladium film thickness, applying 2 bars of H_2 atmosphere and 3 months' illumination by a 650 nm laser beam at 5 mW power level. The obtained results replicated the findings of [7]. The author of [7] noted that the intensity of transmutations depends on the laser wavelength.

Figure 15.2 shows the photo of a circular spot which appeared on the palladium where the laser beam illuminated it. The spot comprises a 100-micron diameter darker core, surrounded by a 200-micron diameter rim. The elemental analysis of the core region reveals a nearly complete transmutation of palladium into lighter elements: N (10.44%), O (68.22%), Na (6.44%), Al (0.42%), Si (0.24%), P (0.07%), S (1.46%), Mn (0.14%), Fe (2.94%), Ni (8.83%), Zn (0.53%), Pd (0.25%). The radiation-free transmutation of palladium and hydrogen into completely different elements heralds a new class of nuclear reactions. We note the high percentage of nitrogen and oxygen: these two elements account for nearly 80% of transmuted atoms. In usual chemical reactions, the oxidation of the 12% Fe-Ni-Zn metal content binds 14–15% oxygen content, and the oxidation of the 6% Na content binds additional 3% oxygen content. That accounts for only 18% oxygen, while the chemical binding of the remaining 50% oxygen and 10% nitrogen is unexpected. A chemist would expect that 60% excess oxygen–nitrogen content to be gaseous. Apparently, the chemical bond-forming environment is different from that it ordinary chemical processes.

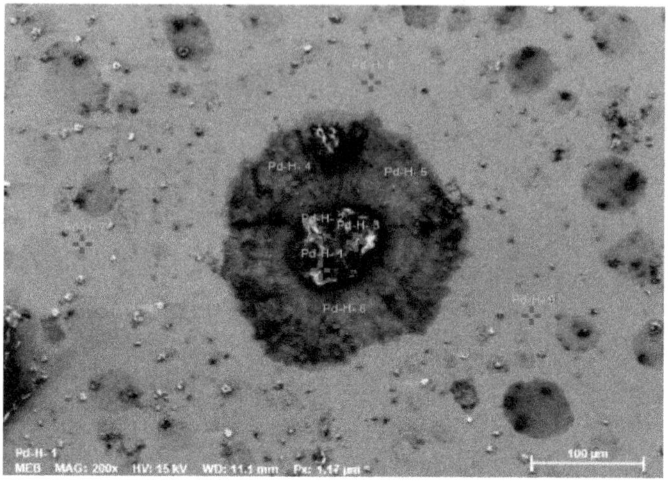

Figure 15.2. A spot which appeared on the palladium-hydride where the laser light was illuminating it. Photo taken by Jean-Paul Biberian.

The transmutations outcomes in these long-duration laser illumination experiments have strong parallels with the fast transmutations in underwater wire explosion experiments: in both cases, the production of oxygen and nitrogen is the primary outcome, and the production of elements around iron is the secondary outcome.

In [8], a reactor comprises a hydrated nickel film, from which hydrogen diffuses through alternating thin layers of nickel and another material. Specifically, the following alternating layer combinations are tested: Ni/Cu, Ni/Cu/Ni/CaO, Ni/Cu/Ni/Y_2O_3. The thickness of interlayers is only 2 nm. This reactor evolved from the deuterated Pd/CaO-type alternating layered system, which was mentioned in the preceding chapter, and thus the data presented in Section 13.4 complements the following observations.

The schematic structure of an interlayer boundary is shown in Figure 15.3. There is a Fermi level difference across the interlayer boundary, causing an accumulation of electrons at the boundary, which generate an electric field across the interface. This electric field prevents electrons from crossing over to the lower Fermi level. Consider diffusing protons, which enter this interface region: since protons are oppositely charged with respect to electrons, they will be accelerated by the electric field in this interface. Thus, diffusing protons gain a few eV energy, which they lose via collisions on the other side of the interface.

After a sufficiently long reactor run, nuclear transmutations are observed at the outer hydrated-nickel surface. Concurrent gamma radiation

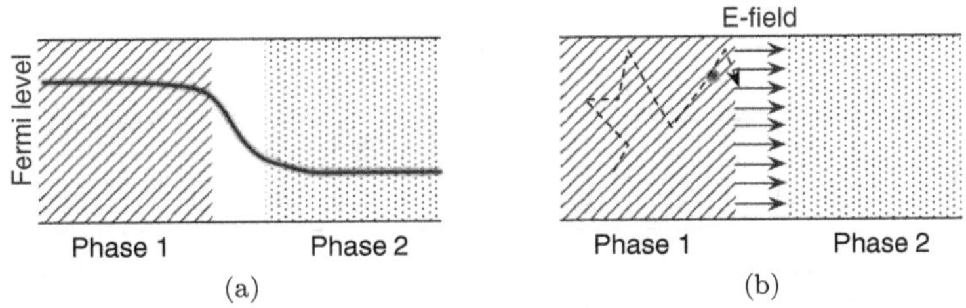

Figure 15.3. Illustration of two metallic phases with Fermi level difference (a), causing hydrogen acceleration to few eV upon diffusion across the phase boundary (b).

monitoring proves that these transmutations are radiatonless events. In the case of Ni/Cu alternating layers, spots appear on the outer nickel surface and have the following composition: O (61.6%), Al (13.8%), Ni (5.9%), K (4.5%), C (2.6%), Cu (2.3%), Na (1.8%). The high concentration of oxygen and the appearance of several new elements is an already familiar transmutation pattern.

Reference [8] also reports that the measured excess power is approximately proportional to the number of interlayers. However, transmutations take place at the outer surface, not at the interlayers. What is being generated in the interlayer structure, which in turn induces transmutations at the outer surface? This question will be investigated in the following section.

The authors of [8] measured 16 keV energy production per diffused proton in case of 6 Ni/Cu alternating layers. For each crossing of a Ni/Cu layer, the average energy release is then 2.6 keV per proton. Such high value makes this setup highly prospective for energy production.

Reference [3] reports the production of tritium in a brightly illuminated solution containing either NaOH or Na_2CO_3 salt, and also a few percentage of D_2O. The illumination sources are red LED lamps, emitting 650 nm light, and the duration of illumination was 1 h. In both NaOH and Na_2CO_3 cases, the solution contained 1.5×10^{24} Na^+ cations, and the number of produced tritium atoms was about 1.5×10^9. Analogous experiments were performed with Li^+ cation containing solutions, but these produced no tritium. Furthermore, the reaction was not accompanied by neutron emission, except for few short bursts of up to 100 neutrons. Therefore, it appears that the produced tritium originates from sodium nuclei.

Reference [10] reports surface transmutations on a Pd electrode, which was under electrolysis for 20 hours. These newly synthesized elements on the Pd surface comprise O (18.5%), Fe (2%), K (2%), Si (1.1 %), Mg (1%). As the Pd electrode was held at negative voltage, the appearing oxygen cannot

Table 15.1. A comparative table of fission-like transmutation experiments.

	Urutskoev *et al.*	Biberian *et al.*	Iwamura *et al.*	Carpinteri *et al.*
Starting material	Ti	Pd (hydrated)	Ni/Cu (hydrated)	Pd (hydrated)
Method	Electric explosion	Laser illumination	Hydrogen diffusion	Electrolysis
Main products	O (71%)	O (68.2%) N (10.4%)	O (61.6%) Al (13.8%)	O (18.5%) Fe (2%)

result from electrolytic oxidation. Moreover, the authors of [10] measured 34% higher neutron count during the electrolysis than the normal neutron background.

The most puzzling aspect of the above discussed fission-like reactions is that they are all endothermic. The destabilization of involved nuclei requires an energy source. Table 15.1 summarizes the observed transmutation reactions, demonstrating that oxygen production is the common pattern. Recalling the clustered structure of nuclei shown in Figure 14.2, it appears that the transmuting nuclei fission into their individual oxygen clusters.

15.3. A Physical Model of Fission-Like Transmutations

Given the highly unexpected characteristics of radiationless transmutations, one must critically examine the measurement data cited in the preceding sections. How reliable are these transmutation measurements? What are the chances of mistaking contamination for transmutation data? For the following reasons, the transmutation measurements cited in this chapter cannot be explained away as contaminations or measurement errors:

- **Independent replications:** The experiment of [7] was replicated by Jean-Paul Biberian, and Alexander Parkhomov's experiment can be considered replicated by [2].
- **Large transmutation percentages:** Nearly 10% of potassium atoms transmuted to calcium atoms in Alexander Parkhomov's experiment, which is too large to be explained as contamination or measurement error. Regarding the experiments of [7] and Jean-Paul Biberian, dark spots appeared exactly where laser light was shining onto palladium-hydride, and palladium almost completely transmuted to other elements in these spots.

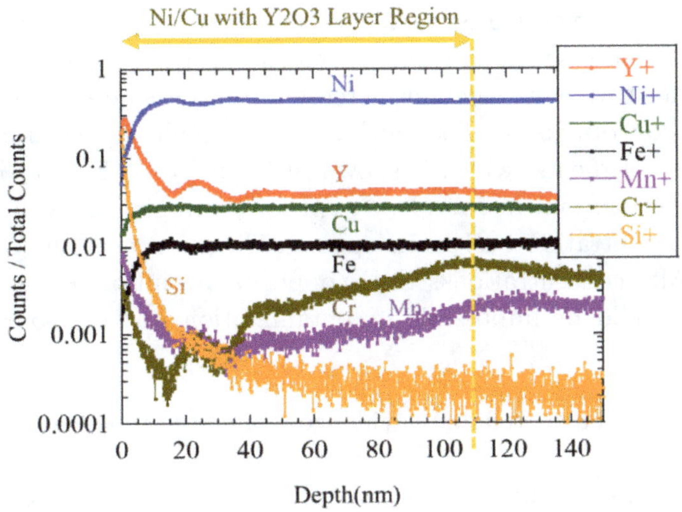

Figure 15.4. The depth profile of a sample after a long-term experiment. The fresh material was hydrated Ni, with Cu and Y_2O_3 nano-layers. Reproduced from [8].

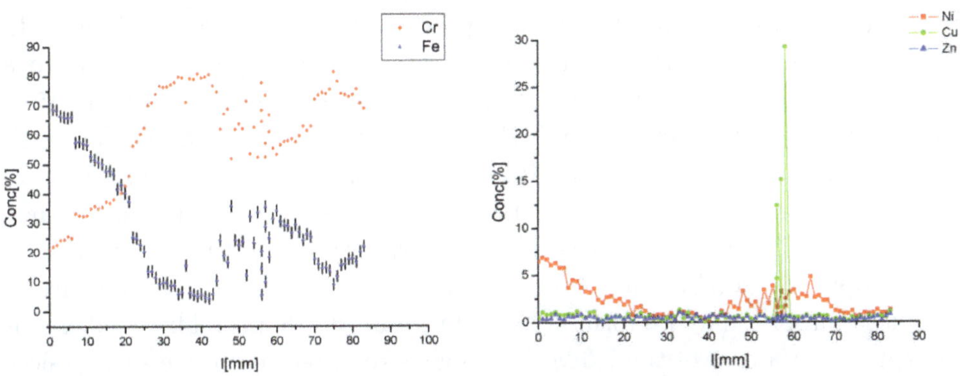

(a) (b)

Figure 15.5. The axial profile of a metal rod surface after elevated-temperature hydrogen treatment. Hydrogen flows from $l = 0$ mm to $l = 100$ mm. The fresh material was hydrated Fe-Cr-Ni-Mn alloy. The alloy composition remained unchanged at $l = 0$ mm. Reproduced from [9].

- **Detection of tritium:** The author of [3] used a scintillator to measure tritium concentration, and such signals are impossible to explain as contamination.
- **Isotope shifts:** The authors of [4–6] consistently measured the depletion of ^{48}Ti during their experiments, relative to other Ti isotopes. Such consistent change in the isotopic distribution cannot result from contamination or measurement error.

- **Time-wise correlations:** The potassium to calcium transmutation report by [2] shows that calcium concentration increases with illumination time, while there is no calcium increase in a potassium-free solution.
- **Spatial correlations:** The author of [8] made a depth profiling of transmuted elements, which is shown in Figure 15.4. Fe, Cr, and Mn are initially absent from the sample, except for contaminating concentrations. The Fe concentration evolves in opposite direction near the surface than Cr and Mn concentrations. Such opposite evolution of concentration gradients would be impossible if contamination was the source of these materials.

Altogether, we have reliable evidence for fusion-like and fission-like transmutations.

To better understand the occurring fission-like transmutations process, we focus on hydrated metal systems, which have a rich literature. Figure 15.4 reveals a large decrease of Fe concentration near the sample surface, along with increasing concentrations of Cr, Si, and Mn. This data indicate that Fe undergoes fission, but the fission product has multiple candidates. In a related experiment, the authors of [9] applied hydrogen treatment to a metal alloy rod, and observed the spatial correlations shown in Figure 15.5. The correlation between Fe and Cr can only be explained as Fe to Cr transmutation. Since the main iron isotope is ^{56}Fe, the fission process occurring in [8, 9] appears to be alpha particle emission, which is endothermic by 7.6 MeV. This proposition is further corroborated by the observation of microscopic "broken bubbles" on the post-experiment metal alloy rod surface [9]; these bubbles may be formed by the produced ^{4}He gas. The challenge is to explain the source of the 7.6 MeV, which is required to induce such fission. This source can be neither gamma radiation nor energetic particle, since the authors of [8] report that all transmutations are radiationless processes.

The key idea is to recall the analysis of longitudinal phonons from Section 2.10; from the macroscopic perspective, a longitudinal phonon is a linearly polarized longitudinal electromagnetic wave. It is mathematically equivalent to a linearly polarized neutrino, but it moves much slower than the speed of light; its propoagation speed is determined by the electromagnetic parameters of the given material. The phonon is internally reflected at the surface of the material and, as discussed in Chapter 2, such **a reflection process is always accompanied by an exponentially decaying tail on the other side of the interface**. Recall from Chapter 2 that the presence of an exponential tail beyond the reflection point directly follows from Maxwell's equation. Figure 15.6 illustrates a longitudinal phonon

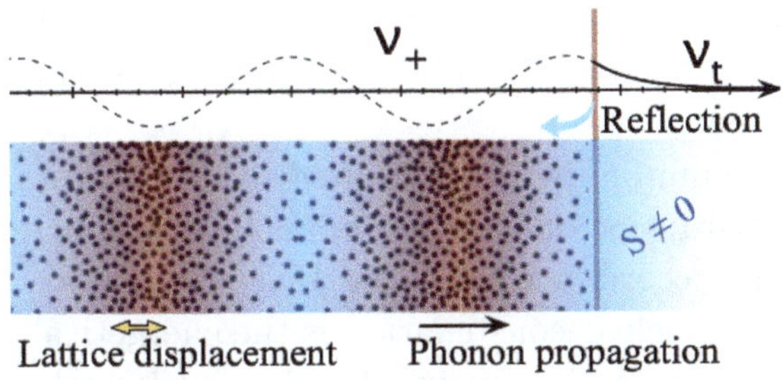

Figure 15.6. An illustration of the evanescent neutrino field, generated by a phonon wave reflecting from the material surface. The horizontal axis is along the z direction and the vertical line represents the material surface.

wave propagating inside the material, and its evanescent tail. In line with Chapter 2 notation, these waves are represented by the ν_+ symbol for the phonon wave and by the ν_t symbol for its evanescent tail. Microscopically, the phonon wave comprises nuclei and electrons; there is no scalar field in-between these particles, and thus the sinusoidal wave is represented by a dashed line in Figure 15.5. The situation is however different for the ν_t tail; it is in a vacuum region and therefore comprises only electromagnetic fields. The mathematical formula of the ν_t field is based on Equation (2.10.2).

The ν_t field is therefore a proper evanescent neutrino field, and it is thus represented by a solid line in Figure 15.5. The lattice particles, which generate a phonon wave, collectively create a non-zero scalar field beyond the material surface.

Now recall the generalized Poynting vector from Chapter 1, given by Equation 1.3.29: it contains a $-S\boldsymbol{E}$ term. Consider the radially oriented electric field around elementary particles. For positively charged protons, the $-S\boldsymbol{E}$ term generates an energy flow into the particle when $S > 0$, while for negatively charged particles the $-S\boldsymbol{E}$ term generates an energy flow out of the particle when $S > 0$. This scalar field-induced energy flow slightly changes a given particle's de Broglie frequency; in opposite direction for oppositely charged particles.

Recall from Chapter 14 that bound nucleons are in a phase-coherent state with respect to neighboring nucleons. Moreover, phase coherent neighbors comprise an oppositely charged proton–nuclear electron pair, which is electrostatically bound. Their phase coherence may be destroyed by an altered de Broglie frequency. When a proton–proton electron pair's phase-coherence is disrupted by a sufficiently strong scalar field, their relative position relative becomes subject to vacuum noise fluctuations. This process can be directly observed in Figure 14.14: the concentrated neutrino radiation at the focal point comprises a concentrated scalar field, which disrupts the metastable constellation of ^{60}Co. Subsequently, vacuum noise fluctuations trigger a beta emission.

The effect of vacuum noise is not limited to terminating only metastable bonds. Upon the disruption of a nuclear bond's phase coherence, the vacuum noise effect acts in accordance with the Heisenberg uncertainty principle. Although the initial nuclear constellation is the energetically most favorable one, the vacuum noise action moves the involved nuclear electron away from its most favorable location, thus being capable of terminating a stable polarized bond between two nuclear fragments. Coulomb repulsion then moves apart the depolarized nuclear fragments.

In summary, we write the experimentally observed ^{56}Fe fission process as follows:

$$^{56}\text{Fe} \quad \underset{\longleftrightarrow}{\nu_t} \quad \left(^{52}\text{Cr} + {}^{4}\text{He}\right)_{\text{polarized}} \quad \underset{\longrightarrow}{v.n.} \; {}^{52}\text{Cr} + {}^{4}\text{He}$$

where "v.n." stands for vacuum noise. The two-way arrow indicates the reversible effect of evanescent neutrinos, while the one-way arrow indicates the irreversible action of vacuum noise. The 7.6 MeV fission energy is thus supplied by the electromagnetic vacuum noise. In other words, a disruption of phase coherent bonding causes the emergence of Heisenberg uncertainty between two constituents that were previously in a phase-coherent composite state, and this process involves energy extraction from the vacuum.

We now compare the predictions of our model against experimental measurements:

- **Transmutations are driven by a longitudinal phonon-generating mechanism:** This prediction follows from the scalar field generating process illustrated in Figure 15.6. The involved phonon-generating mechanism varies from experiment to experiment. In [8], phonons are generated by diffusing protons which gain a few eV energy via the process illustrated in Figure 15.3. In [7] and in Jean-Paul Biberian's replication, phonons are generated by the coherent laser light effect. In Refs. [4–6], phonons are generated by the electric explosion of thin wires or foils. The authors of [10] ascribe the transmutation cause to THz phonon.
- **The intensity of transmutations is proportional to the phonon intensity:** This prediction follows from the phonon reflection process shown in Figure 15.6; the emerging scalar field amplitude is proportional to the amplitude of the reflected phonon wave. Reference [8] indeed reports that the measured non-chemical power is approximately proportional to the number of interlayers, which generate phonons. Several other

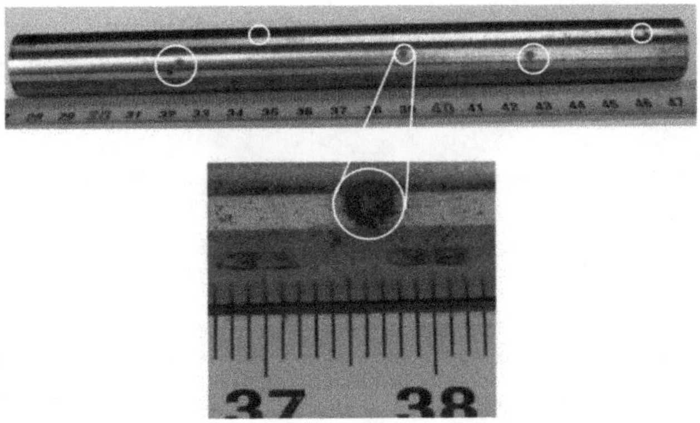

Figure 15.7. Transmutation spots appearing on the surface of an ultra-sonicated steel bar, reproduced from [12]. Approximately half of nuclei in these spots are oxygen and carbon.

experimenters observed phonon-induced nuclear fissions, such as the surface transmutations on ultra-sonicated metal bars shown in Figure 15.7, or THz phonon-induced transmutations on the surface of magnetite and basalt blocks [13].

- **Transmutations occur only near the surface from which phonons reflect:** This prediction follows from the non-zero scalar field value only very near to the material surface. Figure 15.4 indeed demonstrates that transmutations occur only within a few nanometers from the surface, and the same observation is made by the authors of [9, 12]. We now understand why the authors of [8] did not observe transmutations at the embedded interlayers, even though the measured power intensity was determined by the number of interlayers.
- **Endothermic nuclear fissions may occur:** This prediction follows from the effect of vacuum noise, which depolarizes the bound nuclear fragments after their decoherence. A detailed analysis of the electromagnetic vacuum noise was given in Chapter 6.
- **Transmutations are accompanied by neutron emission:** We recall from Section 14.4 that fission reactions are generally accompanied by neutron emission. Indeed the authors of [7, 10, 13] measured excess neutron emission during their experiments.

As far as the author knows, the present model is the first one to account for all the essential experimental features of fission-like transmutations.

Based on the above model, we continue the analysis of observed fission patterns. Regarding the hydrated nickel experiment of [8], it can be seen in

Figure 15.4 that the near-surface silicon concentration increases to a similar level as the initial nickel concentration. The occurring fission processes can therefore be written as follows:

$$^{58}\mathrm{Ni} \quad \underset{\longleftrightarrow}{\nu_t} \quad \left(^{28}\mathrm{Si} + {}^{30}\mathrm{Si}\right)_{\text{polarized}} \quad \underset{\longrightarrow}{v.n.} \quad {}^{28}\mathrm{Si} + {}^{30}\mathrm{Si}$$

$$^{60}\mathrm{Ni} \quad \underset{\longleftrightarrow}{\nu_t} \quad \left(^{30}Si + {}^{30}\mathrm{Si}\right)_{\text{polarized}} \quad \underset{\longrightarrow}{v.n.} \quad {}^{30}\mathrm{Si} + {}^{30}\mathrm{Si}$$

Regarding tritium production from ^{23}Na, we recognize that the sodium nucleus comprises an alpha cluster and an oxygen cluster, which together form a neon core. Besides this neon core, the remaining part of a ^{23}Na nucleus comprises a tritium cluster. The induced fission reaction fragments ^{23}Na along its clusters, and thus corresponds to the following process:

$$^{23}\mathrm{Na} \quad \underset{\longleftrightarrow}{\nu_t} \quad \left(^{20}\mathrm{Ne} + {}^{3}H\right)_{\text{polarized}} \quad \underset{\longrightarrow}{v.n.} \quad {}^{20}\mathrm{Ne} + {}^{3}\mathrm{H}$$

15.4. Are We Observing Fission or Fusion?

In this section we discuss transmutations where it is not immediately clear whether we are observing fission or fusion reactions.

A widely studied reaction is the production of ^{4}He in deuterated palladium electrodes [14], which up to now was assumed to be some type of deuterium fusion. As these reactions are accompanied by only minor neutron emission, the main reaction cannot be deuterium fusion. On the other hand, longitudinal phonons are found to coincide with nuclear heat production [15]. A series of Raman spectroscopy measurements can be seen in Figure 15.8, which show an anti-Stokes line to be present only during nuclear heat production. Based on the ratio of anti-Stokes and Stokes peaks, the author of [15] calculates the effective Stokes temperature, which turns out to be 1650 K. This means that electrolytically accelerated ions generate much more energetic acoustic phonons upon impact than the 300 K ambient thermal fluctuations. During the reflection of these phonons from the PdD surface, the production of evanescent neutrino field is inevitable. Consequently, we anticipate a Pd fission reaction, and propose that the observed ^{4}He is mainly produced during Pd fission. Our proposition is well supported by Table 15.1, which shows Pd fission products after electrolysis. Regarding electron mediated deuterium fusion, Chapter 13 presented its signatures also in electrolytic Pd systems. Nevertheless, the low rate of neutron production means that it is only a minor reaction pathway.

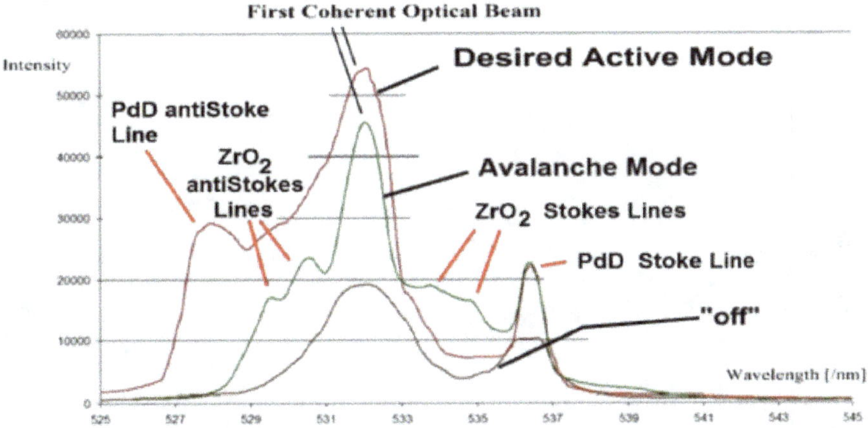

Figure 15.8. Raman scattering spectra of deuterated palladium electrodes during a exothermic reaction ("Desired Active Mode"), during "avalanche mode" electrolysis which suppresses the exothermic reaction ("Avalanche Mode"), and during the inactive state ("off"). The Stokes and anti-Stokes lines are assigned for the two materials comprising the electrode. Reproduced from [15].

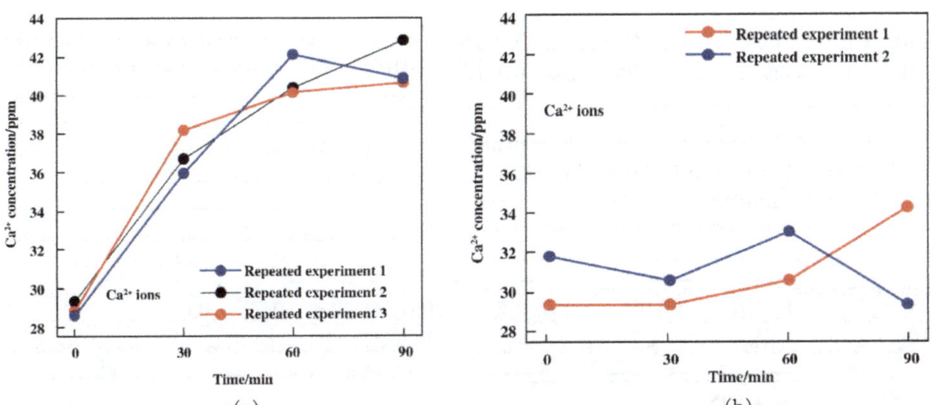

Figure 15.9. Potassium to calcium transmutation in a photoelectrochemical experiment, reproduced from [2]. The figure shows the measurement of calcium concentration during the experiment. (a) Experiment runs with K_2PtCl_6 electrolyte salt. (b) Experiment runs with H_2PtCl_6 electrolyte salt.

There are several reports of potassium-to-calcium transmutation. Alexander Parkhomov immersed a 450 W halogen lamp into a small tank containing 10% KNO_3 in aqueous solution, and this solution was circulated through a cooling device to avoid overheating [1]. The tank's inner surface is reflective, so that illumination is maximized. Table 15.2 lists the transmutations which were registered after 20 hours of this intense illumination.

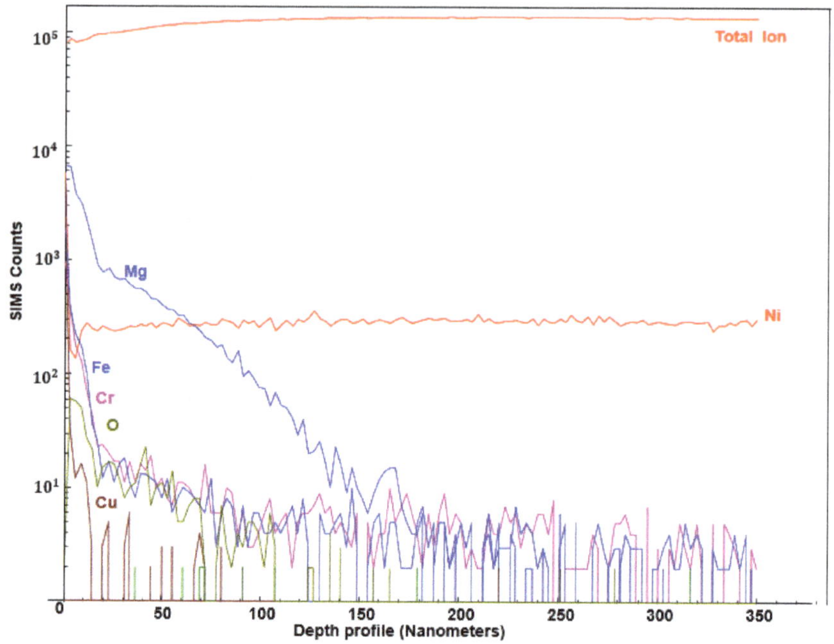

Figure 15.10. The depth profile of a Ni cathode after 10 hours of electrolysis. The fresh material was pure Ni electrode. Reproduced from [16].

Table 15.2. The composition of water-dissolved metal salts before and after the intense illumination of 10% KNO_3 solution.

	K	Ca	Al	Cu	Mg
Before	99.9%	<0.01%	<0.01%	<0.01%	0.05%
After	86%	9%	3%	1%	0.55%

Note: Data provided by Alexander Parkhomov.

Could the main reaction scheme could be $p + \text{K} \rightarrow \text{Ca}$ fusion? Such reaction produces 8.3 MeV heat. Considering that 9–10% of all K nuclei transmutes into Ca, there is enormous reaction heat to be dissipated in case of proton fusion, but there is no plausible explanation for how this heat would be all converted to e.g., neutrino radiation. This prompts us to consider alternative reaction pathways.

A similar result is reported in [2], whose authors observed visible light driven transmutation in the presence of a catalyst (Eosin-Y). The electrolyte salt used by the authors of [2] comprises K^+ cation. As shown in Figure 15.9(a), K to Ca transmutation is reproducibly measured in these experiments. In a potassium-free control experiment, where all other parameters remain unchanged, the Ca concentration remains constant during the experiment runs.

$$^{41}\mathrm{K} \rightarrow {}^{40}\mathrm{Ca} + 2e^{-} + p^{+}$$

The reality of electron-emitting fission was already established in Chapter 14. The above reaction is applicable only for ^{41}K because ^{39}K cannot produce a stable calcium isotope via fission. Therefore an isotopic analysis of calcium would validate whether we are indeed observing fission. Our prediction is that the isotopic analysis will yield only ^{40}Ca.

The authors of [16] electrolyzed a 1 molar K_2CO_3 solution in water, using Ni cathode. The employed voltage is 25 V, and the resulting current is 3.5 A. Figure 15.10 shows the depth profile of the Ni electrode surface after the electrolysis. As expected, the main reaction is a fission-like transmutation of nickel. However, at the same time a fusion-like $\mathrm{Ni} + p \rightarrow \mathrm{Cu}$ reaction also occurs. For both reaction types, the concentration of produced elements decreases exponentially from the electrode surface. While the natural ratio of ^{63}Cu:^{65}Cu isotopes is 69.2:30.8, this fusion-like reaction produced ^{63}Cu:^{65}Cu isotopes in 77.2:22.8 ratio. This ratio is rather close to the natural ratio of ^{62}Ni:^{64}Ni isotopes, which is 79.5:20.5. Therefore, there is a strong evidence for the $\mathrm{Ni} + p \rightarrow \mathrm{Cu}$ fusion reaction, which is a minor reaction in this experiment. Considering that continuous $e^{-} + p^{+} \rightarrow H$ recombination reaction takes place at the negatively charged Ni cathode, we recall from Chapters 12–13 that there will also be Compton-scale electron-proton composite formation. Therefore we ascribe the $\mathrm{Ni} + p \rightarrow \mathrm{Cu}$ reaction to the electron-mediated fusion of protons.

In summary, most transmutation products appear to originate from the fission reactions transmutations discussed in Section 15.3, and for a long time the emitted alpha particles were mistakenly considered to be fusion products. Electron-mediated fusion reactions also occur in some experiments, but at a lower intensity.

15.5. Conclusions

Building onto the results of preceding chapters, we developed a physical model of phonon-triggered nuclear fissions. According to our model, the occurring transmutations are in fact caused by an evanescent neutrino field. The predictions of our model account for all the essential features of this reaction type, as validated across a wide range of experiments. This new insight may lead to a rational engineering of transmutation reactors, designed to produce currently rare materials from more abundant precursors.

Acknowledgments

The author's experimental work is financially supported by the European Union-funded Clean HME project (grant agreement no. 951974). The author thanks Jean-Paul Biberian and Alexander Parkhomov for making their measurement results available, and thanks Giorgio Vassallo for insightful discussions on the generalized Poynting vector.

References

[1] A. Parkhomov *et al.*, LENR as a manifestation of weak nuclear interactions, *RENSIT Journal*, **13**(1) (2021) 45–58.

[2] G. Lu *et al.*, Photocatalytic hydrogen evolution and induced transmutation of potassium to calcium via low energy nuclear reaction (LENR) driven by visible light. *Journal of Molecular Catalysis*, **31**(5) (2017), 401–410.

[3] Y.N. Bazhutov *et al.*, Calorimetric and radiation diagnostics of water solutions under intense light irradiation. *Journal of Condensed Matter Nuclear Science*, **19** (2016) 10–16.

[4] L.I. Urutskoev *et al.*, Detection of abnormal quantity of hydrogen upon electrical explosion of titanium foil in a liquid. *Journal of Condensed Matter Nuclear Science*, **4** (2011) 106–118.

[5] L.I. Urutskoev *et al.*, Observation of transformation of chemical elements during electric discharge. *Annales de la Fondation Louis de Broglie*, **27**(4) (2001).

[6] V.M. Dorovskoj *et al.*, Исследование продуктов электровзрыва титановых фольг с помощью электронного микроскопа. Прикладная физика, **4** (2006) 28–34.

[7] U. Mastromatteo, LENR anomalies in Pd–H_2 systems submitted to laser stimulation. *Journal of Condensed Matter Nuclear Science*, **19** (2016) 173–182.

[8] Y. Iwamura *et al.*, Excess energy generation using a nano-sized multilayer metal composite and hydrogen gas. *Journal of Condensed Matter Nuclear Science*, **33** (2020) 1–13.

[9] E. Campari *et al.*, Surface analysis of hydrogen loaded nickel alloys. In *Proceedings of the ICCF-12 International Conference on Condensed Matter Nuclear Science*, (2006) 414–420.

[10] A. Carpinteri *et al.*, *Hydrogen embrittlement, microcracking, and THz vibrations in the metal electrodes of cold-fusion electrolysis experiments: Nuclear and stoichiometric balances*, 2021.

[11] D. Letts *et al.*, Stimulation of optical phonons in deuterated palladium. In *Proceedings of the ICCF-14 International Conference on Condensed Matter Nuclear Science*, Washington, USA, 2008.

[12] F. Cardone *et al.*, Piezonuclear reactions: Neutrons and transmutations by ultrasonic stress of iron bars. *19th European Conference on Fracture*, 2012.

[13] A. Carpinteri *et al.*, *Acoustic, Electromagnetic, Neutron Emissions from Fracture and Earthquakes.* Springer, Singapore, 2015.

[14] M. McKubre *et al.*, The emergence of a coherent explanation for anomalies observed in D/Pd and H/Pd systems; evidence for ^{4}He and ^{3}He production. In *Proceedings of*

the ICCF-8 International Conference on Condensed Matter Nuclear Science, Bologna, Italy, 2000.

[15] M.R. Swartz *et al.*, Increased PdD anti-stokes peaks are correlated with excess heat mode. *Journal of Condensed Matter Nuclear Science*, **24** (2017), 130–145.

[16] A. Kumar *et al.*, Nuclear transmutations are better facilitated by alloys over pure metal cathodes in electrolysis. Presentation at the *ICCF-23 International Conference on Condensed Matter Nuclear Science*, Xiamen, China, 2021.

https://doi.org/10.1142/9781800611306_0016

Chapter 16

Do Magnetic Monopoles Exist?

András Kovács
BroadBit Energy Technologies, Metallimiehenkuja 8 C
02150 Espoo, Finland
andras.kovacs@broadbit.com

16.1. Introduction

This chapter reviews the observations of helicoidally spiraling particles, which appear to ionize matter despite their low kinetic energy. Such observations are made either during the experiments described in Chapter 15 or during similar experimental conditions. This raises the question whether such particles are somehow related to the evanescent neutrino field which was discussed in Chapter 15. Surprisingly, several experiments suggest that these helicoidally spiraling particles might be magnetically charged. If magnetic monopoles exist, then the relevant theoretic background for their understanding is in Section 1.3.6, where we derived the equations describing magnetic charges and magnetic charge currents, and in Section 2.9, where we pointed out the relationship between magnetic currents and decaying electric charges. The work of Lochak [1] may also be relevant; he proposes that magnetically charged particles are derived from neutrinos.

Despite the above-mentioned theoretical starting points, this chapter remains phenomenological. We don't have yet a clear understanding of the involved physical process, and therefore we leave it to the reader to make conclusions. The title of this chapter contains a question mark for this reason: the confirmation of magnetic monopoles' existence requires the elimination of any alternative interpretations to the measurements described in this chapter.

16.2. Observation of Helicoidal Particle Tracks

Table 16.1 summarizes the observations of helicoidal particle tracks during the transmutation experiments of Chapter 15. The left column shows a track from experiments reported in [1, 3–5], the middle column shows tracks

Table 16.1. Helicoidal particle tracks from transmutation-related experiments.

	Underwater exploded Ti wires	Hydrated Ni after 1350 K heating	230 V arc discharge in KNO_3 solution
Experimenter	L. Urutskoev	V. Chizhov	A. Kovacs
Detector	photographic film	polycarbonate disc	polycarbonate disc
Parallel track			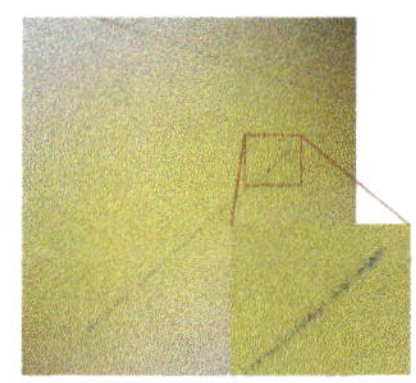
Perpendicular		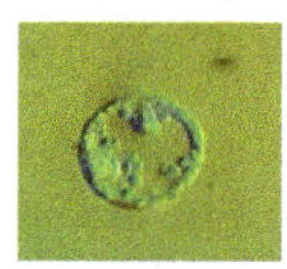	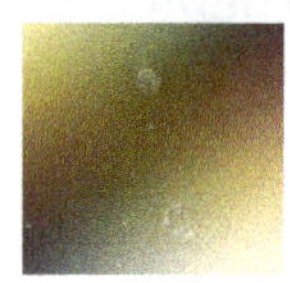

Note: The parallel tracks are along the detector film surface and have 3–5 micron radius, while the perpendicular tracks are orthogonal to the detector film surface and have 100–150 micron radius.

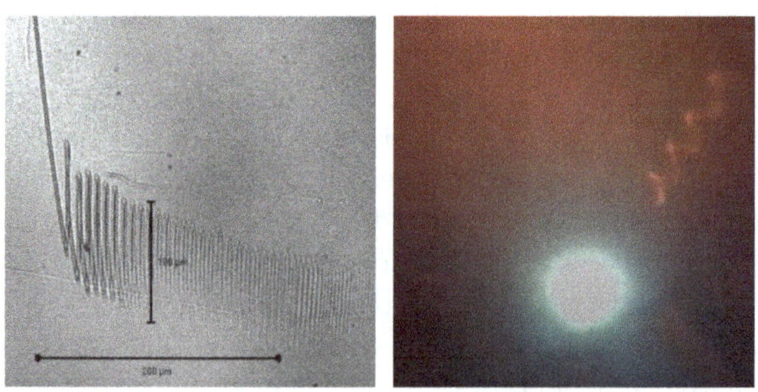

(a) (b)

Figure 16.1. (a) A helicoidal track in a CR-39 detector, reproduced from [10]. (b) A helicoidal track emitted from an electrode into low-pressure gas, presented in [11].

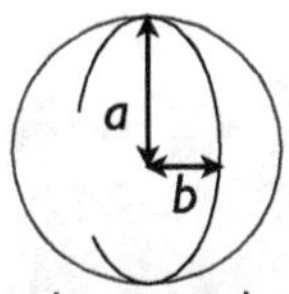

Figure 16.2. The geometry of helicoidal tracks with respect to the detector surface.

recorded by V. Chizhov during a similar experimental setup as in [2], and the right column shows tracks recorded by the author. The helicoidal shape of the parallel tracks was discovered by optical microscopy. The perpendicular tracks have then either circular or elliptical shape, which corresponds to the intersection between the detector film and a helicoidal track.

Other experimenters reported even more direct observation of helicoidal spiraling tracks: these are shown in Figure 16.1. Figure 16.1(a) shows a track recorded in a CR-39 detector, which is emitted from a vacuum electrode illuminated by a laser pulse. Its helicoidal radius is 50 microns. We note that according to Ref. [12] such laser pulse action generates coherent longitudinal phonons in the electrode material, thus creating the conditions described in Chapter 15. Figure 16.1(b) shows the optical camera recording of an emitted particle, which ionizes a low pressure gas along its path, and is emitted from a glowing electrode. Its helicoidal radius is a few millimeters. In that experiment, the glowing electrode is impacted by ions, driven by few hundred volts between the two electrodes, and this ion impact generates phonons in the electrode material.

Suppose that a helicoidally spiraling particle moves perpendicularly to the detector surface. Its track then becomes the arc of a circle: the endpoints of the arc correspond to the particle entry and exit points at the detector edges. In a more general case, the helicoidal movement is at some angle with respect to the detector surface. In this general case, the spiraling particle leaves an elliptic arc track in the detector: the large semi-axis of such an elliptic arc is the helicoidal radius. This geometric concept is illustrated in Figure 16.2.

Reference [9] describes a series of experiments, where elliptic arcs were produced upon intense visible light flashes. These tracks were recorded in photographic emulsion. Figure 16.3 shows such an exemplary elliptical track, which we interpret as a helicoidal track.

Table 16.2 lists the elliptic arc measurements reported in [9]. These data reveal even four nearly equal measurements of the largest a semi-axis value,

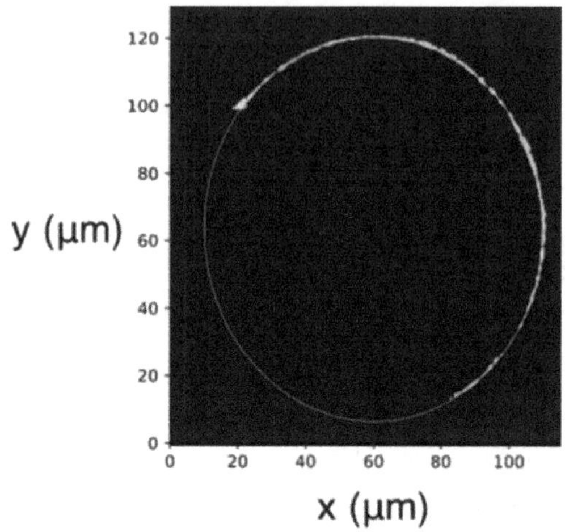

Figure 16.3. An exemplary elliptic arc track, reproduced from [9].

Table 16.2. The table of large (a) and small (b) elliptic semi-axes of the tracks reported in [9].

a (μm)	b (μm)
4091.7 ± 22.5	2241.3 ± 22.5
4078.6 ± 14.6	1996.0 ± 14.6
4077.0 ± 16.1	1129.4 ± 16.1
4076.6 ± 10.3	1053.6 ± 10.3
3141.0 ± 30.5	1474.4 ± 30.5
2391.6 ± 16.1	549.3 ± 16.1
1773.2 ± 24.9	1100.5 ± 24.9
1760.0 ± 23.1	615.3 ± 23.1
908.9 ± 21.8	505.8 ± 21.8
909.0 ± 25.7	367.2 ± 25.7
907.5 ± 11.6	284.3 ± 11.6
622.1 ± 21.1	300.7 ± 21.1
619.9 ± 16.2	192.7 ± 16.2
621.1 ± 7.3	173.8 ± 7.3
254.2 ± 7.4	104.8 ± 7.4
253.6 ± 9.8	202.2 ± 9.8
149.8 ± 13.1	83.0 ± 13.1
111.1 ± 20.6	72.1 ± 20.6
56.9 ± 13.0	49.8 ± 13.0
56.7 ± 9.5	31.5 ± 9.5
38.1 ± 6.2	36.0 ± 6.2
25.6 ± 6.4	15.9 ± 6.4

Note: The circle highlights the largest a values, which reveal the largest helicoidal track radius.

which indicates that we are looking at the limit value of the helicoidal radius. As expected, the b semi-axis values are all different because of the varying directions of monopole movement. This limiting helicoidal radius value is 4.08 mm, which is approximately $4\alpha^3$ times the electron's Zitterbewegung radius value.

16.3. What are the Helicoidally Spiraling Particles?

What kind of particles create the above-discussed helicoidally spiraling tracks, which are nearly macroscopically sized? Coincidentally, many experimentalists who measured helicoidal tracks also claim to have measured magnetic monopole signatures during the same experiment. It is not yet determined whether the helicoidally spiraling particles are magnetic monopoles, or whether there are several related particle types.

In the following paragraphs, we review magnetic monopole claims, which were reported concurrently with helicoidal particle track observations.

In [3], the authors placed ^{57}Fe foils near the reactor with underwater exploding titanium wire, and placed a magnet behind the ^{57}Fe foils. As shown in Figure 16.4, the orientation of the magnet has a measurable effect on the Mössbauer spectrum of ^{57}Fe nuclei, but only in those cases when the foil is placed near the reactor. The author of [3] interprets this effect in the following way: the ^{57}Fe foil captures emitted magnetic monopoles, and opposite magnetic charges shift the effective magnetic field (H_{eff}) on the ^{57}Fe nucleus into opposite directions.

The author of [10] observed that the emitted particle tracks form two beams when a magnetic field was applied in his experiment, and they fly toward the two poles of the applied magnet. A photo of these beam-forming tracks is shown in Figure 16.5. Such tracks are therefore interpreted as magnetic monopole tracks, which separate into two beams according to their magnetic charges.

Vladimir Chizhov placed a container of hydrated nickel powder, which underwent prior heat treatment, into a cloud chamber. There was a magnetic field in the cloud chamber; the red arrows in Figure 16.6 indicate the magnetic field lines. Figure 16.6 contains two frames from a video recording, which shows a particle track emerging from the hydrated nickel powder. The curvature of this track is different from the track of an electrically charged particle, which would spiral around the magnetic field lines. Instead, the emitted particle track may be interpreted as a parabolic curve. The emitted particle initially moves upwards, then the parabola has a peak at a

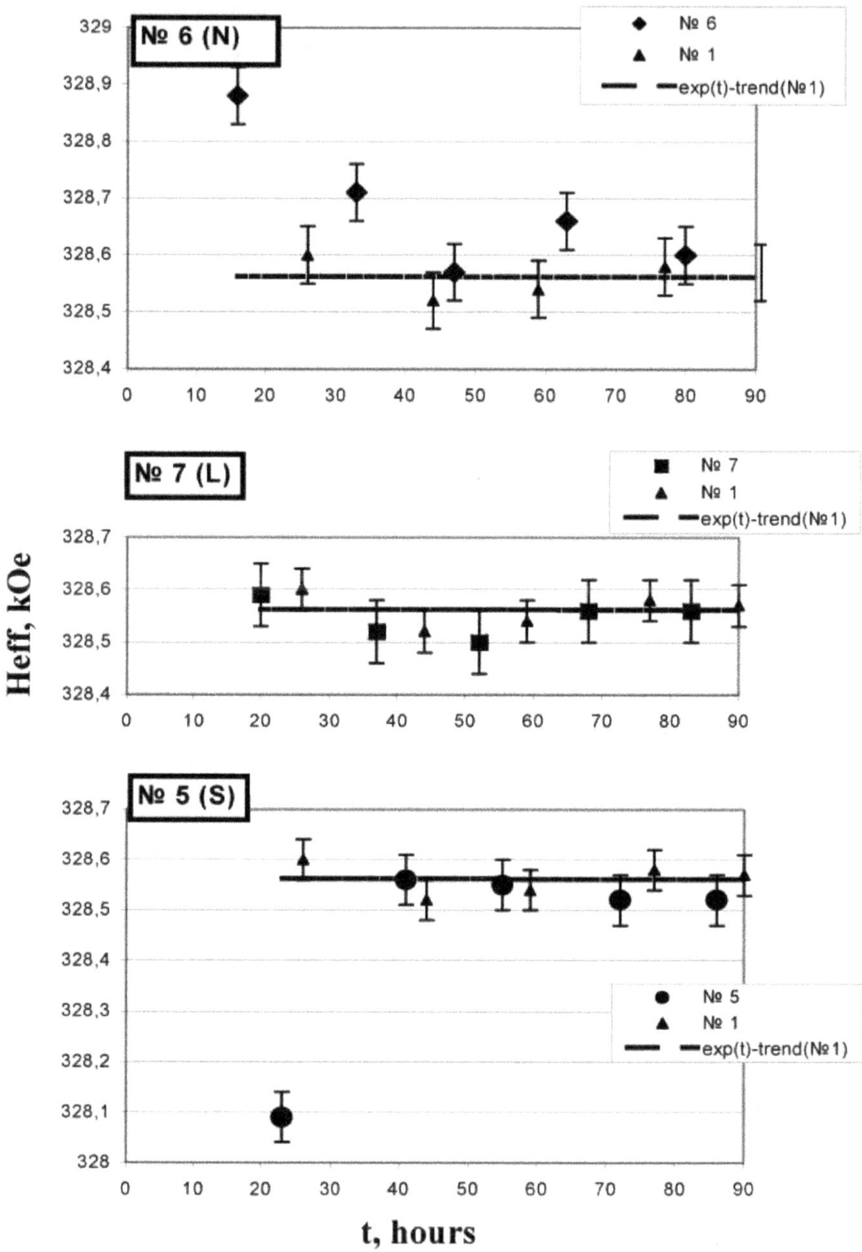

Figure 16.4. Effective magnetic field (H_{eff}) on the ^{57}Fe nucleus, obtained via Mössbauer spectrum measurements. Samples 5, 6, 7 were exposed to magnetic monopoles generated by electric discharge, prior to the first measurement. Sample 1 is the unexposed control sample. The N, S, L symbols indicate the North, South, and Lengthwise parallel orientation of a magnet placed behind the iron foil during discharge, with respect to the electric discharge location. Reproduced from [3].

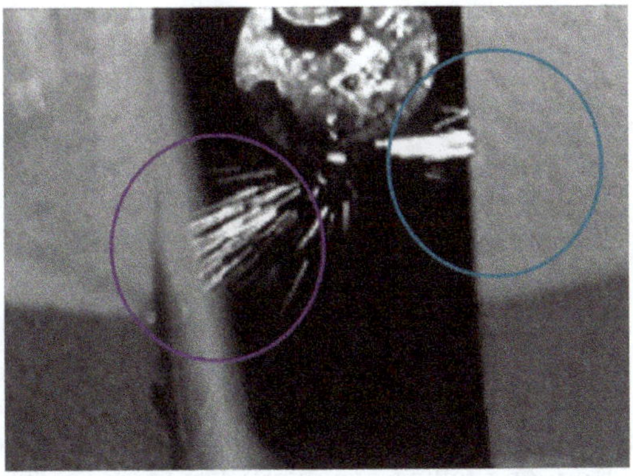

Figure 16.5. Particle tracks emitted from a laser-pulse illuminated electrode. Two beams are formed, which are flying toward the two poles of a magnet. Reproduced from [10].

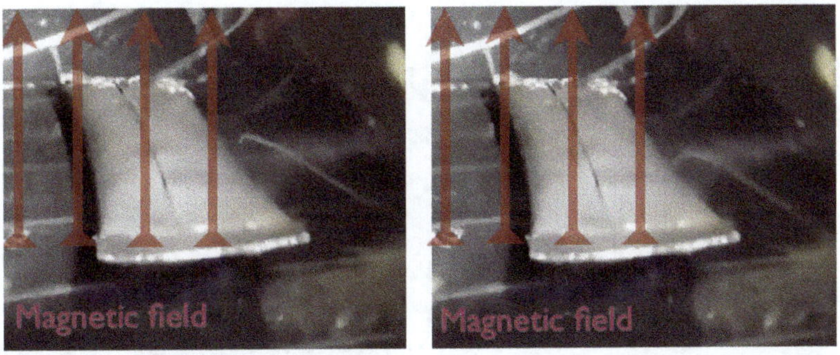

Figure 16.6. Two frames from a video recording which shows a particle track emerging from a container of hydrated nickel powder, that underwent prior heat treatment. This container is in a cloud chamber, and the red arrows indicate the direction of the applied magnetic field. Data provided by V. Chizhov.

certain height, and subsequently the particle track continues downwards. Such track dynamics is consistent with the movement of a magnetically charged particle. An observation of cloud chamber track emission from post-experiment hydrated nickel was also reported in [13].

References [6–8] describe a series of experiments, where magnetic monopole observations were claimed upon illuminating ferromagnetic nanoparticles by intense visible light. In these experiments, the tracks of illuminated and falling nano-particles are well visible. The method of magnetic charge measurement is outlined in Figure 16.7: it is a direct measurement method, and analogous to Millikan's method for measuring the elementary electric charge. Thousands of particle tracks were analyzed in these series of experiments.

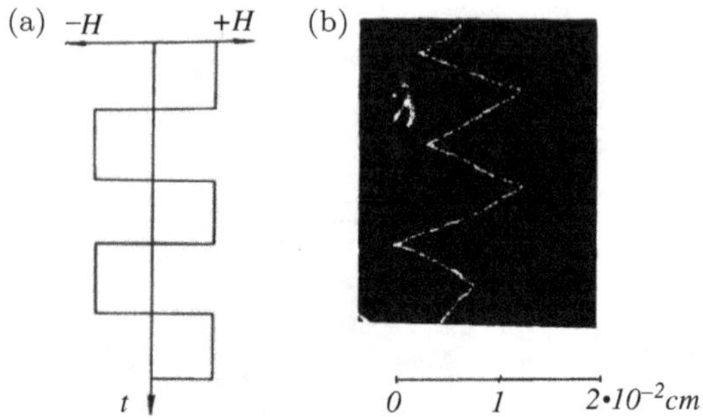

Figure 16.7. Magnetic charge measurement methodology: a time-varying magnetic field (a) causes a zig-zag pattern of a falling nano-particle's track (b), when it is magnetically charged. The magnetic charge is calculated from the analysis of track photographs. Reproduced from [8].

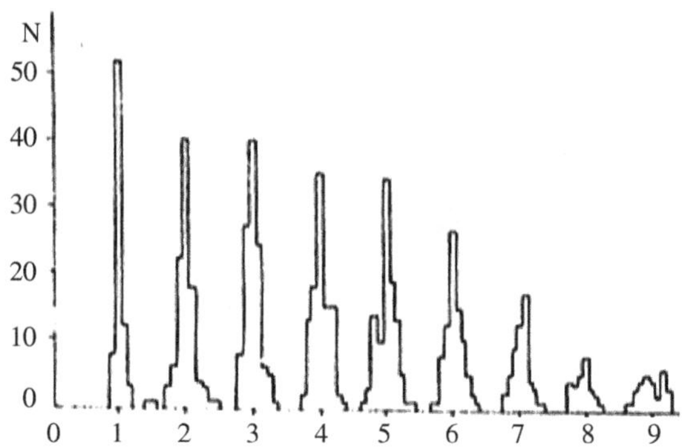

Figure 16.8. A measurement of the magnetic charge quantum. The vertical scale shows the number of samples in the histogram. The horizontal scale shows the magnetic charge value, where the unit 1 corresponds to the magnetic charge quantum value of $g = 5.84 \times 10^{-13}$ Gauss cm^2. Reproduced from [6].

These experiments discovered that magnetic monopoles are produced most effectively by circularly polarized light. The sign of the produced magnetic monopole charge depends on the chirality of the applied circularly polarized light; opposite light chiralities produce opposite magnetic charges in the ferromagnetic nano-particles.

The authors of [6–8] succeeded in measuring the magnetic charge quantum, and found its value to be $g = 5.84 \times 10^{-13}$ Gauss cm^2. Figure 16.8 shows the measured magnetic charge quantization. This magnetic charge quantum value is $g = \frac{\alpha^2}{2.92} g_D = \frac{\alpha^2}{5.84} g_S$, where g_D is the Dirac monopole charge and g_S is the Schwinger monopole charge. These data are unexpected by current mainstream theories, which anticipate the magnetic monopole

Table 16.3. A comparison between the electron and the magnetic monopole candidate.

	$e^{\pm}$	$g^{\pm}$
Charge	$\left[e = \sqrt{\alpha} \approx 8.543 \cdot 10^{-2}\right]_{\mathrm{NU}}$	$\left[g = \frac{\alpha e}{5.84} \approx 2\alpha^2 = 1.065 \cdot 10^{-4}\right]_{\mathrm{NU}}$
Trajectory	386.16 fm Zitterbewegung radius	4.08 mm largest helicoidal radius

	$g^{\pm}/e^{\pm}$ ratio
Charge	$2\alpha^{\frac{3}{2}}$
(Helicoidal radius)$^{-1}$	$4\alpha^3$

Note: The lower table shows the ratio between the monopole and electron values.

charge quantum to be g_D or g_S, which stem from the $\frac{g_D}{e} = \frac{c}{2\alpha}$ and $\frac{g_S}{e} = \frac{c}{\alpha}$ assumptions.

We now compare the above measurement results against the corresponding electron values. The first row of Table 16.3 shows the experimentally measured electron charge and magnetic monopole charge, expressed in natural units. We note this charge ratio to be approximately $2\alpha^{\frac{3}{2}}$. The second row of Table 16.3 compares the electron's largest Zitterbewegung radius, i.e., the rest frame value, against the largest helicoidal track radius measurement. The ratio values of Table 16.3 turn out to be related numbers; $4\alpha^3$ is the square of $2\alpha^{\frac{3}{2}}$.

If the data presented in Table 16.3 are not coincidental, then the next step is to understand the origin and meaning of these numbers.

It is important to note that Figure 16.1 shows a frame from a video recording which demonstrates that the above-discussed helicoidally spiraling particles move much slower than the speed of light. That is in contrast with the electron's Zitterbewegung speed, which is equal to the speed of light.

Different experiments give varying half-life estimates to magnetic monopoles. References [6–8] indicate a half-life of less than 1 second, Figure 16.4 indicates a half-life of several hours, while Figure 16.6 indicates a half-life of several days. Perhaps the half-life value depends on the monopole charge, or some other parameter.

We leave it for the reader to decide whether the herein reviewed magnetic monopole claims altogether prove the existence of monopoles, or whether there is an alternative explanation to these experimental data.

16.4. The Muon's Anomalous Magnetic Moment

Recently, the difference between the electron's and the muon's anomalous magnetic moments received much attention. As discussed in Chapter 6, the muon's g-factor is larger by $\frac{16\pi}{3}\alpha^3$. A magnetic dipole moment may be created either by a current loop, which is the electron case, or by bringing oppositely charged magnetic monopoles close to each other, or by a combination of both. If a magnetic dipole moment is created by bringing oppositely charged monopoles close to each other, its inner field becomes different from a point-like dipole field. At a location $\boldsymbol{r}$, this difference in natural units is

$$\triangle B = \frac{4\pi}{3} m\delta\left(\boldsymbol{r}\right) \tag{16.4.1}$$

where m is the magnetic moment, and $m\delta\left(\boldsymbol{r}\right)$ is the magnetization. Considering that the g-factor is a multiplicative factor of the magnetic moment, the corresponding g-factor difference is $\triangle g = \frac{4\pi}{3}\delta\left(\boldsymbol{r}\right)$, and hence

$$\delta\left(\boldsymbol{r}\right) = 4\alpha^3 \tag{16.4.2}$$

Surprisingly, the above expression matches the value from Table 16.2. If this match is not a coincidence, it raises the following question: is the muon a composite particle, comprising an electron-like constituent and two oppositely charged magnetic monopoles?

Acknowledgments

The author's experimental work is financially supported by the European Union-funded Clean HME project (grant agreement no. 951974). The author thanks György Egely for drawing attention to the literature on magnetic charge measurements, and thanks Vladimir Chizhov for making his measurement results available.

References

[1] G. Lochak, The equation of a light leptonic magnetic monopole and its experimental aspects. *Zeitschrift für Naturforschung*, **62a** (2007) 231–246.

[2] Y. Iwamura *et al.*, Excess energy generation using a nano-sized multilayer metal composite and hydrogen gas. *Journal of Condensed Matter Nuclear Science*, **33** (2020) 1–13.

[3] N.G. Ivoilov *et al.*, The influence of strange radiation on Mössbauer spectrum of Fe57 in metallic foils. *Annales de la Fondation Louis de Broglie*, **29**(3) (2004) 1177–1186.

[4] L.I. Urutskoev *et al.*, Observation of transformation of chemical elements during electric discharge. *Annales de la Fondation Louis de Broglie*, **27**(4) (2002).

[5] V.M. Dorovskoj *et al.*, Исследование продуктов электровзрыва титановых фольг с помощью электронного микроскопа. Прикладная физика, **4** (2006) 28–34.

[6] V.F. Mikhailov, *et al.*, Magnetic charge in the experiments by F. Ehrenhaft and their modern development. *Acta Physica Universitatis Comenianae*, **29** (1989) 97–148.

[7] V.F. Mikhailov, Observation of apparent magnetic charges carried by ferromagnetic particles in water droplets. *Journal of Physics A*, **24**(53) (1991) 53–57.

[8] V.F. Mikhailov, Six experiments with magnetic charge. In *Advanced Electromagnetism* T.W. Barrett and D.M. Grimes (eds.), World Scientific, Singapore, 1995.

[9] K.A. Fredericks, (2019), DOI: 10.20944/preprints201907.0181.v1.

[10] V.A. Skvortsov *et al.*, Main physical properties of exotic quasiparticles. In *Proceedings of EXON-2014 International Symposium*, 2015, pp. 457–463.

[11] R. Greenyer, VEGA — Even more extraordinary traces, https://youtu.be/8a4tnh9yCco.

[12] P. Ruello *et al.*, Physical mechanisms of coherent acoustic phonons generation by ultrafast laser action. *Ultrasonics*, **56** (2015) 21–35.

[13] E. Campari *et al.*, Photon and particle emission, heat production and surface transformation in Ni–H system. In *Proceedings of the ICCF-11 International Conference on Condensed Matter Nuclear Science*, (2005) 405–413.

https://doi.org/10.1142/9781800611306_0017

Chapter 17

Simple Experiments

András Kovács*,§ and Francesco Celani†,‡,¶
*BroadBit Energy Technologies, Metallimiehenkuja 8 C
02150 Espoo, Finland
†Istituto Nazionale di Fisica Nucleare (INFN-LNF)
Via E. Fermi 56, 00044 Frascati, Roma, Italy
‡International Society for Condensed Matter
Nuclear Science (ISCMNS), UK
§andras.kovacs@broadbit.com
¶francesco.celani@lnf.infn.it

17.1. Introduction

In this final chapter, we introduce relatively simple experiments undertaken by the authors, which allow an interested reader to verify the reality of electromagnetically induced nuclear energy production. These experiments don't require overly expensive equipment, and thus may be a good starting point for experimenters.

17.2. Bulk Metal Fueled Energy Production

We use bulk fuel materials, comprising Cu–Ni alloy known as Constantan (97 wt.%), and Li (3 wt.%). Constantan foils and Li pieces were loaded into a steel tube under ambient conditions, and sealed by welding. Because of the ambient conditions, the metallic Li surface contains trace amounts of LiOH. The steel container was then placed into a TZM (99.5% Mo, 0.4% Ti, 0.1% Zr) metal tube, and sealed by welding.

The reactor was operated under a temperature cycling program: constant power heating was used from 1240°C to 1300°C, the heating was turned off at the 1300°C upper temperature threshold, and then the constant power heating was turned back on at the 1240°C lower temperature threshold. Further experimental details are described in [1].

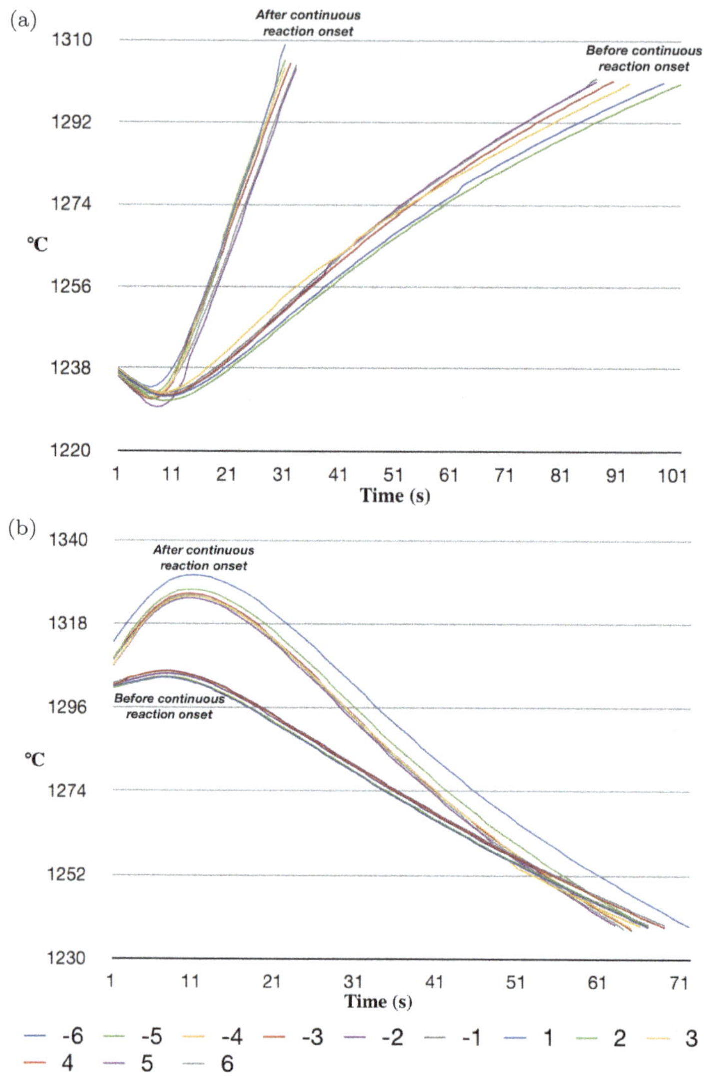

Figure 17.1. Temperature versus time (seconds) charts before and after the continuous reaction process start-up with the Ni-Cu-Li fuel mixture. (a) Heating phases (constant heating power). (b) Cooling phases (zero heating power). The labels indicate cycle no. relative to the starting of continuous reaction.

After about half an hour at the operating temperature regime, a continuous energy-producing reaction starts up and is maintained for approximately three hours. The average reaction power was approximately 30 W/g with respect to the fuel mass during this time. Figure 17.1 shows the temperature evolution of the heating and cooling phases in the six cycles before and after the continuous reaction start-up. The negative (versus positive) numbers of labels indicate cycle numbers before (versus after) the starting of continuous reaction.

The continuous reaction starts up right away at the start of a certain heating cycle (i.e., "cycle 1"), as can be seen from the steeper slope already at the beginning. The fairly constant slope during the heating cycles, as well as the very similar slope among the positive numbered cycles, show that the reaction produces a constant exothermic power during the heating cycles. Signatures of a burst-like reaction process (i.e., sudden temperature jumps) can be seen in some negative numbered preceding cycles, and also simultaneously to the continuous reaction in some positive numbered cycles. Cycles 1, 2, and 5 show very visible temperature jumps.

During the continuous heat production, we detected radio-frequency emission from the reactor with constant spectral power over the measured 1–10 MHz frequency range.

In the cooling phase, the first observable feature is a temperature overshoot in the positive numbered cycles, relative to the temperature evolution of the negative numbered cycles. This indicates that the continuous reaction stays active for some seconds after the heating shut-off. Subsequently, the cooling rate is somewhat faster in the positive numbered cycles than in the negative numbered cycles. This is the consequence of local heat generation during the continuous reaction, which was picked up by the nearby thermocouple. It means that the average reactor temperature at the end of heating is somewhat lower in the positive numbered cycles than in the negative numbered cycles, because the electric heating was turned on for much shorter time. The dynamics of the temperature falling phase indicates little or no exothermic reaction during the temperature fall. Therefore, the reaction dynamics is actually semi-continuous, i.e., the continuous exothermic power is present mainly during the heating phases.

The longer term dynamics of this continuous reaction was analyzed by constructing the frequency spectrum of the temperature signal's autocorrelation function. Figure 17.2 shows this spectrum for the initial part of the operating temperature regime, over the 20 minutes segment prior to the semi-continuous reaction onset, and then the spectrum for the duration of semi-continuous exothermic reaction. Figure 17.2(a) has less sharp autocorrelation peaks, even though the transients at the start of operating temperature regime were excluded. This may be partly caused by remaining transients of the initial operating temperature regime and partly caused by the bursty temperature jumps in the initial phase. The continuous reaction part shows

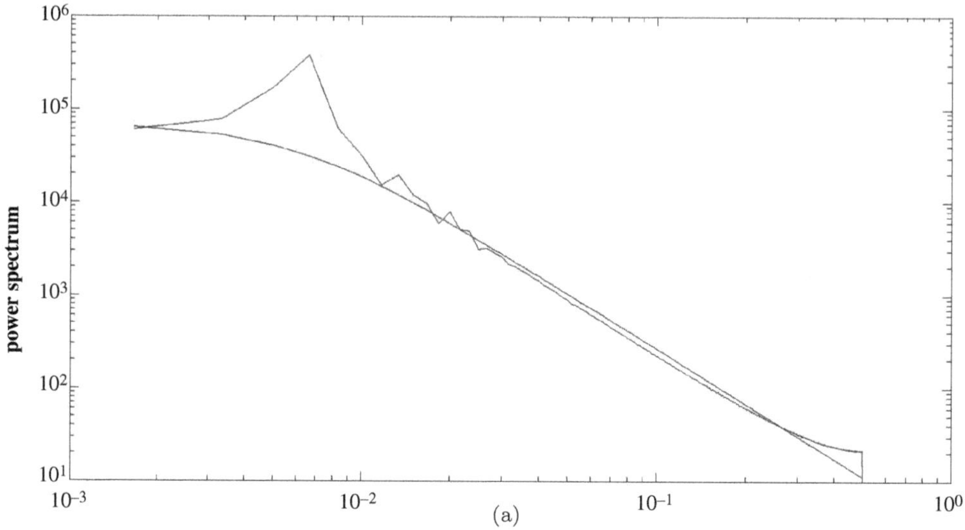

(a)

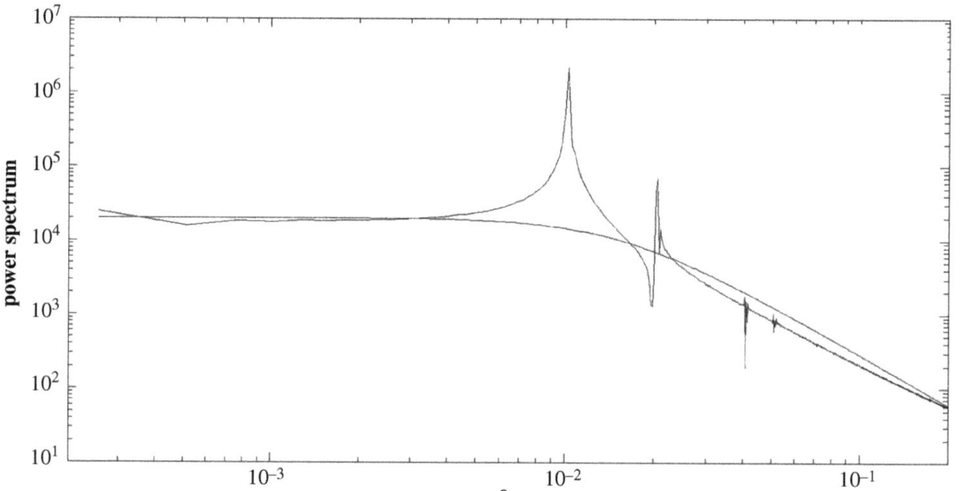

(b)

Figure 17.2. Autocorrelation frequency spectrum during the continuous reaction process with Ni–Cu–Li fuel mixture. (a) The spectrum prior to the continuous reaction. (b) The spectrum during the continuous reaction.

very sharp autocorrelation peaks and the absence of high-frequency noise. Since the heating program is controlled via temperature feedback, these data prove that the reaction dynamics is highly ordered, and remains nearly constant from cycle to cycle. There must be some reaction control parameter which regulates the reaction rate to such constant value. It requires further studies to determine whether the sudden reaction bursts and the semi-continuous reaction process are precursor events to each other or independent phenomena.

17.3. Thin Wire Fueled Energy Production

We use 0.1–0.2 mm diameter Constantan wire with a high surface area. The surface of the wire is coated by a low work function material, thus creating interfaces with a Fermi level difference. As was illustrated in Figure 15.4, diffusing protons gain a few eV energy while crossing this interface, and generate longitudinal phonons upon subsequent collisions. The wire is immersed into hydrogen atmosphere, and heated by a current passing through it. The details of the wire configuration and operating conditions are shown in Figure 17.3. The key requirement is for the Constantan wire to become hydrated, and the progress of this hydration process may be observed by monitoring the resistance value of the wire.

This setup allows a precise investigation of various parameters which affect the energy production rate. It is found that the rate of exothermic power production depends on the following parameters:

- the surface area size of the wire;
- the temperature gradient along the wire;

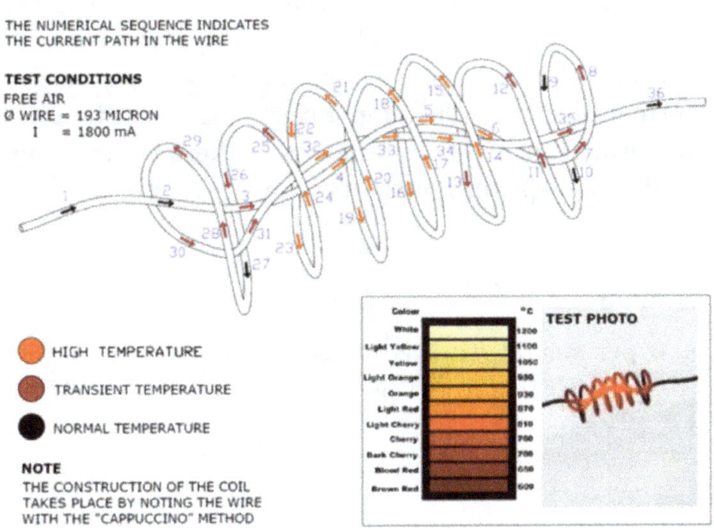

Figure 17.3. Knotted arrangement of the Constantan wire, which produces nuclear energy under hydrogen atmosphere. The inset shows the wire temperature during characterization experiments, in open air conditions.

- the surface coating material on the wire;
- the hydrogen pressure;
- pre-conditioning by high-power electric pulses;
- the electric field effect, which may be created by a high-voltage surface running along the cylindrical axis of the knot structure.

After the analysis of many experimental runs, our group determined that power production is proportional to the hydrogen flux within the wires. Therefore, the above-listed parameters must be set to maximize hydrogen flux. Ideally, the temperature gradient drives a constant flux of hydrogen within the wire, thus supplying a constant flow of protons that become accelerated upon crossing the interface with a Fermi level difference. This stream of accelerated protons then transfers its energy into phonons.

Also, the importance of pre-conditioning by high-power electric pulses was experimentally identified [3]. Such pre-conditioning is applied prior to the normal operating regime.

Altogether, this experimental technique is a relatively straightforward method for reproducible energy production, and facilitates a systematic investigation and optimization of parameters affecting the reaction rate.

Another attractive feature of this technique is that it may allow direct heat-to-electricity conversion. The temperature of the Constantan wire's active region is high enough for this part of the wire to act as a thermionic diode. The evaporated electrons are collected by a colder current collector. If the gas pressure is low enough, self-powered operation may be feasible, wherein the generated thermionic voltage powers an external load and the thermionic current maintains the temperature of the Constantan wire in the suitable temperature range. The production of thermionic voltage has been indeed reported in Ref. [2], although at the moment still at low values.

Acknowledgments

The authors' experimental work is financially supported by the European Union-funded Clean HME project (grant agreement no. 951974).

References

[1] A. Kovacs *et al.*, Exothermic reactions in the partially molten Li–Ni–Cu alloy. *Journal of Condensed Matter Nuclear Science*, **25** (2017), 159–180.

[2] F. Celani *et al.*, Observation of macroscopic current and thermal anomalies, at high temperature, by hetero-structures in thin and long Constantan wires under H_2 gas. *Journal of Condensed Matter Nuclear Science*, **19** (2016), 29–45.

[3] F. Celani *et al.*, Clear and simple evidence of correlations among Open Voltage values and AHE in Constantan wires, DC operations under H2 Gas, after long time conditioning at 70 and 100 W of input power, presentation at the IWAHLM14 conference, Assisi, Italy (2021).

https://doi.org/10.1142/9781800611306_bmatter

Index

D

E

www.ingramcontent.com/pod-product-compliance
Lightning Source LLC
LaVergne TN
LVHW081931200926
848484LV00009B/2405
9781800611290